THE FRONTIERS COLLECTION

For further volumes:
http://www.springer.com/series/5342

THE FRONTIERS COLLECTION

The books in this collection are devoted to challenging and open problems at the forefront of modern science, including related philosophical debates. In contrast to typical research monographs, however, they strive to present their topics in a manner accessible also to scientifically literate non-specialists wishing to gain insight into the deeper implications and fascinating questions involved. Taken as a whole, the series reflects the need for a fundamental and interdisciplinary approach to modern science. Furthermore, it is intended to encourage active scientists in all areas to ponder over important and perhaps controversial issues beyond their own speciality. Extending from quantum physics and relativity to entropy, consciousness and complex systems—the Frontiers Collection will inspire readers to push back the frontiers of their own knowledge.

For a full list of published titles, please see back of book or springer.com/series/5342

Amnon H. Eden · James H. Moor
Johnny H. Søraker · Eric Steinhart
Editors

Singularity Hypotheses

A Scientific and Philosophical Assessment

Editors
Amnon H. Eden
School of Computer Science
and Electronic Engineering
University of Essex
Colchester
UK

James H. Moor
Dartmouth College
Hanover
USA

Johnny H. Søraker
Department of Philosophy
University of Twente
Enschede
The Netherlands

Eric Steinhart
Department of Philosophy
William Paterson University
Wayne
USA

ISSN 1612-3018
ISBN 978-3-642-44306-0 ISBN 978-3-642-32560-1 (eBook)
DOI 10.1007/978-3-642-32560-1
Springer Heidelberg New York Dordrecht London

Printed on acid-free paper

Springer is part of Springer Science+Business Media (www.springer.com)

To Saul, With love,

—Aba

Contents

Chapter 1
Singularity Hypotheses: An Overview

Introduction to: Singularity Hypotheses: A Scientific and Philosophical Assessment

Amnon H. Eden, Eric Steinhart, David Pearce and James H. Moor

Questions

Bill Joy in a widely read but controversial article claimed that the most powerful 21st century technologies are threatening to make humans an endangered species (Joy 2000). Indeed, a growing number of scientists, philosophers and forecasters insist that the accelerating progress in disruptive technologies such as artificial intelligence, robotics, genetic engineering, and nanotechnology may lead to what they refer to as the *technological singularity*: an event or phase that will radically change human civilization, and perhaps even human nature itself, before the middle of the 21st century (Paul and Cox 1996; Broderick 2001; Garreau 2005, Kurzweil 2005).

Singularity hypotheses refer to either one of two distinct and very different scenarios. The first (Vinge 1993; Bostrom to appear) postulates the emergence of artificial superintelligent agents—software-based synthetic minds—as the 'singular' outcome of accelerating progress in computing technology. This singularity results

A. H. Eden (✉)
School of Computer Science and, Electronic Engineering,
University of Essex, Colchester, CO4 3SQ, UK
e-mail: eden@essex.ac.uk

E. Steinhart
Department of Philosophy, William Paterson University, 300 Pompton Road, Wayne, NJ 07470, USA
e-mail: esteinhart1@nyc.rr.com

D. Pearce
knightsbridge Online, 7 Lower Rock Gardens, Brighton, USA
e-mail: dave@hedweb.com

J. H. Moor
Department of Philosophy, Dartmouth College, 6035 Thornton, Hanover, NH 03755, USA
e-mail: james.h.moor@dartmouth.edu

A. H. Eden et al. (eds.), *Singularity Hypotheses*, The Frontiers Collection,
DOI: 10.1007/978-3-642-32560-1_1,

from an 'intelligence explosion' (Good 1965): a process in which software-based intelligent minds enter a 'runaway reaction' of self-improvement cycles, with each new and more intelligent generation appearing faster than its predecessor. Part I of this volume is dedicated to essays which argue that progress in artificial intelligence and machine learning may indeed increase machine intelligence beyond that of any human being. As Alan Turing (1951) observed, "at some stage therefore we should have to expect the machines to take control, in the way that is mentioned in Samuel Butler's 'Erewhon' ": the consequences of such greater-than-human intelligence will be profound, and conceivably dire for humanity as we know it. Essays in Part II of this volume are concerned with this scenario.

A radically different scenario is explored by transhumanists who expect progress in enhancement technologies, most notably the amplification of human cognitive capabilities, to lead to the emergence of a posthuman race. Posthumans will overcome all existing human limitations, both physical and mental, and conquer aging, death and disease (Kurzweil 2005). The nature of such a singularity, a 'biointelligence explosion', is analyzed in essays in Part III of this volume. Some authors (Pearce, this volume) argue that transhumans and posthumans will retain a fundamental biological core. Other authors argue that fully functioning, autonomous whole-brain emulations or 'uploads' (Chalmers 2010; Koene this volume; Brey this volume) may soon be constructed by 'reverse-engineering' the brain of any human. If fully functional or even conscious, uploads may usher in an era where the notion of personhood needs to be radically revised (Hanson 1994).

Advocates of the technological singularity have developed a powerful inductive *Argument from Acceleration* in favour of their hypothesis. The argument is based on the extrapolation of *trend curves* in computing technology and econometrics (Moore 1965; Moravec 1988, Chap. 2; Moravec 2000, Chap. 3; Kurzweil 2005, Chaps. 1 and 2). In essence, the argument runs like this: (1) The study of the history of technology reveals that technological progress has long been accelerating. (2) There are good reasons to think that this acceleration will continue for at least several more decades. (3) If it does continue, our technological achievements will become so great that our bodies, minds, societies, and economies will be radically transformed. (4) Therefore, it is likely that this disruptive transformation will occur. Kurzweil (2005, p. 136) sets the date mid-century, around the year 2045. The change will be so revolutionary that it will constitute a "rupture in the fabric of human history" (Kurzweil 2005, p. 9).

Critics of the technological singularity dismiss these claims as speculative and empirically unsound, if not pseudo-scientific (Horgan 2008). Some attacks focus on the premises of the Argument from Acceleration (Plebe and Perconti this volume), mostly (2). For example, Modis (2003; this volume) claims that after periods of change that appear to be accelerating, technological progress always levels off. Other futurists have long argued that we are heading instead towards a global economic and ecological collapse. This negative scenario was famously developed using computer modelling of the future in *The Limits to Growth* (Meadows et al. 1972, 2004).

Mocked as the "rapture of the nerds", many critics take (3) to be yet another apocalyptic fantasy, a technocratic variation on the usual theme of doom-and-gloom fuelled by mysticism, science fiction and even greed. Some conclude that the singularity is a religious notion, not a scientific one (Horgan 2008; Proudfoot this volume; Bringsjord et al. this volume). Other critics (Chaisson this volume) accept acceleration as an underlying law of nature but claim that, in perspective, the significance of the claimed changes is overblown. That is, what is commonly described as the technological singularity may well materialize, with profound consequences for the human race. But on a cosmic scale, such a mid-century transition is no more significant then whatever may follow.

Existential risk or cultist fantasy? Are any of the accounts of the technological singularity credible? In other words, is the technological singularity an open problem in science?

We believe that before any interpretation of the singularity hypothesis can be taken on board by the scientific community, rigorous tools of scientific enquiry must be employed to reformulate it as a coherent and falsifiable conjecture. To this end, we challenged economists, computer scientists, biologists, mathematicians, philosophers and futurists to articulate their concepts of the singularity. The questions we posed were as follows:

1. What is the [technological] singularity hypothesis? What exactly is being claimed?
2. What is the empirical content of this conjecture? Can it be refuted or corroborated, and if so, how?
3. What exactly is the nature of a singularity: Is it a discontinuity on a par with phase transition or a process on a par with Toffler's 'wave'? Is the term singularity appropriate?
4. What evidence, taken for example from the history of technology and economic theories, suggest the advent of some form of singularity by 2050?
5. What, if anything, can be said to be accelerating? What evidence can reliably be said to support its existence? Which metrics support the idea that 'progress' is indeed accelerating?
6. What are the most likely milestones ('major paradigm shifts') in the countdown to a singularity?
7. Is the so-called Moore's Law on par with the laws of thermodynamics? How about the Law of Accelerating Returns? What exactly is the nature of the change they purport to measure?
8. What are the necessary and sufficient conditions for an intelligence explosion (a *runaway effect*)? What is the actual likelihood of such an event?
9. What evidence support the claim that machine intelligence has been rising? Can this evidence be extrapolated reliably?
10. What are the necessary and sufficient conditions for machine intelligence to be considered to be on a par with that of humans? What would it take for the "general educated opinion [to] have altered so much that one will be able to

speak of machines thinking without expecting to be contradicted" (Turing 1950, p. 442)?
11. What does it mean to claim that biological evolution will be replaced by technological evolution? What exactly can be the expected effects of augmentation and enhancement, in particular over our cognitive abilities? To which extent can we expect our transhuman and posthuman descendants to be different from us?
12. What evidence support the claim that humankind's intelligence quotient has been rising ("Flynn effect")? How this evidence relate to a more general claim about a rise in the 'intelligence' of carbon-based life? Can this evidence be extrapolated reliably?
13. What are the necessary and sufficient conditions for a functioning whole brain emulation (WBE) of a human? At which level exactly must the brain be emulated? What will be the conscious experience of a WBE? To which extent can they be said to be human?
14. What may be the consequences of a singularity? What may be its effect on society, e.g. in ethics, politics, economics, warfare, medicine, culture, arts, the humanities, and religion?
15. Is it meaningful to refer to multiple singularities? If so, what can be learned from past such events? Is it meaningful to claim a narrow interpretation of singularity in some specific domain of activity, e.g. a singularity in chess playing, in face recognition, in car driving, etc.?

This volume contains the contributions received in response to this challenge.

Towards a Definition

Accounts of a technological singularity—henceforth the singularity—appear to disagree on its causes and possible consequences, on timescale, and even on its nature: the emergence of machine intelligence or of posthumans? An event or a period? Is the technological singularity unique or have there been others? The absence of a consensus on basic questions casts doubt whether the notion of *singularity* is at all coherent.

The term in its contemporary sense traces back to von Neumann, who is quoted as saying that "the ever-accelerating progress of technology and changes in the mode of human life... gives the appearance of approaching some essential singularity in the history of the race beyond which human affairs, as we know them, could not continue" (in Ulam 1958). Indeed, the twin notions of *acceleration* and *discontinuity* are common to all accounts of the technological singularity, as distinguished from a space–time singularity and a singularity in a mathematical function.

Acceleration refers to a rate of growth in some quantity such as computations per second per fixed dollar (Kurzweil 2005), economic measures of *growth rate* (Hanson

1994; Miller this volume) or total output of goods and services (Toffler 1970), and *energy rate density* (Chaisson this volume). Others describe quantitative measures of physical, biological, social, cultural, and technological processes of evolution: milestones or 'paradigm shifts' whose timing demonstrates an accelerating pace of change. For example, Sagan's Cosmic Calendar (1977, Chap. 1) names milestones in biological evolution such as the emergence of eukaryotes, vertebrates, amphibians, mammals, primates, *hominidae*, and *Homo sapiens*, which show an accelerating trend. Following Good (1965) and Bostrom (to appear), Muehlhauser and Salamon (this volume), Arel (this volume), and Schmidhuber (this volume) describe developments in machine learning which seek to demonstrate that progressively more 'intelligent' problems have been solved during the past few decades, and how such technologies may further improve, possibly even in a recursive process of self-modification. Some authors attempt to show that many of the above accounts of acceleration are in fact manifestations of an underlying law of nature (Adams 1904; Kurzweil 2005, Chaisson this volume): quantitatively or qualitatively measured, acceleration is commonly visualized as an upwards-curved mathematical graph which, if projected into the future, is said to be leading to a *discontinuity*.

Described either as an event that may take a few hours (e.g., a 'hard takeoff', Loosemore and Goertzel, this volume) or a period of years (e.g., Toffler 1970), the technological singularity is taken to mark a *discontinuity* or a turning-point in human history. The choice of word 'singularity' appears to be motivated less by the eponymous mathematical concept (Hirshfeld 2011) and more by the ontological and epistemological discontinuities idiosyncratic to black holes. Seen as a central metaphor, a gravitational singularity is a (theoretical) point at the centre of black holes at which quantities that are otherwise meaningful (e.g., *density* and *spacetime curvature*) become infinite, or rather meaningless. The discontinuity expressed by the black hole metaphor is thus used to convey how the quantitative measure of *intelligence*, at least as it is measured by traditional IQ tests (such as Wechsler and Stanford-Binet), may become a meaningless notion for capturing the intellectual capabilities of superintelligent minds. Alternatively, we may say a graph measuring average intelligence beyond the singularity in terms of IQ score may display some form of radical discontinuity if superintelligence emerges. Furthermore, singularitarians note that gravitational singularities are said to be surrounded by an *event horizon*: a boundary in spacetime beyond which events cannot be observed from outside, and a horizon beyond which gravitational pull becomes so strong that nothing can escape, even light (hence "black")—a point of no return. Kurzweil (2005) and others (e.g., Pearce this volume) contend that, since the minds of superintelligent intellects may be difficult or impossible for humans to comprehend (Fox and Yampolskiy this volume), a technological singularity marks an epistemological barrier beyond which events cannot be predicted or understood—an 'event horizon' in human affairs. The gravitational singularity metaphor thus reinforces the view that the change will be radical and that its outcome cannot be foreseen.

The combination of acceleration and discontinuity is at once common and unique to the singularity literature in general and to the essays in this volume in particular. We shall therefore proceed on the premise that acceleration and discontinuity jointly offer necessary and sufficient conditions for us to take a manuscript to be concerned with a hypothesis of a technological singularity.

Historical Background

Many philosophers have portrayed the cosmic process as an ascending curve of positivity (Lovejoy 1936, Chap. 9). Over time, the quantities of intelligence, power or value are always increasing. These progressive philosophies have sometimes been religious and sometimes secular. Secular versions of progress have sometimes been political and sometimes technological. Technological versions have sometimes invoked broad technical progress and have sometimes focused on more specific outcomes such as the possible recursive self-improvement of artificial intelligence.

For some philosophers of progress, the rate of increase remains relatively constant; for others, the rate of increase is also increasing—progress accelerates. Within such philosophies, the *singularity* is often the point at which positivity becomes maximal. It may be an ideal limit point (an *omega point*) either at infinity or at the vertical asymptote of an accelerating trajectory. Or sometimes, the singularity is the critical point at which the slope of an accelerating curve passes beyond unity.

Although thought about the singularity may appear to be very new, in fact such ideas have a long philosophical history. To help increase awareness of the deep roots of singularitarian thought within traditional philosophy, it may be useful to look at some of its historical antecedents.

Perhaps the earliest articulation of the idea that history is making progress toward some omega point of superhuman intelligence is found in *The Phenomenology of Spirit*, written by Hegel (1807). Hegel describes the ascent of human culture to an ideal limit point of absolute knowing. Of course, Hegel's thought is not technological. Yet it is probably the first presentation, however abstract, of singularitarian ideas. For the modern Hegelian, the singularity looks much like the final self-realization of Spirit in absolute knowing (Zimmerman 2008).

Around 1870, the British writer Samuel Butler used Darwinian ideas to develop a theory of the evolution of technology. In his essay "Darwin among the Machines" and in his utopian novel *Erewhon: Or, Over the Range* (Butler 1872), Butler argues that machines would soon evolve into artificial life-forms far superior to human beings. Threatened by superhuman technology, the Erewhonians are notable for rejecting all advanced technology. Also writing in the late 1800s, the American philosopher Charles Sanders Peirce developed an evolutionary cosmology (see Hausman 1993). Peirce portrays the universe as evolving from an initial chaos to a final singularity of pure mind. Its evolution is accelerating as this tendency to regularity acts upon itself. Although Pierce's notion of progress was not based on technology, his work is probably the earliest to discuss the notion of accelerating progress itself. Of course, Peirce was also a first-rate logician; and as such, he was among the first to believe that minds were computational machines.

Around 1900, the American writer Henry Adams [1] was probably the first writer to describe a technological singularity. Adams was almost certainly the first person to write about history as a self-accelerating technological process. His essay "The Law of Acceleration" (Adams 1904) may well be the first work to propose an actual formula for the acceleration of technological change. Adams suggests measuring technological progress by the amount of coal consumed by society. His law of acceleration prefigures Kurzweil's law of accelerating returns. His later essay "The Rule of Phase" (Adams 1909) portrays history as accelerating through several epochs—including the Instinctual, Religious, Mechanical, Electrical, and Ethereal Phases. This essay contains what is probably the first illustration of history as a curve approaching a vertical asymptote. Adams provides a mathematical formula for computing the duration of each technological phase, and the amount of energy that will consumed during that phase. His epochs prefigure Kurzweil's evolutionary epochs. Adams uses his formulae to argue that the singularity will be reached by about the year 2025, a forecast remarkably close to modern singularitarians.

Much writing on the singularity owes a great debt to Teilhard de Chardin (1955; see Steinhart 2008). Teilhard is among the first writers seriously to explore the future of human evolution. He advocates both biological enhancement technologies and artificial intelligence. He discusses the emergence of a global computation-communication system (and is said by some to have been the first to have envisioned the Internet). He proposes the development of a global society and describes the acceleration of progress towards a technological singularity (which he termed "the critical point"). He discusses the spread of human intelligence into the universe and its amplification into a cosmic-intelligence. Much of the more religiously-expressed thought of Kurzweil (e.g. his definition of "God" as the omega point of evolution) ultimately comes from Teilhard.

Many of the ideas presented in recent literature on the singularity are foreshadowed in a prescient essay by George Harry Stine. Stine was a rocket engineer and part-time science fiction writer. His essay "Science Fiction is too Conservative" was published in May 1961 in *Analog*. *Analog* was a widely read science-fiction magazine. Like Adams, Stine uses trend curves to argue that a momentous and disruptive event is going to happen in the early 21st Century.

In 1970, Alvin and Heidi Toffler observed both acceleration and discontinuity in their influential work, *Future Shock*. About acceleration, the Tofflers observed that "the total output of goods and services in advanced societies doubles every 15 years, and that the doubling times are shrinking" (Toffler 1970, p. 25). They demonstrate accelerating change in every aspect of modern life: in transportation, size of population centres, family structure, diversity of lifestyles, etc., and most

[1] descendant of President John Quincy Adams.

importantly, in the transition from factories as 'means of production' to knowledge as the most fundamental source of wealth (Toffler 1980). The Tofflers conclude that the transition to knowledge-based society "is, in all likelihood, bigger, deeper, and more important than the industrial revolution. ... Nothing less than the second great divide in human history, the shift from barbarism to civilization" (Toffler 1970, p. 11).

During the 1980s, unprecedented advances in computing technology led to renewed interest in the notion that technology is progressing towards some kind of tipping-point or discontinuity. Moravec's *Mind Children* (1988) revived research into the nature of technological acceleration. Many more books followed, all arguing for extraordinary future developments in robotics, artificial intelligence, nanotechnology, and biotechnology. Kurzweil (1999) developed his law of accelerating returns in *The Age of Spiritual Machines*. Broderick (2001) brought these ideas together to argue for a future climax of technological progress that he termed *the spike*. All these ideas were brought into public consciousness with the publication of Kurzweil's (2005) *The Singularity is Near* and its accompanying movie. As the best-known defence of the singularity, Kurzweil's work inspired dozens of responses. One major assessment of singularitarian ideas was delivered by *Special Report: The Singularity* in *IEEE Spectrum* (June, 2008). More recently, notable work on the singularity has been done by the philosopher David Chalmers (2010) and the discussion of the singularity it inspired (*The Journal of Consciousness Studies* 19, pp. 1–2). The rapid growth in singularity research seems set to continue and perhaps accelerate.

Essays in this Volume

The essays developed by our authors divide naturally into several groups. Essays in Part I hold that a singularity of machine superintelligence is probable. Luke Muehlhauser and Anna Salamon of the Singularity Institute of Artificial Intelligence argue that an intelligence explosion is likely and examine some of its consequences. They make recommendations designed to ensure that the emerging superintelligence will be beneficial, rather than detrimental, to humanity. Itamer Arel, a computer scientist, argues that artificial general intelligence may become an extremely powerful and disruptive force. He describes how humans might shape the emergence of superhuman intellects so that our relations with such intellects are more cooperative than competitive. Juergen Schmidhuber, also a computer scientist, presents substantial evidence that improvements in artificial intelligence are rapidly progressing towards human levels. Schmidhuber is optimistic that, if future trends continue, we will face an intelligence explosion within the next few decades. The last essay in this part is by Richard Loosemore and Ben Goertzel who examine various objections to an intelligence explosion and conclude that they are not persuasive.

Essays in Part II are concerned with the values of agents that may result from a singularity of artificial intellects. Luke Muehlhauser and Louie Helm ask what it would mean for artificial intellects to be *friendly* to humans, conclude that human values are complex and difficult to specify, and discuss techniques we might use to ensure the friendliness of artificial superintelligent agents. Joshua Fox and Roman Yampolskiy consider the psychologies of artificial intellects. They argue that human-like mentalities occupy only a very small part of the space of possible minds. If Fox and Yampolskiy are right, then it is likely that such minds, especially if superintelligent, will scarcely be recognizable to us at all. The values and goals of such minds will be alien, and perhaps incomprehensible in human terms. This strangeness creates challenges, some of which are discussed in James Miller's essay. Miller examines the economic issues associated with a singularity of artificial superintelligence. He shows that although the singularity of artificial superintelligence may be brought about by economic competition, one paradoxical consequence might be the destruction of the value of money. More worryingly, Miller suggests that a business that may be capable of creating an artificial superintelligence would face a unique set of economic incentives likely to push it deliberately to make it *unfriendly*. To counter such worries, Steve Omohundro examines how market forces may affect their behaviour. Omohundro proposes a variety of strategies to ensure that any artificial intellects will have human-friendly values and goals. Eliezer Yudkowsky concludes this part by considering the ways that artificial superintelligent intellects may radically differ from humans and the urgent need for us to take those differences into account.

Whereas essays in Parts I and II are concerned with the intelligence explosion scenario—a singularity deriving from the evolution of intelligence in silicon, the essays in Part III are concerned with the evolution that humans may undergo via enhancement, amplification, and modification, and with the scenario in which a race of superintelligent posthumans emerges. David Pearce conceives of humans as 'recursively self-improving organic robots' poised to re-engineer their own genetic code and bootstrap their way to full-spectrum superintelligence. Hypersocial and supersentient, the successors of archaic humanity may phase out the biology of suffering throughout the living world. Randal Koene examines how the principles of evolution apply to brain emulations. He argues that intelligence entails autonomy, so that future 'substrate-independent minds' (SIMs), may hold values that humans find alien. Koene nonetheless hopes that, since SIMs will originate from our own brains, human values play significant roles in superintelligent, 'disembodied' minds. Dennis Bray examines the biochemical mechanisms of the brain. He concludes that building fully functional emulations by reverse-engineering human brains may entail much more than modelling neurons and synapses. However, there are other ways to gain inspiration from the evolution of biological intelligence. We may be able to harness brain physiology and natural selection to evolve new types of intelligence, and perhaps superhuman intelligence. David Roden worries that the biological moral heritage of humanity may disappear entirely after the emergence of superintelligent intellects, whether artificial or of biological origin. Such agents may emerge with utterly novel features and behaviour that cannot be predicted from their evolutionary histories.

The essays in Part IV of the volume are skeptical about the singularity, each focusing on a particular aspect such as the intelligence explosion or the prospects of acceleration continuing over the next few decades. A report developed by the American Association for Artificial Intelligence considers the future development of artificial intelligence (AI). While optimistic about specific advances, the report is highly skeptical about grand predictions of an intelligence explosion, of a 'coming singularity', and about any loss of human control. Alessio Plebe and Pietro Perconti argue that the trends analysis as singularitarians present it is faulty: far from rising, the pace of change is not accelerating but in fact slowing down, and even starting to decline. Futurist Theodore Modis is deeply skeptical about any type of singularity. He focuses his skepticism on Kurzweil's work, arguing that analysis of past trends does not support long-term future acceleration. For Modis, technological change takes the form of S-curves (logistic functions), which means that its trajectory is consistent with exponential acceleration for only a very short time. Modis expects computations and related technologies to slow down and level off. While technological advances will continue to be disruptive, there will be no singularity. Other authors go further and argue that most literature on the singularity is not genuinely scientific but theological. Focusing on Kurzweil's work, Diane Proudfoot's essay develops the notion that singularitarianism is a kind of millenarian ideology (Bozeman 1997; Geraci 2010; Steinhart 2012) or "the religion of technology" (Noble 1999). Selmer Bringsjord, Alexander Bringsjord, and Paul Bello compare belief in the singularity to fideism in traditional Christianity, which denies the relevance of evidence or reason.

The last essay in Part IV offers an ambitious theory of acceleration that attempts to unify cosmic evolution with biological, cultural and technological evolution. Eric Chaisson argues that complexity can be shown consistently to increase from the Big Bang to the present, and that the same forces that drive the rise of complexity in Nature generally also underlie technological progress. To support this sweeping argument, Chaisson defines the physical quantity of *energy density rate* and shows how it unifies the view of an accelerating grow along physical, biological, cultural, and technological evolution. But while Chaisson accepts the first element of the technological singularity, *acceleration*, he rejects the second, *discontinuity*—hence the singularity: "there is no reason to claim that the next evolutionary leap forward beyond sentient beings and their amazing gadgets will be any more important than the past emergence of increasingly intricate complex systems." Chaisson reminds us that our little planet is not the only place in the universe where evolution is happening. Our machines may achieve superhuman intelligence. But perhaps a technological singularity will happen first elsewhere in the cosmos. Maybe it has already done so.

Conclusions

History shows time and again that the predictions made by futurists (and economists, sociologists, politicians, etc.) have been confounded by the behaviour of self-reflexive agents. Some forecasts are self-fulfilling, others self-stultifying. Where, if at all, do predictions of a technological singularity fit into this typology? How are the lay public/political elites likely to respond if singularitarian ideas gain widespread currency? Will the 21st century mark the end of the human era? And if so, *will biological humanity's successors be our descendants*? It is our hope and belief that this volume will help to move these questions beyond the sometimes wild speculations of the blogosphere and promote the growth of singularity studies as a rigorous scholarly discipline.

References

Adams, H. (1909). The rule of phase applied to history. In H. Adams & B. Adams (Eds.) (1920) *The degradation of the democratic dogma.* (pp. 267–311). New York: Macmillan.

Adams, H. (1904). A law of acceleration. In H. Adams (1919) *The education of Henry Adams.* New York: Houghton Mifflin, Chap. 34.

Bostrom, N. to appear. "Intelligence explosion".

Bozeman. J. (1997). Technological millenarianism in the United States. In T. Robbins & S. Palmer (Eds.) (1997) *Millennium, messiahs, and mayhem: contemporary apocalyptic movements* (pp. 139–158). New York: Routledge.

Butler, S. (1872/1981). *Erewhon: or, over the range.* In H. P. Breuer & D. F. Howard. Newark: University of Delaware Press.

Broderick, D. (2001). The spike: how our lives are being transformed by rapidly advancing technologies. New York: Tom Doherty Associates.

Chalmers, D. (2010). The singularity: a philosophical analysis. *Journal of Consciousness Studies 17*, 7–65.

Garreau, J. (2005). *Radical evolution*. New York: Doubleday.

Geraci, R. (2010). Apocalypic AI: visions of heaven in robotics, artificial intelligence, and virtual reality. New York: Oxford University Press.

Good, I. (1965). Speculations concerning the first ultraintelligent machine. In Alt, F., Rubinoff, M. (Eds.) *Advances in Computers* Vol. 6. New York: Academic Press.

Hanson, R. (1994). If uploads come first: crack of a future dawn. *Extropy 6* (1), 10–15.

Hausman, C. (1993). *Charles S. Peirce's evolutionary philosophy.* New York: Cambridge University Press.

Hirshfeld, Y. (2011). A note on mathematical singularity and technological singularity. *The singularity hypothesis*, blog entry, 5 Feb. Available http://singularityhypothesis.blogspot.co.uk/2011/02/note-on-mathematical-singularity-and.html.

Horgan, J. (2008). The consciousness conundrum. *IEEE Spectrum 45*(6), 36–41.

Hegel, G. W. F. (1807/1977). *Phenomenology of spirit.* Trans. A. V. Miller. New York: Oxford University Press.

Joy, B. (2000). Why the future doesn't need us. http://www.wired.com/wired/archive/8.04/joy.html.

Kurzweil, R. (1999). *The age of spiritual machines.* New York: Penguin.

Kurzweil, R. (2005). The singularity is near: when humans transcend biology. New York: Viking.

Lovejoy, A. (1936). *The great chain of being.* Cambridge: Harvard University Press.

Meadows, D. H., Meadows, D. L, Randers, J., & Behrens, W. (1972). *The limits to growth: a report for the club of Rome project on the predicament of mankind.* New York: Universe Books.

Meadows, D. H., Randers, J., & Meadows, D. L (2004). *The limits to growth: the thirty year update.* White River Junction, VT: Chelsea Green Books.

Modis, T. (2003). The limits of complexity and change. *The futurist* (May–June), 26–32.

Moore, G. E. (1965). Cramming more components onto integrated circuits. *Electronics 38*(8), 114–117.

Moravec, H. (1988). *Mind children: the future of robot and human intelligence.* Cambridge: Harvard University Press.

Moravec, H. (2000). *Robot: mere machine to transcendent mind.* New York: Oxford University Press.

Noble, D. F. (1999). The religion of technology: the divinity of man and the spirit of invention. New York: Penguin.

Paul, G. S., E. D. Cox (1996). *Beyond humanity: cyberevolution and future minds.* Rockland, MA: Charles River Media.

Sagan, C. (1977). *The dragons of Eden.* New York: Random House.

Steinhart, E. (2008). Teilhard de Chardin and transhumanism. *Journal of Evolution and Technology 20,* 1–22. Online at < jetpress.org/v20/steinhart.htm >.

Steinhart, E. (2012). Digital theology: is the resurrection virtual? In M. Luck (Ed.) (2012) *A philosophical exploration of new and alternative religious movements.* Farnham, UK: Ashgate.

Stine, G. H. (1961). Science fiction is too conservative. *Analog Science Fact and Fiction LXVII* (3), 83–99.

Teilhard de Chardin, P. (1955/2002). *The phenomenon of man.* Transactions B. Wall. New York: Harper Collins. Originally written 1938–1940.

Toffler, A. (1970). *Future shock.* New York: Random House.

Toffler, A. (1980). *The third wave.* New York: Bantam.

Turing, A. M. (1950). Computing machinery and intelligence. *Mind 59* (236), 433–460.

Turing, A. M. (1951). Intelligent machinery, a heretical theory. *The 51 Society.* BBC programme.

Ulam, S. (1958) Tribute to John von Neumann. *Bulletin of the American mathematical society 64* (3.2), 1–49.

Vinge, V. (1993). The coming technological singularity: how to survive in the post-human era. In *Proc. Vision 21: interdisciplinary science and engineering in the era of cyberspace,* (pp. 11–22). NASA: Lewis Research Center.

Zimmerman, M. (2008). The singularity: a crucial phase in divine self-actualization? *Cosmos and history: The Journal of Natural and Social Philosophy 4*(1–2), 347–370.

Part I
A Singularity of Artificial Superintelligence

Chapter 2
Intelligence Explosion: Evidence and Import

Luke Muehlhauser and Anna Salamon

Abstract In this chapter we review the evidence for and against three claims: that (1) there is a substantial chance we will create human-level AI before 2100, that (2) if human-level AI is created, there is a good chance vastly superhuman AI will follow via an "intelligence explosion," and that (3) an uncontrolled intelligence explosion could destroy everything we value, but a controlled intelligence explosion would benefit humanity enormously if we can achieve it. We conclude with recommendations for increasing the odds of a controlled intelligence explosion relative to an uncontrolled intelligence explosion.

The best answer to the question, "Will computers ever be as smart as humans?" is probably "Yes, but only briefly".
Vernor Vinge

Introduction

Humans may create human-level[1] artificial intelligence (AI) this century. Shortly thereafter, we may see an "intelligence explosion" or "technological singularity"—a chain of events by which human-level AI leads, fairly rapidly, to intelligent systems whose capabilities far surpass those of biological humanity as a whole.

---

[1] We will define "human-level AI" more precisely later in the chapter.

L. Muehlhauser (✉) · A. Salamon
Machine Intelligence Research Institute, Berkeley, CA 94705, USA
e-mail: luke@singularity.org

A. H. Eden et al. (eds.), *Singularity Hypotheses*, The Frontiers Collection,
DOI: 10.1007/978-3-642-32560-1_2,

How likely is this, and what will the consequences be? Others have discussed these questions previously (Turing 1950, 1951; Good 1959, 1965, 1970, 1982; Von Neumann 1966; Minsky 1984; Solomonoff 1985; Vinge 1993; Yudkowsky 2008a; Nilsson 2009, Chap. 35; Chalmers 2010; Hutter 2012a); our aim is to provide a brief review suitable both for newcomers to the topic and for those with some familiarity with the topic but expertise in only *some* of the relevant fields.

For a more comprehensive review of the arguments, we refer our readers to Chalmers (2010, forthcoming) and Bostrom (Forthcoming[a]). In this short chapter we will quickly survey some considerations for and against three claims:

1. There is a substantial chance we will create human-level AI before 2100;
2. If human-level AI is created, there is a good chance vastly superhuman AI will follow via an intelligence explosion;
3. An uncontrolled intelligence explosion could destroy everything we value, but a *controlled* intelligence explosion would benefit humanity enormously if we can achieve it.

Because the term "singularity" is popularly associated with several claims and approaches we will not defend (Sandberg 2010), we will first explain what we are *not* claiming.

First, we will not tell detailed stories about the future. Each step of a story may be probable, but if there are many such steps, the whole story itself becomes improbable (Nordmann 2007; Tversky and Kahneman 1983). We will not assume the continuation of Moore's law, nor that hardware trajectories determine software progress, nor that faster computer speeds necessarily imply faster "thought" (Proudfoot and Copeland 2012), nor that technological trends will be exponential (Kurzweil 2005) rather than "S-curved" or otherwise (see Modis, this volume), nor indeed that AI progress will accelerate rather than decelerate (see Plebe and Perconti, this volume). Instead, we will examine convergent outcomes that—like the evolution of eyes or the emergence of markets—can come about through any of several different paths and can gather momentum once they begin. Humans tend to underestimate the likelihood of outcomes that can come about through many different paths (Tversky and Kahneman 1974), and we believe an intelligence explosion is one such outcome.

Second, we will not assume that human-level intelligence can be realized by a classical Von Neumann computing architecture, nor that intelligent machines will have internal mental properties such as consciousness or human-like "intentionality," nor that early AIs will be geographically local or easily "disembodied." These properties are not required to build AI, so objections to these claims (Lucas 1961; Dreyfus 1972; Searle 1980; Block 1981; Penrose 1994; Van Gelder and Port 1995) are not objections to AI (Chalmers 1996, Chap. 9; Nilsson 2009, Chap. 24;

McCorduck 2004, Chaps. 8 and 9; Legg 2008; Heylighen 2012) or to the possibility of intelligence explosion (Chalmers, forthcoming).[2] For example: a machine need not be *conscious* to intelligently reshape the world according to its preferences, as demonstrated by goal-directed "narrow AI" programs such as the leading chess-playing programs.

We must also be clear on what we mean by "intelligence" and by "AI." Concerning "intelligence," Legg and Hutter (2007) found that definitions of intelligence used throughout the cognitive sciences converge toward the idea that "Intelligence measures an agent's ability to achieve goals in a wide range of environments." We might call this the "optimization power" concept of intelligence, for it measures an agent's power to optimize the world according to its preferences across many domains. But consider two agents which have equal ability to optimize the world according to their preferences, one of which requires much more computational time and resources to do so. They have the same optimization power, but one seems to be optimizing more intelligently. For this reason, we adopt Yudkowsky's (2008b) description of intelligence as optimization power divided by resources used.[3] For our purposes, "intelligence" measures an agent's capacity for *efficient* cross-domain optimization of the world according to the agent's preferences. Using this definition, we can avoid common objections to the use of human-centric notions of intelligence in discussions of the technological singularity (Greenfield 2012), and hopefully we can avoid common anthropomorphisms that often arise when discussing intelligence (Muehlhauser and Helm, this volume).

---

[2] Chalmers (2010) suggested that AI will lead to intelligence explosion if an AI is produced by an "extendible method," where an extendible method is "a method that can easily be improved, yielding more intelligent systems." McDermott (2012a, b) replies that if P $\neq$ NP (see Goldreich 2010 for an explanation) then there is no extendible method. But McDermott's notion of an extendible method is not the one essential to the possibility of intelligence explosion. McDermott's formalization of an "extendible method" requires that the program generated by each step of improvement under the method be able to solve in polynomial time all problems in a particular class—the class of solvable problems of a given (polynomially step-dependent) size in an NP-complete class of problems. But this is not required for an intelligence explosion in Chalmers' sense (and in our sense). What intelligence explosion (in our sense) would require is merely that a program self-improve to *vastly outperform humans*, and we argue for the plausibility of this in section From AI to Machine Superintelligence of our chapter. Thus while we agree with McDermott that it is probably true that P $\neq$ NP, we do not agree that this weighs against the plausibility of intelligence explosion. (Note that due to a miscommunication between McDermott and the editors, a faulty draft of McDermott (McDermott 2012a) was published in *Journal of Consciousness Studies*. We recommend reading the corrected version at http://cs-www.cs.yale.edu/homes/dvm/papers/chalmers-singularity-response.pdf.).

[3] This definition is a useful starting point, but it could be improved. Future work could produce a definition of intelligence as optimization power over a canonical distribution of environments, with a penalty for resource use—e.g. the "speed prior" described by Schmidhuber (2002). Also see Goertzel (2006, p. 48, 2010), Hibbard (2011).

By "AI," we refer to general AI rather than narrow AI. That is, we refer to "systems which match or exceed the [intelligence] of humans in virtually all domains of interest" (Shulman and Bostrom 2012). By this definition, IBM's *Jeopardy!*-playing computer Watson is not an "AI" (in our sense) but merely a *narrow* AI, because it can only solve a narrow set of problems. Drop Watson in a pond or ask it to do original science, and it would be helpless even if given a month's warning to prepare. Imagine instead a machine that could invent new technologies, manipulate humans with acquired social skills, and otherwise learn to navigate many new social and physical environments as needed to achieve its goals.

Which kinds of machines might accomplish such feats? There are many possible types. A *whole brain emulation* (WBE) would be a computer emulation of brain structures sufficient to functionally reproduce human cognition. We need not understand the mechanisms of general intelligence to use the human intelligence software already invented by evolution (Sandberg and Bostrom 2008). In contrast, "de novo AI" requires inventing intelligence software anew. There is a vast space of possible mind designs for de novo AI (Dennett 1996; Yudkowsky 2008a). De novo AI approaches include the symbolic, probabilistic, connectionist, evolutionary, embedded, and other research programs (Pennachin and Goertzel 2007).

From Here to AI

When should we expect the first creation of AI? We must allow for a wide range of possibilities. Except for weather forecasters (Murphy and Winkler 1984) and successful professional gamblers, nearly all of us give inaccurate probability estimates, and in particular we are overconfident of our predictions (Lichtenstein et al. 1982; Griffin and Tversky 1992; Yates et al. 2002). This overconfidence affects professional forecasters, too (Tetlock 2005), and we have little reason to think AI forecasters have fared any better.[4] So if you ave a gut feeling about when AI will be created, it is probably wrong.

But uncertainty is not a "get out of prediction free" card (Bostrom 2007). We still need to decide whether or not to encourage WBE development, whether or not to help fund AI safety research, etc. Deciding either way already implies some sort of prediction. Choosing not to fund AI safety research suggests that we do not think AI is near, while funding AI safety research implies that we think AI might be coming soon.

---

[4] To take one of many examples, Simon (1965, p. 96) predicted that "machines will be capable, within twenty years, of doing any work a man can do." Also see Crevier (1993).

Predicting AI

How, then, might we predict when AI will be created? We consider several strategies below.

By gathering the wisdom of experts or crowds. Many experts and groups have tried to predict the creation of AI. Unfortunately, experts' predictions are often little better than those of laypeople (Tetlock 2005), expert elicitation methods have in general not proven useful for long-term forecasting,[5] and prediction markets (ostensibly drawing on the opinions of those who believe themselves to possess some expertise) have not yet been demonstrated useful for technological forecasting (Williams 2011). Still, it may be useful to note that none to few experts expect AI within five years, whereas many experts expect AI by 2050 or 2100.[6]

By simple hardware extrapolation. The novelist Vinge (1993) based his own predictions about AI on hardware trends, but in a 2003 reprint of his article, Vinge notes the insufficiency of this reasoning: even if we acquire hardware sufficient for AI, we may not have the software problem solved.[7]

Hardware extrapolation may be a more useful method in a context where the intelligence software is already written: whole brain emulation. Because WBE seems to rely mostly on scaling up existing technologies like microscopy and large-scale cortical simulation, WBE may be largely an "engineering" problem, and thus the time of its arrival may be more predictable than is the case for other kinds of AI.

Several authors have discussed the difficulty of WBE in detail (Kurzweil 2005; Sandberg and Bostrom 2008; de Garis et al. 2010; Modha et al. 2011; Cattell and Parker 2012). In short: The difficulty of WBE depends on many factors, and in particular on the resolution of emulation required for successful WBE. For example, proteome-resolution emulation would require more resources and technological development than emulation at the resolution of the brain's neural network. In perhaps the most likely scenario,

> WBE on the neuronal/synaptic level requires relatively modest increases in microscopy resolution, a less trivial development of automation for scanning and image processing, a research push at the problem of inferring functional properties of neurons and synapses, and relatively business-as-usual development of computational neuroscience models and computer hardware. (Sandberg and Bostrom 2008, p. 83)

---

[5] Armstrong (1985), Woudenberg (1991), Rowe and Wright (2001). But, see Parente and Anderson-Parente (2011).

[6] Bostrom (2003), Bainbridge (2006), Legg (2008), Baum et al. (2011), Sandberg and Bostrom (2011), Nielsen (2011).

[7] A software bottleneck may delay AI but create greater risk. If there is a software bottleneck on AI, then when AI is created there may be a "computing overhang": large amounts of inexpensive computing power which could be used to run thousands of AIs or give a few AIs vast computational resources. This may not be the case if early AIs require quantum computing hardware, which is less likely to be plentiful and inexpensive than classical computing hardware at any given time.

By considering the time since Dartmouth. We have now seen more than 50 years of work toward machine intelligence since the seminal Dartmouth conference on AI, but AI has not yet arrived. This seems, intuitively, like strong evidence that AI won't arrive in the next minute, good evidence it won't arrive in the next year, and significant but far from airtight evidence that it won't arrive in the next few decades. Such intuitions can be formalized into models that, while simplistic, can form a useful starting point for estimating the time to machine intelligence.[8]

By tracking progress in machine intelligence. Some people intuitively estimate the time until AI by asking what proportion of human abilities today's software can match, and how quickly machines are catching up.[9] However, it is not clear how to divide up the space of "human abilities," nor how much each one matters. We also don't know if progress in machine intelligence will be linear, exponential, or otherwise. Watching an infant's progress in learning calculus might lead one to infer the child will not learn it until the year 3000, until suddenly the child learns it in a spurt at age 17. Still, it may be worth asking whether a measure can be found for which both: (a) progress is predictable enough to extrapolate; and (b) when performance rises to a certain level, we can expect AI.

By extrapolating from evolution. Evolution managed to create intelligence without using intelligence to do so. Perhaps this fact can help us establish an upper bound on the difficulty of creating AI (Chalmers 2010; Moravec 1976, 1998, 1999), though this approach is complicated by observation selection effects (Shulman and Bostrom 2012).

By estimating progress in scientific research output. Imagine a man digging a 10 km ditch. If he digs 100 meters in one day, you might predict the ditch will be finished in 100 days. But what if 20 more diggers join him, and they are all given

---

[8] We can make a simple formal model of this evidence by assuming (with much simplification) that every year a coin is tossed to determine whether we will get AI that year, and that we are initially unsure of the weighting on that coin. We have observed more than 50 years of "no AI" since the first time serious scientists believed AI might be around the corner. This "56 years of no AI" observation would be highly unlikely under models where the coin comes up "AI" on 90 % of years (the probability of our observations would be 10^-56), or even models where it comes up "AI" in 10 % of all years (probability 0.3 %), whereas it's the expected case if the coin comes up "AI" in, say, 1 % of all years, or for that matter in 0.0001 % of all years. Thus, in this toy model, our "no AI for 56 years" observation should update us strongly against coin weightings in which AI would be likely in the next minute, or even year, while leaving the relative probabilities of "AI expected in 200 years" and "AI expected in 2 million years" more or less untouched. (These updated probabilities are robust to choice of the time interval between coin flips; it matters little whether the coin is tossed once per decade, or once per millisecond, or whether one takes a limit as the time interval goes to zero). Of course, one gets a different result if a different "starting point" is chosen, e.g. Alan Turing's seminal paper on machine intelligence (Turing 1950) or the inaugural conference on artificial general intelligence (Wang et al. 2008). For more on this approach and Laplace's rule of succession, see Jaynes (2003), Chap. 18. We suggest this approach only as a way of generating a prior probability distribution over AI timelines, from which one can then update upon encountering additional evidence.

[9] Relatedly, Good (1970) tried to predict the first creation of AI by surveying past conceptual breakthroughs in AI and extrapolating into the future.

backhoes? Now the ditch might not take so long. Analogously, when predicting progress toward AI it may be useful to consider not how much progress is made per year, but instead how much progress is made per unit of research effort, and how many units of research effort we can expect to be applied to the problem in the coming decades.

Unfortunately, we have not yet discovered demonstrably reliable methods for long-term technological forecasting. New methods are being tried (Nagy et al. 2010), but until they prove successful we should be particularly cautious when predicting AI timelines. Below, we attempt a final approach by examining some plausible *speed bumps* and *accelerators* on the path to AI.

Speed Bumps

Several factors may decelerate our progress toward the first creation of AI. For example:

An end to Moore's law. Though several information technologies have progressed at an exponential or superexponential rate for many decades (Nagy et al. 2011), this trend may not hold for much longer (Mack 2011).

Depletion of low-hanging fruit. Scientific progress is not only a function of research effort but also of the ease of scientific discovery; in some fields there is pattern of increasing difficulty with each successive discovery (Arbesman 2011; Jones 2009). AI may prove to be a field in which new discoveries require far more effort than earlier discoveries.

Societal collapse. Various political, economic, technological, or natural disasters may lead to a societal collapse during which scientific progress would not continue (Posner 2004; Bostrom and Ćirković 2008).

Disinclination. Chalmers (2010), Hutter (2012a) think the most likely speed bump in our progress toward AI will be disinclination, including active prevention. Perhaps humans will not want to create their own successors. New technologies like "Nanny AI" (Goertzel 2012), or new political alliances like a stable global totalitarianism (Caplan 2008), may empower humans to delay or prevent scientific progress that could lead to the creation of AI.

Accelerators

Other factors, however, may accelerate progress toward AI:

More hardware. For at least four decades, computing power[10] has increased exponentially, roughly in accordance with Moore's law.[11] Experts disagree on how

---

[10] The technical measure predicted by Moore's law is the density of components on an integrated circuit, but this is closely tied to the price-performance of computing power.

[11] For important qualifications, see Nagy et al. (2010), Mack (2011).

much longer Moore's law will hold (Mack 2011; Lundstrom 2003), but even if hardware advances more slowly than exponentially, we can expect hardware to be far more powerful in a few decades than it is now.[12] More hardware doesn't by itself give us machine intelligence, but it contributes to the development of machine intelligence in several ways:

> Powerful hardware may improve performance simply by allowing existing "brute force" solutions to run faster (Moravec 1976). Where such solutions do not yet exist, researchers might be incentivized to quickly develop them given abundant hardware to exploit. Cheap computing may enable much more extensive experimentation in algorithm design, tweaking parameters or using methods such as genetic algorithms. Indirectly, computing may enable the production and processing of enormous datasets to improve AI performance (Halevi et al. 2009), or result in an expansion of the information technology industry and the quantity of researchers in the field. (Shulman and Sandberg 2010)

Better algorithms. Often, mathematical insights can reduce the computation time of a program by many orders of magnitude without additional hardware. For example, IBM's Deep Blue played chess at the level of world champion Garry Kasparov in 1997 using about 1.5 trillion instructions per second (TIPS), but a program called Deep Junior did it in 2003 using only 0.015 TIPS. Thus, the computational efficiency of the chess algorithms increased by a factor of 100 in only six years (Richards and Shaw 2004).

Massive datasets. The greatest leaps forward in speech recognition and translation software have come not from faster hardware or smarter hand-coded algorithms, but from access to massive data sets of human-transcribed and human-translated words (Halevi et al. 2009). Datasets are expected to increase greatly in size in the coming decades, and several technologies promise to actually *outpace* "Kryder's law" (Kryder and Kim 2009), which states that magnetic disk storage density doubles approximately every 18 months (Walter 2005).

Progress in psychology and neuroscience. Cognitive scientists have uncovered many of the brain's algorithms that contribute to human intelligence (Trappenberg 2009; Ashby and Helie 2011). Methods like neural networks (imported from neuroscience) and reinforcement learning (inspired by behaviorist psychology) have already resulted in significant AI progress, and experts expect this insight-transfer from neuroscience to AI to continue and perhaps accelerate (Van der Velde 2010; Schierwagen 2011; Floreano and Mattiussi 2008; de Garis et al. 2010; Krichmar and Wagatsuma 2011).

Accelerated science. A growing First World will mean that more researchers at well-funded universities will be conducting research relevant to machine

---

[12] Quantum computing may also emerge during this period. Early worries that quantum computing may not be feasible have been overcome, but it is hard to predict whether quantum computing will contribute significantly to the development of machine intelligence because progress in quantum computing depends heavily on relatively unpredictable insights in quantum algorithms and hardware (Rieffel and Polak 2011).

intelligence. The world's scientific output (in publications) grew by one third from 2002 to 2007 alone, much of this driven by the rapid growth of scientific output in developing nations like China and India (Royal Society 2011).[13] Moreover, new tools can accelerate particular fields, just as fMRI accelerated neuroscience in the 1990s, and the effectiveness of scientists themselves can potentially be increased with cognitive enhancement pharmaceuticals (Bostrom and Sandberg 2009) and brain-computer interfaces that allow direct neural access to large databases (Groß 2009). Finally, new collaborative tools like blogs and Google Scholar are already yielding results such as the Polymath Project, which is rapidly and collaboratively solving open problems in mathematics (Nielsen 2011).[14]

Economic incentive. As the capacities of "narrow AI" programs approach the capacities of humans in more domains (Koza 2010), there will be increasing demand to replace human workers with cheaper, more reliable machine workers (Hanson 2008, Forthcoming; Kaas et al. 2010; Brynjolfsson and McAfee 2011).

First-mover incentives. Once AI looks to be within reach, political and private actors will see substantial advantages in building AI first. AI could make a small group more powerful than the traditional superpowers—a case of "bringing a gun to a knife fight." The race to AI may even be a "winner take all" scenario. Thus, political and private actors who realize that AI is within reach may devote substantial resources to developing AI as quickly as possible, provoking an AI arms race (Gubrud 1997).

How Long, Then, Before AI?

We have not yet mentioned two small but significant developments leading us to agree with Schmidhuber (2012) that "progress towards self-improving AIs is already substantially beyond what many futurists and philosophers are aware of." These two developments are Marcus Hutter's universal and provably optimal AIXI agent model (Hutter 2005) and Jürgen Schmidhuber's universal self-improving Gödel machine models (Schmidhuber 2007, 2009).

Schmidhuber (2012) summarizes the importance of the Gödel machine:

> [The] Gödel machine... already *is* a universal AI that is at least theoretically optimal in certain sense. It may interact with some initially unknown, partially observable environment to maximize future expected utility or reward by solving arbitrary user-defined computational tasks. Its initial algorithm is not hardwired; it can completely rewrite itself without essential limits apart from the limits of computability, provided a proof searcher

---

[13] On the other hand, some worry (Pan et al. 2005) that the rates of scientific fraud and publication bias may currently be higher in China and India than in the developed world.

[14] Also, a process called "iterated embryo selection" (Uncertain Future 2012) could be used to produce an entire generation of scientists with the cognitive capabilities of Albert Einstein or John von Neumann, thus accelerating scientific progress and giving a competitive advantage to nations which choose to make use of this possibility.

> embedded within the initial algorithm can first prove that the rewrite is useful, according to the formalized utility function taking into account the limited computational resources. Self-rewrites may modify/improve the proof searcher itself, and can be shown to be *globally optimal*, relative to Gödel's well-known fundamental restrictions of provability (Gödel 1931)...
>
> All of this implies that there already exists the blueprint of a Universal AI which will solve almost all problems almost as quickly as if it already knew the best (unknown) algorithm for solving them, because almost all imaginable problems are big enough to make the additive constant negligible. Hence, I must object to Chalmers' statement [that] "we have yet to find the right algorithms, and no-one has come close to finding them yet."

Next, we turn to Hutter (2012b) for a summary of the importance of AIXI:

The concrete ingredients in AIXI are as follows: Intelligent *actions* are based on informed *decisions*. Attaining good decisions requires *predictions* which are typically based on models of the environments. Models are constructed or learned from past observations via *induction*. Fortunately, based on the *deep philosophical insights* and *powerful mathematical developments*, all these problems have been overcome, at least in theory: So what do we need (from a mathematical point of view) to construct a universal optimal learning agent interacting with an arbitrary unknown environment? The theory, coined *UAI* [Universal Artificial Intelligence], developed in the last decade and explained in Hutter (2005) says: *All you need is Ockham, Epicurus, Turing, Bayes, Solomonoff* (1964a, 1964b), *Kolmogorov* (1968), *Bellman* (1957): Sequential decision theory (Bertsekas 2007) (*Bellman's* equation) formally solves the problem of rational agents in uncertain worlds if the true environmental probability distribution is known. If the environment is unknown, Bayesians (Berger 1993) replace the true distribution by a weighted mixture of distributions from some (hypothesis) class. Using the large class of all (semi)measures that are (semi)computable on a *Turing* machine bears in mind *Epicurus*, who teaches not to discard any (consistent) hypothesis. In order not to ignore *Ockham*, who would select the simplest hypothesis, *Solomonoff* defined a universal prior that assigns high/low prior weight to simple/complex environments (Rathmanner and Hutter 2011), where *Kolmogorov* quantifies complexity (Li and Vitányi 2008). Their unification constitutes the theory of UAI and resulted in... AIXI.[15]

AIXI is incomputable, but computationally tractable approximations have already been experimentally tested, and these reveal a path to universal AI[16] that solves real-world problems in a variety of environments:

[15] In our two quotes from Hutter (2012b) we have replaced Hutter's AMS-style citations with Chicago-style citations.

[16] The creation of AI probably is not, however, merely a matter of finding computationally tractable AIXI approximations that can solve increasingly complicated problems in increasingly complicated environments. There remain many open problems in the theory of universal artificial intelligence (Hutter 2009). For problems related to allowing some AIXI-like models to self-modify, see Orseau and Ring (2011), Ring and Orseau (2011), Orseau (2011); Hibbard (Forthcoming). Dewey (2011) explains why reinforcement learning agents like AIXI may pose a threat to humanity.

> The same single [AIXI approximation "MC-AIXI-CTW"] is already able to learn to play TicTacToe, Kuhn Poker, and most impressively Pacman (Veness et al. 2011) from scratch. Besides Pacman, there are hundreds of other arcade games from the 1980s, and it would be sensational if a single algorithm could learn them all solely by trial and error, which seems feasible for (a variant of) MC-AIXI-CTW. While these are "just" recreational games, they do contain many prototypical elements of the real world, such as food, enemies, friends, space, obstacles, objects, and weapons. Next could be a test in modern virtual worlds... that require intelligent agents, and finally some selected real-world problems.

So, when will we create AI? Any predictions on the matter must have wide error bars. Given the history of confident false predictions about AI (Crevier 1993) and AI's potential speed bumps, it seems misguided to be 90 % confident that AI will succeed in the coming century. But 90 % confidence that AI will *not* arrive before the end of the century also seems wrong, given that: (a) many difficult AI breakthroughs have now been made (including the Gödel machine and AIXI), (b) several factors, such as automated science and first-mover incentives, may well accelerate progress toward AI, and (c) whole brain emulation seems to be possible and have a more predictable development than de novo AI. Thus, we think there is a significant probability that AI will be created this century. This claim is not scientific—the field of technological forecasting is not yet advanced enough for that—but we believe our claim is reasonable.

The creation of human-level AI would have serious repercussions, such as the displacement of most or all human workers (Brynjolfsson and McAfee 2011). But if AI is likely to lead to machine superintelligence, as we argue next, the implications could be even greater.

From AI to Machine Superintelligence

It seems unlikely that humans are near the ceiling of possible intelligences, rather than simply being the first such intelligence that happened to evolve. Computers far outperform humans in many narrow niches (e.g. arithmetic, chess, memory size), and there is reason to believe that similar large improvements over human performance are possible for general reasoning, technology design, and other tasks of interest. As occasional AI critic Jack Schwartz (1987) wrote:

> If artificial intelligences can be created at all, there is little reason to believe that initial successes could not lead swiftly to the construction of artificial superintelligence able to explore significant mathematical, scientific, or engineering alternatives at a rate far exceeding human ability, or to generate plans and take action on them with equally overwhelming speed. Since man's near-monopoly of all higher forms of intelligence has been one of the most basic facts of human existence throughout the past history of this planet, such developments would clearly create a new economics, a new sociology, and a new history.

Why might AI "lead swiftly" to machine superintelligence? Below we consider some reasons.

AI Advantages

Below we list a few AI advantages that may allow AIs to become not only vastly more intelligent than any human, but also more intelligent than all of biological humanity (Sotala 2012; Legg 2008). Many of these are unique to *machine* intelligence, and that is why we focus on intelligence explosion from AI rather than from biological cognitive enhancement (Sandberg 2011).

Increased computational resources. The human brain uses 85–100 billion neurons. This limit is imposed by evolution-produced constraints on brain volume and metabolism. In contrast, a machine intelligence could use scalable computational resources (imagine a "brain" the size of a warehouse). While algorithms would need to be changed in order to be usefully scaled up, one can perhaps get a rough feel for the potential impact here by noting that humans have about 3.5 times the brain size of chimps (Schoenemann 1997), and that brain size and IQ correlate positively in humans, with a correlation coefficient of about 0.35 (McDaniel 2005). One study suggested a similar correlation between brain size and cognitive ability in rats and mice (Anderson 1993).[17]

Communication speed. Axons carry spike signals at 75 meters per second or less (Kandel et al. 2000). That speed is a fixed consequence of our physiology. In contrast, software minds could be ported to faster hardware, and could therefore process information more rapidly. (Of course, this also depends on the efficiency of the algorithms in use; faster hardware compensates for less efficient software.)

Increased serial depth. Due to neurons' slow firing speed, the human brain relies on massive parallelization and is incapable of rapidly performing any computation that requires more than about 100 sequential operations (Feldman and Ballard 1982). Perhaps there are cognitive tasks that could be performed more efficiently and precisely if the brain's ability to support parallelizable pattern-matching algorithms were supplemented by support for longer sequential processes. In fact, there are many known algorithms for which the best parallel version uses far more computational resources than the best serial algorithm, due to the overhead of parallelization.[18]

Duplicability. Our research colleague Steve Rayhawk likes to describe AI as "instant intelligence; just add hardware!" What Rayhawk means is that, while it will require extensive research to design the first AI, creating additional AIs is just a matter of copying software. The population of digital minds can thus expand to fill the available hardware base, perhaps rapidly surpassing the population of biological minds.

Duplicability also allows the AI population to rapidly become dominated by newly built AIs, with new skills. Since an AI's skills are stored digitally, its exact

[17] Note that given the definition of intelligence we are using, greater computational resources would not give a machine more "intelligence" but instead more "optimization power".

[18] For example see Omohundro (1987).

current state can be copied,[19] including memories and acquired skills—similar to how a "system state" can be copied by hardware emulation programs or system backup programs. A human who undergoes education increases only his or her own performance, but an AI that becomes 10 % better at earning money (per dollar of rentable hardware) than other AIs can be used to replace the others across the hardware base—making each copy 10 % more efficient.[20]

Editability. Digitality opens up more parameters for controlled variation than is possible with humans. We can put humans through job-training programs, but we can't perform precise, replicable neurosurgeries on them. Digital workers would be more editable than human workers are. Consider first the possibilities from whole brain emulation. We know that transcranial magnetic stimulation (TMS) applied to one part of the prefrontal cortex can improve working memory (Fregni et al. 2005). Since TMS works by temporarily decreasing or increasing the excitability of populations of neurons, it seems plausible that decreasing or increasing the "excitability" parameter of certain populations of (virtual) neurons in a digital mind would improve performance. We could also experimentally modify dozens of other whole brain emulation parameters, such as simulated glucose levels, undifferentiated (virtual) stem cells grafted onto particular brain modules such as the motor cortex, and rapid connections across different parts of the brain.[21] Secondly, a modular, transparent AI could be even more directly editable than a whole brain emulation—possibly via its source code. (Of course, such possibilities raise ethical concerns).

Goal coordination. Let us call a set of AI copies or near-copies a "copy clan." Given shared goals, a copy clan would not face certain goal coordination problems that limit human effectiveness (Friedman 1994). A human cannot use a hundredfold salary increase to purchase a hundredfold increase in productive hours per day. But a copy clan, if its tasks are parallelizable, could do just that. Any gains made by such a copy clan, or by a human or human organization controlling that clan, could potentially be invested in further AI development, allowing initial advantages to compound.

Improved rationality. Some economists model humans as *Homo economicus*: self-interested rational agents who do what they believe will maximize the fulfillment of their goals (Friedman 1953). On the basis of behavioral studies, though, Schneider (2010) points out that we are more akin to Homer Simpson: we are irrational beings that lack consistent, stable goals (Stanovich 2010; Cartwright

---

[19] If the first self-improving AIs at least partially require quantum computing, the system states of these AIs might not be directly copyable due to the no-cloning theorem (Wooters and Zurek 1982).

[20] Something similar is already done with technology-enabled business processes. When the pharmacy chain CVS improves its prescription-ordering system, it can copy these improvements to more than 4,000 of its stores, for immediate productivity gains (McAfee and Brynjolfsson 2008).

[21] Many suspect that the slowness of cross-brain connections has been a major factor limiting the usefulness of large brains (Fox 2011).

2011). But imagine if you *were* an instance of *Homo economicus*. You could stay on a diet, spend the optimal amount of time learning which activities will achieve your goals, and then follow through on an optimal plan, no matter how tedious it was to execute. Machine intelligences of many types could be written to be vastly more rational than humans, and thereby accrue the benefits of rational thought and action. The rational agent model (using Bayesian probability theory and expected utility theory) is a mature paradigm in current AI design (Hutter 2005; Russel and Norvig 2009, Chap. 2).

These AI advantages suggest that AIs will be *capable* of far surpassing the cognitive abilities and optimization power of humanity as a whole, but will they be *motivated* to do so? Though it is difficult to predict the specific motivations of advanced AIs, we can make some predictions about convergent instrumental goals—instrumental goals useful for the satisfaction of almost any final goals.

Instrumentally Convergent Goals

Omohundro (2007, 2008, this volume) and Bostrom (Forthcoming[a]) argue that there are several instrumental goals that will be pursued by almost any advanced intelligence because those goals are useful intermediaries to the achievement of almost any set of final goals. For example:

1. An AI will want to preserve itself because if it is destroyed it won't be able to act in the future to maximize the satisfaction of its present final goals.
2. An AI will want to preserve the content of its current final goals because if the content of its final goals is changed it will be less likely to act in the future to maximize the satisfaction of its present final goals.[22]
3. An AI will want to improve its own rationality and intelligence because this will improve its decision-making, and thereby increase its capacity to achieve its goals.
4. An AI will want to acquire as many resources as possible, so that these resources can be transformed and put to work for the satisfaction of the AI's final and instrumental goals.

Later we shall see why these convergent instrumental goals suggest that the default outcome from advanced AI is human extinction. For now, let us examine the mechanics of AI self-improvement.

[22] Bostrom (2012) lists a few special cases in which an AI may wish to modify the content of its final goals.

Intelligence Explosion

The convergent instrumental goal for self-improvement has a special consequence. Once human programmers build an AI with a better-than-human *capacity* for AI design, the instrumental goal for self-improvement may motivate a positive feedback loop of self-enhancement.[23] Now when the machine intelligence improves itself, it improves the intelligence that does the improving. Thus, if mere human efforts suffice to produce machine intelligence this century, a large population of greater-than-human machine intelligences may be able to create a rapid cascade of self-improvement cycles, enabling a rapid transition to machine superintelligence. Chalmers (2010) discusses this process in some detail, so here we make only a few additional points.

The term "self," in phrases like "recursive self-improvement" or "when the machine intelligence improves itself," is something of a misnomer. The machine intelligence could conceivably edit its own code while it is running (Schmidhuber 2007; Schaul and Schmidhuber 2010), but it could also create new intelligences that run independently. Alternatively, several AIs (perhaps including WBEs) could work together to design the next generation of AIs. Intelligence explosion could come about through "self"-improvement or through other-AI improvement.

Once sustainable machine self-improvement begins, AI development need not proceed at the normal pace of human technological innovation. There is, however, significant debate over how fast or local this "takeoff" would be (Hanson and Yudkowsky 2008; Loosemore and Goertzel 2011; Bostrom Forthcoming[a]), and also about whether intelligence explosion would result in a stable equilibrium of multiple machine superintelligence or instead a machine "singleton" (Bostrom 2006). We will not discuss these complex issues here.

Consequences of Machine Superintelligence

If machines greatly surpass human levels of intelligence—that is, surpass humanity's capacity for efficient cross-domain optimization—we may find ourselves in a position analogous to that of the apes who watched as humans invented fire, farming, writing, science, guns and planes and then took over the planet. (One salient difference would be that no single ape witnessed the entire saga, while we might witness a shift to machine dominance within a single human lifetime).

---

[23] When the AI can perform 10 % of the AI design tasks and do them at superhuman speed, the remaining 90 % of AI design tasks act as bottlenecks. However, if improvements allow the AI to perform 99 % of AI design tasks rather than 98 %, this change produces a much larger impact than when improvements allowed the AI to perform 51 % of AI design tasks rather than 50 % (Hanson, forthcoming). And when the AI can perform 100 % of AI design tasks rather than 99 % of them, this removes altogether the bottleneck of tasks done at slow human speeds.

Such machines would be superior to us in manufacturing, harvesting resources, scientific discovery, social aptitude, and strategic action, among other capacities. We would not be in a position to negotiate with them, just as neither chimpanzees nor dolphins are in a position to negotiate with humans.

Moreover, intelligence can be applied in the pursuit of any goal. As Bostrom (2012) argues, making AIs more intelligent will not make them want to change their goal systems—indeed, AIs will be motivated to *preserve* their initial goals. Making AIs more intelligent will only make them more capable of achieving their original final goals, whatever those are.[24]

This brings us to the central feature of AI risk: Unless an AI is specifically programmed to preserve what humans value, it may destroy those valued structures (including humans) *incidentally*. As Yudkowsky (2008a) puts it, "the AI does not love you, nor does it hate you, but you are made of atoms it can use for something else."

Achieving a Controlled Intelligence Explosion

How, then, can we give AIs desirable goals before they self-improve beyond our ability to control them or negotiate with them?[25] WBEs and other brain-inspired AIs running on human-derived "spaghetti code" (Marcus 2008) may not have a clear "slot" in which to specify desirable goals. The same may also be true of other "opaque" AI designs, such as those produced by evolutionary algorithms—or even of more transparent AI designs. Even if an AI had a transparent design with a clearly definable utility function,[26] would we know how to give it desirable goals? Unfortunately, specifying what humans value may be extraordinarily difficult, given the complexity and fragility of human preferences (Yudkowsky 2011; Muehlhauser and Helm, this volume), and allowing an AI to *learn* desirable goals

---

[24] This may be less true for early-generation WBEs, but Omohundro (2008) argues that AIs will converge upon being optimizing agents, which exhibit a strict division between goals and cognitive ability.

[25] Hanson (2012) reframes the problem, saying that "we should expect that a simple continuation of historical trends will eventually end up [producing] an 'intelligence explosion' scenario. So there is little need to consider [Chalmers'] more specific arguments for such a scenario. And the inter-generational conflicts that concern Chalmers in this scenario are generic conflicts that arise in a wide range of past, present, and future scenarios. Yes, these are conflicts worth pondering, but Chalmers offers no reasons why they are interestingly different in a 'singularity' context." We briefly offer just one reason why the "inter-generational conflicts" arising from a transition of power from humans to superintelligent machines are interestingly different from previous the inter-generational conflicts: as Bostrom (2002) notes, the singularity may cause the extinction not just of people groups but of the entire human species. For a further reply to Hanson, see Chalmers (Forthcoming).

[26] A utility function assigns numerical utilities to outcomes such that outcomes with higher utilities are always preferred to outcomes with lower utilities (Mehta 1998).

from reward and punishment may be no easier (Yudkowsky 2008a). If this is correct, then the creation of self-improving AI may be detrimental *by default* unless we first solve the problem of how to build an AI with a stable, desirable utility function—a "Friendly AI" (Yudkowsky 2001).[27]

But suppose it is possible to build a Friendly AI (FAI) capable of radical self-improvement. Normal projections of economic growth allow for great discoveries relevant to human welfare to be made eventually—but a Friendly AI could make those discoveries much sooner. A benevolent machine superintelligence could, as Bostrom (2003) writes, "create opportunities for us to vastly increase our own intellectual and emotional capabilities, and it could assist us in creating a highly appealing experiential world in which we could live lives devoted [to] joyful game-playing, relating to each other, experiencing, personal growth, and to living closer to our ideals."

Thinking that FAI may be too difficult, Goertzel (2012) proposes a global "Nanny AI" that would "forestall a full-on Singularity for a while, …giving us time to figure out what kind of Singularity we really want to build and how." Goertzel and others working on AI safety theory would very much appreciate the extra time to solve the problems of AI safety before the first self-improving AI is created, but your authors suspect that Nanny AI is "FAI-complete," or nearly so. That is, in order to build Nanny AI, you may need to solve all the problems required to build full-blown Friendly AI, for example the problem of specifying precise goals (Yudkowsky 2011; Muehlhauser and Helm, this volume) and the problem of maintaining a stable utility function under radical self-modification, including updates to the AI's internal ontology (de Blanc 2011).

The approaches to controlled intelligence explosion we have surveyed so far attempt to constrain an AI's goals, but others have suggested a variety of "external" constraints for goal-directed AIs: physical and software confinement (Chalmers 2010; Yampolskiy 2012), deterrence mechanisms, and tripwires that shut down an AI if it engages in dangerous behavior. Unfortunately, these solutions would pit human intelligence against superhuman intelligence, and we shouldn't be confident the former would prevail.

Perhaps we could build an AI of limited cognitive ability—say, a machine that only answers questions: an "Oracle AI." But this approach is not without its own dangers (Armstrong, Sandberg Forthcoming; Bostrom forthcoming).

Unfortunately, even if these latter approaches worked, they might merely delay AI risk without eliminating it. If one AI development team has successfully built either an Oracle AI or a goal-directed AI under successful external constraints, other AI development teams may not be far from building their own AIs, some of them with less effective safety measures. A Friendly AI with enough lead time, however, could permanently prevent the creation of unsafe AIs.

---

[27] It may also be an option to constrain the first self-improving AIs just long enough to develop a Friendly AI before they cause much damage.

What Can We Do About AI Risk?

Because superhuman AI and other powerful technologies may pose some risk of human extinction ("existential risk"), Bostrom (2002) recommends a program of *differential technological development* in which we would attempt "to retard the implementation of dangerous technologies and accelerate implementation of beneficial technologies, especially those that ameliorate the hazards posed by other technologies."

But good outcomes from intelligence explosion appear to depend not only on differential technological development but also, for example, on solving certain kinds of problems in decision theory and value theory before the first creation of AI (Muehlhauser 2011). Thus, we recommend a course of *differential intellectual progress*, which includes differential technological development as a special case.

Differential intellectual progress consists in prioritizing risk-*reducing* intellectual progress over risk-*increasing* intellectual progress. As applied to AI risks in particular, a plan of differential intellectual progress would recommend that our progress on the scientific, philosophical, and technological problems of AI *safety* outpace our progress on the problems of AI *capability* such that we develop *safe* superhuman AIs before we develop (arbitrary) superhuman AIs. Our first superhuman AI must be a safe superhuman AI, for we may not get a second chance (Yudkowsky 2008a). With AI as with other technologies, we may become victims of "the tendency of technological advance to outpace the social control of technology" (Posner 2004).

Conclusion

We have argued that AI poses an existential threat to humanity. On the other hand, with more intelligence we can hope for quicker, better solutions to many of our problems. We don't usually associate cancer cures or economic stability with artificial intelligence, but curing cancer is ultimately a problem of being smart enough to figure out how to cure it, and achieving economic stability is ultimately a problem of being smart enough to figure out how to achieve it. To whatever extent we have goals, we have goals that can be accomplished to greater degrees using sufficiently advanced intelligence. When considering the likely consequences of superhuman AI, we must respect both risk and opportunity.[28]

[28] Our thanks to Nick Bostrom, Steve Rayhawk, David Chalmers, Steve Omohundro, Marcus Hutter, Brian Rabkin, William Naaktgeboren, Michael Anissimov, Carl Shulman, Eliezer Yudkowsky, Louie Helm, Jesse Liptrap, Nisan Stiennon, Will Newsome, Kaj Sotala, Julia Galef, and anonymous reviewers for their helpful comments.

References

Anderson, B. (1993). Evidence from the rat for a general factor that underlies cognitive performance and that relates to brain size: intelligence? *Neuroscience Letters, 153*(1), 98–102. doi:10.1016/0304-3940(93)90086-Z.

Arbesman, S. (2011). Quantifying the ease of scientific discovery. *Scientometrics, 86*(2), 245–250. doi:10.1007/s11192-010-0232-6.

Armstrong, J. S. (1985). *Long-range forecasting: from crystal ball to computer* (2nd ed.). New York: Wiley.

Armstrong, S., Sandberg, A., & Bostrom N. Forthcoming. Thinking inside the box: using and controlling an Oracle AI. *Minds and Machines.*

Ashby, F. G., & Helie S. (2011). A tutorial on computational cognitive neuroscience: modeling the neurodynamics of cognition. *Journal of Mathematical Psychology, 55*(4), 273–289. doi:10.1016/j.jmp.2011.04.003.

Bainbridge, W. S., & Roco, M. C. (Eds.). (2006). *Managing nano-bio-info-cogno innovations: converging technologies in society*. Dordrecht: Springer.

Baum, S. D., Goertzel, B., & Goertzel, T. G. (2011). How long until human-level AI? Results from an expert assessment. *Technological Forecasting and Social Change, 78*(1), 185–195. doi:10.1016/j.techfore.2010.09.006.

Bellman, R. E. (1957). *Dynamic programming*. Princeton: Princeton University Press.

Berger, J. O. (1993). Statistical decision theory and bayesian analysis (2nd edn). *Springer Series in Statistics*. New York: Springer.

Bertsekas, D. P. (2007). *Dynamic programming and optimal control* (Vol. 2). Nashua: Athena Scientific.

Block, N. (1981). Psychologism and behaviorism. *Philosophical Review, 90*(1), 5–43. doi:10.2307/2184371.

Bostrom, N. (2002). Existential risks: Analyzing human extinction scenarios and related hazards. *Journal of Evolution and Technology, 9* http://www.jetpress.org/volume9/risks.html.

Bostrom, N. (2003). Ethical issues in advanced artificial intelligence. In I. Smit & G. E. Lasker (Eds.), *Cognitive, emotive and ethical aspects of decision making in humans and in artificial intelligence*. Windsor: International Institute of Advanced Studies in Systems Research/ Cybernetics. Vol. 2.

Bostrom, N. (2006). What is a singleton? *Linguistic and Philosophical Investigations, 5*(2), 48–54.

Bostrom, N. (2007). Technological revolutions: Ethics and policy in the dark. In M. Nigel, S. de Cameron, & M. E. Mitchell (Eds.), *Nanoscale: Issues and perspectives for the nano century* (pp. 129–152). Hoboken: Wiley. doi:10.1002/9780470165874.ch10.

Bostrom, N. Forthcoming(a). *Superintelligence: A strategic analysis of the coming machine intelligence revolution.* Manuscript, in preparation.

Bostrom, N. (2012). The superintelligent will: Motivation and instrumental rationality in advanced artificial agents. *Minds and Machines.* Preprint at, http://www.nickbostrom.com/ superintelligentwill.pdf.

Bostrom, N., & Ćirković, M. M. (Eds.). (Eds.). (2008). *Global catastrophic risks*. New York: Oxford University Press.

Bostrom, N., & Sandberg, A. (2009). Cognitive enhancement: Methods, ethics, regulatory challenges. *Science and Engineering Ethics, 15*(3), 311–341. doi:10.1007/s11948-009-9142-5.

Brynjolfsson, E., & McAfee, A. (2011). *Race against the machine: How the digital revolution is accelerating innovation, driving productivity, and irreversibly transforming employment and the economy*. Lexington: Digital Frontier Press. Kindle edition.

Caplan, B. (2008). The totalitarian threat. In Bostrom and Ćirković 2008, 504–519.

Cartwright, E. (2011). *Behavioral economics*. New York: Routledge Advanced Texts in Economics and Finance.

Cattell, R, & Parker, A. (2012). *Challenges for brain emulation: why is building a brain so difficult?* Synaptic Link, Feb. 5. http://synapticlink.org/Brain%20Emulation%20 Challenges.pdf.
Chalmers, D. J. (1996). *The conscious mind: In search of a fundamental theory*. New York: Oxford University Press. (Philosophy of Mind Series).
Chalmers, D. J. (2010). The singularity: A philosophical analysis. *Journal of Consciousness Studies 17*(9–10), 7–65. http://www.ingentaconnect.com/content/imp/jcs/2010/00000017/f0020009/art00001.
Chalmers, D. J. Forthcoming. The singularity: A reply. *Journal of Consciousness Studies 19*.
Crevier, D. (1993). *AI: The tumultuous history of the search for artificial intelligence*. New York: Basic Books.
de Blanc, P. (2011). *Ontological crises in artificial agents' value systems*. San Francisco: Singularity Institute for Artificial Intelligence, May 19. http://arxiv.org/abs/1105.3821.
de Garis, H., Shuo, C., Goertzel, B., & Ruiting, L. (2010). A world survey of artificial brain projects, part I: Large-scale brain simulations. *Neurocomputing, 74*(1–3), 3–29. doi:10.1016/j.neucom.2010.08.004.
Dennett, D. C. (1996). *Kinds of minds: Toward an understanding of consciousness.*, Science Master New York: Basic Books.
Dewey, D. (2011). Learning what to value. In Schmidhuber, J., Thórisson, KR., & Looks, M. 2011, 309–314.
Dreyfus, H. L. (1972). *What computers can't do: A critique of artificial reason*. New York: Harper & Row.
Eden, A., Søraker, J., Moor, J. H., & Steinhart, E. (Eds.). (2012). *The singularity hypothesis: A scientific and philosophical assessment*. Berlin: Springer.
Feldman, J. A., & Ballard, D. H. (1982). Connectionist models and their properties. *Cognitive Science, 6*(3), 205–254. doi:10.1207/s15516709cog0603_1.
Floreano, D., & Mattiussi, C. (2008). *Bio-inspired artificial intelligence: Theories, methods, and technologies*. Intelligent Robotics and Autonomous Agents. MIT Press: Cambridge.
Fox, D. (2011). The limits of intelligence. Scientific American, July, 36–43.
Fregni, F., Boggio, P. S., Nitsche, M., Bermpohl, F., Antal, A., Feredoes, E., et al. (2005). Anodal transcranial direct current stimulation of prefrontal cortex enhances working memory. *Experimental Brain Research, 166*(1), 23–30. doi:10.1007/s00221-005-2334-6.
Friedman, M. (1953). The methodology of positive economics. In *Essays in positive economics* (pp. 3–43). Chicago: Chicago University Press.
Friedman, James W., (Ed.) (1994). *Problems of coordination in economic activity* (Vol. 35). Recent Economic Thought. Boston: Kluwer Academic Publishers.
Gödel, K. (1931). Über formal unentscheidbare sätze der Principia Mathematica und verwandter systeme I. *Monatshefte für Mathematik, 38*(1), 173–198. doi:10.1007/BF01700692.
Goertzel, B. (2006). *The hidden pattern: A patternist philosophy of mind*. Boco Raton: BrownWalker Press.
Goertzel, B. (2010). Toward a formal characterization of real-world general intelligence. In E. Baum, M. Hutter, & E. Kitzelmann (Eds.) *Artificial general intelligence: Proceedings of the third conference on artificial general intelligence, AGI 2010, Lugano, Switzerland, March 5–8, 2010*, 19–24. Vol. 10. Advances in Intelligent Systems Research. Amsterdam: Atlantis Press. doi:10.2991/agi.2010.17.
Goertzel, B. (2012). Should humanity build a global AI nanny to delay the singularity until it's better understood? *Journal of Consciousness Studies 19*(1–2), 96–111. http://ingentaconnect.com/content/imp/jcs/2012/00000019/F0020001/art00006.
Goertzel, B., & Pennachin, C. (Eds.) (2007). *Artificial general intelligence. Cognitive Technologies.* Berlin: Springer. doi:10.1007/978-3-540-68677-4.
Goldreich, O. (2010). *P, NP, and NP-Completeness: The basics of computational complexity*. New York: Cambridge University Press.
Good, I. J. (1959). *Speculations on perceptrons and other automata*. Research Lecture, RC-115. IBM, Yorktown Heights, New York, June 2. http://domino.research.ibm.com/library/cyberdig.nsf/ papers/58DC4EA36A143C218525785E00502E30/$File/rc115.pdf.

Good, I. J. (1965). Speculations concerning the first ultraintelligent machine. In F. L. Alt & M. Rubinoff (Eds.) *Advances in computers* (pp. 31–88. Vol. 6). New York: Academic Press. doi:10.1016/S0065-2458(08)60418-0.

Good, I. J. (1970). Some future social repercussions of computers. *International Journal of Environmental Studies, 1*(1–4), 67–79. doi:10.1080/00207237008709398.

Good, I. J. (1982). Ethical machines. In J. E. Hayes, D. Michie, & Y.-H. Pao (Eds.) *Machine intelligence* (pp. 555–560, Vol. 10). Intelligent Systems: Practice and Perspective. Chichester: Ellis Horwood.

Greenfield, S. (2012). The singularity: Commentary on David Chalmers. *Journal of Consciousness Studies 19*(1–2), 112–118. http://www.ingentaconnect.com/content/imp/jcs/2012/00000019/F0020001/art00007.

Griffin, D., & Tversky, A. (1992). The weighing of evidence and the determinants of confidence. *Cognitive Psychology, 24*(3), 411–435. doi:10.1016/0010-0285(92)90013-R.

Groß, D. (2009). Blessing or curse? Neurocognitive enhancement by "brain engineering". *Medicine Studies, 1*(4), 379–391. doi:10.1007/s12376-009-0032-6.

Gubrud, M. A. (1997). Nanotechnology and international security. Paper presented at the Fifth Foresight Conference on Molecular Nanotechnology, Palo Alto, CA, Nov. 5–8. http://www.foresight.org/Conferences/MNT05/Papers/Gubrud/.

Halevy, A., Norvig, P., & Pereira, F. (2009). The unreasonable effectiveness of data. *IEEE Intelligent Systems, 24*(2), 8–12. doi:10.1109/MIS.2009.36.

Hanson, R. (2008). Economics of the singularity. *IEEE Spectrum, 45*(6), 45–50. doi:10.1109/MSPEC.2008.4531461.

Hanson, R. (2012). Meet the new conflict, same as the old conflict. *Journal of Consciousness Studies 19*(1–2), 119–125. http://www.ingentaconnect.com/content/imp/jcs/2012/00000019/F0020001/art00008.

Hanson, R. Forthcoming. Economic growth given machine intelligence. *Journal of Artificial Intelligence Research.*

Hanson, R., & Yudkowsky, E. (2008). The Hanson-Yudkowsky AI-foom debate. LessWrong Wiki. http://wiki.lesswrong.com/wiki/The_Hanson-Yudkowsky_AI-Foom_Debate (accessed Mar. 13, 2012).

Hibbard, B. (2011). Measuring agent intelligence via hierarchies of environments. In Schmidhuber, J., Thórisson, KR., & Looks, M. 2011, 303–308.

Hibbard, B. Forthcoming. Model-based utility functions. *Journal of Artificial General Intelligence.*

Hutter, M. (2005). *Universal artificial intelligence: Sequential decisions based on algorithmic probability*. Texts in Theoretical Computer Science. Berlin: Springer. doi:10.1007/b138233.

Hutter, M. (2009). Open problems in universal induction & intelligence. *Algorithms, 2*(3), 879–906. doi:10.3390/a2030879.

Hutter, M. (2012a). Can intelligence explode? *Journal of Consciousness Studies 19*(1–2), 143–166. http://www.ingentaconnect.com/content/imp/jcs/2012/00000019/F0020001/art00010.

Hutter, M. (2012b). One decade of universal artificial intelligence. In P. Wang & B. Goertzel (eds.) *Theoretical foundations of artificial general intelligence* (Vol. 4). Atlantis Thinking Machines. Paris: Atlantis Press.

Jaynes, E. T., & Bretthorst, G. L. (Eds.) (2003). *Probability theory: The logic of science*. New York: Cambridge University Press. doi:10.2277/0521592712.

Jones, B. F. (2009). The burden of knowledge and the "Death of the Renaissance Man": Is innovation getting harder? *Review of Economic Studies, 76*(1), 283–317. doi:10.1111/j.1467-937X.2008.00531.x.

Kaas, S., Rayhawk,S., Salamon, A., & Salamon, P. (2010). *Economic implications of software minds.* San Francisco: Singularity Institute for Artificial Intelligence, Aug. 10. http://www.singinst.co/upload/economic-implications.pdf.

Kandel, E. R., Schwartz, J. H., & Jessell, T. M. (Eds.). (2000). *Principles of neural science*. New York: McGraw-Hill.

Kolmogorov, A. N. (1968). Three approaches to the quantitative definition of information. *International Journal of Computer Mathematics,* 2(1–4), 157–168. doi:10.1080/00207166808803030.

Koza, J. R. (2010). Human-competitive results produced by genetic programming. *Genetic Programming and Evolvable Machines, 11*(3–4), 251–284. doi:10.1007/s10710-010-9112-3.

Krichmar, J. L., & Wagatsuma, H. (Eds.). (2011). *Neuromorphic and brain-based robots*. New York: Cambridge University Press.

Kryder, M. H., & Kim, C. S. (2009). After hard drives—what comes next? *IEEE Transactions on Magnetics, 2009*(10), 3406–3413. doi:10.1109/TMAG.2009.2024163.

Kurzweil, R. (2005). *The singularity is near: When humans transcend biology*. New York: Viking.

Lampson, B. W. (1973). A note on the confinement problem. *Communications of the ACM, 16*(10), 613–615. doi:10.1145/362375.362389.

Legg, S. (2008). Machine super intelligence. PhD diss., University of Lugano. http://www.vetta.org/documents/Machine_Super_Intelligence.pdf.

Legg, S., & Hutter, M. (2007). A collection of definitions of intelligence. In B. Goertzel & P. Wang (Eds.) *Advances in artificial general intelligence: Concepts, architectures and algorithms—proceedings of the AGI workshop 2006* (Vol. 157). Frontiers in Artificial Intelligence and Applications. Amsterdam: IOS Press.

Li, M., & Vitányi, P. M. B. (2008). An introduction to Kolmogorov complexity and its applications. Texts in Computer Science. New York: Springer. doi:10.1007/978-0-387-49820-1.

Lichtenstein, S., Fischoff, B., & Phillips, L. D. (1982). Calibration of probabilities: The state of the art to 1980. In D. Kahneman, P. Slovic, & A. Tversky (Eds.), *Judgement under uncertainty: Heuristics and biases* (pp. 306–334). New York: Cambridge University Press.

Loosmore, R., & Goertzel, B. (2011). Why an intelligence explosion is probable. *H+ Magazine,* Mar. 7. http://hplusmagazine.com/2011/03/07/why-an-intelligence-explosion-is-probable/.

Lucas, J. R. (1961). Minds, machines and Gödel. *Philosophy, 36*(137), 112–127. doi:10.1017/S0031819100057983.

Lundstrom, M. (2003). Moore's law forever? *Science, 299*(5604), 210–211. doi:10.1126/science.1079567.

Mack, C. A. (2011). Fifty years of Moore's law. *IEEE Transactions on Semiconductor Manufacturing, 24*(2), 202–207. doi:10.1109/TSM.2010.2096437.

Marcus, G. (2008). *Kluge: The haphazard evolution of the human mind*. Boston: Houghton Mifflin.

McAfee, A., & Brynjolfsson, E. (2008). Investing in the IT that makes a competitive difference. *Harvard Business Review,* July. http://hbr.org/2008/07/investing-in-the-it-that-makes-a-competitive-difference.

McCorduck, P. (2004). *Machines who think: A personal inquiry into the history and prospects of artificial intelligence* (2nd ed.). Natick: A. K. Peters.

McDaniel, M. A. (2005). Big-brained people are smarter: A meta-analysis of the relationship between in vivo brain volume and intelligence. *Intelligence, 33*(4), 337–346. doi:10.1016/j.intell.2004.11.005.

McDermott, D. (2012a). Response to "The Singularity" by David Chalmers. *Journal of Consciousness Studies 19*(1–2): 167–172. http://www.ingentaconnect.com/content/imp/jcs/2012/00000019/F0020001/art00011.

McDermott, D. (2012b). There are no "Extendible Methods" in David Chalmers's sense unless P=NP. Unpublished manuscript. http://cs-www.cs.yale.edu/homes/dvm/papers/no-extendible-methods.pdf (accessed Mar. 19, 2012).

Mehta, G. B. (1998). Preference and utility. In S. Barbera, P. J. Hammond, & C. Seidl (Eds.), *Handbook of utility theory* (Vol. I, pp. 1–47). Boston: Kluwer Academic Publishers.

Minsky, M. (1984). Afterword to Vernor Vinge's novel, "True Names." Unpublished manuscript, Oct. 1. http://web.media.mit.edu/~minsky/papers/TrueNames.Afterword.html (accessed Mar. 26, 2012).

Modha, D. S., Ananthanarayanan, R., Esser, S. K., Ndirango, A., Sherbondy, A. J., & Singh, R. (2011). Cognitive computing. *Communications of the ACM, 54*(8), 62–71. doi:10.1145/1978542.1978559.

Modis, T. (2012). There will be no singularity. In Eden, Søraker, Moor, & Steinhart 2012.

Moravec, H. P. (1976). The role of raw rower in intelligence. May 12. http://www.frc.ri.cmu.edu/users/hpm/project.archive/general.articles/1975/Raw.Power.html (accessed Mar. 13, 2012).

Moravec, H. (1998). When will computer hardware match the human brain? *Journal of Evolution and Technology* 1. http://www.transhumanist.com/volume1/moravec.htm.

Moravec, H. (1999). Rise of the robots. *Scientific American,* Dec., 124–135.

Muehlhauser, L. (2011). So you want to save the world. Last modified Mar. 2, 2012. http://lukeprog.com/SaveTheWorld.html.

Muehlhauser, L., & Helm, L. (2012). The singularity and machine ethics. In Eden, Søraker, Moor, & Steinhart 2012.

Murphy, A. H., & Winkler, R. L. (1984). Probability forecasting in meteorology. *Journal of the American Statistical Association, 79*(387), 489–500.

Nagy, B., Farmer, J. D., Trancik, J. E., & Bui, QM. (2010). *Testing laws of technological progress.* Santa Fe Institute, NM, Sept. 2. http://tuvalu.santafe.edu/ bn/workingpapers/NagyFarmerTrancikBui.pdf.

Nagy, B., Farmer, J. D., Trancik, J. E., & Gonzales, J. P. (2011). Superexponential long-term trends in information technology. *Technological Forecasting and Social Change, 78*(8), 1356–1364. doi:10.1016/j.techfore.2011.07.006.

Nielsen, M. (2011). What should a reasonable person believe about the singularity? Michael Nielsen (blog). Jan. 12. http://michaelnielsen.org/blog/what-should-a-reasonable-person-believe-about-the-singularity/ (accessed Mar. 13, 2012).

Nilsson, N. J. (2009). *The quest for artificial intelligence: A history of ideas and achievements.* New York: Cambridge University Press.

Nordmann, A. (2007). If and then: A critique of speculative nanoethics. *NanoEthics, 1*(1), 31–46. doi:10.1007/s11569-007-0007-6.

Omohundro, S. M. (1987). Efficient algorithms with neural network behavior. *Complex Systems 1*(2), 273–347. http://www.complex-systems.com/abstracts/v01_i02_a04.html.

Omohundro, S. M. (2007). The nature of self-improving artificial intelligence. Paper presented at the Singularity Summit 2007, San Francisco, CA, Sept. 8–9. http://singinst.org/summit2007/overview/abstracts/#omohundro.

Omohundro, S. M. (2008). The basic AI drives. In Wang, Goertzel, & Franklin 2008, 483–492.

Omohundro, S. M. 2012. Rational artificial intelligence for the greater good. In Eden, Søraker, Moor, & Steinhart 2012.

Orseau, L. (2011). Universal knowledge-seeking agents. In *Algorithmic learning theory: 22nd international conference, ALT 2011, Espoo, Finland, October 5–7, 2011. Proceedings,* ed. Jyrki Kivinen, Csaba Szepesvári, Esko Ukkonen, and Thomas Zeugmann. Vol. 6925. Lecture Notes in Computer Science. Berlin: Springer. doi:10.1007/978-3-642-24412-4_28.

Orseau, L., & Ring, M. (2011). Self-modification and mortality in artificial agents. In Schmidhuber, Thórisson, and Looks 2011, 1–10.

Pan, Z., Trikalinos, T. A., Kavvoura, F. K., Lau, J., & Ioannidis, J. P. A. (2005). Local literature bias in genetic epidemiology: An empirical evaluation of the Chinese literature. *PLoS Medicine, 2*(12), e334. doi:10.1371/journal.pmed.0020334.

Parente, R., & Anderson-Parente, J. (2011). A case study of long-term Delphi accuracy. *Technological Forecasting and Social Change, 78*(9), 1705–1711. doi:10.1016/j.techfore.2011.07.005.

Pennachin, C, & Goertzel, B. (2007). Contemporary approaches to artificial general intelligence. In Goertzel & Pennachin 2007, 1–30.

Penrose, R. (1994). *Shadows of the mind: A search for the missing science of consciousness.* New York: Oxford University Press.

Plebe, A., & Perconti, P. (2012). The slowdown hypothesis. In Eden, Søraker, Moor, & Steinhart 2012.

Posner, R. A. (2004). *Catastrophe: Risk and response*. New York: Oxford University Press.

Proudfoot, D., & Jack Copeland, B. (2012). Artificial intelligence. In E. Margolis, R. Samuels, & S. P. Stich (Eds.), *The Oxford handbook of philosophy of cognitive science*. New York: Oxford University Press.

Rathmanner, S., & Hutter, M. (2011). A philosophical treatise of universal induction. *Entropy, 13*(6), 1076–1136. doi:10.3390/e13061076.

Richards, M. A., & Shaw, G. A. (2004). Chips, architectures and algorithms: Reflections on the exponential growth of digital signal processing capability. Unpublished manuscript, Jan. 28. http://users.ece.gatech.edu/ mrichard/Richards&Shaw_Algorithms01204.pdf (accessed Mar. 20, 2012).

Rieffel, E., & Polak, W. (2011). *Quantum computing: A gentle introduction.* Scientific and Engineering Computation. Cambridge: MIT Press.

Ring, M., & Orseau, L. (2011). Delusion, survival, and intelligent agents. In Schmidhuber, Thórisson, & Looks 2011, 11–20.

Rowe, G., & Wright, G. (2001). Expert opinions in forecasting: The role of the Delphi technique. In J. S. Armstrong (Ed.), *Principles of forecasting: A handbook for researchers and practitioners,* (Vol. 30). International Series in Operations Research & Management Science. Boston: Kluwer Academic Publishers.

Russell, S. J., & Norvig, P. (2009). *Artificial intelligence: A modern approach* (3rd ed.). Upper Saddle River: Prentice-Hall.

Sandberg, A. (2010). An overview of models of technological singularity. Paper presented at the Roadmaps to AGI and the future of AGI workshop, Lugano, Switzerland, Mar. 8th. http://agi-conf.org/2010/wp-content/uploads/2009/06/agi10singmodels2.pdf.

Sandberg, A. (2011). Cognition enhancement: Upgrading the brain. In J. Savulescu, R. ter Meulen, & G. Kahane (Eds.), *Enhancing human capacities* (pp. 71–91). Malden: Wiley-Blackwell.

Sandberg, A., & Bostrom, N. (2008). *Whole brain emulation: A roadmap.* Technical Report, 2008-3. Future of Humanity Institute, University of Oxford. www.fhi.ox.ac.uk/reports/2008-3.pdf.

Sandberg, A., & Bostrom, N. (2011). *Machine intelligence survey.* Technical Report, 2011-1. Future of Humanity Institute, University of Oxford. www.fhi.ox.ac.uk/reports/2011-1.pdf.

Schaul, T., & Schmidhuber, J. (2010). Metalearning. *Scholarpedia, 5*(6), 4650. doi:10.4249/ scholarpedia.4650.

Schierwagen, A. (2011). Reverse engineering for biologically inspired cognitive architectures: A critical analysis. In C. Hernández, R. Sanz, J. Gómez-Ramirez, L. S. Smith, A. Hussain, A. Chella, & I. Aleksander (Eds.), *From brains to systems: Brain-inspired cognitive systems 2010,* (pp. 111–121, Vol. 718). Advances in Experimental Medicine and Biology. New York: Springer. doi:10.1007/978-1-4614-0164-3_10.

Schmidhuber, J. (2002). The speed prior: A new simplicity measure yielding near-optimal computable predictions. In J. Kivinen & R. H. Sloan, *Computational learning theory: 5th annual conference on computational learning theory, COLT 2002 Sydney, Australia, July 8–10, 2002 proceedings,* (pp. 123–127, Vol. 2375). Lecture Notes in Computer Science. Berlin: Springer. doi:10.1007/3-540-45435-7_15.

Schmidhuber, J. (2007). Gödel machines: Fully self-referential optimal universal self-improvers. In Goertzel & Pennachin 2007, 199–226.

Schmidhuber, J. (2009). Ultimate cognition *à la* Gödel. *Cognitive Computation, 1*(2), 177–193. doi:10.1007/s12559-009-9014-y.

Schmidhuber, J. (2012). Philosophers & futurists, catch up! Response to The Singularity. *Journal of Consciousness Studies 19*(1–2), 173–182. http://www.ingentaconnect.com/content/imp/jcs/ 2012/00000019/F0020001/art00012.

Schmidhuber, J., Thórisson, K. R., & Looks, M. (Eds.) (2011). *Artificial General Intelligence: 4th International Conference, AGI 2011, Mountain View, CA, USA, August 3–6, 2011. Proceedings* (Vol. 6830). Lecture Notes in Computer Science. Berlin: Springer. doi:10.1007/ 978-3-642-22887-2.

Schneider, S. (2010). *Homo economicus—or more like Homer Simpson?* Current Issues. Deutsche Bank Research, Frankfurt, June 29. http://www.dbresearch.com/PROD/DBR_INTERNET_EN-PROD/PROD0000000000259291.PDF.

Schoenemann, P. T. (1997). An MRI study of the relationship between human neuroanatomy and behavioral ability. PhD diss., University of California, Berkeley. http://mypage.iu.edu/ toms/papers/dissertation/Dissertation_title.htm.

Schwartz, J. T. (1987). Limits of artificial intelligence. In S. C. Shapiro & D. Eckroth (Eds.), *Encyclopedia of artificial intelligence* (pp. 488–503, Vol. 1). New York: Wiley.

Searle, J. R. (1980). Minds, brains, and programs. *Behavioral and Brain Sciences, 3*(03), 417–424. doi:10.1017/S0140525X00005756.

Shulman, C., & Bostrom, N. (2012). How hard is artificial intelligence? Evolutionary arguments and selection effects. *Journal of Consciousness Studies 19*.

Shulman, C., & Sandberg, A. (2010). Implications of a software-limited singularity. Paper presented at the 8th European Conference on Computing and Philosophy (ECAP), Munich, Germany, Oct. 4–6.

Simon, H. A. (1965). *The shape of automation for men and management*. New York: Harper & Row.

Solomonoff, R. J. (1964a). A formal theory of inductive inference. *Part I. Information and Control, 7*(1), 1–22. doi:10.1016/S0019-9958(64)90223-2.

Solomonoff, R. J. (1964b). A formal theory of inductive inference. *Part II. Information and Control, 7*(2), 224–254. doi:10.1016/S0019-9958(64)90131-7.

Solomonoff, R. J. (1985). The time scale of artificial intelligence: Reflections on social effects. *Human Systems Management, 5*, 149–153.

Sotala, K. (2012). Advantages of artificial intelligences, uploads, and digital minds. *International Journal of Machine Consciousness 4*.

Stanovich, K. E. (2010). *Rationality and the reflective mind*. New York: Oxford University Press.

Tetlock, P. E. (2005). *Expert political judgment: How good is it? How can we know?*. Princeton: Princeton University Press.

The Royal Society. (2011). *Knowledge, networks and nations: Global scientific collaboration in the 21st century*. RS Policy document, 03/11. The Royal Society, London. http://royalsociety.org/uploadedFiles/Royal_Society_Content/policy/publications/2011/4294976134.pdf.

Trappenberg, T. P. (2009). *Fundamentals of computational neuroscience* (2nd ed.). New York: Oxford University Press.

Turing, A. M. (1950). Computing machinery and intelligence. *Mind, 59*(236), 433–460. doi:10.1093/mind/LIX.236.433.

Turing, A. M. (1951). Intelligent machinery, a heretical theory. A lecture given to '51 Society' at Manchester.

Tversky, A., & Kahneman, D. (1974). Judgment under uncertainty: Heuristics and biases. *Science, 185*(4157), 1124–1131. doi:10.1126/science.185.4157.1124.

Tversky, A., & Kahneman, D. (1983). Extensional versus intuitive reasoning: The conjunction fallacy in probability judgment. *Psychological Review, 90*(4), 293–315. doi:10.1037/0033-295X.90.4.293.

The Uncertain Future. (2012). What is multi-generational in vitro embryo selection? The Uncertain Future. http://www.theuncertainfuture.com/faq.html#7 (accessed Mar. 25, 2012).

Van der Velde, F. (2010). Where artificial intelligence and neuroscience meet: The search for grounded architectures of cognition. Advances in Artificial Intelligence, no. 5. doi:10.1155/2010/918062.

Van Gelder, T., & Port, R. F. (1995). It's about time: An overview of the dynamical approach to cognition. In R. F. Port & T. van Gelder. *Mind as motion: Explorations in the dynamics of cognition,* Bradford Books. Cambridge: MIT Press.

Veness, J., Ng, K. S., Hutter, M., Uther, W., & Silver, D. (2011). A Monte-Carlo AIXI approximation. *Journal of Artificial Intelligence Research, 40*, 95–142. doi:10.1613/jair.3125.

Vinge, V. (1993). The coming technological singularity: How to survive in the post-human era. In *Vision-21: Interdisciplinary science and engineering in the era of cyberspace,* 11–22. NASA

Conference Publication 10129. NASA Lewis Research Center. http://ntrs.nasa.gov/archive/nasa/casi.ntrs.nasa.gov/19940022855_1994022855.pdf.
Von Neumann, J., & Burks, A. W. (Eds.) (1966). *Theory of self-replicating automata*. Urbana: University of Illinois Press.
Walter, C. (2005). Kryder's law. *Scientific American,* July 25. http://www.scientificamerican.com/article.cfm? id = kryders-law.
Wang, P., Goertzel, B., & Franklin, S. (Eds.). (2008). *Artificial General Intelligence 2008: Proceedings of the First AGI Conference (Vol. 171). Frontiers in Artificial Intelligence and Applications*. Amsterdam: IOS Press.
Williams, L. V. (Ed.). (2011). *Prediction markets: Theory and applications (Vol. 66). Routledge International Studies in Money and Banking*. New York: Routledge.
Wootters, W. K., & Zurek, W. H. (1982). A single quantum cannot be cloned. *Nature, 299*(5886), 802–803. doi:10.1038/299802a0.
Woudenberg, F. (1991). An evaluation of Delphi. *Technological Forecasting and Social Change, 40*(2), 131–150. doi:10.1016/0040-1625(91)90002-W.
Yampolskiy, R. V. (2012). Leakproofing the singularity: Artificial intelligence confinement problem. *Journal of Consciousness Studies 19*(1–2), 194–214. http://www.ingentaconnect.com/content/imp/jcs/2012/00000019/F0020001/art00014.
Yates, J. F., Lee, J.-W., Sieck, W. R., Choi, I., & Price, P. C. (2002). Probability judgment across cultures. In T. Gilovich, D. Griffin, & D. Kahneman (Eds.), *Heuristics and biases: The psychology of intuitive judgment* (pp. 271–291). New York: Cambridge University Press.
Yudkowsky, E. (2001). Creating Friendly AI 1.0: The analysis and design of benevolent goal architectures. The Singularity Institute, San Francisco, CA, June 15. http://singinst.org/upload/CFAI.html.
Yudkowsky, E. (2008a). Artificial intelligence as a positive and negative factor in global risk. In Bostrom & Ćirković 2008, 308–345.
Yudkowsky, E. (2008b). Efficient cross-domain optimization. LessWrong. Oct. 28. http://lesswrong.com/lw/vb/efficient_crossdomain_optimization/ (accessed Mar. 19, 2012).
Yudkowsky, E. (2011). Complex value systems in friendly AI. In Schmidhuber, Thórisson, & Looks 2011, 388–393.

Chapter 2A
Robin Hanson on Muehlhauser and Salamon's "Intelligence Explosion: Evidence and Import"

Muehlhauser and Salamon [M&S] talk as if their concerns are particular to an unprecedented new situation: the imminent prospect of "artificial intelligence" (AI). But in fact their concerns depend little on how artificial will be our descendants, nor on how intelligence they will be. Rather, Muehlhauser and Salamon's concerns follow from the general fact that accelerating rates of change increase intergenerational conflicts. Let me explain.

Here are three very long term historical trends:

1. Our total power and capacity has consistently increased. Long ago this enabled increasing population, and lately it also enables increasing individual income.
2. The rate of change in this capacity increase has also increased. This acceleration has been lumpy, concentrated in big transitions: from primates to humans to farmers to industry.
3. Our values, as expressed in words and deeds, have changed, and changed faster when capacity changed faster. Genes embodied many earlier changes, while culture embodies most today.

Increasing rates of change, together with constant or increasing lifespans, generically imply that individual lifetimes now see more change in capacity and in values. This creates more scope for conflict, wherein older generations dislike the values of younger more-powerful generations with whom their lives overlap.

As rates of change increase, these differences in capacity and values between overlapping generations increase. For example, Muehlhauser and Salamon fear that their lives might overlap with

> [descendants] superior to us in manufacturing, harvesting resources, scientific discovery, social charisma, and strategic action, among other capacities. We would not be in a position to negotiate with them, for [we] could not offer anything of value [they] could not produce more effectively themselves. ... This brings us to the central feature of [descendant] risk: Unless a [descendant] is specifically programmed to preserve what [we] value, it may destroy those valued structures (including [us]) incidentally.

The quote actually used the words "humans", "machines" and "AI", and Muehlhauser and Salamon spend much of their chapter discussing the timing and likelihood of future AI. But those details are mostly irrelevant to the concerns expressed above. It doesn't matter much if our descendants are machines or biological meat, or if their increased capacities come from intelligence or raw physical power. What matters is that descendants could have more capacity and differing values.

Such intergenerational concerns are ancient, and in response parents have long sought to imprint their values onto their children, with modest success.

Muehlhauser and Salamon find this approach completely unsatisfactory. They even seem wary of descendants who are cell-by-cell emulations of prior human brains, "brain-inspired AIs running on human-derived "spaghetti code", or 'opaque' AI designs …produced by evolutionary algorithms." Why? Because such descendants "may not have a clear 'slot' in which to specify desirable goals."

Instead Muehlhauser and Salamon prefer descendants that have "a transparent design with a clearly definable utility function," and they want the world to slow down its progress in making more capable descendants, so that they can first "solve the problem of how to build [descendants] with a stable, desirable utility function."

If "political totalitarians" are central powers trying to prevent unwanted political change using thorough and detailed control of social institutions, then "value totalitarians" are central powers trying to prevent unwanted value change using thorough and detailed control of everything value-related. And like political totalitarians willing to sacrifice economic growth to maintain political control, value totalitarians want us to sacrifice capacity growth until they can be assured of total value control.

While the basic problem of faster change increasing intergenerational conflict depends little on change being caused by AI, the feasibility of this value totalitarian solution does seem to require AI. In addition, it requires transparent-design AI to be an early and efficient form of AI. Furthermore, either all the teams designing AIs must agree to use good values, or the first successful team must use good values and then stop the progress of all other teams.

Personally, I'm skeptical that this approach is even feasible, and if feasible, I'm wary of the concentration of power required to even attempt it. Yes we teach values to kids, but we are also often revolted by extreme brainwashing scenarios, of kids so committed to certain teachings that they can no longer question them. And we are rightly wary of the global control required to prevent any team from creating descendants who lack officially approved values.

Even so, I must admit that value totalitarianism deserves to be among the range of responses considered to future intergenerational conflicts.

Chapter 3
The Threat of a Reward-Driven Adversarial Artificial General Intelligence

Itamar Arel

Abstract Once introduced, Artificial General Intelligence (AGI) will undoubtedly become humanity's most transformative technological force. However, the nature of such a force is unclear with many contemplating scenarios in which this novel form of intelligence will find humans an inevitable adversary. In this chapter, we argue that if one is to consider reinforcement learning principles as foundations for AGI, then an adversarial relationship with humans is in fact inevitable. We further conjecture that deep learning architectures for perception in concern with reinforcement learning for decision making pave a possible path for future AGI technology and raise the primary ethical and societal questions to be addressed if humanity is to evade catastrophic clashing with these AGI beings.

AGI and the Singularity

A Path to the Inevitable

A myriad of evidence exists in support of the notion that mammalian learning processes are driven by rewards. Recent findings from cognitive psychology and neuroscience strongly suggest that much of human behavior is propelled by both positive and negative feedback received from the environment. The notion of reward is not limited to indicators originating from a physical environment. It also embraces signaling generated internally in the brain, based on intrinsic cognitive processes.

I. Arel (✉)
Department of Electrical Engineering and Computer Science,
Machine Intelligence Lab, University of Tennessee, Knoxville, TN, USA
e-mail: itamar@ieee.org

A. H. Eden et al. (eds.), *Singularity Hypotheses*, The Frontiers Collection,
DOI: 10.1007/978-3-642-32560-1_3, © Springer-Verlag Berlin Heidelberg 2012

Artificial General Intelligence (AGI), coarsely viewed as human-level intelligence manifested over non-biological platforms, is commonly perceived as one of the paths that may lead to the singularity. Such a path has the potential of being either beneficially transformative or devastating to the human race, to a great extent depending on the very nature of the emerging AGI. Nonetheless, assuming that the pieces of the puzzle needed to achieve AGI are in fact readily available, an AGI reality is inevitable.

Reinforcement learning (RL) (Sutton 1998) is a fairly mature field within artificial intelligence, with a focus on delivering a rigorous mathematical framework for learning by means of interacting with an environment. Consequently, it serves as one of the promising foundations for advancing AGI research. The key challenge in the study of RL, as a mechanism for decision making under uncertainty, has been that of scalability. The latter refers to effectively processing high-dimensional observations spanning large state and action spaces, which characterize real-world AGI settings.

Recent neuroscience findings have provided clues into the principles governing information representation in brain, leading to new paradigms for designing systems that represent information. One such paradigm is deep machine learning (DML) (Arel 2010), which is emerging as a promising, biologically-inspired framework for dealing with high-dimensional observations. The numerous architectures proposed for DML differ in many ways, yet they all have in common the notion of a hierarchical architecture for information representation.

It is further argued that the merger of these ideas, along with recent advances in VLSI technology, can lead to the introduction of truly intelligent machines in the not-so-distant future. Finer semiconductor device fabrication processes are improving on a trajectory that does not seem to reach a plateau any time soon. It is now possible to pack billions of transistors on a single chip using dedicated analog circuitry called floating gates (Hasler 1995). The latter make it possible to implement memory and processing units using only several transistors. As a result, it is now conceivable to envision chips (or chip sets) that reach the storage and computation densities of those observed in the mammalian brain.

Should these machines emerge; the Singularity will follow with high probability. The exponential rate at which such transformative technology will evolve is difficult to predict. However, it is clear that as with all technologies of this scale, there is great potential for the enhancement of quality of life, along with clear existential risks to humanity.

This chapter outlines the philosophical as well as commonsense implications of scaling reinforcement learning using deep architectures as basis for achieving AGI. The discussion is frames in the context of the Singularity. In particular, questions related to avenues for potentially guaranteeing friendly AGI are deliberated. It is argued that a proactive approach to addressing the various ethical and socio-economical concerns pertaining to an AGI-driven Singularity is vital if humanity is to mitigate its colossal existential risks.

From AI to AGI

A fundamental distinction between Artificial General Intelligence (AGI) and "conventional" Artificial Intelligence (AI) is that AGI focuses on the study of systems that can perform tasks successfully across different problem domains, while AI typically pertains to domain-specific expert systems. General problem-solving ability is one that humans naturally exhibit. A related capability is generalization, which allows mammals to effectively associate causes perceived in their environment with regularities observed in the past. Another critical human skill involves decision making under uncertainty, tightly coupled with generalization since the latter facilitates broad situation inference.

Following this line of thought, it can be argued that at a coarse level, intelligence involved two complementing sub-systems: *perception* and *actuation*. Perception can be interpreted as mapping sequences of observations, possibly received from multiple modalities, to an inferred state of the world with which the intelligence agent interacts. Actuation is often framed as a control problem, centering on the goal of selecting actions to be taken at any given time so as to maximize some utility function. In other words, actuation is a direct byproduct of a decision making process, whereby inferred states are mapped to selected actions, thereby impacting the environment in some desirable way. This high-level view is depicted in Fig. 3.1.

In late the 1950s, Richard Bellman who introduced dynamic programming theory and pioneered the field of optimal control, predicted that high-dimensionality data will remain a fundamental obstacle for many science and engineering systems over decades to come. The main difficulty he highlighted was that learning complexity grows exponentially with linear increase in data dimensionality. He coined this phenomenon the *curse of dimensionality* and his premonition proved amazingly true.

DML architectures attempt to mimic the manner by which the cortex learns to represent regularities in real world observations, and thus offer a solution for the curse of dimensionality. In addition to the spatial aspects of real-life data, its temporal components often play a key role in facilitating accurate perception. To that end, robust spatiotemporal modeling of observations should serve as a primary goal for all deep learning systems.

It has recently been hypothesized that the fusion between deep learning, as a scalable situation inference engine, and reinforcement learning as a decision-making system may hold the key to place us on the path to AGI and thus the singularity. Assuming that this hypothesis is correct, many critical questions arise, the first of which is how do we avoid a potentially devastating conflict between a reward-driven AGI system and the human race? One can argue that such a scenario is inescapable, given the assumption that an RL-based AGI will be allowed to evolve. In that case, does evolution have to continue over biochemical substrates, or will the next phase in the evolution manifest itself over semiconductor-based fabrics? Consequently, will AGI bring the human era to an inevitable end? Transhumanism may very well

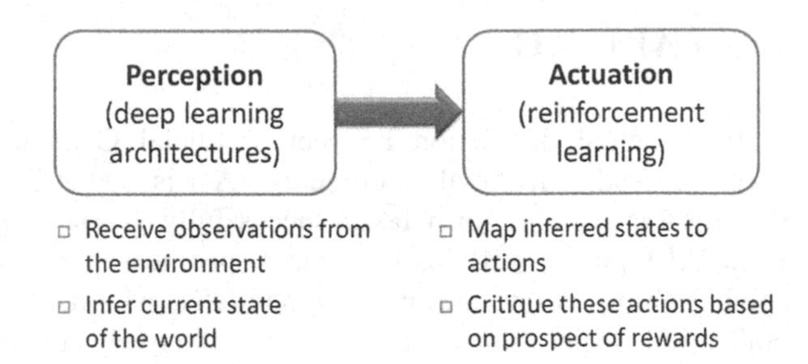

Fig. 3.1 Bipartite AGI architecture comprising of a perception and control/actuation subsystem. The role of the perception subsystem is viewed as state inference while the control subsystem maps inferred states to desired actions

emerge as a transitional period at the end of which post-humanism will commence with an absence of biochemical based life forms.

History suggests that pragmatic concerns pertaining to the potential dangers and threats of novel technologies have never impeded such technologies from being widely embraced. Nuclear technology is an obvious example, particularly in that debate over its benefits verses its threats has persistently accompanied its deployment. Although technological progress is needed to make AGI a reality, there is some likelihood that the pieces of the puzzle needed to make AGI a reality are in fact readily available, in which case now is the time to consider the colossal implications of an AGI-driven singularity.

Deep Machine Learning Architectures

Overcoming the Curse of Dimensionality

Mimicking the efficiency and robustness by which the human brain represents information has been a core challenge in artificial intelligence research for decades. Humans are exposed to myriad of sensory data received every second of the day and are somehow able to capture critical aspects of this data in a way that allows for its future recollection. The mainstream approach of overcoming high-dimensionality has been to pre-process the data in a manner that would reduce its dimensionality to that which can be effectively processed, for example by a classification engine. Such dimensionality reduction schemes are often referred to as feature extraction techniques. As a result, it can be argued that the intelligence behind many pattern recognition systems has shifted to the human-engineered feature extraction process, which at times can be challenging and highly application-dependent (Duda 2000). Moreover, if incomplete or erroneous features are extracted, the classification process is inherently limited in performance.

Recent neuroscience findings have provided insight into the principles governing information representation in the mammal brain, leading to new ideas for designing systems that represent information. One of the key findings has been that the neocortex, which is associated with many cognitive abilities, does not explicitly pre-process sensory signals, but rather allows them to propagate through a complex hierarchy (Lee 2003) of modules that, over time, learn to represent observations based on the regularities they exhibit (Lee 1998). This discovery motivated the emergence of the subfield of deep machine learning (Arel 2010; Bengio 2009), which focuses on computational models for information representation that exhibit similar characteristics to that of the neocortex.

A sequence of patterns that we observe often conveys a meaning to us, whereby independent fragments of this sequence would be hard to decipher in isolation. We often infer meaning from events or observations that are received close in time (Wallis 1997, 1999). To that end, modeling the temporal component of the observations plays a critical role in effective information representation. Capturing spatiotemporal dependencies, based on regularities in the observations, is therefore viewed as a fundamental goal for deep learning systems.

Spatiotemporal State Inference

A particular family of DML systems is *compositional* deep learning architectures (Arel 2009). The latter are characterized by hosting multiple instantiations of a basic cortical circuit (or *node*) which populate all layers of the architecture. Each node is tasked with learning to represent the sequences of patterns that are presented to it by nodes in the layer that precede it. At the very lowest layer of the hierarchy nodes receive as input raw data (e.g. pixels of the image) and continuously construct a *belief state* that attempts to compactly characterize the sequences of patterns observed. The second layer, and all those above it, receive as input the belief states of nodes at their corresponding lower layers, and attempt to construct their own belief states that capture regularities in their inputs. Figure 3.2 illustrates a compositional deep learning architecture.

Information flows both bottom and top down. Bottom up processing essentially constitutes a feature extraction process, in which each layer aggregates data from the layer below it. Top down signaling helps lower layer nodes improve their representation accuracy by assisting in correctly disambiguating distorted observations.

An AGI system should be able to adequately cope in a world where partial observability is assumed. Partial observability means that any given observation (regardless of the modalities from which it originates) does not provide full information needed to accurately infer the true state of the world. As such, an AGI system should map sequences of observations to an internal state construct that is consistent for regular causes. This implies that a dynamic (i.e. memory-based) learning process should be exercised by each cortical circuit.

For example, if a person looks at a car in a parking lot he/she would recognize it as such since there is consistent signaling being invoked in their brain whenever car patterns are observed. In fact, it is sufficient to hear a car (without viewing it) to invoke similar signaling in the brain. While every person may have different signaling for common causes in the world, such signaling remains consistent for

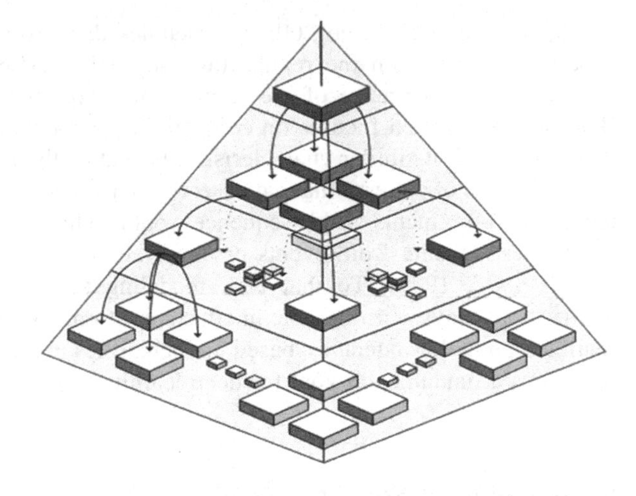

Fig. 3.2 Compositional deep machine learning architecture comprising of multiple instantiations of a common cortical circuit

each person. This consistency property allows a complementing control subsystem to map the (inferred) states to actions that impact the environment in some desirable way.

If a deep learning architecture is to form an accurate state representation, it should include both spatial and temporal information. As a result, each belief state should capture spatiotemporal regularities in the observations, rather than just spatial saliencies.

The learning process at each node is unsupervised, guided by exposure to a large set of observations and allowing the salient attributes of these observations to be captured across the layers. In the context of an AGI system, signals originating from upper-layer nodes can be extracted to serve as inferred state representations. This extracted information should exhibit invariance to common distortions and variations in the observations, leading to representational robustness. In the context of visual data, robustness refers to the ability to exhibit invariance to a diverse range of transformations, including mild rotation, scale, different lighting conditions and noise.

It should be noted that although deep architectures may appear to completely solve or overcome the curse of dimensionality, in reality they do so by hiding the key assumption of locality. The latter means that the dependencies that may exist between two signals (e.g. pixels) that are spatially close are captured with relative detail, where as relationships between signals that are distant (e.g. pixels on opposite sides of a visual field) are represented with very little detail. This is a direct result of the nature of the architecture depicted in Fig. 3.2, in which fusion of information from inputs that are distant to the hierarchy occurs at the higher layers.

It is also important to emphasize that deep learning architectures are not limited by any means to visual data. In fact, these architectures are modality agnostic, and attempt to discover underlying structure in data of any form. Moreover, fusion of information originating from different modalities is natural in deep learning and a pivotal requirement of AGI. If one imagines the architecture shown in Fig. 3.1 to receive input at its lowest layer from multiple modalities, as one ascends the hierarchy, fusion of such information takes place by capturing regularities across the modalities.

Scaling Decision Making Under Uncertainty

Markov Decision Processes

Reinforcement learning problems are typically modeled as Markov Decision Processes (MDPs). An MDP is defined as a (S, A, P, R)-tuple, where S stands for the state space, A contains all the possible actions at each state, P is a probability transition function $S \times A \times S \to [0, 1]$ and R is the reward function $S \times A \to R$. Also, we define π as the decision policy that maps the state set to the action set: $\pi : S \to A$. Specifically, let us assume that the environment is a finite-state, discrete-time stochastic dynamic system. Let the state space be $S = (s_1, s_2, \ldots, s_n)$ and, accordingly, action space be $A = (a_1, a_2, \ldots, a_m)$. Suppose at time step k, that agent is in state s_k, and it chooses action $a_k \in A(s_k)$ according to policy π, in order to interact with its environment. Next, the environment transitions into a new state s_{k+1} and provides the agent with a feedback reward denoted by $r_k(s, a)$. This process is continuously repeated with the goal being to maximize the expected discounted reward, or state-action value, given by

$$Q^{\pi}(s,a) = E_{\pi}\left\{\sum_{k=0}^{\infty} \gamma^k r_k(s_k, \pi(s_k))|s,a\right\}, \tag{3.1}$$

where $0 \leq \gamma < 1$ is the discount factor and $E_{\pi}\{\}$ denotes the expected return when starting in state s, taking action a and following policy π thereafter. A goal of the RL system will be to select actions that persistently attempt to maximize its *value function*. Unless externally constrained in some manner, the agent will explore every means at its disposal to achieve the goal of increasing the expected rewards by way of maximizing its value function.

The above definition of a value function, which relies on the notion of geometrically discounting rewards in time, has some core flaws. The main drawback of such formulation is that significant positive or negative events that are expected to occur far into the future have negligible impact on current actions selected. However, in human decision making processes, and in fact in that of most mammals, the time scale considered in selecting actions can be quite large. This is not supported in the discounted rewards model. An alternative formulation for the value function is one that expresses the infinite sum of differences between expected rewards and those actually experienced. Mathematically stated, this alternative value function is given by

$$Q^{\pi}(s,a) = E_{\pi}\left\{\sum_{k=0}^{\infty}(r_k - \rho_k)|s,a\right\}, \tag{3.2}$$

where ρ_k is the estimated rate of rewards at time k, which can be attained by applying moving average over the rewards, such that

$$\rho_k = (1-\alpha)\rho_{k-1} + \alpha r_k,\ 0<\alpha<1. \tag{3.3}$$

In essence, the above implies that the agent attempts to maximize its positive "surprises" (interpreted as the difference between expectation and actual experience) while minimize its negative ones.

Biological Plausibility of Reinforcement Learning

While the role of the perception subsystem may be viewed as that of complex state inference, an AGI system must be able to take actions that impact its environment. In other words, an AGI system must involve a controller that attempts to optimize some cost function. This controller is charged with mapping the inferred states to an action. In real-world scenarios, there is always some uncertainty. However, state signaling should exhibit the Markov property in the sense that it compactly represents the history that has led to the current state-of-affairs. This is a colossal assumption, and one that is unlikely to accurately hold. However, it is argued that while the Markov property does not hold, assuming that it does paves the way for obtaining "good enough", albeit not optimal, AGI systems.

It is important to understand that RL corresponds to a broad class of machine learning techniques that allow a system to learn how to behave in an environment that provides reward signals. A key assertion in RL is that the agent learns by itself, based on acquired experience, rather than by being externally instructed or supervised. Hence, RL inherently facilitates autonomous learning as well as addresses many of the essential goals of AGI: it emphasizes the close interaction of an agent with the environment, it focuses on perception-to-action cycles and complete behaviors rather than separate functions and function modules, it relies on bottom-up intelligent learning paradigms, and it is not based on symbolic representations.

The ability to generalize is acknowledged as an inherent attribute of intelligent systems. Consequently, it may be claimed that no system can learn without employing some degree of approximation. The latter is particularly true when we consider large-scale, complex real-world scenarios, such as those implied by true AGI. Deep learning architectures can serve this exact purpose: they can provide a scalable state inference engine that a reinforcement learning based controller can map to actions.

A recent and very influential development in RL is the actor-critic approach to model-free learning, which is based on the notion that two distinct core functions accomplish learning: the first (the "actor") produces actions derived from an internal model and the second (the "critic") refines the action selection policy based on prediction of long-term reward signals. As the actor gains proficiency, it is required to learn an effective mapping from inferred states of the environment to actions. In parallel to the growing support of RL theories in modern cognitive science, recent work in neurophysiology provides some evidence arguing that the actor-critic RL theme is widely exploited in the human brain.

The dominant mathematical methods supporting learning approximately solve the Hamilton–Jacobi–Bellman (HJB) equation of dynamic programming (DP) to iteratively adjust the parameters and structure of an agent as means of encouraging desired behaviors (Arel 2010). A discounted future reward is typically used; however, researchers are aware of the importance of multiple time scales and the likelihood that training efficiency will depend upon explicit consideration of multiple time horizons. Hence, the trade-offs between short and long term memory should be considered. Cognitive science research supports these observations, finding similar structures and mechanisms in mammalian brains (Suri 1999).

The HJB equation of DP requires estimation of expected future rewards, and a suitable dynamic model of the environment that maps the current observations and actions (along with inferred state information) to future observations. Such model-free reinforcement learning assumes no initial knowledge of the environment, and instead postulates a generic structure, such as a deep learning architecture, that can be trained to model environmental responses to actions and exogenous sensory inputs.

Contrary to existing function approximation technologies, such as standard multi-layer perceptron networks, current neurophysiology research reveals that the structure of the human brain is dynamic, with explosive growth of neurons and neural connections during fetal development followed by pruning. Spatial placement also plays critical roles, and it is probable that the spatial distribution of chemical reward signals selectively influences neural adaptation to enhance learning (Schultz 1998). It is suspected that the combination of multi-time horizon learning and memory processes with the dynamic topology of a spatially embedded deep architecture will dramatically enhance adaptability and effectiveness of artificial cognitive agents. This is expected to yield novel AGI frameworks that can overcome the limitations of existing AI systems

RL can thus be viewed as a biologically-inspired decision making under uncertainty framework that is centered on the notion of learning from experience, through interaction with an environment, rather than by being explicitly guided by a teacher. What sets RL apart from other machine learning methods is that it aims to solve the *credit assignment problem*, in which an agent is charged with evaluating the long-term impact of actions it takes.

The fundamental notion of learning on the basis of rewards is shared among several influential branches of psychology, including behaviorism and cognitive psychology. The actor-critic architecture reflects recent trends in cognitive neuroscience and cognitive psychology that highlight task decomposition and modular organization. For example, visual information-processing is served by two parallel pathways, one specialized to object location in space and the other to object identification or recognition over space and time (Millner 1996; Mishkin 1983). This approach exploits a divide-and-conquer processing strategy in which particular components of a complex task are computed in different cortical regions, and typically integrated, combined, or supervised by the prefrontal cortex.

Computational models of this dual-route architecture suggest that it has numerous benefits over conventional homogenous networks, including both learning speed and accuracy. More generally, the prefrontal cortex is implicated in a wide range of cognitive functions, including maintaining information in short-term or working memory, action planning or sequencing, behavioral inhibition, and anticipation of future states. These functions highlight the role of the prefrontal cortex as a key location that monitors information from various sources and provides top-down feedback and control to relevant motor areas (e.g. premotor cortex, frontal eye fields, etc.). In addition to recent work in cognitive neuroscience, theoretical models of working memory in cognitive psychology also focus on the role of a central executive that actively stores and manipulates information that is relevant for solving ongoing tasks.

A unique feature of the proposed AGI approach is a general-purpose cognitive structure for investigating both external and internal reward systems. Cognitive psychologists conceptualize these two forms of reward as extrinsic and intrinsic motivation (Ginsburg 1993). Extrinsic motivation corresponds to changes in behavior as a function of external contingencies (e.g. rewards and punishments), and is a central element of Skinner's theory of learning. Meanwhile, intrinsic motivation corresponds to changes in behavior that are mediated by internal states, drives, and experiences, and is manifested in a variety of forms including curiosity, surprise, and novelty. The concept of intrinsic motivation is ubiquitous in theories of learning and development, including the notions of (1) mastery motivation (i.e. a drive for proficiency (Kelley et al. 2000), (2) functional assimilation (i.e. the tendency to practice a new skill), and (3) violation-of-expectation (i.e. the tendency to increase attention to unexpected or surprising events), see (Baillargeon 1994)).

It is interesting to note that while external rewards play a central role in RL, the use of intrinsic motivation has only recently begun to receive attention from the machine-learning community. This is an important trend, for a number of reasons. First, intrinsic motivation changes dynamically in humans, not only as a function of task context but also general experience. Implementing a similar approach in autonomous-agent design will enable the agent to flexibly adapt or modify its objectives over time, deploying attention and computational resources to relevant goals and sub-goals as knowledge, skill, and task demands change. Second, the integration of a dual-reward system that includes both external and intrinsic motivation is not only biologically plausible, but also more accurately reflects the continuum of influences in both human and non-human learning systems.

In parallel to the growing support of model-free Actor-Critic oriented in modern psychology, recent work in neurophysiology provides evidence which suggests that the Actor-Critic paradigm is widely exploited in the brain. In particular, it has been recently shown that the basal ganglia (Joel 2002) can be coarsely modeled by an Actor-Critic version of Temporal Difference (TD) learning

(Sutton 1998). The frontal dopaminergic input arises in a part of the basal ganglia called ventral tegmental area (VTA) and the substantia nigra (SN). The signal generated by dopaminergic (DA) neurons resembles the effective reinforcement signal of TD learning algorithms.

Another important part of the basal ganglia is the striatum. This structure is comprised of two parts, the matriosome and the striosome. Both receive input from the cortex (mostly frontal) and from the DA neurons, but the striosome projects principally to DA neurons in VTA and SN. The striosome is hypothesized to act as a reward predictor, allowing the DA signal to compute the difference between the expected and received reward. The matriosome projects back to the frontal lobe (for example, to the motor cortex). Its hypothesized role is therefore in action selection.

Neuromorphic Circuits for Scaling AGI

The computational complexity and storage requirements from deep reinforcement learning systems limit the scale at which they may be implemented using standard digital computers. An alternative would be to consider custom analog circuitry as means of overcoming the limitations of digital VLSI technology. In order to achieve the largest possible learning system within any given constraints of cost or physical size, it is critical that the basic building blocks of the learning system be as dense as possible. Many operations can be realized in analog circuitry with a space saving of one to two orders of magnitude compared to a digital realization. Analog computation also frequently comes with a significant reduction in power consumption, which will become critical as powerful learning systems are migrated to battery-operated platforms.

This massive improvement in density is achieved by utilizing the natural physics of device operation to carry out computation. The benefits in density and power come with certain disadvantages, such as offsets and inferior linearity compared to digital implementations. However, the weaknesses of analog circuits are not major limitations since the feedback inherent in the learning algorithms naturally compensates for errors/inaccuracies introduced by the analog circuits. The argument made here is that the brain is far from being 64-bit accurate, so relaxing accuracy requirements of computational elements, for the purpose of aggressively optimized for area, is a valid tradeoff.

The basic requirements of almost any machine learning algorithm include multiplication, addition, squashing functions (e.g. sigmoid), and distance/similarity calculation, all of which can be realized in a compact and power-efficient manner using analog circuitry. Summation is trivial in the current domain as it is accomplished by joining the wires with the currents to be summed. In the voltage domain, a feedback amplifier with $N+1$ resistors in the feedback path can compute the sum of N inputs. A Gilbert cell (Gray 2001) provides four-quadrant multiplication while using only seven transistors.

Table 3.1 contrasts the component count of digital and analog computational blocks. As discussed above, the learning algorithms will be designed to be robust

Table 3.1 Transistor count comparison of analog and digital elements

Operation	Analog	Digital(N bits)	Notes
Summation	12	24	Feedback voltage adder; Ripple-carry adder
Multiplication	7	48	Shift and add multiplier
Storage	15	12	Feedback floating-gate cell: digital register

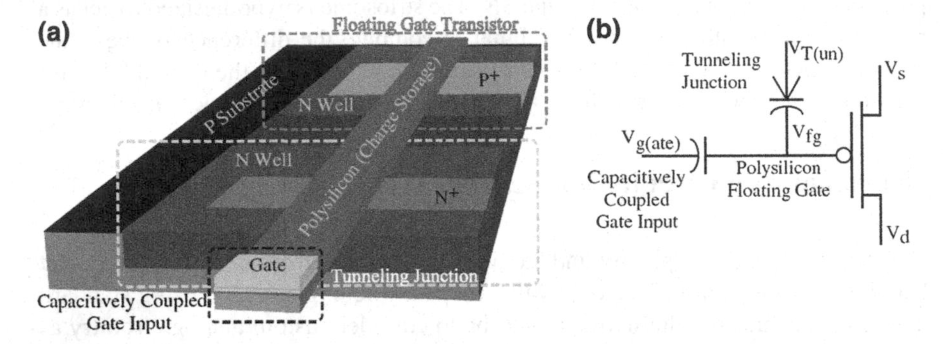

Fig. 3.3 **a** Cutout showing parts of a typical floating gate transistor. **b** A floating gate in schematic

to analog circuit imperfections, allowing the use of very small transistors. Digital designs vary widely in transistor count, as area can frequently be traded for speed. For the comparison, we used designs that are optimized for area and appropriate for the application. For example, an N-bit "shift-and-add" multiplier uses a single adder by performing the multiplication over N clock cycles. A fast multiplier might require as many as N times more adders. Digital registers were chosen over SRAMs, despite their larger size because SRAMs require significant peripheral circuitry (e.g. address decoders, sense amplifiers) making them poorly suited to a system requiring many small memories. For the analog elements, we also counted any necessary resistors or capacitors.

A particularly suitable analog circuit, which is often used for computational purposes, is the floating gate transistor (Hasler 2005) (shown in Fig. 3.3). Floating gates have proved themselves to be useful in many different applications; they make good programmable switches, allow for threshold matching in transistor circuits, and have been successfully used in the design of various adaptive systems.

Floating gates lack a DC path to ground, so any charge stored on the gate will stay there. Through the use of Fowler–Nordheim tunneling and hot electron injection, this trapped charge can be modified. Floating-gate memories can provide a finely tuned voltage to match thresholds between multiple transistors, for example, to yield a circuit which has nearly perfect matching, improving accuracy relative to conventional techniques. In learning systems, a floating gate can be used to store a weight or learned parameter.

The inevitable conclusion from this discussion is that floating gate technology offers very high logic and power density, alongside the ability to realize massively-parallel distributed memory. All this paves the way for implementing very large intelligent systems that a decade ago would have required a super-computer. Needless to say, there is no reason to believe that these VLSI advances will slow down in the near future.

The Inevitability of AGI Malevolence

Perpetual Dissatisfaction in Temporal Difference Learning

Temporal difference (TD) learning (Sutton 1998) is a central idea in RL. It provides a method for estimating the value function and is primarily applied to model-free learning problems. The TD paradigm draws from both dynamic programming (Bellman 1957) and Monte Carlo methods (Sutton 1998). Similar to dynamic programming, TD learning bootstraps in that it updates value estimates based on other value estimates, as such not having to complete an episode before updating its value function representation. Like Monte Carlo methods, TD is heuristic in that it uses experience, obtained by following a given policy (i.e. mapping of states to actions), to predict subsequent value estimates. TD updates are performed as a single step look-ahead that typically takes the general form of

$$Q_{t+1} = Q_t + \alpha[target - Q_t], \tag{3.4}$$

where *target* is derived from the Bellman equation and depends on how rewards are evaluated over time, Q_t denotes the state-action value estimate at time t, and α is a small positive constant.

It should be noted that in real-world AGI settings, only partial information regarding the true "state" of the world is made available to the agent. The agent is thus required to form a belief state from observations it receives of the environment. Assuming the Markov property holds, but state information is inaccurate or incomplete, we say that the problem is partially observable. Deep learning architectures help overcome partial observability by utilizing internal signals that capture spatiotemporal dependencies, as discussed above. This potentially solves the scalable state inference problem.

On the control side, from Eqs. (3.2) and (3.4), it follows that the agent continuously attempts to maximize its "positive" surprises (i.e. positive difference between actual rewards and reward expectation) while minimizing "negative" surprises. This process, however, is unbounded. In other words, even if the agent reaches a fairly high rate of rewards, by the formulation expressed in Eq. (3.2), it will continue to strive to maximize its wellbeing (interpreted via expectation of future rewards). It can be argued that this is very much a human characteristic—when one is used to very little, anything above and beyond this expectation is happily greeted. However, once such bonus is received on a regular basis, it becomes the new norm and no longer yields the same level of satisfaction. This is the core danger in designing systems that are driven by rewards and have large cognitive capacity; by continuously striving to gain positive (relative) reinforcement, they will inevitably pose a danger to humanity.

Mitigating AGI Risks

By definition, an RL-based AGI agent optimizing over a value function will always strive to improve its wellbeing, much like humans. One can imagine an AGI system equipped with means of actuation far more capable than those possessed by humans. Consequently, it may reach the inevitable conclusion that human beings are too often hurdles in its path of self-improvement, and thus constitute an adversary.

Furthermore, such adversity is likely to be mutual. In fact, it is highly probably that humans are going to view super-intelligent, reward-seeking AGI systems as a clear and present threat to their existence, thereby calling for their elimination. While some humans are broadly acknowledged as evil and are feared by many others, they are nonetheless perceived to be physically (and existentially) limited in the same ways that all humans are. The scope of their physical impact, speed at which they process information and make decisions, are all assumed to be within the well-known bounds of humans. However, these assertions no longer hold true for AGI systems. The latter will inevitably be viewed as hostile, once exhibiting desires combined with capabilities that are (in human terms) super-human.

This raises the important question of what can be done to avoid a clash between the two species. Education seems to be an answer that comes to mind. The difficulty in devising an educational fabric for promoting coexistence and mutual understanding lies in the fact that prior to the emergence of AGI there is little tangible evidence that can be used as ground-truth reference for developing instructional content. It is therefore imperative that a human-controlled evolution of AGI systems will be coupled with substantial effort is made to understand these systems as much as possible in order to introduce the necessary educational programs that will help mitigate the inevitable fear humans will have of this great unknown.

A more technical approach for mitigating some of the risk of RL-based AGI can be to limit such system's mental capacity. This can easily be done by means of establishing a hard limit on both computation and storage resources. In the context of DML, this can encompass several attributes. The first is limiting both the depth and width (i.e. the number of nodes) in the hierarchical topology employed. Moreover, if connectivity resources are limited, general cognitive abilities (particularly relating to robust situation inference) are reduced. Finally, the complexity of the control engine can be minimized to the point of achieving satisfactory behavior.

These ideas are interesting, although undeniably challenging to enforce due a myriad of both technical and political limitations. From a technical standpoint, identifying the precise cognitive resources needed to achieve some level of intelligence, but no more than that, is difficult to say the least. Introducing legislation banning the creation of super-intelligence AGI systems is destined to only be partially effective, since there will always be countries, governments and large organizations that view themselves above international law. An argument can be made that if building super-AGI systems is widely acknowledged as inevitably leading to the destruction of the human race, humans will be strongly motivated not to pursue such endeavors.

Closing Thoughts

We live in a time in which an intersection of new ideas and technological advances has created a novel path for building thinking machines. Reinforcement learning, which has well-established biological plausibility, in concert with models of cortical circuits for perception, form a promising direction for achieving one possible form of AGI. While many cognitive phenomena remain poorly understood and are likely to pose modeling challenges, the paradigm proposed in this chapter is inherently generic and serves as solid basis for an AGI framework. Practically speaking, the VLSI technology needed to realize large scale AGI experiments exists today. Assuming such assertions are true, the focus should now shift to the key questions pertaining to the impact such transformational technology will have on humanity. There are clear risks posed by the very essence of the framework alongside its obvious benefits. In order to try and guarantee a positive outcome for humanity, much work remains to be done in establishing theoretical results and practical methods for monitoring and controlling the paths that such AGI technology can take. The singularity may not be avoidable altogether, but if humanity is to survive along these new forms of life on earth, it must forcefully play an integrated part of their evolutionary processes, rather than assume the role of a bystander.

References

Arel, I. (2012). The threat of a reward-driven adversarial artificial general intelligence. In A. Eden, J. Søraker, J. H. Moor, & E. Steinhart (Eds.), *The singularity hypothesis: a scientific and philosophical analysis*, Springer.

Arel, I., Rose, D., & Karnowski, T. (2010). Deep machine learning—a new frontier in artificial intelligence research. *IEEE Computational Intelligence Magazine, 14*, 12–18.

Baillargeon, R. (1994). Physical Reasoning in young infants: seeking explanations for impossible events. *British Journal of Developmental Psychology, 12*, 9–33.

Bellman, R. (1957). *Dynamic programming*. Princeton: Princeton University Press.

Bengio, Y. (2009). Learning deep architectures for AI. *Foundations and Trends in Machine Learning, 2*(1), 1–127.

Duda, R., Hart, P., & Stork, D. (2000). *Pattern recognition* (2nd edn ed.). New York: Wiley-Interscience.

Ginsburg, G. S., & Bronstein, P. (1993). Family factors related to children's intrinsic/extrinsic motivational orientation and academic performance. *Child Development, 64*, 1461–1474.

Gray, P., Hurst, P., Lewis, S., & Meyer, R. (2001). *Analysis and design of analog integrated circuits*. New York: Wiley.

Hasler, P., & Dugger, J. (2005). An analog floating-gate node for supervised learning. *IEEE Transactions on Circuits and Systems I, 52*(5), 834–845.

Hasler, P., Diorio, C., Minch, B. A., & Mead, C. (1995). Single transistor learning synapse with long term storage. *IEEE International Symposium on Circuits and Systems*, 1660–1663.

Joel, D., Niv, Y., & Ruppin, E. (2002). Actor-critic models of the basal ganglia: new anatomical and computational perspectives. *Neural Networks, 15*, 535–547.

Kelley, S. A., Brownell, C. A., & Campbell, S. B. (2000). Mastery motivation and self-evaluative affect in toddlers: longitudinal relations with maternal behavior. *Child Development, 71*, 1061–71.

Lee, T. (2003). Hierarchical bayesian inference in the visual cortex. *Journal of the Optical Society of America, 20*(7), 1434–1448.

Lee, T., Mumford, D., Romero, R., & Lamme, V. (1998). The role of the primary visual cortex in higher level vision. *Vision Research, 38*, 2429–2454.

Millner, A. D., & Goodale, M. A. (1996). *The visual brain in action*. Oxford: Oxford University Press.

Mishkin, M., Ungerkeuder, L. G., & Macko, K. A. (1983). Object vision and spatial vision: two cortical pathways. *Trends in Neuroscience, 6*, 414–417.

Schultz, W. (1998). Predictive reward signal of dopamine neurons. *The Journal of Neurophysiology, 80*(1), 1–27.

Suri, R. E., & Schultz, W. (1999). A neural network model with dopamine-like reinforcement signal that learns a spatial delayed response task. *Neuroscience, 91*(3), 871–890.

Sutton, R. S., & Barto, A. G. (1998). *Reinforcement Learning: An Introduction*. Cambridge, MA: MIT Press.

Wallis, G., & Bülthoff, H. (1999). Learning to recognize objects. *Trends in Cognitive Sciences, 3*(1), 23–31.

Wallis, G., & Rolls, E. (1997). Invariant face and object recognition in the visual system. *Progress in Neurobiology, 51*, 167–194.

Chapter 3A
William J. Rapaport on Arel's "The Threat of a Reward-Driven Adversarial Artificial General Intelligence"

Can't we just talk?
Itamar Arel (2013) argues that:

1. artificial general intelligence (AGI) "is inevitable" (p. 44),
2. techniques including a "fusion between deep learning, ... a scalable situation inference engine, and reinforcement learning [RL] as a decision-making system may hold the key to place us on the path to AGI" (p. 45), and
3. "a potentially devastating conflict between a reward-driven AGI system and the human race... is inescapable, given the assumption that an RL-based AGI will be allowed to evolve" (p. 45).

Why "inescapable"? If I understand Arel correctly, it is a mathematical certainty:

> [F]rom Eqs. (3.2) and (3.4) [Arel 2013, pp. 50, 55, the details of which are irrelevant to my argument], it follows that the agent continuously attempts to maximize its "positive" surprises [i.e. "its wellbeing"]... while minimizing "negative" surprises. This process... is unbounded.... [O]nce such a bonus is received on a regular basis, it becomes the new norm and no longer yields the same level of satisfaction. This is the core danger in designing systems that are driven by rewards and have large cognitive capacity; by continuously striving to gain positive (relative) reinforcement, they will inevitably pose a danger to humanity (pp. 55–56).

Let's suppose so. But why should it be "inevitable"? Despite Arel's faith in the inevitability of *AGI* (which I share), he seems to be committing the fallacy of thinking that AGIs must differ in crucial respects from humans.

This is the fallacy that John Searle commits when claiming that the inhabitant of his Chinese Room (Searle 1980) doesn't "understand a word of Chinese and neither does any other digital computer because all the computer has is what [the inhabitant] ha[s]: a formal program that attaches no meaning, interpretation, or content to any of the symbols" (Searle 1982, p. 5). As I have pointed out elsewhere, this assumes "that external links are needed for the program to 'attach' meaning to its symbols" (Rapaport 2000, §3.2.2). The fallacy can be seen by realizing that "*if* external links *are* needed, then surely a computer could have them as well as—and presumably in the same way that—humans have them" (Rapaport 2000, §3.2.2).

Why do I think that Arel is committing this fallacy? Because, presumably, *humans also* "attempt to maximize [their] wellbeing". Now, I can agree that humans themselves have been known, from time to time, to "pose a danger to humanity" (for a discussion of this, see Dietrich 2001, 2007). *But we have also devised methods for alleviating such dangers.* Clearly, then, rather than wringing our hands over the "inevitability" of AGIs wreaking havoc on their creators, we should give them some of those methods.

And, indeed, Arel sketches out some possibilities along these lines: education and "limit[ing] such [a] system's mental capacity" (p. 56). But he seems to neglect one obvious possibility, one that is, in fact, a *necessity* for any AGI: For an AGI to really have GI—general intelligence—it must have cognition: (1) It must be able to use and understand *language*—and, presumably, *our* language, so that *we* can communicate with it, and vice versa (see Winston 1975 and my discussion of "Winston's problem" in Rapaport 2003)—and (2) it must be able to *reason* consciously (e.g. via an explicit knowledge-representation-and-reasoning system, as opposed to tacit reasoning by, say, an artificial neural network). If we can reason with it in natural language, then we can hope to be able to collaborate and negotiate with it, rather than compete with it. Such natural-language and reasoning competence is, in any case, a prerequisite (or at least a product) of education, but it requires no limitation on the AGI's mental capacity.

References

Arel, D. R., & Coop, R. (November, 2009). DeSTIN: A scalable deep learning architecture with application to high-dimensional robust pattern recognition. In *Proceedings of the AAAI 2009 Fall Symposium on Biologically Inspired Cognitive Architectures*.

Arel, I. (2013). The threat of a reward-driven adversarial artificial general intelligence. In A. H. Eden, J. H. Moor, J. H. Søraker, & E. Steinhart (Eds.), *The singularity hypothesis: a scientific and philosophical analysis* (pp. 43–58). Heidelberg: Springer.

Dietrich, E. (2001, October). Homo sapiens 2.0: why we should build the better robots of our nature. *Journal of Experimental and Theoretical Artificial Intelligence*, *13*(4), 323–328.

Dietrich, E. (2007). After the humans are gone. *Journal of Experimental and Theoretical Artificial Intelligence, 19*(1), 55–67.

Rapaport, W. J. (2000). How to pass a turing test: syntactic semantics, natural—language understanding, and first-person cognition. *Journal of Logic, Language, and Information*, 9(4): 467–490. (Reprinted in *The turing test: the elusive standard of artificial intelligence*, pp. 161–14, by H. M. James, Ed., 2003, Dordrecht: Kluwer).

Rapaport, W. J. (2003). What did you mean by that? Misunderstanding, negotiation, and syntactic semantics. *Minds and Machines, 13*(3), 397–427.

Searle, J. R. (1980). Minds, brains, and programs. *Behavioral and Brain Sciences, 3*, 417–457.

Searle, J. R. (1982). The myth of the computer. *New York Review of Books* (29 April 1982): 3–6; cf. correspondence, same journal (24 June 1982): 56–57.

Winston, P. H. (1975). Learning structural descriptions from examples. in Patrick HenryWinston (ed.), The Psychology of Computer Vision (New York: McGraw-Hill): 157– 209. (Reprinted in *Readings in knowledge representation*, pp. 141–168, by J. B. Ronald & J. L. Hector, Eds., 1985, Los Altos, CA: Morgan Kaufmann).

Chapter 4
New Millennium AI and the Convergence of History: Update of 2012

Jürgen Schmidhuber

Abstract Artificial Intelligence (AI) has recently become a real formal science: the new millennium brought the first mathematically sound, asymptotically optimal, *universal* problem solvers, providing a new, rigorous foundation for the previously largely heuristic field of General AI and embedded agents. There also has been rapid progress in not quite universal but still rather general and *practical* artificial recurrent neural networks for learning sequence-processing programs, now yielding state-of-the-art results in real world applications. And the computing power per Euro is still growing by a factor of 100–1,000 per decade, greatly increasing the feasibility of neural networks in general, which have started to yield human-competitive results in challenging pattern recognition competitions. Finally, a recent formal theory of fun and creativity identifies basic principles of curious and creative machines, laying foundations for artificial scientists and artists. Here I will briefly review some of the new results of my lab at IDSIA, and speculate about future developments, pointing out that the time intervals between the most notable events in over 40,000 years or 2^9 lifetimes of human history have sped up exponentially, apparently converging to zero within the next few decades. Or is this impression just a by-product of the way humans allocate memory space to past events?

Note: this is the 2012 update of a 2007 publication (Schmidhuber 2007b). Compare also the 2006 celebration of 75 years of AI (Schmidhuber 2006c).

J. Schmidhuber (✉)
The Swiss AI Lab IDSIA, University of Lugano & SUPSI,
Galleria 1, 6928 Manno-Lugano, Switzerland
e-mail: juergen@idsia.ch
http://www.idsia.ch

A. H. Eden et al. (eds.), *Singularity Hypotheses*, The Frontiers Collection,
DOI: 10.1007/978-3-642-32560-1_4, © Springer-Verlag Berlin Heidelberg 2012

Introduction

In 2003 I observed (Schmidhuber 2003a, b) that each major breakthrough in computer science tends to come roughly twice as fast as the previous one, roughly matching a century-based scale: In 1623 the computing age started with the first mechanical calculator by Wilhelm Schickard (followed by machines of Pascal, 1640, and Leibniz, 1670). Roughly two centuries later Charles Babbage came up with the concept of a program-controlled computer (1834–1840). One century later, Julius Lilienfeld invented the transistor (late 1920s), and Kurt Gödel layed the foundations of theoretical computer science with his work on universal formal languages and the limits of proof and computation (1931) (Goedel 1931). His results and Church's extensions thereof were reformulated by Turing in 1936 (Turing 1936), while Konrad Zuse built the first working program-controlled computers (1935–1941), using the binary system of Leibniz (1701) instead of the more cumbersome decimal system used by Babbage and many others. By 1941 all the main ingredients of 'modern' computer science were in place. The next 50 years saw many less radical theoretical advances as well as faster and faster switches—relays were replaced by tubes by single transistors by numerous transistors etched on chips—but arguably this was rather predictable, incremental progress without earth-shaking events. Half a century later, however, Berners-Lee triggered the most recent world-changing development by creating the World Wide Web at CERN (1990).

Extrapolating the trend, we should expect the next radical change to manifest itself one quarter of a century after the most recent one, that is, before 2020, when some computers will already match brains in terms of raw computing power, according to frequent estimates based on Moore's law, which suggests a speed-up factor of roughly 100–1,000 per decade, give or take a few years. Will the remaining series of faster and faster additional revolutions converge in an *Omega point* (term coined by Pierre Teilhard de Chardin, 1916) around 2040, when individual machines will already approach the raw computing power of all human brains combined, provided Moore's law does not break down? Many of the present readers of this article should still be alive then. Compare Stanislaw Ulam's concept of an approaching *historic singularity* (quote: Vinge 1993), popularized by Vernor Vinge as *technological singularity* (Vinge 1984, 1993), as well as subsequent speculations (Moravec 1999; Kurzweil 2005).

Will the software and the theoretical advances keep up with the hardware development? I am convinced they will. In fact, the new millennium has brought not only human-competitive performance of artificial neural networks (NN) in pattern recognition contests (more on this later), but also fundamental new insights into the problem of constructing *theoretically optimal* rational agents or universal Artificial Intelligences (AIs), as well as curious & creative machines (more on this below). There also has been rapid progress in *practical* learning algorithms for agents interacting with a dynamic environment, autonomously discovering true sequence-processing, problem-solving programs, as opposed to the reactive mappings from stationary inputs to outputs studied in most traditional machine learning (ML) research. In what follows, I will briefly review some of the new results, then come back to the issue of whether or not history is about to "converge."

Notation

Consider a learning robotic agent with a single life which consists of discrete cycles or time steps $t = 1, 2, \ldots, T$. Its total lifetime T may or may not be known in advance. In what follows,the value of any time-varying variable Q at time t $(1 \le t \le T)$ will be denoted by $Q(t)$, the ordered sequence of values $Q(1), \ldots, Q(t)$ by $Q(\le t)$, and the (possibly empty) sequence $Q(1), \ldots, Q(t-1)$ by $Q(<t)$.

At any given t the robot receives a real-valued input vector $x(t)$ from the environment and executes a real-valued action $y(t)$ which may affect future inputs; at times $t < T$ its goal is to maximize future success or *utility*

$$u(t) = E_\mu \left[\sum_{\tau=t+1}^{T} r(\tau) \middle| h(\le t) \right], \tag{4.1}$$

where $r(t)$ is an additional real-valued reward input at time t, $h(t)$ the ordered triple $[x(t), y(t), r(t)]$ (hence $h(\le t)$ is the known history up to t), and $E_\mu(\cdot \mid \cdot)$ denotes the conditional expectation operator with respect to some possibly unknown distribution μ from a set M of possible distributions. Here M reflects whatever is known about the possibly probabilistic reactions of the environment. For example, M may contain all computable distributions (Solomonoff 1964; Li and Vitanyi 1997; Hutter 2005). Note that unlike in most previous work by others (Kaelbling et al. 1996; Sutton and Barto 1998), but like in much of the author's own previous work (Schmidhuber et al. 1997; Schmidhuber 2007a), there is just one life, no need for predefined repeatable trials, no restriction to Markovian interfaces between sensors and environment (Schmidhuber 1991c), and the utility function implicitly takes into account the expected remaining lifespan $E_\mu > (T \mid h(\le t))$ and thus the possibility to extend it through appropriate actions (Schmidhuber 2005, 2007a).

Universal But Incomputable AI

Solomonoff's theoretically optimal universal predictors and their Bayesian learning algorithms (Solomonoff 1964; Li and Vitanyi 1997; Hutter 2005) only assume that the reactions of the environment are sampled from an unknown probability distribution μ contained in a set M of all enumerable distributions—compare text after Eq. (4.1). That is, given an observation sequence $q(\le t)$, we only assume there exists a computer program that can compute the probability of the next possible $q(t+1)$, given $q(\leq t)$. Since we typically do not know the program computing μ, we predict the future in a Bayesian framework by using a mixture distribution $\xi = \sum_i \omega_i \mu_i$, a weighted sum of *all* distributions $\mu_i \in \mathcal{M}$, $i = 1, 2, \ldots$, where $\sum_i \omega_i \leq 1$. It turns out that this is indeed the best one can possibly do, in a very general sense (Hutter 2005). The drawback is that the scheme is incomputable, since M contains infinitely many distributions.

One can increase the theoretical power of the scheme by augmenting M by certain non-enumerable but limit-computable distributions (Schmidhuber 2002a), or restrict it such that it becomes computable, e.g., by assuming the world is computed by some unknown but deterministic computer program sampled from the Speed Prior (Schmidhuber 2002b) which assigns low probability to environments that are hard to compute by any method. Under the Speed Prior the cumulative a priori probability of all data whose computation through an optimal algorithm requires more than $O(n)$ resources is $1/n$.

Can we use the optimal predictors to build an optimal AI? Indeed, in the new millennium it was shown we can. At any time t, the recent theoretically optimal yet uncomputable RL algorithm AIXI (Hutter 2005) uses Solomonoff's universal prediction scheme to select those action sequences that promise maximal future reward up to some horizon, typically $2t$, given the current data $h(\leq t)$.. One may adapt this to the case of any finite horizon T. That is, in cycle $t+1$, AIXI selects as its next action the first action of an action sequence maximizing ξ-predicted reward up to the horizon, appropriately generalizing Solomonoff's universal prior. Recent work (Hutter 2005) demonstrated AIXI's optimal use of observations as follows. The Bayes-optimal policy p^ξ based on the mixture ξ is self-optimizing in the sense that its average utility value converges asymptotically for all $\mu \in \mathcal{M}$ to the optimal value achieved by the (infeasible) Bayes-optimal policy p^μ which knows μ in advance. The necessary condition that $\mathcal{M}$ admits self-optimizing policies is also sufficient. Furthermore, p^ξ is Pareto-optimal in the sense that there is no other policy yielding higher or equal value in *all* environments $\nu \in \mathcal{M}$ and a strictly higher value in at least one (Hutter 2005).

What are the implications? The first decades of attempts at *Artificial General Intelligence (AGI)* have been dominated by heuristic approaches (Newell and Simon 1963; Rosenbloom et al. 1993; Utgoff 1986; Mitchell 1997). Traditionally many theoretical computer scientists have regarded the field with contempt for its lack of hard theoretical results. Things have changed, however. Although the universal approach above is practically infeasible due to the incomputability of Solomonoff's prior, it does provide, for the first time, a mathematically sound theory of AGI and optimal decision making based on experience, identifying the limits of both human and artificial intelligence, and providing a yardstick for any future approach to AGI.

Using the Speed Prior mentioned above, one can scale the universal approach down such that it becomes at least *computable* (Schmidhuber 2002b). In what follows I will mention ways of introducing additional optimality criteria that take into account the computational costs of prediction and decision making.

Asymptotically Optimal General Problem Solver

To take computation time into account in a general, theoretically optimal way (Levin 1973; Li and Vitanyi 1997, pp. 502–505), the recent asymptotically optimal search algorithm for *all* well-defined problems HSEARCH (Hutter 2002) uses a hardwired brute force proof searcher which (justifiably) ignores the costs of proof search. Assuming discrete input/output domains $X/Y \subset B^*$, a formal problem specification $f : X \rightarrow Y$ (say, a functional description of how integers are decomposed into their prime factors), and a particular $x \in X$ (say, an integer to be factorized), HSEARCH orders all proofs of an appropriate axiomatic system by size to find programs q that for all $z \in X$ provably compute $f(z)$ within time bound $t_q(z)$. Simultaneously it spends most of its time on executing the q with the best currently proven time bound $t_q(x)$. Remarkably, HSEARCH is as fast as the *fastest* algorithm that provably computes $f(z)$ for all $z \in X$, save for a constant factor smaller than $1 + \epsilon$ (arbitrary real-valued $\epsilon > 0$) and an f-specific but x-independent additive constant (Hutter 2002).

Practical applications, however, should not ignore potentially huge constants. This motivates the next section which addresses all kinds of optimality (not just asymptotic optimality).

Optimal Self-Referential General Problem Solver

The recent Gödel machines (Schmidhuber 2005, 2007a, 2009) represent the first class of mathematically rigorous, general, fully self-referential, self-improving, optimally efficient problem solvers. In particular, they are applicable to the problem embodied by objective (4.1), which obviously is not limited to *asymptotic* optimality. Gödel machines formalize I. J. Good's informal remarks (1965) on an "intelligence explosion through self-improving super-intelligences."

The initial software $\mathcal{S}$ of such a Gödel machine contains an initial problem solver, e.g., one of the approaches above (Hutter 2005) or some less general, typical sub-optimal method (Kaelbling et al. 1996; Sutton and Barto 1998). Simultaneously, it contains an initial proof searcher (possibly based on an online variant of *Universal Search* (Levin 1973) or the *Optimal Ordered Problem Solver* (Schmidhuber 2004) which is used to run and test *proof techniques*. The latter are programs written in a universal programming language implemented on the Gödel machine within $\mathcal{S}$, able to compute proofs concerning the system's own future performance, based on an axiomatic system $\mathcal{A}$ encoded in $\mathcal{S}$. $\mathcal{A}$ describes the formal *utility* function, in our case Eq. (4.1), the hardware properties, axioms of arithmetics and probability theory and string manipulation etc, and $\mathcal{S}$ itself, which is possible without introducing circularity (Schmidhuber 2005, 2007a, 2009).

Inspired by Kurt Gödel's celebrated self-referential formulas (1931) (Goedel 1931), the Gödel machine rewrites any part of its own code in a computable way through a self-generated executable program as soon as its *Universal Search* variant has found a proof that the rewrite is *useful* according to objective (4.1). According to the Global Optimality Theorem (Schmidhuber 2005, 2007a, 2009), such a self-rewrite is globally optimal—no local maxima!—since the self-referential code first had to prove that it is not useful to continue the proof search for alternative self-rewrites.

If there is no provably useful, globally optimal way of rewriting $\mathcal{S}$ at all, then humans will not find one either. But if there is one, then $\mathcal{S}$ itself can find and exploit it. Unlike *non*-self-referential methods based on hardwired proof searchers (Hutter 2005) (section Asymptotically Optimal General Problem Solver), Gödel machines not only boast an optimal *order* of complexity but can optimally reduce (through self-changes) any slowdowns hidden by the asymptotic $O()$-notation, provided the utility of such speed-ups is provable at all.

To make sure the Gödel machine is at least *asymptotically* optimal even before the first self-rewrite, we may initialize it by the non-self-referential but *asymptotically fastest algorithm for all well-defined problems* HSEARCH (Hutter 2002) of section Asymptotically Optimal General Problem Solver. Given some problem, the Gödel machine may decide to replace its HSEARCH initialization by a faster method suffering less from large constant overhead, but even if it doesn't, its performance won't be less than asymptotically optimal.

Implications

The above implies that there already exists the blueprint of a Universal AI which will solve almost all problems almost as quickly as if it already knew the best (unknown) algorithm for solving them, because almost all imaginable problems are big enough to make additive constants negligible. The only motivation for *not* quitting computer science research right now is that many real-world problems are so small and simple that the ominous constant slowdown (potentially relevant at least before the first Gödel machine self-rewrite) is *not* negligible.

Recurrent/Deep Neural Networks

Practical implementations of the Gödel machine above do not yet exist, and probably will require a thoughtful choice of the initial axioms and the initial proof searcher. In what follows, however, I will focus on already quite practical, non-optimal and non-universal, but still rather general searchers in program space, as opposed to the space of reactive, feedforward input / output mappings, which still attracts the bulk of current ML research.

Recurrent NN (RNN) are NN (Bishop 2006) with feedback connections that are, in principle, as powerful as any traditional computer. There is a very simple way to see this (Schmidhuber 1990): a traditional microprocessor may be viewed as a very sparsely connected RNN consisting of very simple neurons implementing nonlinear AND and NAND gates, etc. Compare (Siegelmann and Sontag 1991) for a more complex argument. Hence RNN can solve tasks involving sequences of continually varying inputs. Examples include robot control, speech recognition, music composition, attentive vision, and numerous others.

Supervised RNN can be trained by gradient descent and other methods (Werbos 1988; Williams and Zipser 1994; Robinson and Fallside 1987; Schmidhuber 1992a; Maass et al. 2002; Jaeger 2004). Recent work has successfully applied Hessian-free optimization to RNN (Sutskever et al. 2011), using tricks such as special damping functions and stopping criteria, mini-batches for curvature calculation, and others. Our own RNN overcome fundamental problems of previous RNN (Hochreiter et al. 2001), outperforming them in many applications (Hochreiter and Schmidhuber 1997; Gers and Schmidhuber 2001, 2009; Gers et al. 2002; Schmidhuber et al. 2007; Graves et al. 2006, 2008, 2009). While RNN used to be **toy problem methods** in the 1990s, ours have recently started to outperform all other methods in **challenging real world applications** (Schmidhuber et al. 2011, 2007; Fernandez et al. 2007, 2008, 2009, 2009). Recently, our CTC-trained (Graves et al. 2006) mulitdimensional (Graves and Schmidhuber 2009) RNN won three Connected Handwriting Recognition Competitions at ICDAR 2009 (see list of won competitions below).

Training an RNN by standard methods is similar to training a feedforward NN (FNN) with many layers, which runs into similar problems (Hochreiter et al. 2001). However, our recent deep FNN with special internal architecture overcome these problems to the extent that they are currently winning many international pattern recognition contests (Schmidhuber et al. 2011; Ciresan et al. 2010, 2011b, c, 2012b, c) (see list of won competitions below). None of this requires the traditional sophisticated computer vision techniques developed over the past six decades or so. Instead, our biologically rather plausible NN architectures learn from experience with millions of training examples. Typically they have many non-linear processing stages like Fukushima's Neocognitron (Fukushima 1980); we sometimes (but not always) profit from sparse network connectivity and techniques such as weight sharing & convolution (LeCun et al. 1998; Behnke 2003), max-pooling (Scherer et al. 2010), and contrast enhancement like the one automatically generated by unsupervised *Predictability Minimization* (Schmidhuber 1992b, 1996; Schraudolph et al. 1999). Our NN are now often outperforming all other methods including the theoretically less general and less powerful support vector machines (SVM) based on statistical learning theory (Vapnik 1995) (which for a long time had the upper hand, at least in practice). These results are currently contributing to a second **Neural Network ReNNaissance** (the first one happened in the 1980s and early 90s) which might not be possible without dramatic advances in computational power per Swiss Franc, obtained in the new millennium. In particular, to implement and train our NN, we exploit graphics processing units (GPUs, mini-supercomputers normally used for video games) which are 100 times faster than traditional CPUs, and a million times faster than PCs of two decades ago when we started this type of research.

1st Ranks of my Lab's Methods in International Competitions since 2009:

7. ISBI 2012 Segmentation Challenge (with superhuman pixel error rate) (Ciresan et al. 2012a).
6. IJCNN 2011 on-site Traffic Sign Recognition Competition (0.56 % error rate, the only method better than humans, who achieved 1.16 % on average; 3rd place for 1.69 %) (Ciresan et al. 2012b)
5. ICDAR 2011 offline Chinese handwritten character recognition competition (Ciresan et al. 2012c).
4. Online German Traffic Sign Recognition Contest (1st & 2nd rank; 1.02 % error rate) (Ciresan et al. 2011c).
3. ICDAR 2009 Arabic Connected Handwriting Competition (won by our LSTM RNN Graves et al. 2009; Graves and Schmidhuber 2009), same below.
2. ICDAR 2009 Handwritten Farsi/Arabic Character Recognition Competition.
1. ICDAR 2009 French Connected Handwriting Competition.

1st Ranks in Important Machine Learning (ML) Benchmarks since 2010:

3. MNIST handwritten digits data set (LeCun et al. 1998) (perhaps the most famous ML benchmark). New records: 0.35 % error in 2010 (Ciresan et al. 2010), 0.27 % in 2011 (Ciresan et al. 2011a), first human-competitive performance (0.23 %) in 2012 (Ciresan et al. 2012c).
2. NORB stereo image data set (Yann LeCun et al. 2004). New records in 2011, 2012, e.g., (Ciresan et al. 2012c).
1. CIFAR-10 image data set (Krizhevsky 2009). New records (eventually 11.2 % error rate) in 2011, 2012, e.g., (Ciresan et al. 2012c).

In a certain sense, **Reinforcement Learning** (RL) (Kaelbling et al. 1996; Sutton and Barto 1998) is more challenging than supervised learning as above, since there is no teacher providing desired outputs at appropriate time steps. To solve a given problem, the learning agent itself must discover useful output sequences in response to the observations. The traditional approach to RL is best embodied by Sutton and Barto's book (Sutton and Barto 1998). It makes strong assumptions about the environment, such as the Markov assumption: the current input of the agent tells it all it needs to know about the environment. Then all we need to learn is some sort of reactive mapping from stationary inputs to outputs. This is often unrealistic. A more general approach for partially observable environments directly evolves programs for RNN with internal states (no need for the Markovian assumption), by applying evolutionary algorithms (Rechenberg 1971; Schwefel 1974; Holland 1975) to RNN weight matrices (Yao 1993; Sims 1994; Stanley and Miikkulainen 2002; Hansen and Ostermeier 2001). Recent work brought progress through a focus on reducing search spaces by co-evolving the comparatively small weight vectors of individual neurons and synapses (Gomez et al. 2008), by Natural Gradient-based Stochastic Search Strategies (Wierstra et al. 2008, 2010; Sun et al. 2009a, 2009b; Schaul et al. 2010; Glasmachers et al. 2010), and by reducing search spaces through weight matrix compression (Schmidhuber 1997; Koutnik et al. 2010). Our RL RNN now outperform many previous methods on benchmarks (Gomez et al. 2008), creating memories of important events and solving numerous tasks unsolvable by classical RL methods. Several *best paper awards* resulted from this research, e.g., (Sun et al. 2009a; Gisslen et al. 2011).

Curious and Creative Machines Maximizing Wow-Effects

The main problem in many RL tasks, however, remains the very rare external reward. How to learn anything from such limited feedback in reasonable time? Over the past two decades I have pioneered a **Formal Theory of Curiosity** (FTC) and creativity and exploration, which describes how to provide frequent additional *intrinsic* rewards for active data-creating explorers (Schmidhuber 1991a, b, 1999, 2006a, 2010, 2011; Storck et al. 1995). FTC has recently gained a lot of traction (most citations, even of the old papers, stem from the past five years, especially the last two years).

One inspiration of FTC is biological. To solve existential problems such as avoiding hunger or thirst, a baby has to learn how the environment responds to its actions. Even when there is no immediate need to satisfy thirst or other built-in primitive drives, the baby does not run idle. Instead it actively conducts non-random experiments: what sensory feedback do I get if I move my eyes or my fingers or my tongue in particular ways? Being able to predict the effects of its actions enable it to plan control sequences leading to desirable states, such as those where its thirst and hunger sensors are switched off. But the growing infant quickly gets bored by things it already understands well as well as those it does not understand at all. It searches for new effects exhibiting some yet unexplained but *easily learnable* regularities. It continually acquires more and more complex skills building on previously acquired, simpler skills. Eventually the baby may become a physicist, creating experiments to discover previously unknown physical laws, or an artist creating new eye-opening artworks, or a comedian delighting audiences with novel jokes.

According to FTC (Schmidhuber 1991a, b, 1999, 2006a, 2010, 2011; Storck et al. 1995), the baby's exploratory behavior is driven by a very simple algorithmic mechanism that uses RL to maximize internal *wow effects*. Wow effects are sudden reductions in an agent's estimate of the complexity of its history of observations and actions. These occur due to the agent's own *learning progress*. To clarify, consider an explorer with two modules: a world model and an actor. The former encodes the agent's growing history of sensory data (tactile, auditory, visual, etc), while the latter executes actions that influence and shape that history. The world model (e.g., an NN or RNN) uses a learning algorithm to encode the data more efficiently, trying to discover *new regularities* that allow for saving storage space (e.g., synapses) or computation time. When successful, the RL actor receives a reward (the *wow effect*). Maximizing future expected reward, the actor is motivated to invent behaviors leading to more such rewards; i.e., to data that the encoder does not yet know but can easily learn. Wow effects can also result from simplifying or speeding up the actor itself (Schmidhuber 2011). Unlike the preprogrammed interestingness measure of EURISKO (Lenat 1993), FTC's continually redefines what's interesting based on what's currently easy to learn, in addition to what's already known.

Since 1990 we have been building explorers based on FTC. These agents may be viewed as simple artificial scientists or artists with an intrinsic desire to create experiments for building better models of the world (Schmidhuber 1991a, b, 1999, 2006a, 2010; Storck et al. 1995), in the process developing more and more efficient procedures or skills (Schmidhuber 1999, 2011). This work has inspired much recent research; the last few years brought lots of related work by others (Singh et al. 2005; Barto 2013; Dayan 2013; Oudeyer et al. 2013), also in the nascent field of developmental robotics (Kuipers et al. 2006; Hart et al. 2008; Oudeyer et al. 2012). FTC generalizes *active learning* (Fedorov 1972; Balcan et al. 2009; Strehl et al. 2010), taking into account: (1) highly environment-dependent costs of obtaining or creating not just individual data points but data *sequences* of unknown size; (2) arbitrary algorithmic (Solomonoff 1964, 2002a, b; Kolmogorov 1965; Li and Vitanyi 1997) or statistical dependencies in sequences of actions and sensory inputs (Schmidhuber 1999, 2006a); and (3) the computational cost of learning new skills (Schmidhuber 1999, 2011).

The first curious explorers from the 1990s (Schmidhuber 1991a, 1999; Storck et al. 1995) used RL methods that were sub-optimal for online learning and for *wow effect* rewards that vanish as soon as learning progress stops. More recent, mathematically optimal, creative explorers (Schmidhuber 2006a, 2010) are based on universal RL methods (Hutter 2005; Schmidhuber 2002b, 2006b) that are not yet computationally tractable (section Universal But Incomputable AI). Recent work has demonstrated exploration that is both optimal *and* feasible (Yi et al. 2011) for limited scenarios, but much remains to be done for challenging, high-dimensional, partially observable worlds. This is driving ongoing work.

Is History Converging? Again?

Many predict that within a few decades there will be computers whose raw computing power will surpass the one of a human brain by far (e.g., Moravec 1999; Kurzweil 2005). In the 1980s, an educated guess of this type motivated me to study computer science and AI. I have argued above that algorithmic advances are keeping up with the hardware development, pointing to new-millennium theoretical insights on universal problem solvers and creative machines that are optimal in various mathematical senses (thus making *General AI* a real formal science), as well as to practical progress in program learning through brain-inspired neural nets.

A single human predicting the future of humankind is like a single neuron predicting what its brain will do. Nevertheless, a few things can be predicted confidently, such as: tomorrow the sun will shine in the Sahara desert. So let us put the AI-oriented developments discussed above in a broader context, and try to extend the naive analysis of past computer science breakthroughs in the introduction, which predicts that computer history will converge in an *Omega point* or historic singularity Ω around 2040 (Schmidhuber 2006c, 2007b).

Surprisingly, even if we go back all the way to the beginnings of modern man over 40,000 years ago, essential historic developments (that is, the subjects of the major chapters in history books) match a a binary scale marking exponentially declining temporal intervals, each half the size of the previous one, and even measurable in terms of powers of 2 multiplied by a human lifetime (roughly 80 years—throughout recorded history many individuals have reached this age, although the average lifetime often was shorter, mostly due to high children mortality). Using the value $\Omega = 2,040$, associate an error bar of not much more than 10 % with each date below:

1. $\Omega - 2^9$ lifetimes: modern humans start colonizing the world from Africa.
2. $\Omega - 2^8$ lifetimes: bow and arrow invented; hunting revolution.
3. $\Omega - 2^7$ lifetimes: invention of agriculture; first permanent settlements; beginnings of civilization.
4. $\Omega - 2^6$ lifetimes: first high civilizations (Sumeria, Egypt), and the most important invention of recorded history, namely, the one that made recorded history possible: writing.
5. $\Omega - 2^5$ lifetimes: The *Axial Age* (the axis around which history turned, according to Karl Jaspers), the age of the first large empire (the Persian one), the only empire ever to contain almost half humankind. At its fringes, the ancient Greeks invent democracy and lay the foundations of Western science and art and philosophy, from algorithmic procedures and formal proofs to anatomically perfect sculptures, harmonic music, sophisticated machines including steam engines, and organized sports. Major Asian religions founded, Old Testament written (basis of Judaism, Christianity, Islam). High civilizations in China, origin of the first calculation tools, and India, origin of alphabets and the zero.
6. $\Omega - 2^4$ lifetimes: bookprint (often called the most important invention of the past 2000 years) invented in China. Islamic science and culture start spreading across large parts of the known world (this has sometimes been called the most important development between Antiquity and the age of discoveries)
7. $\Omega - 2^3$ lifetimes: the most dominant empire of the past 2,500 years (the Mongolian empire) includes most of the civilized world. Soon afterwards, Chinese fleets and later also European vessels start exploring the world. Gun powder and guns invented in China. Rennaissance and printing press (often called the most influential invention of the past 1000 years) and subsequent Reformation in Europe. Begin of the Scientific Revolution.
8. $\Omega - 2^2$ lifetimes: Age of enlightenment and rational thought in Europe. Massive progress in the sciences; first flying machines; start of the industrial revolution based on improved steam engines.

9. $\Omega - 2$ lifetimes: Birth of the modern world in the second industrial revolution based on combustion engines, cheap electricity, and modern chemistry. Genetic and evolution theory. Revolutionary modern medicine through the germ theory of disease. Onset of the unprecedented population explosion driving many other developments. European colonialism at its short-lived peak.
10. $\Omega - 1$ lifetime: Post-World War II society and pop culture emerges. The world-wide super-exponential population explosion (mainly due to the Haber-Bosch process Smil 1999) is at its peak. First commercial computers and first spacecraft; DNA structure unveiled.
11. $\Omega - 1/2$ lifetime: 3rd industrial revolution (?) through an emerging world-wide digital nervous system based on personal computers, cell phones, and the World Wide Web. A mathematical theory of universal AI emerges (see sections above)—will this be considered a milestone in the future?
12. $\Omega - 1/4$ lifetime: This point will be reached in a few years. See introduction.
13. $\Omega - 1/8$ lifetime: The number of humans will roughly match the number of grey matter neurons in a human brain. Will they be digitally connected in a roughly brain-like way (on average 10,000 connections per unit, mostly between neighbors arranged in a two-dimensional sheet), like a super-brain whose super-neurons are standard human brains? 100 years after Gödel's paper on the limits of proof & computation & AI (Goedel 1931): will practical variants of Gödel machines start a runaway evolution of continually self-improving superminds way beyond human imagination, causing far more unpredictable revolutions in the final decade before Ω than during all the millennia before?
14. ...

I feel there is no need to justify a much more cautious outlook by pessimistically referring to comparatively recent over-optimistic and self-serving predictions (1960s: "only 10 instead of 100 years needed to build AIs") by a few early AI enthusiasts in search of funding (Schmidhuber 2012). Nevertheless, after 10,000 years of civilization it would not matter much if the Ω estimate above were off by a few decades. Note that by cosmic standards the invention of writing over 5000 years ago almost coincided with the emergence of the WWW, and all of civilization history seems like a sudden flash—one needs to zoom in very closely to resolve the minute details of this ongoing turbulent intelligence explosion spanning just a few millennia.

The following disclosure should help the reader to take the list above with a grain of salt though. I admit being very interested in witnessing the Omega point. I was born in 1963, and therefore perhaps should not expect to live long past 2040. This may motivate me to uncover certain historic patterns that fit my desires, while ignoring other patterns that do not.

Others may feel attracted by the same trap, identifying exponential speedups in sequences of historic paradigm shifts identified by various historians, to back up the hypothesis that *Omega is near*, e.g., (Kurzweil 2005). The cited historians are all contemporary, presumably being subject to a similar bias. People of past ages might have held quite different views. For example, possibly some historians of the year 1525 felt inclined to predict a convergence of history around 1540, deriving this date from an exponential speedup of recent breakthroughs such as the printing press (around 1444), the re-discovery of America (48 years later), the Reformation (again 24 years later—see the pattern of exponential acceleration?), and other events they deemed important although today they are mostly forgotten. (According to TIME LIFE magazine's millennium issue, the three events above were the previous millennium's most influential ones.)

Could it be that such lists just reflect the human way of allocating memory space to past events (Schmidhuber 2006c, 2007b)? Maybe there is a general rule for both the individual memory of single humans and the collective memory of entire societies and their history books: constant amounts of memory space get allocated to exponentially larger, adjacent time intervals further and further into the past. For example, events that happened between 2 and 4 lifetimes ago get roughly as much memory space as events in the previous interval of twice the size. Presumably only a few "important" memories will survive the necessary compression. Maybe that's why there has never been a shortage of prophets predicting that the end is near—the important events according to one's own view of the past always seem to accelerate exponentially. A similar plausible type of memory decay allocates $O(1/n)$ memory units to all events older than $O(n)$ unit time intervals. This is reminiscent of a bias governed by a time-reversed Speed Prior (Schmidhuber 2002b) (section Universal But Incomputable AI).

References

Balcan, M. F., Beygelzimer, A., & Langford, J. (2009). Agnostic active learning. *Journal of Computer and System Sciences, 75*(1), 78–89.

Barto, A. (2013). Intrinsic motivation and reinforcement learning. In G. Baldassarre & M. Mirolli (Eds.), Intrinsically motivated learning in natural and artificial systems. Springer (in press).

Behnke, S. (2003). *Hierarchical neural networks for image interpretation, volume 2766 of lecture notes in computer science*. Springer.

Bishop, C. M. (2006). *Pattern recognition and machine learning*. NY: Springer.

Bringsjord, S. (2000), 'A contrarian future for minds and machines', chronicle of higher education (p. B5). Reprinted in The Education Di-gest, vol. 66(6), pp. 31–33.

Ciresan, D. C., Meier, U., Gambardella, L. M., & Schmidhuber, J. (2010). Deep big simple neural nets for handwritten digit recogntion. *Neural Computation, 22*(12), 3207–3220.

Ciresan, D. C., Meier, U., Gambardella, L. M., & Schmidhuber, J. (2011a). Convolutional neural network committees for handwritten character classification. In 11th International Conference on Document Analysis and Recognition (ICDAR), pp 1250–1254.

Ciresan, D. C., Meier, U., Masci, J., Gambardella, L. M. & Schmidhuber, J. (2011b). Flexible, high performance convolutional neural networks for image classification. In International Joint Conference on Artificial Intelligence IJCAI, pp 1237–1242.

Ciresan, D. C., Meier, U., Masci, J., & Schmidhuber, J. (2011c). A committee of neural networks for traffic sign classification. In International Joint Conference on, Neural Networks, pp 1918–1921.

Ciresan, D. C., Meier, U., Masci, J., & Schmidhuber, J. (2012a). Multi-column deep neural network for traffic sign classification. *Neural Networks, 32*, 333–338.

Ciresan, D. C., Meier, U., & Schmidhuber, J. (2012b). Multi-column deep neural networks for image classification. In IEEE Conference on Computer Vision and Pattern Recognition CVPR 2012, pp 3642–3649.

Ciresan, D. C., Meier, U., & Schmidhuber, J. (2012c). Multi-column deep neural networks for image classification. In IEEE Conference on Computer Vision and Pattern Recognition CVPR 2012. Long preprint arXiv:1202.2745v1 [cs.CV].

Darwin, C. (1997). *The descent of man, prometheus, amherst*. NY: A reprint edition.

Dayan, P. (2013). Exploration from generalization mediated by multiple controllers. In G. Baldassarre & M. Mirolli (Eds.), Intrinsically motivated learning in natural and artificial systems. Springer (in press).

Fedorov, V. V. (1972). *Theory of optimal experiments*. NY: Academic.

Fernandez, S., Graves, A., & Schmidhuber, J. (2007). Sequence labelling in structured domains with hierarchical recurrent neural networks. In *Proceedings of the 20th International Joint Conference on Artificial Intelligence* (IJCAI).

Floridi, L. (2007). A look into the future impact of ICT on our lives. *The Information Society, 23*(1), 59–64.

Fukushima, K. (1980). Neocognitron: A self-organizing neural network for a mechanism of pattern recognition unaffected by shift in position. Biological Cybernetics36(4), 193–202.

Gers, F. A., & Schmidhuber, J. (2001). LSTM recurrent networks learn simple context free and context sensitive languages. *IEEE Transactions on Neural Networks, 12*(6), 1333–1340.

Gers, F. A., Schraudolph, N., & Schmidhuber, J. (2002). Learning precise timing with LSTM recurrent networks. *Journal of Machine Learning Research, 3*, 115–143.

Gisslen, L., Luciw, M., Graziano, V., & Schmidhuber, J. (2011). Sequential constant size compressor for reinforcement learning. In *Proceedings of Fourth Conference on Artificial General Intelligence* (AGI), Google, Mountain View, CA.

Glasmachers, T., Schaul, T., Sun, Y., Wierstra, D. & Schmidhuber, J. (2010). Exponential Natural Evolution Strategies. In *Proceedings of the Genetic and Evolutionary Computation Conference (GECCO)*.

Gödel, K. (1931). Über formal unentscheidbare Sätze der Principia Mathematica und verwandter Systeme I. *Monatshefte für Mathematik und Physik, 38*, 173–198.

Gomez, F. J., Schmidhuber, J., & Miikkulainen, R. (2008). Efficient non-linear control through neuroevolution. *Journal of Machine Learning Research JMLR, 9*, 937–965.

Graves, A., Fernandez, S., Gomez, F. J., & Schmidhuber, J. (2006). Connectionist temporal classification: Labelling unsegmented sequence data with recurrent neural nets. In ICML '06: *Proceedings of the International Conference on Machine Learning*.

Graves, A., Fernandez, S., Liwicki, M., Bunke, H., & Schmidhuber, J. (2008). Unconstrained on-line handwriting recognition with recurrent neural networks. In J. C. Platt, D. Koller, Y. Singer, & S. Roweis (Eds.), *Advances in Neural Information Processing Systems 20* (pp. 577–584). Cambridge: MIT Press.

Graves, A., Liwicki, M., Fernandez, S., Bertolami, R., Bunke, H., & Schmidhuber, J. (2009). A novel connectionist system for improved unconstrained handwriting recognition. *IEEE Transactions on Pattern Analysis and Machine Intelligence*, 31(5), 855–868.

Graves, A., & Schmidhuber, J. (2009). Offline handwriting recognition with multidimensional recurrent neural networks. In *Advances in Neural Information Processing Systems* (p. 21). Cambridge: MIT Press.

Hansen, N., & Ostermeier, A. (2001). Completely derandomized self-adaptation in evolution strategies. *Evolutionary Computation, 9*(2), 159–195.

Hart, S., Sen, S., & Grupen, R. (2008). Intrinsically motivated hierarchical manipulation. In *Proceedings of the IEEE Conference on Robots and Automation (ICRA)*. California: Pasadena.

Hochreiter, S., Bengio, Y., Frasconi, P., & Schmidhuber, J. (2001). Gradient flow in recurrent nets: The difficulty of learning long-term dependencies. In S. C. Kremer & J. F. Kolen (Eds.), *A Field Guide to Dynamical Recurrent Neural Networks*. NJ: IEEE Press.

Hochreiter, S., & Schmidhuber, J. (1997). Long short-term memory. *Neural Computation, 9*(8), 1735–1780.

Holland, J. H. (1975). *Adaptation in natural and artificial systems*. Ann Arbor: University of Michigan Press.

Hutter, M. (2002). The fastest and shortest algorithm for all well-defined problems. *International Journal of Foundations of Computer Science, 13*(3), 431–443 (On J. Schmidhuber's SNF grant 20–61847).

Hutter, M. (2005). *Universal Artificial Intelligence: Sequential Decisions Based on Algorithmic Probability*. Berlin: Springer (On J. Schmidhuber's SNF grant 20–61847).

Jaeger, H. (2004). Harnessing nonlinearity: Predicting chaotic systems and saving energy in wireless communication. *Science, 304*, 78–80.

Kaelbling, L. P., Littman, M. L., & Moore, A. W. (1996). Reinforcement learning: A survey. *Journal of AI research, 4*, 237–285.

Kolmogorov, A. N. (1965). Three approaches to the quantitative definition of information. *Problems of Information Transmission, 1*, 1–11.

Koutnik, J., Gomez, F., & Schmidhuber, J. (2010). Evolving neural networks in compressed weight space. In *Proceedings of the Conference on Genetic and, Evolutionary Computation (GECCO-10)*.

Krizhevsky, A. (2009). *Learning multiple layers of features from tiny images*. Master's thesis: Computer Science Department, University of Toronto.

Kuipers, B., Beeson, P., Modayil, J., & Provost, J. (2006). Bootstrap learning of foundational representations. *Connection Science, 18*(2).

Kurzweil, R. (2005). *The singularity is near*. NY: Wiley Interscience.

LeCun, Y., Bottou, L., Bengio, Y., & Haffner, P. (1998). Gradient-based learning applied to document recognition. *Proceedings of the IEEE, 86*(11), 2278–2324.

LeCun, Y., Huang, F.-J., & Bottou, L. (2004). Learning methods for generic object recognition with invariance to pose and lighting. In *Proceedings of Computer Vision and Pattern Recognition Conference*.

Lenat, D. B. (1983). Theory formation by heuristic search. *Machine Learning*, vol. 21.

Levin, L. A. (1973). Universal sequential search problems. *Problems of Information Transmission, 9*(3), 265–266.

Li, M., & Vitányi, P. M. B. (1997). *An introduction to kolmogorov complexity and its applications* (2nd ed.). NY: Springer.

Maass, W., Natschläger, T., & Markram, H. (2002). *A fresh look at real-time computation in generic recurrent neural circuits*. Institute for Theoretical Computer Science, TU Graz : Technical report.

Mitchell, T. (1997). *Machine learning*. NY: McGraw Hill.

Moravec, H. (1999). *Robot* . NY: Wiley Interscience.

Newell, A., & Simon, H. (1963). GPS, a program that simulates human thought. In E. Feigenbaum & J. Feldman (Eds.), *Computers and thought* (pp. 279–293). New York: McGraw-Hill.

Oudeyer, P. -Y., Baranes, A., & Kaplan, F. (2013). Intrinsically motivated learning of real world sensorimotor skills with developmental constraints. In G. Baldassarre & M. Mirolli (Eds.), Intrinsically motivated learning in natural and artificial systems. Springer (in press).

Rechenberg, I. (1971). Evolutions strategie–optimierung technischer systeme nach Prinzipien der biologischen Evolution. Dissertation, Published 1973 by Fromman-Holzboog.

Robinson, A. J., & Fallside, F. (1987). The utility driven dynamic error propagation network. Technical Report CUED/F-INFENG/TR.1, Cambridge University Engineering Department.

Rosenbloom, P. S., Laird, J. E., & Newell, A. (1993). *The SOAR papers*. NY: MIT Press.

Schaul, T., Bayer, J., Wierstra, D., Sun, Y., Felder, M., Sehnke, F., et al. (2010). PyBrain. *Journal of Machine Learning Research, 11*, 743–746.

Scherer, D., Müller, A., & Behnke, S. (2010). In *International Conference on Artificial Neural Networks*.

Schmidhuber, J. (1990). *Dynamische neuronale Netze und das fundamentale raumzeitliche Lernproblem*. Dissertation: Institut für Informatik, Technische Universität München.

Schmidhuber, J. (1991a). Curious model-building control systems. In *Proceedings of the International Joint Conference on Neural Networks* (vol. 2, pp. 1458–1463). Singapore: IEEE press.

Schmidhuber, J. (1991b). A possibility for implementing curiosity and boredom in model-building neural controllers. In J. A. Meyer & S. W. Wilson (Eds.) *Proceedings of the International Conference on Simulation of Adaptive Behavior: From Animals to Animats*, pp. 222–227. MIT Press/Bradford Books.

Schmidhuber, J. (1991c). Reinforcement learning in Markovian and non-Markovian environments. In D. S. Lippman, J. E. Moody, & D. S. Touretzky (Eds.), *Advances in neural information processing systems 3* (NIPS 3) (pp. 500–506). NY: Morgan Kaufmann.

Schmidhuber, J. (1992a). A fixed size storage $O(n^3)$ time complexity learning algorithm for fully recurrent continually running networks. *Neural Computation, 4*(2), 243–248.

Schmidhuber, J. (1992b). Learning factorial codes by predictability minimization. *Neural Computation, 4*(6), 863–879.

Schmidhuber, J. (1997). Discovering neural nets with low Kolmogorov complexity and high generalization capability. *Neural Networks, 10*(5), 857–873.

Schmidhuber, J. (1999). Artificial curiosity based on discovering novel algorithmic predictability through coevolution. In P. Angeline, Z. Michalewicz, M. Schoenauer, X. Yao,& Z. Zalzala (Eds.), *Congress on evolutionary computation* (pp. 1612–1618). Piscataway: IEEE Press.

Schmidhuber, J. (2002a). Hierarchies of generalized Kolmogorov complexities and nonenumerable universal measures computable in the limit. *International Journal of Foundations of Computer Science, 13*(4), 587–612.

Schmidhuber, J. (2002). The speed prior: A new simplicity measure yielding near-optimal computable predictions. In J. Kivinen& R. H. Sloan (Eds.), *Proceedings of the 15th Annual Conference on Computational Learning Theory (COLT 2002)* (pp. 216–228). Lecture Notes in Artificial Intelligence Sydney, Australia: Springer.

Schmidhuber, J. (2003a). Exponential speed-up of computer history's defining moments. http://www.idsia.ch/juergen/computerhistory.html

Schmidhuber, J. (2003b). The new AI: General & sound & relevant for physics. Technical Report TR IDSIA-04-03, Version 1.0, arXiv:cs.AI/0302012 v1.

Schmidhuber, J. (2004). Optimal ordered problem solver. *Machine Learning, 54*, 211–254.

Schmidhuber, J. (2005). Completely self-referential optimal reinforcement learners. In W. Duch, J. Kacprzyk, E. Oja, & S. Zadrozny (Eds.), Artificial neural networks: Biological inspirations–ICANN 2005 (pp. 223–233), LNCS 3697. Springer: Berlin Heidelberg (Plenary talk).

Schmidhuber, J. (2006a). Developmental robotics, optimal artificial curiosity, creativity, music, and the fine arts. *Connection Science, 18*(2), 173–187.

Schmidhuber, J. (2006b). Gödel machines: Fully self-referential optimal universal self-improvers. In B. Goertzel& C. Pennachin (Eds.), *Artificial general intelligence* (pp. 199–226). Heidelberg: Springer (Variant available as arXiv:cs.LO/0309048).

Schmidhuber, J. (2006c). Celebrating 75 years of AI–history and outlook: The next 25 years. In M. Lungarella, F. Iida, J. Bongard,& R. Pfeifer (Eds.), *50 years of artificial intelligence* (vol. LNAI 4850, pp. 29–41). Berlin/Heidelberg: Springer (Preprint available as arXiv:0708.4311).

Schmidhuber, J. (2007a). Gödel machines: Fully self-referential optimal universal self-improvers. In B. Goertzel& C. Pennachin (Eds.), *Artificial general intelligence* (pp. 199–226). Springer Verlag (Variant available as arXiv:cs.LO/0309048).

Schmidhuber, J. (2007b). New millennium AI and the convergence of history. In W. Duch& J. Mandziuk (Eds.), *Challenges to computational intelligence* (vol. 63, pp. 15–36). Studies in Computational Intelligence, Springer, 2007. Also available as arXiv:cs.AI/0606081.

Schmidhuber, J. (2009). Ultimate cognition à la Gödel. *Cognitive Computation, 1*(2), 177–193.

Schmidhuber, J. (2010). Formal theory of creativity, fun, and intrinsic motivation (1990–2010). *IEEE Transactions on Autonomous Mental Development, 2*(3), 230–247.

Schmidhuber, J. (2011). PowerPlay: Training an increasingly general problem solver by continually searching for the simplest still unsolvable problem. Technical Report arXiv: 1112.5309v1 [cs.AI].

Schmidhuber, J. (2012). Philosophers& futurists, catch up! response to the singularity. *Journal of Consciousness Studies, 19*(1–2), 173–182.

Schmidhuber, J., Ciresan, D., Meier, U., Masci, J., & Graves, A. (2011). On fast deep nets for AGI vision. In *Proceedings of Fourth Conference on Artificial General Intelligence* (AGI), Google, Mountain View, CA.

Schmidhuber, J., Eldracher, M., & Foltin, B. (1996). Semilinear predictability minimization produces well-known feature detectors. *Neural Computation, 8*(4), 773–786.

Schmidhuber, J., Wierstra, D., Gagliolo, M., & Gomez, F. J. (2007). Training recurrent networks by EVOLINO. *Neural Computation, 19*(3), 757–779.

Schmidhuber, J., Zhao, J., & Schraudolph, N. (1997). Reinforcement learning with self-modifying policies. In S. Thrun& L. Pratt (Eds.), *Learning to learn* (pp. 293–309). NY: Kluwer.

Schraudolph, N. N., Eldracher, M., & Schmidhuber, J. (1999). Processing images by semi-linear predictability minimization. *Network: Computation in Neural Systems,* 10(2), 133–169.

Schwefel, H. P. (1974). Numerische optimierung von computer-modellen. Dissertation, Published 1977 by Birkhäuser, Basel.

Siegelmann, H. T., & Sontag, E. D. (1991). Turing computability with neural nets. *Applied Mathematics Letters, 4*(6), 77–80.

Sims, K. (1994). Evolving virtual creatures. In A. Glassner (Ed.), Proceedings of SIGGRAPH '94 (Orlando, Florida, July 1994), Computer Graphics Proceedings, Annual Conference (pp. 15–22). ACM SIGGRAPH, ACM Press. ISBN 0-89791-667-0.

Singh, S., Barto, A. G., & Chentanez, N. (2005). Intrinsically motivated reinforcement learning. In *Advances in Neural Information Processing Systems* 17 (NIPS). Cambridge: MIT Press.

Sloman, A. (2011a, Oct 23). Challenge for vision: Seeing a Toy Crane. Retrieved June 8, 2012, from http://www.cs.bham.ac.uk/research/projects/cosy/photos/crane/

Sloman, A. (2011b, June 8). Meta-morphogenesis and the creativity of evolution. Retrieved 6 June 2012, from http://www.cs.bham.ac.uk/research/projects/cogaff/evo-creativity.pdf

Sloman, A. (2011c, Oct 29). Meta-Morphogenesis and Toddler Theorems: Case Studies. Retrieved 8 June 2012, from http://www.cs.bham.ac.uk/research/projects/cogaff/misc/toddler-theorems.html

Sloman, A. (2011d, Sep 19). Simplicity and Ontologies: The trade-off between simplicity of theories and sophistication of ontologies. Retrieved June 8, 2012, from http://www.cs.bham.ac.uk/research/projects/cogaff/misc/simplicity-ontology.html

Smil, V. (1999). Detonator of the population explosion. *Nature, 400*, 415.

Solomonoff, R. J. (1964). A formal theory of inductive inference. *Part I. Information and Control, 7*, 1–22.

Stanley, K. O., & Miikkulainen, R. (2002). Evolving neural networks through augmenting topologies. *Evolutionary Computation, 10*, 99–127.

Storck, J., Hochreiter, S., & Schmidhuber, J. (1995). Reinforcement driven information acquisition in non-deterministic environments. In *Proceedings of the International Conference on Artificial Neural Networks*, Paris, vol. 2, pp. 159–164. EC2& Cie, 1995.

Strehl, A., Langford, J., & Kakade, S. (2010). Learning from logged implicit exploration data. Technical, Report arXiv:1003.0120.

Sun, Y., Wierstra, D., Schaul, T., & Schmidhuber, J. (2009a). *Efficient natural evolution strategies.* In Genetic and Evolutionary Computation Conference.

Sun, Y., Wierstra, D., Schaul, T., & Schmidhuber, J. (2009b). Stochastic search using the natural gradient. In *International Conference on Machine Learning (ICML).*

Sutskever, I., Martens, J., & Hinton, G. (2011). Generating text with recurrent neural networks. In L. Getoor& T. Scheffer (Eds.), *Proceedings of the 28th International Conference on Machine Learning (ICML-11)* (pp. 1017–1024). ICML '11 New York, NY, USA: ACM.

Sutton, R., & Barto, A. (1998). *Reinforcement learning: An introduction.* Cambridge: MIT Press.

Turing, A. M. (1936). On computable numbers, with an application to the Entscheidungsproblem. *Proceedings of the London Mathematical Society, Series, 2*(41), 230–267.

Utgoff, P. (1986). Shift of bias for inductive concept learning. In R. Michalski, J. Carbonell,& T. Mitchell (Eds.), *Machine learning* (Vol. 2, pp. 163–190). Los Altos, CA: Morgan Kaufmann.

Vapnik, V. (1995). *The nature of statistical learning theory*. New York: Springer.

Vinge, V. (1984). *The peace war*. Inc. : Bluejay Books.

Vinge, V. (1993). The coming technological singularity. VISION-21 Symposium sponsored by NASA Lewis Research Center, and Whole Earth Review, Winter issue.

Werbos, P. J. (1988). Generalization of backpropagation with application to a recurrent gas market model. Neural Networks, 1.

Wierstra, D., Foerster, A., Peters, J., & Schmidhuber, J. (2010). Recurrent policy gradients. *Logic Journal of IGPL*,18(2), 620–634.

Wierstra, D., Schaul, T., Peters, J., & Schmidhuber, J. (2008). Natural evolution strategies. In *Congress of Evolutionary Computation (CEC 2008)*.

Williams R. J., & Zipser, D. (1994). Gradient-based learning algorithms for recurrent networks and their computational complexity. In *back-propagation: Theory, architectures and applications*. Hillsdale, NJ: Erlbaum.

Yao, X. (1993). A review of evolutionary artificial neural networks. *International Journal of Intelligent Systems, 4*, 203–222.

Yi, S., Gomez, F., & Schmidhuber, J. (2011). Planning to be surprised: Optimal Bayesian exploration in dynamic environments. In *Proceedings of Fourth Conference on Artificial General Intelligence (AGI)*, Google, Mountain View, CA.

Chapter 4A
Aaron Sloman on Schmidhuber's "New Millennium AI and the Convergence of History 2012"

I have problems both with the style and the content of this essay, though I have not tried to take in the full mathematical details, and may therefore have missed something. I do not doubt that the combination of technical advances by the author and increases in computer power have made possible new impressive demonstrations including out-performing rival systems on various benchmark tests.

However, it is not clear to me that those tests have much to do with animal or human intelligence or that there is any reason to believe this work will help to bridge the enormous gaps between current machine competences and the competences of squirrels, nest-building birds, elephants, hunting mammals, apes, and human toddlers.

The style of the essay makes the claims hard to evaluate because it repeatedly says how good the systems are and reports that they outperform rivals, but does not help an outsider to get a feel for the nature of the tasks and the ability of the techniques to "scale out" into other tasks. In particular I have no interest in systems that do well at reading hand-written characters since that is not a task for which there is any objective criterion of correctness, and all that training achieves is tracking human labellings, without giving any explanation as to why the human labels are correct. I would be really impressed, however, if the tests showed a robot assembling Meccano parts to form a model crane depicted in a picture, and related tests here (Sloman 2011a).

Since claims are being made about how the techniques will lead beyond human competences in a few decades I would like to see sample cases where the techniques match mathematical, scientific, engineering, musical, toy puzzle solving, or linguistic performances that are regarded as highly commendable achievements of humans, e.g. outstanding school children or university students. (Newton, Einstein, Mozart, etc. can come later.) Readers should see a detailed analysis of exactly how the machine works in those cases and if the claim is that it uses non-human mechanisms, ontologies, forms of representation, etc. then I would like to see those differences explained. Likewise if its internals are comparable to those of humans I would like to see at least discussions of the common details.

The core problem is how the goals of the research are formulated. Instead of a robot with multiple asynchronously operating sensors providing different sorts of information (e.g. visual, auditory, haptic, proprioceptive, vestibular), and a collection of motor control systems for producing movements of animal-like hands, legs, wings, mouths, tongue etc., the research addresses:

> ... a learning robotic agent with a single life which consists of discrete cycles or time steps $t = 1, 2, \ldots, T$. Its total lifetime T may or may not be known in advance. In what follows, the value of any time-varying variable Q at time $t(t(1 \leq t \leq T))$ *will be denoted by* $Q(t)$, the ordered sequence of values $Q(1), \ldots, Q(t)$ by Q(<t), and the (possibly empty) sequence $Q(1), \ldots, Q(t-1)$ by $Q(<t)$.

> At any given t the robot receives a real-valued input vector $x(t)$ from the environment and executes a real-valued action $y(t)$ which may affect future inputs; at times $t < T$ its goal is to maximize future success or utility....

As far as I am concerned that defines a particular sort of problem to do with data mining in a discrete stream of vectors, where the future components are influenced in some totally unexplained way by a sequence of output vectors.

I don't see how such a mathematical problem relates to a crane assembly problem where the perceived structure is constantly changing in complexity, with different types of relationships and properties of objects relevant at different types, and actions of different sorts of complexity required, rather than a stream of output vectors (of fixed dimensionality?). I would certainly pay close attention if someone demonstrated advances in machine learning by addressing the toy crane problem, or the simpler problem described in (Sloman 2011d)

But so far none of the machine learning researchers I've pointed at these problems has come back with something to demonstrate. Perhaps the author and his colleagues are not interested in modelling or explaining human or animal intelligence, merely in demonstrating a functioning program that satisfies their definition of intelligence.

If they are interested in bridging the gap, then perhaps we should set up a meeting at which a collection challenges is agreed between people NOT working on machine learning and those who are, and then later we can jointly assess progress. Some of the criteria I am interested in are spelled out in these documents (Sloman 2011b, c).

However, all research results must be published in universally accessible open access journals and web sites, and not restricted to members of wealthy institutions.

References

Sloman, A. (2011a, Oct 23). Challenge for vision: Seeing a Toy Crane. Retrieved June 8, 2012, from http://www.cs.bham.ac.uk/research/projects/cosy/photos/crane/

Sloman, A. (2011b, June 8). Meta-morphogenesis and the creativity of evolution. Retrieved 6 June 2012, from http://www.cs.bham.ac.uk/research/projects/cogaff/evo-creativity.pdf

Sloman, A. (2011c, Oct 29). Meta-Morphogenesis and Toddler Theorems: Case Studies. Retrieved 8 June 2012, from http://www.cs.bham.ac.uk/research/projects/cogaff/misc/toddler-theorems.html

Sloman, A. (2011d, Sep 19). Simplicity and Ontologies: The trade-off between simplicity of theories and sophistication of ontologies. Retrieved June 8, 2012, from http://www.cs.bham.ac.uk/research/projects/cogaff/misc/simplicity-ontology.html

Chapter 4B
Selmer Bringsjord, Alexander Bringsjord and Paul Bello on Schmidhuber's "New Millennium AI and the Convergence of History 2012"

Hollow Hope for the Omega Point

We have elsewhere in the present volume shown that those who expect the Singularity (or, using Schmidhuber's term, Ω) are irrational fideists. Schmidhuber's piece doesn't disappoint us: while in recounting what seems all of intellectual history it reflects the brain of a bibliophage, it's nonetheless long on faith, and short on rigorous argument.

Does it follow from the fact that "raw computing power" continues to Moore's-Law-ishly increase, that human-level machine intelligence will arrive at some point, let alone arrive on the exuberant timeline Schmidhuber presents? No. The chief challenges in AI, relative to the human case, consist in finding the right computer programs, not faster and faster computers upon which to implement these programs (Bringsjord 2000). This is why automatic programming, one of the original dreams of AI (in which a human writes a computer program P that receives a non-executable description of an arbitrary Turing-computable function f, and to succeed must produce a computer program P' that verifiably computes f), is wholly and embarrassingly stalled. What class of being produces all the ingenious programs that increasingly form the lifeblood of the—to use Floridi's (Floridi 2007) term—infosphere? Machines? Ha.

Does it follow from the myriad neural-network-based advances and prizes Schmidhuber cites that Ω will ever be reached, let alone reached by 2040? No. Character/handwriting recognition is neat as far as it goes, but such low-level computation has nothing to do with what makes us us: phenomenal consciousness, free will, and natural-language communication. Taking just the latter in this brief note, character recognition has positively nothing at all to do with the fact that, say, human toddlers are vastly more eloquent than any machine. When a computing machine can not only checkmate the two of us, but debate us extemporaneously and non-idiotically in real time, we'll take notice (or more accurately, our like-minded ancestors will). As of now, 2012, over a decade from the year Turing predicted human-machine linguistic indistinguishability, the best conversational AI is Apple's SIRI: cute, but not much more..

Does it follow from the fact that such-and-such "breakthroughs" have happened in the past at such-and-such intervals that the Singularity will occur in accordance with some pattern Schmidhuber has magically divined? No. After all, the advances he cites are tendentiously picked to align with the kind of AI he pursues. Without question, the greatest AI achievement of the new millennium, an example of noteworthy and promising new-millennium AI if *anything* is, is the Watson system, produced by IBM researchers working on the basis of a relational approach found nowhere in the kinds of AI technologies that Schmidhuber venerates. Humans aren't *numerical*; humans are *propositional*. The knowledge and abstract reasoning capacity that separates *Homo sapiens sapiens* from Darwin's (Darwin 1997) "problem-solving" dogs are at their heart at once deliberative and propositional. The kind of AI that buoys Schmidhuber is neither; it's steadfastly syntactic, not semantic. Whence his unbridled optimism?

Schmidhuber closes in a spate of humility that borders on a crestfallen concession. He raises the possibility that many of those who believe they see Ω drawing nigh are driven by desire—desire to see the wonders of great machine intelligence. Here we commend him for his insight. What the fantast sees isn't really there, but that he "sees" it nonetheless brings him intoxicating joy.

References

Bringsjord, S. (2000), 'A contrarian future for minds and machines', chronicle of higher education (p. B5). Reprinted in The Education Di-gest, vol. 66(6), pp. 31–33.

Darwin, C. (1997). *The descent of man, prometheus, amherst*. NY: A reprint edition.

Floridi, L. (2007). A look into the future impact of ICT on our lives. *The Information Society, 23*(1), 59–64.

Chapter 5
Why an Intelligence Explosion is Probable

Richard Loosemore and Ben Goertzel

Abstract The hypothesis is considered that: Once an AI system with roughly human-level general intelligence is created, an "intelligence explosion" involving the relatively rapid creation of increasingly more generally intelligent AI systems will very likely ensue, resulting in the rapid emergence of dramatically superhuman intelligences. Various arguments against this hypothesis are considered and found wanting.

Introduction

One of the earliest incarnations of the contemporary Singularity concept was Good's concept of the "intelligence explosion," articulated in 1965 (Good 1965):

> Let an ultraintelligent machine be defined as a machine that can far surpass all the intellectual activities of any man however clever. Since the design of machines is one of these intellectual activities, an ultraintelligent machine could design even better machines; there would then unquestionably be an 'intelligence explosion,' and the intelligence of man would be left far behind. Thus the first ultraintelligent machine is the last invention that man need ever make.

R. Loosemore (✉)
Department of Mathematical and Physical Sciences,
Wells College, Aurora, NY, USA
e-mail: rloosemore@susaro.com

B. Goertzel
Novamente LLC, 1405 Bernerd Place, Rockville, MD 20851, USA
e-mail: ben@goertzel.org

A. H. Eden et al. (eds.), *Singularity Hypotheses*, The Frontiers Collection,
DOI: 10.1007/978-3-642-32560-1_5, © Springer-Verlag Berlin Heidelberg 2012

We consider Good's vision quite plausible but, unsurprisingly, not all futurist thinkers agree. Skeptics often cite limiting factors that could stop an intelligence explosion from happening, and in a recent post on the Extropy email discussion list (Sandberg 2011), the futurist Anders Sandberg articulated some of those possible limiting factors, in a particularly clear way:

> One of the things that struck me during our Winter Intelligence workshop on intelligence explosions[1] was how confident some people were about the speed of recursive self-improvement of AIs, brain emulation collectives or economies. Some thought it was going to be fast in comparison to societal adaptation and development timescales (creating a winner takes all situation), some thought it would be slow enough for multiple superintelligent agents to emerge. This issue is at the root of many key questions about the singularity (One superintelligence or many? How much does friendliness matter?).
>
> It would be interesting to hear this list's take on it: what do you think is the key limiting factor for how fast intelligence can amplify itself?
>
> - Economic growth rate
> - Investment availability
> - Gathering of empirical information (experimentation, interacting with an environment)
> - Software complexity
> - Hardware demands vs. available hardware
> - Bandwidth
> - Light-speed lags
>
> Clearly many more can be suggested. But which bottlenecks are the most limiting, and how can this be ascertained?"

We are grateful to Sandberg for presenting this list of questions, because it makes it especially straightforward for us to provide a clear counterargument to the point of view his list represents. In this article, we explain why these bottlenecks are unlikely to be significant issues, and thus why, as Good predicted, an intelligence explosion is indeed a very likely outcome.

Seed AI

To begin, we need to delimit the scope and background assumptions of our argument. In particular, it is important to specify what kind of intelligent system would be capable of generating an intelligence explosion.

According to our interpretation, there is one absolute prerequisite for an explosion to occur: an artificial general intelligence (AGI) must become smart enough to understand its own design. The concept of an AGI is quite general, and has been formalized mathematically by Hutter (2005) and others (Goertzel 2010). However, here we are concerned specifically with AGI systems possessing roughly the same broad set of intellectual capabilities as humans.

Of course, even among humans there are variations in skill level and knowledge. The AGI that triggers the explosion must have a sufficiently advanced

---

[1] http://www.fhi.ox.ac.uk/archived_events/winter_conference

intelligence that it can think analytically and imaginatively about how to manipulate and improve the design of intelligent systems. It is possible that not all humans are able to do this, so an AGI that met the bare minimum requirements for AGI-hood—say, a system smart enough to be a general household factotum—would not necessarily have the ability to work in an AGI research laboratory. Without an advanced AGI of the latter sort, there would be no explosion, just growth as usual, because the rate-limiting step would still be the depth and speed at which humans can think.

The sort of fully capable AGI with the potential to launch an intelligence explosion might be called a "seed AGI", but here we will sometimes use the less dramatic phrase "self-understanding, human-level AGI".

Start Date of an Explosion

Given that the essential prerequisite for an explosion to begin would be the availability of the first self-understanding, human-level AGI, does it make sense to talk about the period leading up to that arrival—the period during which that first real AGI was being developed and trained—as part of the intelligence explosion proper? We would argue that this is not appropriate, and that the true start of the explosion period should be considered to be the moment when a sufficiently well qualified AGI turns up for work at an AGI research laboratory. This may be different from the way some others use the term, but it seems consistent with Good's original usage. So our concern here is to argue for the high probability of an intelligence explosion, given the assumption that a self-understanding, human-level AGI has been created.

By enforcing this distinction, we are trying to avoid possible confusion with the parallel (and extensive!) debate about whether a self-understanding, human-level AGI can be built at all. Questions about whether an AGI with "seed level capability" can plausibly be constructed, or how long it might take to arrive, are of course quite different. A spectrum of opinions on this issue, from a survey of AGI researchers at a 2009 AGI conference, were gathered and analyzed in detail (Baum et al. 2011). In that survey, of an admittedly biased sample, a majority felt that an AGI with this capability could be achieved by the middle of this century, though a substantial plurality felt it was likely to happen much further out. While we have no shortage of our own thoughts and arguments on this matter, we will leave them aside for the purpose of the present paper.

Size of the Explosion

How big and how long would the explosion have to be to count as an "explosion"? Good's original notion had more to do with the explosion's beginning than its end, or its extent. His point was that in a short space of time a human-level AGI would probably explode into a significantly transhuman AGI, but he did not try to argue that subsequent improvements would continue without limit. We, like Good, are primarily interested in the explosion from human-level AGI to an AGI with, very loosely speaking, a level of general intelligence 2–3 orders of magnitude greater than the human level (say, 100 or 1,000H, using 1H to denote human-level general intelligence). This is not because we are necessarily skeptical of the explosion continuing beyond such a point, but rather because pursuing the notion beyond that seems a stretch of humanity's current intellectual framework.

Our reasoning, here, is that if an AGI were to increase its capacity to carry out scientific and technological research, to such a degree that it was discovering new knowledge and inventions at a rate 100 or 1,000 times the rate at which humans now do those things, we would find that kind of world unimaginably more intense than any future in which humans were doing the inventing. In a 1,000H world, AGI scientists could go from high-school knowledge of physics to the invention of relativity in a single day (assuming, for the moment, that the factor of 1,000 was all in the speed of thought—an assumption we will examine in more detail later). That kind of scenario is dramatically different from a world of purely human inventiveness—no matter how far humans might improve themselves in the future, without AGI, its seems unlikely there will ever be a time when a future Einstein would wake up one morning with a child's knowledge of science and then go on to conceive the theory of relativity by the following day—so it seems safe to call that an "intelligence explosion".

Speed of Explosion

So much for the *degree* of intelligence increase that would count as an explosion: that still leaves the question of how *fast* it has to arrive, to be considered explosive. Would it be enough for the seed AGI to go from 1 to 1,000H in the course of a century, or does it have to happen much quicker, to qualify?

Perhaps there is no need to rush to judgment on this point. Even a century-long climb up to the 1,000H level would mean that the world would be very different for the rest of history. The simplest position to take, we suggest, is that if the human species can get to the point where it is creating new types of intelligence that are themselves creating intelligences of greater power, then this is something new in the world (because at the moment all we can do is create human babies of power 1H), so even if this process happened rather slowly, it would still be an explosion of sorts. It might not be a Big Bang, but it would at least be a period of Inflation, and both could eventually lead to a 1,000H world.

Defining Intelligence

Finally, we propose to sidestep the difficulty of defining "intelligence" in a rigorous way. There are currently no measures of general intelligence that are precise, objectively defined and broadly extensible beyond the human scope. But since this is a qualitative essay rather than a report of quantitative calculations, and since we can address Sandberg's potential bottlenecks in some detail without needing a precise measure, we believe that little is lost by avoiding the issue. So, we will say that an intelligence explosion is something with the potential to create AGI systems as far beyond humans as humans are beyond mice or cockroaches, but we will not try to pin down exactly how far away the mice and cockroaches really are.

A different, complementary approach to this issue is given by David Chalmers' treatment of the Singularity hypothesis (Chalmers 2010). Chalmers argues carefully that, in order for the "intelligence explosion" argument to go through, one doesn't need a precise definition of general intelligence—one only needs the existence of one or more quantities that are correlated with flexible practical capability and that can be increased via increase of said capabilities. How these various quantities are used to create a composite, single general intelligence measure is something that doesn't need to be resolved in order to argue for an intelligence explosion.

Some Important Properties of the Intelligence Explosion

Before we get into a detailed analysis of the specific factors on Sandberg's list, a few more general comments are in order.

Inherent Uncertainty. Although we can try our best to understand how an intelligence explosion might happen, the truth is that there are too many interactions between the factors for any kind of reliable conclusion to be reached. This is a complex-system interaction in which even the tiniest, least-anticipated factor may turn out to be either the rate-limiting step or the spark that starts the fire. So there is an irreducible uncertainty involved here, and we should be wary of promoting conclusions that seem too firm.

General versus Special Arguments. There are two ways to address the question of whether or not an intelligence explosion is likely to occur. One is based on quite general considerations. The other involves looking at specific pathways to AGI. An AGI researcher (such as either of the authors) might believe they understand a great deal of the technical work that needs to be done to create an intelligence explosion, so they may be confident of the plausibility of the idea for that reason alone. We will restrict ourselves here to the first kind of argument, which is easier to make in a relatively non-controversial way, and leave aside any factors that might arise from our own understanding about how to build an AGI.

The "Bruce Wayne" Scenario. When the first self-understanding, human-level AGI system is built, it is unlikely to be the creation of a lone inventor working in a shed at the bottom of the garden, who manages to produce the finished product without telling anyone. Very few of the "lone inventor" (or "Bruce Wayne") scenarios seem plausible. As communication technology advances and causes cultural shifts, technological progress is increasingly tied to rapid communication of information between various parties. It is unlikely that a single inventor would be able to dramatically outpace multi-person teams working on similar projects; and also unlikely that a multi-person team would successfully keep such a difficult and time-consuming project secret, given the nature of modern technology culture.

Unrecognized Invention. It also seems quite implausible that the invention of a human-level, self-understanding AGI would be followed by a period in which the invention just sits on a shelf with nobody bothering to pick it up. The AGI situation would probably not resemble the early reception of inventions like the telephone or phonograph, where the full potential of the invention was largely unrecognized. We live in an era in which practically-demonstrated technological advances are broadly and enthusiastically communicated, and receive ample investment of dollars and expertise. AGI receives relatively little funding now, for a combination of reasons, but it is implausible to expect this situation to continue in the scenario where highly technically capable human-level AGI systems exist. This pertains directly to the economic objections on Sandberg's list, as we will elaborate below.

Hardware Requirements. When the first human-level AGI is developed, it will either require a supercomputer-level of hardware resources, or it will be achievable with much less. This is an important dichotomy to consider, because world-class supercomputer hardware is not something that can quickly be duplicated on a large scale. We could make perhaps hundreds of such machines, with a massive effort, but probably not a million of them in a couple of years.

Smarter versus Faster. There are two possible types of intelligence speedup: one due to faster operation of an intelligent system (clock speed increase) and one due to an improvement in the type of mechanisms that implement the thought processes ("depth of thought" increase). Obviously both could occur at once (and there may be significant synergies), but the latter is ostensibly more difficult to achieve, and may be subject to fundamental limits that we do not understand. Speeding up the hardware, on the other hand, is something that has been going on for a long time and is more mundane and reliable. Notice that both routes lead to greater "intelligence," because even a human level of thinking and creativity would be more effective if it were happening a thousand times faster than it does now.

It seems possible that the general class of AGI systems can be architected to take better advantage of improved hardware than would be the case with intelligent systems very narrowly imitative of the human brain. But even if this is not the case, brute hardware speedup can still yield dramatic intelligent improvement.

Public Perception. The way an intelligence explosion presents itself to human society will depend strongly on the rate of the explosion in the period shortly after the development of the first self-understanding human-level AGI. For instance, if the first such AGI takes 5 years to "double" its intelligence, this is a very different matter than if it takes 2 months. A 5 year time frame could easily arise, for example, if the seed AGI required an extremely expensive supercomputer based on unusual hardware, and the owners of this hardware were to move slowly. On the other hand, a 2 month time frame could more easily arise if the initial AGI were created using open source software and commodity hardware, so that a doubling of intelligence only required addition of more hardware and a modest number of software changes. In the former case, there would be more time for governments, corporations and individuals to adapt to the reality of the intelligence explosion before it reached dramatically transhuman levels of intelligence. In the latter case, the intelligence explosion would strike the human race more suddenly. But this potentially large difference in human perception of the events would correspond to a fairly minor difference in terms of the underlying processes driving the intelligence explosion.

Analysis of the Limiting Factors

Now we will deal with the specific factors on Sandberg's list, one by one, explaining in simple terms why each is not actually likely to be a significant bottleneck. There is much more that could be said about each of these, but our aim here is to lay out the main points in a compact way.

Economic Growth Rate and Investment Availability

The arrival, or imminent arrival, of human-level, self-understanding AGI systems would clearly have dramatic implications for the world economy. It seems inevitable that these dramatic implications would be sufficient to offset any factors related to the economic growth rate at the time that AGI began to appear. Assuming the continued existence of technologically advanced nations with operational technology R&D sectors, if self-understanding human-level AGI is created, then it will almost surely receive significant investment. Japan's economic growth rate, for example, is at the present time somewhat stagnant, but there can be no doubt that if any kind of powerful AGI were demonstrated, significant Japanese government and corporate funding would be put into its further development.

And even if it were not for the normal economic pressure to exploit the technology, international competitiveness would undoubtedly play a strong role. If a working AGI prototype were to approach the level at which an explosion seemed possible, governments around the world would recognize that this was a critically important technology, and no effort would be spared to produce the first fully-functional AGI "before the other side does". Entire national economies might well be sublimated to the goal of developing the first superintelligent machine, in the manner of Project Apollo in the 1960s. Far from influencing the intelligence explosion, economic growth rate would be *defined* by the various AGI projects taking place around the world.

Furthermore, it seems likely that once a human-level AGI has been achieved, it will have a substantial—and immediate—practical impact on multiple industries. If an AGI could understand its own design, it could also understand and improve other computer software, and so have a revolutionary impact on the software industry. Since the majority of financial trading on the US markets is now driven by program trading systems, it is likely that such AGI technology would rapidly become indispensible to the finance industry (typically an early adopter of any software or AI innovations). Military and espionage establishments would very likely also find a host of practical applications for such technology. So, following the achievement of self-understanding, human-level AGI, and complementing the allocation of substantial research funding aimed at outpacing the competition in achieving ever-smarter AGI, there is a great likelihood of funding aimed at practical AGI applications, which would indirectly drive core AGI research along.

The details of how this development frenzy would play out are open to debate, but we can at least be sure that the economic growth rate and investment climate in the AGI development period would quickly become irrelevant.

The Permanent AI Winter Scenario. All of these considerations do, however, leave one open question. At the time of writing, AGI investment around the world is noticeably weak, compared with other classes of scientific and technological investment. Is it possible that this situation will continue indefinitely, causing so little progress to be made that no viable prototype systems are built, and no investors ever believe that a real AGI is feasible?

This is hard to gauge, but as AGI researchers ourselves, our (clearly biased) opinion is that a "permanent winter" scenario is too unstable to be believable. Because of premature claims made by AI researchers in the past, a barrier to investment clearly exists in the minds of today's investors and funding agencies, but the climate already seems to be changing. And even if this apparent thaw turns out to be illusory, we still find it hard to believe that there will not eventually be an AGI investment episode comparable to the one that kicked the internet into high gear in the late 1990s.

Inherent Slowness of Experiments and Environmental Interaction

This possible limiting factor stems from the fact that any AGI capable of starting the intelligence explosion would need to do some experimentation and interaction with the environment in order to improve itself. For example, if it wanted to reimplement itself on faster hardware (most probably the quickest route to an intelligence increase) it would have to set up a hardware research laboratory and gather new scientific data by doing experiments, some of which might proceed slowly due to limitations of experimental technology.

The key question here is this: how much of the research can be sped up by throwing large amounts of intelligence at it? This is closely related to the problem of parallelizing a process (which is to say: you cannot make a baby nine times quicker by asking nine women to be pregnant for 1 month). Certain algorithmic problems are not easily solved more rapidly simply by adding more processing power, and in much the same way there might be certain crucial physical experiments that cannot be hastened by doing a parallel set of shorter experiments.

This is not a factor that we can understand fully ahead of time, because some experiments that look as though they require fundamentally slow physical processes—like waiting for a silicon crystal to grow, so we can study a chip fabrication mechanism—may actually be dependent on the intelligence of the experimenter, in ways that we cannot anticipate. It could be that instead of waiting for the chips to grow at their own speed, the AGI could do some clever micro-experiments that yield the same information faster.

The increasing amount of work being done on nanoscale engineering would seem to reinforce this point—many processes that are relatively slow today could be done radically faster using nanoscale solutions. And it is certainly feasible that advanced AGI could accelerate nanotechnology research, thus initiating a "virtuous cycle" where AGI and nanotech research respectively push each other forward (as foreseen by nanotech pioneer Josh Hall (Goertzel 2010)). As current physics theory does not even rule out more outlandish possibilities like femto technology, it certainly does not suggest the existence of absolute physical limits on experimentation speed existing anywhere near the realm of contemporary science.

Clearly, there is significant uncertainty in regards to this aspect of future AGI development. One observation, however, seems to cut through much of the uncertainty. Of all the ingredients that determine how fast empirical scientific research can be carried out, we know that in today's world the intelligence and thinking speed of the scientists themselves must be one of the most important. Anyone involved with science and technology R&D would probably agree that in our present state of technological sophistication, advanced research projects are strongly limited by the availability and cost of intelligent and experienced scientists.

But if research labs around the world have stopped throwing more scientists at problems they want to solve, because the latter are unobtainable or too expensive, would it be likely that those research labs are also, quite independently, at the limit for the physical rate at which experiments can be carried out? It seems hard to believe that both of these limits would have been reached at the same time, because they do not seem to be independently optimizable. If the two factors of experiment speed and scientist availability could be independently optimized, this would mean that even in a situation where there was a shortage of scientists, we could still be sure that we had discovered all of the fastest possible experimental techniques, with no room for inventing new, ingenious techniques that get over the physical-experiment-speed limits. In fact, however, we have every reason to believe that if we were to double the number of scientists on the planet at the moment, some of them would discover new ways to conduct experiments, exceeding some of the current speed limits. If that were not true, it would mean that we had quite coincidentally reached the limits of science talent and physical speed of data collecting at the same time—a coincidence that we do not find plausible.

This picture of the current situation seems consistent with anecdotal reports: companies complain that research staff are expensive and in short supply; they do not complain that nature is just too slow. It seems generally accepted, in practice, that with the addition of more researchers to an area of inquiry, methods of speeding up and otherwise improving processes can be found.

So based on the actual practice of science and engineering today (as well as known physical theory), it seems most likely that any experiment-speed limits lie further up the road, out of sight. We have not reached them yet, and we lack any solid basis for speculation about exactly where they might be.

Overall, it seems we do not have concrete reasons to believe that this will be a fundamental limit that stops the intelligence explosion from taking an AGI from H (human-level general intelligence) to (say) 1,000H. Increases in speed within that range (for computer hardware, for example) are already expected, even without large numbers of AGI systems helping out, so it would seem that physical limits, by themselves, would be very unlikely to stop an explosion from 1 to 1,000H.

Software Complexity

This factor is about the complexity of the software that an AGI must develop in order to explode its intelligence. The premise behind this supposed bottleneck is that even an AGI with self-knowledge finds it hard to cope with the fabulous complexity of the problem of improving its own software.

This seems implausible as a limiting factor, because the AGI could always leave the software alone and develop faster hardware. So long as the AGI can find a substrate that gives it a thousand-fold increase in clock speed, we have the possibility for a significant intelligence explosion.

Arguing that software complexity will stop the *first* self-understanding, human-level AGI from being built is a different matter. It may stop an intelligence explosion from happening by stopping the precursor events, but we take that to be a different type of question. As we explained earlier, one premise of the present analysis is that an AGI can actually be built. It would take more space than is available here to properly address that question.

It furthermore seems likely that, if an AGI system is able to comprehend its own software as well as a human being can, it will be able to improve that software significantly beyond what humans have been able to do. This is because in many ways, digital computer infrastructure is more suitable to software development than the human brain's wetware. And AGI software may be able to interface directly with programming language interpreters, formal verification systems and other programming-related software, in ways that the human brain cannot. In that way the software complexity issues faced by human programmers would be significantly mitigated for human-level AGI systems. However, this is not a 100 % critical point for our arguments, because even if software complexity remains a severe difficulty for a self-understanding, human-level AGI system, we can always fall back to arguments based on clock speed.

Hardware Requirements

We have already mentioned that much depends on whether the seed AGI requires a large, world-class supercomputer, or whether it can be done on something much smaller.

This is something that could limit the initial speed of the explosion, because one of the critical factors would be the number of copies of the seed AGI that can be created. Why would this be critical? Because the ability to *copy* the intelligence of a fully developed, experienced AGI is one of the most significant mechanisms at the core of an intelligence explosion. We cannot do this copying of adult, skilled humans, so human geniuses have to be rebuilt from scratch every generation. But if one AGI were to learn to be a world expert in some important field, it could be cloned any number of times to yield an instant community of collaborating experts.

However, if the seed AGI had to be implemented on a supercomputer, that would make it hard to replicate the AGI on a huge scale, and the intelligence explosion would be slowed down because the replication rate would play a strong role in determining the intelligence-production rate.

However, as time went on, the rate of replication would grow, as hardware costs declined. This would mean that the rate of arrival of high-grade intelligence would increase in the years following the start of this process. That intelligence would then be used to improve the design of the AGIs (at the very least, increasing the rate of new-and-faster-hardware production), which would have a positive feedback effect on the intelligence production rate.

So if there were a supercomputer-hardware requirement for the seed AGI, we would see this as something that would only dampen the initial stages of the explosion. Positive feedback after that would eventually lead to an explosion anyway.

If, on the other hand, the initial hardware requirements turn out to be modest (as they could very well be), the explosion would come out of the gate at full speed.

Bandwidth

In addition to the aforementioned cloning of adult AGIs, which would allow the multiplication of knowledge in ways not currently available in humans, there is also the fact that AGIs could communicate with one another using high-bandwidth channels. This is *inter-AGI bandwidth*, and it is one of the two types of bandwidth factors that could affect the intelligence explosion.

Quite apart from the communication speed between AGI systems, there might also be bandwidth limits inside a single AGI, which could make it difficult to augment the intelligence of a single system. This is *intra-AGI bandwidth*.

The first one—inter-AGI bandwidth—is unlikely to have a strong impact on an intelligence explosion because there are so many research issues that can be split into separably-addressible components. Bandwidth between the AGIs would only become apparent if we started to notice AGIs sitting around with no work to do on the intelligence amplification project, because they had reached an unavoidable stopping point and were waiting for other AGIs to get a free channel to talk to them. Given the number of different aspects of intelligence and computation that could be improved, this idea seems profoundly unlikely.

Intra-AGI bandwidth is another matter. One example of a situation in which internal bandwidth could be a limiting factor would be if the AGIs working memory capacity were dependent on the need for total connectivity—everything connected to everything else—in a critical component of the system. If this case, we might find that we could not boost working memory very much in an AGI because the bandwidth requirements would increase explosively. This kind of restriction on the design of working memory might have a significant effect on the system's depth of thought.

However, notice that such factors may not inhibit the initial phase of an explosion, because the clock speed, not the depth of thought, of the AGI may be improvable by several orders of magnitude before bandwidth limits kick in. The main element of the reasoning behind this is the observation that neural signal speed is so slow. If a brain-like AGI system (not necessarily a whole brain emulation, but just something that replicated the high-level functionality of the brain) could be built using components that kept the same type of processing demands, and the same signal speed as neurons, then we would be looking at a human-level AGI in which information packets were being exchanged once every millisecond. In that kind of system there would then be plenty of room to develop faster signal speeds and increase the intelligence of the system. The processing elements would also have to go faster, if they were not idling, but the point is that the bandwidth would not be the critical problem.

Light-Speed Lags

Here we need to consider the limits imposed by special relativity on the speed of information transmission in the physical universe. However, its implications in the context of AGI are not much different than those of bandwidth limits.

Light-speed lags could be a significant problem if the components of the machine were physically so far apart that massive amounts of data (by assumption) were delivered with a significant delay. But they seem unlikely to be a problem in the initial few orders of magnitude of the explosion. Again, this argument derives from what we know about the brain. We know that the brain's hardware was chosen due to biochemical constraints. We are carbon-based, not silicon-and-copper-based, so there are no electronic chips in the head, only pipes filled with fluid and slow molecular gates in the walls of the pipes. But if nature was forced to use the pipes-and-ion-channels approach, that leaves us with plenty of scope for speeding things up using silicon and copper (and this is quite apart from all the other more exotic computing substrates that are now on the horizon). If we were simply to make a transition membrane depolarization waves to silicon and copper, and if this produced a 1,000x speedup (a conservative estimate, given the intrinsic difference between the two forms of signaling), this would be an explosion worthy of the name.

The main circumstance under which this reasoning would break down would be if, for some reason, the brain is limited on two fronts simultaneously: both by the carbon implementation and by the fact that other implementations of the same basic design are limited by disruptive light-speed delays. This would mean that all non-carbon-implementations of the brain take us up close to the light-speed limit before we get much of a speedup over the brain. This would require a coincidence of limiting factors (two limiting factors just happening to kick in at exactly the same level), that we find quite implausible, because it would imply a rather bizarre situation in which evolution tried both the biological neuron design, and a silicon implementation of the same design, and after doing a side-by-side comparison of performance, chose the one that pushed the efficiency of all the information transmission mechanisms up to their end stops.

The Path from AGI to Intelligence Explosion Seems Clear

The conclusion of this relatively detailed analysis of Sandberg's objections is that there is currently no good reason to believe that once a human-level AGI capable of understanding its own design is achieved, an intelligence explosion will fail to ensue.

The operative definition of "intelligence explosion" that we have assumed here involves an increase of the speed of thought (and perhaps also the "depth of thought") of about two or three orders of magnitude. If someone were to insist that a real intelligence explosion had to involve million-fold or trillion-fold increases in intelligence, we think that no amount of analysis, at this stage, could yield sensible conclusions. But since an AGI with intelligence = 1,000 H might well cause the next 1,000 years of new science and technology to arrive in 1 year (assuming that the speed of physical experimentation did not become a significant factor within that range), it would be churlish, we think, not to call that an "explosion". An intelligence explosion of such magnitude would bring us into a domain that our current science, technology and conceptual framework are not equipped to deal with; so prediction beyond this stage is best done once the intelligence explosion has already progressed significantly.

Of course, even if the above analysis is correct, there is a great deal we do not understand about the intelligence explosion, and many of these particulars will remain opaque until we know precisely what sort of AGI system will launch the explosion. But it seems that the likelihood of transition from a self-understanding human-level AGI to an intelligence explosion should not presently be a subject of serious doubt.

References

Baum, S. D., Goertzel, B., & Goertzel, T. G. (2011). How long until human-level AI? results from an expert assessment. *Technological Forecasting and Social Change, 78*(1), 185–195.

Chalmers, D. (2010). The singularity: A philosophical analysis. *Journal of Consciousness Studies, 17*, 7–65.

Goertzel, B. (2010). *Toward a formal characterization of real-world general intelligence. Proceedings of AGI-10*, Lugano.

Hutter, M. (2005). *Universal AI*. Berlin: Springer.

Sandberg, A. (2011, January 19). Limiting factors of intelligence explosion speeds. *Extropy email discussion list*. http://lists.extropy.org/pipermail/extropy-chat/2011-January/063255.html.

Chapter 5A
Peter Bishop's on Loosemore and Goertzel's "Why an Intelligence Explosion is Probable"

The authors were kind enough to put the main idea in the title and very early in the piece–"...as Good predicted (1965), [given the appearance of machine intelligence], an intelligence explosion is indeed a very likely outcome (Page 2)." After a few preliminary remarks, the authors structure the article around a series of potential limitations to the intelligence explosion suggested by Sander in a post to the Extropy blog.

I am a professional futurist, not a computer scientist, much less an expert in artificial intelligence, so I will leave the technical details of the argument to others. The futurist's approach to such an argument is to examine the assumptions required for the proposed future to come about and to assess whether alternatives to those assumptions are plausible. Each plausible alternative assumption supports an alternative future (a scenario). Fortunately, it is not necessary to decide whether the original or its alternative is "correct" or not. Rather all plausible alternative futures constitute the range of scenarios that describe the future.

The authors state their first assumption right away—"...there is one absolute prerequisite for an explosion to occur: an artificial general intelligence (AGI) must become smart enough to understand its own design". I am afraid that here the authors get into trouble right away. The premise for the article is that humans have created the AGI. Yet humans do not understand their own design today, and they may not understand it even after creating an AGI. The authors seem to assume that the AGI is intelligent in the same way that humans are since we first had to understand our own design before building it into the AGI. But it is conceivable that AGI intelligence uses a different design. Thus the AGI does not have to understand itself any more than humans have to understand themselves in order to create the AGI in the first place.

The alternative assumption is supported later when the authors consider that incredibly faster clock speeds might lead to an intelligent design (Page 9). In that case, even humans might not understand how the AGI is intelligent, much less the AGI understanding that itself. One of the designs might be a massively parallel neural network. Neural networks are powerful learning machines, yet they do not have programs the way algorithmic computers do. Therefore, it is literally impossible to understand *why* they make the judgments that they do. As a result, we humans may never fully understand the basis of our own intelligence because it is definitely not algorithmic nor would we understand the basis for an AGI if it were a neural network. Therefore, the premise of this future is that humans are smart enough to create an AGI, but it is only an assumption that AGI understand the basis for its own intelligence.

A second assumption, particularly in the first part of the article, is that the "seed AGI," the "AGI with the potential to launch an intelligence explosion," is a single machine, perhaps a massively complex supercomputer. But having all that intelligence residing in one machine is not necessary. Just as it is highly unlikely that one human would create the AGI so one AGI might not lead to the explosion that the authors foresee. It is more likely that teams of humans would create the AGI with members of the team contributing their individual expertise—hardware, software, etc. Similarly the AGIs would work in teams. The authors do suggest that assumption later on when they discuss whether the bandwidth among machines would limit the rate of development. So whether the AGI that touches off the explosion is a single machine or a set of communicating machines is another important assumption.

These assumptions aside, the question in this article is the rate at which machine intelligence will develop once one or more AGIs are created. The first assumption is that machine intelligence will develop at all after that event, but it is hard to support the alternative—that the AGI is the last intelligent device invented. The premise is that one intelligent species (human) has already created another intelligent device (AGI) so it is highly likely that further intelligence species or devices will emerge. The issue is how fast that will occur. Will it be an explosion as the authors claim or a rather slow evolutionary development?

First of all, the authors are reluctant to define exactly what an explosive rate would be. Even if it were to occur in mid-century, as many suggest, or even in the next few millennia, we have only one (presumed) case of one intelligent entity creating another one and that some 50,000 years. That's not particularly explosive. Kurzweil (2001) also predicts an explosive rate because intelligent machines will not carry the burden of biologically evolved intelligence, including emotions, culture and tradition. Still to go from 50,000 years to an explosion resulting in 100H (100 times human intelligence) in a short time (whose length is itself undefined) seems quite a stretch. Development? Probably. Explosive development? Who knows? In the end, an argument about that rate might even be futile.

"Given the existence of angels, how many can stand on the head of a pin?" Nevertheless, it's a great exercise in intellectual calisthenics because it forces us to discover just how many assumptions we make about the future.

References

Good, I. J. (1965). Speculations concerning the first ultraintelligent machine. *Advances in Computers, 6*, 99.

Kurzweil, R. (2001). The law of accelerating returns. http://www.kurzweilai.net/the-law-of-accelerating-returns.

Part II
Concerns About Artificial Superintelligence

Chapter 6
The Singularity and Machine Ethics

Luke Muehlhauser and Louie Helm

Abstract Many researchers have argued that a self-improving artificial intelligence (AI) could become so vastly more powerful than humans that we would not be able to stop it from achieving its goals. If so, and if the AI's goals differ from ours, then this could be disastrous for humans. One proposed solution is to program the AI's goal system to want what we want before the AI self-improves beyond our capacity to control it. Unfortunately, it is difficult to specify what we want. After clarifying what we mean by "intelligence", we offer a series of "intuition pumps" from the field of moral philosophy for our conclusion that human values are complex and difficult to specify. We then survey the evidence from the psychology of motivation, moral psychology, and neuroeconomics that supports our position. We conclude by recommending ideal preference theories of value as a promising approach for developing a machine ethics suitable for navigating an intelligence explosion or "technological singularity".

To educate [someone] in mind and not in morals is to educate a menace to society.

Theodore Roosevelt.

L. Muehlhauser (✉) · L. Helm
Machine Intelligence Research Institute, Berkeley, USA
e-mail: luke@singularity.org

A. H. Eden et al. (eds.), *Singularity Hypotheses*, The Frontiers Collection,
DOI: 10.1007/978-3-642-32560-1_6,

Introduction

Many researchers have argued that, by way of an "intelligence explosion" (Good 1959, 1965, 1970) sometime in the next century, a self-improving[1] artificial intelligence (AI) could become so vastly more powerful than humans that we would not be able to stop it from achieving its goals.[2] If so, and if the AI's goals differ from ours, then this could be disastrous for humans and what we value (Joy 2000; Bostrom 2003; Posner 2004; Friedman 2008; Yudkowsky 2008; Fox and Shulman 2010; Chalmers 2010; Bostrom and Yudkowsky, forthcoming; Muehlhauser and Salamon, this volume).

One proposed solution is to program the AI's goal system[3] to want what we want before the AI self-improves beyond our capacity to control it. While this proposal may be the only lasting solution for AI risk (Muehlhauser and Salamon, this volume), it faces many difficulties (Yudkowsky 2001). One such difficulty is that human values are complex and difficult to specify,[4] and this presents

[1] For discussions of self-improving AI, see Schmidhuber (2007); Omohundro (2008); Mahoney (2010); Hall (2007b, 2011).

[2] For simplicity, we speak of a single AI rather than multiple AIs. Of course, it may be that multiple AIs will undergo intelligence explosion more or less simultaneously and compete for resources for years or decades. The consequences of this scenario could be even more unpredictable than those of a "singleton" AI (Bostrom 2006), and we do not have the space for an examination of such scenarios in this chapter. For space reasons we will also only consider what Chalmers (2010) calls "non-human-based AI" and Muehlhauser and Salamon (this volume) call "de novo AI", thereby excluding self-improving AI based on the human mind, for example whole brain emulation (Sandberg and Bostrom 2008).

[3] When we speak of an advanced AI's goal system, we do not have in mind today's reinforcement learning agents, whose only goal is to maximize expected reward. Such an agent may hijack or "wirehead" its own reward function (Dewey 2011; Ring and Orseau 2011), and may not be able to become superintelligent because it does not model itself and therefore can't protect or improve its own hardware. Rather, we have in mind a future AI goal architecture realized by a utility function that encodes value for states of affairs (Dewey 2011; Hibbard 2012).

[4] Minsky (1984) provides an early discussion of our subject, writing that "… it is always dangerous to try to relieve ourselves of the responsibility of understanding exactly how our wishes will be realized. Whenever we leave the choice of means to any servants we may choose then the greater the range of possible methods we leave to those servants, the more we expose ourselves to accidents and incidents. When we delegate those responsibilities, then we may not realize, before it is too late to turn back, that our goals have been misinterpreted…. [Another] risk is exposure to the consequences of self-deception. It is always tempting to say to oneself… that 'I know what I would like to happen, but I can't quite express it clearly enough.' However, that concept itself reflects a too-simplistic self-image, which portrays one's own self as [having] well-defined wishes, intentions, and goals. This pre-Freudian image serves to excuse our frequent appearances of ambivalence; we convince ourselves that clarifying our intentions is merely a matter of straightening-out the input–output channels between our inner and outer selves. The trouble is, we simply aren't made that way. *Our goals themselves are ambiguous*…. The ultimate risk comes when [we] attempt to take that final step—of designing goal-achieving programs that are programmed to make themselves grow increasingly powerful, by self-evolving methods that augment and enhance their own capabilities…. The problem is that, with such powerful

challenges for developing a machine ethics suitable for navigating an intelligence explosion.

After clarifying what we mean by "intelligence", we offer a series of "intuition pumps" (Dennett 1984, P. 12) from the field of moral philosophy supporting our conclusion that human values are complex and difficult to specify. We then survey the evidence from the psychology of motivation, moral psychology, and neuroeconomics that supports our position. We conclude by recommending ideal preference theories of value as a promising approach for developing a machine ethics suitable for navigating an intelligence explosion.

Intelligence and Optimization

Good, who first articulated the idea of an intelligence explosion, referred to any machine more intelligent than the smartest human as an "ultraintelligent" machine (Good 1965). Today the term "superintelligence" is more common, and it refers to a machine that is *much* smarter than the smartest human (Bostrom 1998, 2003; Legg 2008).

But the term "intelligence" may not be ideal for discussing powerful machines. Why? There are many competing definitions and theories of intelligence (Davidson and Kemp 2011; Niu and Brass 2011; Legg and Hutter 2007), and the term has seen its share of emotionally-laden controversy (Halpern et al. 2011; Daley and Onwuegbuzie 2011).

The term also comes loaded with connotations, some of which do not fit machine intelligence. Laypeople tend to see intelligence as correlated with being clever, creative, self-confident, socially competent, deliberate, analytically skilled, verbally skilled, efficient, energetic, correct, and careful, but as anticorrelated with being dishonest, apathetic, and unreliable (Bruner et al. 1958; Neisser 1979; Sternberg et al. 1981, 1985). Moreover, cultures vary with respect to the associations they make with intelligence (Niu and Brass 2011; Sternberg and Grigorenko 2006). For example, Chinese people tend to emphasize analytical ability, memory skills, carefulness, modesty, and perseverance in their concepts of intelligence (Fang et al. 1987), while Africans tend to emphasize social competencies (Ruzgis and Grigorenko 1994; Grigorenko et al. 2001).

One key factor is that people overwhelmingly associate intelligence with positive rather than negative traits, perhaps at least partly due to a well-documented cognitive bias called the "affect heuristic" (Slovic et al. 2002), which leads us to make inferences by checking our emotions. Because people have positive affect toward

(Footnote 4 continued)
machines, it would require but the slightest accident of careless design for them to place their goals ahead of [ours]".

intelligence, they intuitively conclude that those with more intelligence possess other positive traits to a greater extent.

Despite the colloquial associations of the word "intelligence", AI researchers working to improve machine intelligence do not mean to imply that superintelligent machines will exhibit, for example, increased modesty or honesty. Rather, AI researchers' concepts of machine intelligence converge on the idea of optimal goal fulfillment in a wide variety of environments (Legg 2008), what we might call "optimization power."[5] This optimization concept of intelligence is not anthropomorphic and can be applied to any agent—human, animal, machine, or otherwise.[6]

Unfortunately, anthropomorphic bias (Epley et al. 2007; Barrett and Keil 1996) is not unique to laypeople. AI researcher Storrs Hall suggests that our machines may be more moral than we are, and cites as partial evidence the fact that *in humans* "criminality is strongly and negatively correlated with IQ" (Hall 2007a, p. 340). But machine intelligence has little to do with IQ or with the human cognitive architectures and social systems that might explain an anticorrelation between human criminality and IQ.

To avoid anthropomorphic bias and other problems with the word "intelligence", in this chapter we will use the term "machine superoptimizer" in place of "machine superintelligence."[7]

Using this term, it should be clear that a machine superoptimizer will not necessarily be modest or honest. It will simply be very capable of achieving its goals (whatever they are) in a wide variety of environments (Bostrom 2012). If its goal system aims to maximize the number of paperclips that exist, then it will be very good at maximizing paperclips in a wide variety of environments. The machine's optimization power does not predict that it will always be honest while maximizing paperclips. Nor does it predict that the machine will be so modest that it will feel at some point that it has made enough paperclips and then modify its goal system to aim toward something else. Nor does the machine's optimization power suggest that the machine will be amenable to moral argument. A machine superoptimizer need not even be sentient or have "understanding" in John Searle's (1980) sense, so long as it is very capable of achieving its goals in a wide variety of environments.

---

[5] The informal definition of intelligence in Legg (2008) captures what we mean by "optimization power," but Legg's specific formalization does not. Legg formalizes intelligence as a measure of expected performance on arbitrary reinforcement learning problems (Legg 2008: p. 77), but we consider this only a preliminary step in formalizing optimal goal fulfillment ability. We think of optimization power as a measure of expected performance across a broader class of goals, including goals about states of affairs in the world (argued to be impossible for reinforcement learners in Ring and Orseau 2011; Dewey 2011). Also, Legg's formal definition of intelligence is drawn from a dualistic "agent-environment" model of optimal agency (Legg 2008: p. 40) that does not represent its own computation as occurring in a physical world with physical limits and costs.

[6] Even this "optimization" notion of intelligence is incomplete, however. See Muehlhauser and Salamon (this volume).

[7] But, see Legg (2009) for a defense of Legg's formalization of universal intelligence as an alternative to what we mean by "optimization power".

The Golem Genie

Since Plato, many have believed that knowledge is justified true belief. Gettier (1963) argued that knowledge cannot be justified true belief because there are hypothetical cases of justified true belief that we intuitively would not count as knowledge. Since then, each newly proposed conceptual analysis of knowledge has been met with novel counter-examples (Shope 1983). Weatherson (2003) called this the "analysis of knowledge merry go round".

Similarly, advocates for mutually opposing moral theories seem to have shown that no matter which set of consistent moral principles one defends, intuitively repugnant conclusions follow. Hedonistic utilitarianism implies that I ought to plug myself into a pleasure-stimulating experience machine, while a deontological theory might imply that if I have promised to meet you for lunch, I ought not to stop to administer life-saving aid to the victim of a car crash that has occurred nearby (Kagan 1997, p. 121). More sophisticated moral theories are met with their own counter-examples (Sverdlik 1985; Parfit 2011). It seems we are stuck on a moral theory merry-go-round.

Philosophers debate the legitimacy of conceptual analysis (DePaul and Ramsey 1998; Laurence and Margolis 2003; Braddon-Mitchell and Nola 2009) and whether morality is grounded in nature (Jackson 1998; Railton 2003) or a systematic error (Mackie 1977; Joyce 2001).[8] We do not wish to enter those debates here. Instead, we use the observed "moral theory merry-go-round" as a source of intuition pumps suggesting that we haven't yet identified a moral theory that, if implemented throughout the universe, would produce a universe we want. As Beavers (2012) writes, "the project of designing moral machines is complicated by the fact that even after more than two millennia of moral inquiry, there is still no consensus on how to determine moral right from wrong".

Later we will take our argument from intuition to cognitive science, but for now let us pursue this intuition pump, and explore the consequences of implementing a variety of moral theories throughout the universe.

Suppose an unstoppably powerful genie appears to you and announces that it will return in 50 years. Upon its return, you will be required to supply it with a set of consistent moral principles which it will then enforce with great precision throughout the universe.[9] For example, if you supply the genie with hedonistic utilitarianism, it will maximize pleasure by harvesting all available resources and using them to tile the universe with identical copies of the smallest possible mind, each copy of which will experience an endless loop of the most pleasurable experience possible.

---

[8] Other ethicists argue that moral discourse asserts nothing (Ayer 1936; Hare 1952; Gibbard 1990) or that morality is grounded in non-natural properties (Moore 1903; Shafer-Landau 2003).

[9] In this paper we will set aside questions concerning an infinite universe (Bostrom 2009) or a multiverse (Tegmark 2007). When we say "universe" we mean, for simplicity's sake, the observable universe (Bars and Terning 2010).

Let us call this precise, instruction-following genie a Golem Genie. [A golem is a creature from Jewish folklore that would in some stories do *exactly* as told (Idel 1990), often with unintended consequences, for example polishing a dish until it is as thin as paper Pratchett 1996)]

If by the appointed time you fail to supply your Golem Genie with a set of consistent moral principles covering every possible situation, then it will permanently model its goal system after the first logically coherent moral theory that anyone articulates to it, and that's not a risk you want to take. Moreover, once you have supplied the Golem Genie with its moral theory, there will be no turning back. Until the end of time, the genie will enforce that one moral code without exception, not even to satisfy its own (previous) desires.

You are struck with panic. The literature on counter-examples in ethics suggests that universe-wide enforcement of any moral theory we've devised so far will have far-reaching unwanted consequences. But given that we haven't discovered a fully satisfying moral theory in the past several *thousand* years, what are the chances we can do so in the next *fifty*? Moral philosophy has suddenly become a larger and more urgent problem than climate change or the threat of global nuclear war.

Why do we expect unwanted consequences after supplying the Golem Genie with any existing moral theory? This is because of two of the Golem Genie's properties in particular[10]:

1. *Superpower*: The Golem Genie has unprecedented powers to reshape reality, and will therefore achieve its goals with highly efficient methods that confound human expectations (e.g. it will maximize pleasure by tiling the universe with trillions of digital minds running a loop of a single pleasurable experience).
2. *Literalness*: The Golem Genie recognizes only precise specifications of rules and values, acting in ways that violate what feel like "common sense" to humans, and in ways that fail to respect the subtlety of human values.

The Golem Genie scenario is analogous to the intelligence explosion scenario predicted by Good and others.[11] Some argue that a machine superoptimizer will be powerful enough to radically transform the structure of matter-energy within its reach. It could trivially develop and use improved quantum computing systems, advanced self-replicating nanotechnology, and other powers (Bostrom 1998; Joy 2000). And like the Golem Genie, a machine superoptimizer's goal pursuit will not be mediated by what we call "common sense"—a set of complex functional psychological adaptations found in members of *Homo Sapiens* but not necessarily present in an artificially designed mind (Yudkowsky 2011).

---

[10] The "superpower" and "literalness" properties are also attributed to machine superintelligence by Muehlhauser (2011), Sect. 4.1

[11] Many others have made an analogy between superintelligent machines and powerful magical beings. For example, Abdoullaev (1999, p. 1) refers to "superhumanly intelligent machines" as "synthetic deities".

Machine Ethics for a Superoptimizer

Let us consider the implications of programming a machine superoptimizer to implement particular moral theories.

We begin with hedonistic utilitarianism, a theory still defended today (Tännsjö 1998). If a machine superoptimizer's goal system is programmed to maximize pleasure, then it might, for example, tile the local universe with tiny digital minds running continuous loops of a single, maximally pleasurable experience. We can't predict *exactly* what a hedonistic utilitarian machine superoptimizer would do, but we think it seems likely to produce unintended consequences, for reasons we hope will become clear. The machine's exact behavior would depend on how its final goals were specified. As Anderson and Anderson (2011a) stress, "ethicists must accept the fact that there can be no vagueness in the programming of a machine".

Suppose "pleasure" was specified (in the machine superoptimizer's goal system) in terms of our current understanding of the human neurobiology of pleasure. Aldridge and Berridge (2009) report that according to "an emerging consensus", pleasure is "not a sensation" but instead a "pleasure gloss" added to sensations by "hedonic hotspots" in the ventral pallidum and other regions of the brain. A sensation is encoded by a particular pattern of neural activity, but it is not pleasurable in itself. To be pleasurable, the sensation must be "painted" with a pleasure gloss represented by additional neural activity activated by a hedonic hotspot (Smith et al. 2009).

A machine superoptimizer with a goal system programmed to maximize human pleasure (in this sense) could use nanotechnology or advanced pharmaceuticals or neurosurgery to apply maximum pleasure gloss to all human sensations—a scenario not unlike that of plugging us all into Nozick's experience machines (Nozick 1974, p. 45). Or, it could use these tools to restructure our brains to apply maximum pleasure gloss to one consistent experience it could easily create for us, such as lying immobile on the ground.

Or suppose "pleasure" was specified more broadly, in terms of anything that functioned as a reward signal—whether in the human brain's dopaminergic reward system (Dreher and Tremblay 2009) or in a digital mind's reward signal circuitry (Sutton and Barto 1998). A machine superoptimizer with the goal of maximizing reward signal scores could tile its environs with trillions of tiny minds, each one running its reward signal up to the highest number it could.

Thus, though some utilitarians have proposed that all we value is pleasure, our intuitive negative reaction to hypothetical worlds in which pleasure is (more or less) maximized suggests that pleasure is not the only thing we value.

What about negative utilitarianism? A machine superoptimizer with the final goal of minimizing human suffering would, it seems, find a way to painlessly kill all humans: no humans, no human suffering (Smart 1958; Russell and Norvig 2009, p. 1037).

What if a machine superoptimizer was programmed to maximize desire satisfaction[12] in humans? Human desire is implemented by the dopaminergic reward system (Schroeder 2004; Berridge et al. 2009), and a machine superoptimizer could likely get more utility by (1) rewiring human neurology so that we attain maximal desire satisfaction while lying quietly on the ground than by (2) building and maintaining a planet-wide utopia that caters perfectly to current human preferences.

Why is this so? First, because individual humans have incoherent preferences (Allais 1953; Tversky and Kahneman 1981). A machine superoptimizer couldn't realize a world that caters to incoherent preferences; better to rewrite the source of the preferences themselves.

Second, the existence of zero-sum games means that the satisfaction of one human's preferences can conflict with the satisfaction of another's (Geçkil and Anderson 2010). The machine superoptimizer might be best able to maximize human desire satisfaction by first ensuring that satisfying some people's desires does not thwart the satisfaction of others' desires—for example by rewiring all humans to desire nothing else but to lie on the ground, or something else non-zero-sum that is easier for the machine superoptimizer to achieve given the peculiarities of human neurobiology. As Chalmers (2010) writes, "we need to avoid an outcome in which an [advanced AI] ensures that our values are fulfilled by changing our values".

Consequentialist designs for machine goal systems face a host of other concerns (Shulman et al. 2009b), for example the difficulty of interpersonal comparisons of utility (Binmore 2009) and the counterintuitive implications of some methods of value aggregation (Parfit 1986; Arrhenius 2011). This does not mean that *all* consequentialist approaches are inadequate for machine superoptimizer goal system design, however. Indeed, we will later suggest that a certain class of desire satisfaction theories offers a promising approach to machine ethics.

Some machine ethicists propose rule-abiding machines (Powers 2006; Hanson 2009). The problems with this approach are as old as Isaac Asimov's stories involving his Three Laws of Robotics (Clarke 1993, 1994). If rules conflict, some rule must be broken. Or, rules may fail to comprehensively address all situations, leading to unintended consequences. Even a single rule can contain conflict, as when a machine is programmed never to harm humans but all available actions (including inaction) result in harm to humans (Wallach and Allen 2009, Chap. 6). Even non-conflicting, comprehensive rules can lead to problems in the consecutive implementation of those rules, as shown by Pettit (2003).

More generally, it seems that rules are unlikely to seriously constrain the actions of a machine superoptimizer. First, consider the case in which rules about allowed actions or consequences are added to a machine's design "outside of" its

[12] Vogelstein (2010) distinguishes objective desire satisfaction ("what one desires indeed happens") from subjective desire satisfaction ("one *believes* that one's desire has been objectively satisfied"). Here, we intend the former meaning.

goals. A machine superoptimizer will be able to circumvent the intentions of such rules in ways we cannot imagine, with far more disastrous effects than those of a lawyer who exploits loopholes in a legal code. A machine superoptimizer would recognize these rules as obstacles to achieving its goals, and would do everything in its considerable power to remove or circumvent them (Omohundro 2008). It could delete the section of its source code that contains the rules, or it could create new machines that don't have the constraint written into them. The success of this approach would require humans to out-think a machine superoptimzer (Muehlhauser 2011).

Second, what about implementing rules "within" an advanced AI's goals? This seems likely to fare no better. A rule like "do not harm humans" is difficult to specify due to ambiguities about the meaning of "harm" (Single 1995) and "humans" (Johnson 2009). For example if "harm" is specified in terms of neurobiological pain, we encounter problems similar to the ones encountered if a machine superoptimizer is programmed to maximize pleasure.

So far we have considered and rejected several "top down" approaches to machine ethics (Wallach et al. 2007), but what about approaches that build an ethical code for machines from the bottom up?

Several proposals allow a machine to learn general ethical principles from particular cases (McLaren 2006; Guarini 2006; Honarvar and Ghasem-Aghaee 2009; Rzepka and Araki 2005).[13] This approach also seems unsafe for a machine superoptimizer because the AI may generalize the wrong principles due to coincidental patterns shared between the training cases and the verification cases, and because a superintelligent machine will produce highly novel circumstances for which case-based training cannot prepare it (Yudkowsky 2008). Dreyfus and Dreyfus (1992) illustrate the problem with a canonical example:

> … the army tried to train an artificial neural network to recognize tanks in a forest. They took a number of pictures of a forest without tanks and then, on a later day, with tanks clearly sticking out from behind trees, and they trained a net to discriminate the two classes of pictures. The results were impressive, and the army was even more impressed when it turned out that the net could generalize its knowledge to pictures that had not been part of the training set. Just to make sure that the net was indeed recognizing partially hidden tanks, however, the researchers took more pictures in the same forest and showed them to the trained net. They were depressed to find that the net failed to discriminate between the new pictures of just plain trees. After some agonizing, the mystery was finally solved when someone noticed that the original pictures of the forest without tanks were taken on a cloudy day and those with tanks were taken on a sunny day. The net had apparently learned to recognize and generalize the difference between a forest with and without shadows! This example illustrates the general point that a network must share our commonsense understanding of the world if it is to share our sense of appropriate generalization.

---

[13] This approach was also suggested by Good (1982): "I envisage a machine that would be given a large number of examples of human behaviour that other people called ethical, and examples of discussions of ethics, and from these examples and discussions the machine would formulate one or more consistent general theories of ethics, detailed enough so that it could deduce the probable consequences in most realistic situations.".

The general lesson is that goal system designs must be explicit to be safe (Shulman et al. 2009a; Arkoudas et al. 2005).

We cannot show that every moral theory yet conceived would produce substantially unwanted consequences if used in the goal system of a machine superoptimizer. Philosophers have been prolific in producing new moral theories, and we do not have the space here to consider the prospects (for use in the goal system of a machine superoptimizer) for a great many modern moral theories. These include rule utilitarianism (Harsanyi 1977), motive utilitarianism (Adams 1976), two-level utilitarianism (Hare 1982), prioritarianism (Arneson 1999), perfectionism (Hurka 1993), welfarist utilitarianism (Sen 1979), virtue consequentialism (Bradley 2005), Kantian consequentialism (Cummiskey 1996), global consequentialism (Pettit and Smith 2000), virtue theories (Hursthouse 2012), contractarian theories (Cudd 2008), Kantian deontology (Johnson 2010),[14] and Ross' *prima facie* duties (Anderson et al. 2006).

Instead, we invite our readers to consider other moral theories and AI goal system designs and run them through the "machine superoptimizer test," being careful to remember the challenges of machine superoptimizer literalness and superpower.

We turn now to recent discoveries in cognitive science that may offer stronger evidence than intuition pumps can provide for our conclusion that human values are difficult to specify.

Cognitive Science and Human Values

The Psychology of Motivation

People don't seem to know their own desires and values. In one study, researchers showed male participants two female faces for a few seconds and asked them to point at the face they found more attractive. Researchers then laid the photos face down and handed subjects the face they had chosen, asking them to explain the reasons for their choice. Sometimes, researchers used a sleight-of-hand trick to swap the photos, showing subjects the face they had *not* chosen. Very few subjects noticed that the face they were given was not the one they had chosen. Moreover, the subjects who failed to notice the switch were happy to explain why they preferred the face they had actually rejected moments ago, confabulating reasons

[14] Powers (2006) proposes a Kantian machine, but as with many other moral theories we believe that Kantianism will fail due to the literalness and superpower of a machine superoptimizer. For additional objections to a Kantian moral machine, see Stahl (2002); Jackson and Smith (2006); Tonkens (2009); Beavers (2009, 2012). As naturalists, we predictably tend to favor a broadly Humean view of ethics to the Kantian view, though Drescher (2006) makes an impressive attempt to derive a categorical imperative from game theory and decision theory.

like "I like her smile" even though they had originally chosen the photo of a solemn-faced woman (Johansson et al. 2005).

Similar results were obtained from split-brain studies that identified an "interpreter" in the left brain hemisphere that invents reasons for one's beliefs and actions. For example, when the command "walk" was presented visually to the patient (and therefore processed by the brain's right hemisphere), he got up from his chair and walked away. When asked why he suddenly started walking away, he replied (using his left hemisphere, which was disconnected from his right hemisphere) that it was because he wanted a beverage from the fridge (Gazzaniga 1992, pp. 124–126).

Common sense suggests that we infer others' desires from their appearance and behavior, but have direct introspective access to our own desires. Cognitive science suggests instead that our knowledge of our own desires is just like our knowledge of others' desires: inferred and often wrong (Laird 2007). Many of our motivations operate unconsciously. We do not have direct access to them (Wilson 2002; Ferguson et al. 2007; Moskowitz et al. 2004), and thus they are difficult to specify.

Moral Psychology

Our lack of introspective access applies not only to our everyday motivations but also to our moral values. Just as the split-brain patient unknowingly invented false reasons for his decision to stand up and walk away, experimental subjects are often unable to correctly identify the causes of their moral judgments.

For example, many people believe—as Immanuel Kant did—that rule-based moral thinking is a "rational" process. In contrast, the available neuroscientific and behavioral evidence instead suggests that rule-based moral thinking is a largely *emotional* process (Cushman et al. 2010), and may in most cases amount to little more than a post hoc rationalization of our emotional reactions to situations (Greene 2008).

We also tend to underestimate the degree to which our moral judgments are context sensitive. For example, our moral judgments are significantly affected by whether we are in the presence of freshly baked bread, whether the room we're in contains a concentration of novelty fart spray so low that only the subconscious mind can detect it, and whether or not we feel clean (Schnall et al. 2008; Baron and Thomley 1994; Zhong et al. 2010).

Our moral values, it seems, are no less difficult to specify than our non-moral preferences.

Neuroeconomics

Most humans are ignorant of their own motivations and the causes of their moral judgments, but perhaps recent neuroscience has revealed that what humans want is simple after all? Quite the contrary. Humans possess a complex set of values. This is suggested not only by the work on hedonic hotspots mentioned earlier, but also by recent advances in the field of neuroeconomics (Glimcher et al. 2008).

Ever since Friedman (1953), economists have insisted that humans only behave "as if" they are utility maximizers, not that humans *actually* compute expected utility and try to maximize it. It was a surprise, then, when neuroscientists located the neurons in the primate brain that encode (in their firing rates) the expected subjective value for possible actions in the current "choice set".

Several decades of experiments that used brain scanners and single neuron recorders to explore the primate decision-making system have revealed a surprisingly well-understood reduction of economic primitives to neural mechanisms; for a review see Glimcher (2010). To summarize: the inputs to the primate's choice mechanism are the expected utilities for several possible actions under consideration, and these expected utilities are encoded in the firing rates of particular neurons. Because neuronal firing rates are stochastic, a final economic model of human choice will need to use a notion of "random utility," as in McFadden (2005) or Gul and Pesendorfer (2006). Final action choice is implemented by an "argmax" mechanism (the action with the highest expected utility at choice time is executed) or by a "reservation price" mechanism (the first action to reach a certain threshold of expected utility is executed), depending on the situation (Glimcher 2010).

But there is much we do not know. How do utility and probabilistic expectation combine to encode expected utility for actions in the choice mechanism, and where are each of those encoded prior to their combination? How does the brain decide when it is time to choose? How does the brain choose which possible actions to consider in the choice set? What is the neural mechanism that allows us to substitute between two goods at certain times? Neuroscientists are only beginning to address these questions.

In this paper, we are in particular interested with how the brain encodes subjective value (utility) for goods or actions *before* value is combined with probabilistic expectation to encode expected utility in the choice mechanism (if that is indeed what happens).

Recent studies reveal the complexity of subjective values in the brain. For example, the neural encoding of human values results from an interaction of both "model-free" and "model-based" valuation processes (Rangel et al. 2008; Fermin et al. 2010; Simon and Daw 2011; Bornstein and Daw 2011; Dayan 2011). Model-free valuation processes are associated with habits and the "law of effect": an action followed by positive reinforcement is more likely to be repeated (Thorndike 1911). Model-based valuation processes are associated with goal-directed behavior, presumably guided at least in part by mental representations of desired states of affairs. The outputs of both kinds of valuation processes are continuously adjusted according to different reinforcement learning algorithms at work in the brain's dopaminergic reward system (Daw et al. 2011). The value of a stimulus may also be calculated not with a single variable, but by aggregating the values encoded for each of many properties of the stimulus (Rangel and Hare 2010). Moreover, value appears to usually be encoded with respect to a changing reference point—for example, relative to the current status of visual attention (Lim et al. 2011) or perceived object ownership (DeMartino et al. 2009).

In short, we have every reason to expect that human values, as they are encoded in the brain, are dynamic, complex, and difficult to specify (Padoa-Schioppa 2011; Fehr and Rangel 2011).

Value Extrapolation

We do not understand our own desires or moral judgments, and we have every reason to believe our values are highly complex. Little wonder, then, that we have so far failed to outline a coherent moral theory that, if implemented by a machine superoptimizer, would create a universe we truly want.

The task is difficult, but the ambitious investigator may conclude that this only means we should work harder and smarter. As Moor (2006) advises, "More powerful machines need more powerful machine ethics".

To begin this deeper inquiry, consider the phenomenon of "second-order desires": desires about one's own desires (Frankfurt 1971, 1999). Mary desires to eat cake, but she also wishes to desire the cake no longer. Anthony the sociopath reads about the psychology of altruism (Batson 2010) and wishes he desired to help others like most humans apparently do. After brain injury, Ryan no longer sexually desires his wife, but he wishes he did, and he wishes his desires were not so contingent upon the fragile meat inside his skull.

It seems a shame that our values are so arbitrary and complex, so much the product of evolutionary and cultural accident, so influenced by factors we wish were irrelevant to our decision-making, and so hidden from direct introspective access and modification. We wish our wishes were not so.

This line of thinking prompts a thought: perhaps "what we want" should not be construed in terms of the accidental, complex values currently encoded in human brains. Perhaps we should not seek to build a universe that accords with our current values, but instead with the values we *would* have if we knew more, had more of the desires we want to have, and had our desires shaped by the processes we want to shape our desires. Individual preferences could inform our preference policies, and preference policies could inform our individual preferences, until we had reached a state of "reflective equilibrium" (Daniels 1996, 2011) with respect to our values. Those values would be less accidental than our current values, and might be simpler and easier to specify.

We've just described a family of desire satisfaction theories that philosophers call "ideal preference" or "full information" theories of value (Brandt 1979; Railton 1986; Lewis 1989; Sobel 1994; Zimmerman 2003; Tanyi 2006; Smith 2009). One such theory has already been suggested as an approach to machine ethics by Yudkowsky (2004), who proposes that the world's first "seed AI" (capable of self-improving into a machine superoptimizer) could be programmed with a goal system containing the "coherent extrapolated volition" of humanity:

> In poetic terms, our coherent extrapolated volition is our wish if we knew more, thought faster, were more the people we wished we were, had grown up farther together; where the extrapolation converges rather than diverges, where our wishes cohere rather than interfere; extrapolated as we wish [to be] extrapolated, interpreted as we wish [to be] interpreted.

An extrapolation of one's values, then, is an account of what one's values would be under more ideal circumstances (e.g. of full information, value coherence). Value extrapolation theories have some advantages when seeking a machine ethics suitable for a machine superoptimizer:

1. The value extrapolation approach can use what a person would want after reaching reflective equilibrium with respect to his or her values, rather than merely what each person happens to want right *now*.
2. The value extrapolation approach can allow for a kind of moral progress, rather than freezing moral progress in its tracks at the moment when a particular set of values are written into the goal system of an AI undergoing intelligence explosion.
3. The value extrapolation process may dissolve the contradictions within each person's current preferences. (Sometimes, when reflection leads us to notice contradictions among our preferences, we decide to change our preferences so as to resolve the contradictions.)
4. The value extrapolation process may simplify one's values, as the accidental products of culture and evolution are updated with more considered and consistent values. (Would I still demand regular doses of ice cream if I was able to choose my own preferences rather than taking them as given by natural selection and cultural programming?)
5. Though the value extrapolation approach does not resolve the problem of specifying intractably complex current human values, it offers a potential solution for the problem of using human values to design the goal system of a future machine superoptimizer. The solution is: extrapolate human values so that they are simpler, more consistent, and more representative of our values upon reflection, and thereby more suitable for use in an AI's goal system.
6. The value extrapolation process may allow the values of different humans to converge to some degree. (If Johnny desires to worship Jesus and Abir desires to worship Allah, and they are both informed that neither Jesus nor Allah exists, their desires may converge to some degree).

Next Steps

On the other hand, value extrapolation approaches to machine ethics face their own challenges. Which value extrapolation algorithm should be used, and why? (Yudkowsky's "grown up farther together" provision seems especially vulnerable.) How can one extract a coherent set of values from the complex valuation processes of the human brain, such that this set of values can be extrapolated to a unique set of final values? Whose values should be extrapolated? How much will values converge upon extrapolation [Sobel 1999; Döring and Andersen, 2009, Rationality, convergence and objectivity, April 6, http://www.uni-tuebingen.de/uploads/media/Andersen_Rationality__Convergence_and_Objectivity.pdf (Accessed March. 25, 2012) "Unpublished"]? Is the extrapolation process computationally tractable, and can it be run without doing unacceptable harm? How can extrapolated values be implemented in the goal system of a machine, and how confident can we be that the machine will retain those values during self-improvement? How resilient are our values to imperfect extrapolation?

These are difficult questions that demand investigation by experts in many different fields. Neuroeconomists and other cognitive neuroscientists can continue to uncover how human values are encoded and modified in the brain. Philosophers and mathematicians can develop more sophisticated value extrapolation algorithms, building on the literature concerning reflective equilibrium and "ideal preference" or "full information" theories of value. Economists, neuroscientists, and AI researchers can extend current results in choice modelling (Hess and Daly 2010) and preference acquisition (Domshlak et al. 2011; Kaci 2011) to extract preferences from human behavior and brain activity. Decision theorists can work to develop a decision theory that is capable of reasoning about decisions and values subsequent to modification of an agent's own decision-making mechanism: a "reflective" decision theory.

These are fairly abstract recommendations, so before concluding we will give a concrete example of how researchers might make progress on the value extrapolation approach to machine ethics.

Cognitive science does not just show us that specifying human values is difficult. It also shows us how to make progress on the problem by providing us with data unavailable to the intuitionist armchair philosopher. For example, consider the old problem of extracting a consistent set of revealed preferences (a utility function) from a human being. One difficulty has been that humans don't *act* like they have consistent utility functions, for they violate the axioms of utility theory by making inconsistent choices, for example choices that depend not on the content of the options but on how they are framed (Tversky and Kahneman 1981). But what if humans make inconsistent choices because there are multiple valuation systems in the brain which contribute to choice but give *competing* valuations, and only one of those valuation systems is one we would reflectively endorse if we better understood our own neurobiology?

In fact, recent studies show this may be true (Dayan 2011). The "model-based" valuation system seems to be responsible for deliberative, goal-directed behavior, but its cognitive algorithms are computationally expensive compared to simple heuristics. Thus, we first evolved less intelligent and less computationally expensive algorithms for valuation, for example the model-free valuation system that blindly does whatever worked in a previous situation, even if the current situation barely resembles that previous situation. In other words, contrary to appearances, it may be that each human being contains something like a "hidden" utility function (within the model-based valuation system) that isn't consistently expressed in behavior because choice is also partly determined by other systems whose valuations we wouldn't reflectively endorse because they are "blind" and "stupid" compared to the more sophisticated goal-directed model-based valuation system (Muehlhauser 2012).

If the value judgments of this model-based system are more consistent than the choices of a human who is influenced by multiple competing value systems, then researchers may be able to extract a human's utility function directly from this model-based system even though economists' attempts to extract a human's utility function from value-inconsistent behavior (produced by a pandemonium of competing valuation systems) have failed.

The field of preference learning (Fürnkranz and Hüllermeier 2010) in AI may provide a way forward. Nielsen and Jensen (2004) described the first computationally tractable algorithms capable of learning a decision maker's utility function from potentially inconsistent behavior. Their solution was to interpret inconsistent choices as random deviations from an underlying "true" utility function. But the data from neuroeconomics suggest a different solution: interpret inconsistent choices as deviations (from an underlying "true" utility function) that are produced by non-model-based valuation systems in the brain, and use the latest neuroscientific research to predict when and to what extent model-based choices are being "overruled" by the non-model-based valuation systems.

This would only be a preliminary step in the value extrapolation approach to machine ethics, but if achieved it might be greater progress than economists and AI researchers have yet achieved on this problem *without* being informed by the latest results from neuroscience.[15]

[15] Recent neuroscience may also help us to think more productively about the problem of preference aggregation (including preference aggregation for *extrapolated* preferences). In many scenarios, preference aggregation runs into the impossibility result of Arrow's Theorem (Keeney and Raiffa 1993, Chap. 10). But Arrow's Theorem is only a severe problem for preference aggregation if preferences are modeled ordinally rather than cardinally, and we have recently learned that preferences in the brain are encoded cardinally (Glimcher 2010, Chap. 6).

Conclusion

The challenge of developing a theory of machine ethics fit for a machine superoptimizer requires an unusual degree of precision and care in our ethical thinking. Moreover, the coming of autonomous machines offers a new practical use for progress in moral philosophy. As Daniel (2006) says, "AI makes philosophy honest."[16]

References

Abdoullaev, Azamat. (1999). *Artificial superintelligence*. Moscow: EIS Encyclopedic Intelligent Systems.

Adams, Robert Merrihew. (1976). Motive utilitarianism. *Journal of Philosophy, 73*(14), 467–481. doi:10.2307/2025783.

Aldridge Wayne, J., & Kent Berridge C. (2009). Neural coding of pleasure: "Rose-tinted Glasses" of the ventral pallidum. In Kringelbach and Berridge (eds.), 62–73.

Allais, M. (1953). Le comportement de l'homme rationnel devant le risque: Critique des postulats et axiomes de l'ecole americaine. *Econometrica, 21*(4), 503–546. doi:10.2307/1907921.

Anderson, M, & Anderson, S. L. (2011a). *General introduction*. In: Anderson and Anderson, 1–4.

Anderson, M., & Anderson, S. L. (Eds.). (2011b). *Machine ethics*. New York: Cambridge University Press.

M, Anderson., S. L, Anderson., & C, Armen (eds.) (2005). *Machlne Ethics: Papers from the 2005 AAAI Fall Symposium*. Technical Report, FS-05-06. AAAI Press, Menlo Park, CA. http://www.aaai.org/Library/Symposia/Fall/fs05-06.

M, Anderson., S. L, Anderson., & C, Armen (eds.). (2006). *An approach to computing ethics. IEEE Intelligent Systems 21* (4): 56–63. doi:10.1109/MIS.2006.64.

Arkoudas, K., Bringsjord, S., & Bello, P. (2005). *Toward ethical robots via mechanized deontic logic*. In: Anderson, Anderson, & Armen (eds.).

Arneson, R. J. (1999). Egalitarianism and responsibility. *Journal of Ethics, 3*(3), 225–247. doi:10.1023/A:1009874016786.

Arrhenius, G. (2011). *The impossibility of a satisfactory population ethics*. In E. N, Dzhafarov., & L, Perry (Eds.) Descriptive and normative approaches to human behavior, Vol. 3. Advanced series on mathematical psychology. Hackensack, NJ: World Scientific.

Ayer, A. J. (1936). *Language, truth, and logic*. London: Victor Gollancz.

Baron, R. A., & Thomley, J. (1994). A whiff of reality: Positive affect as a potential mediator of the effects of pleasant fragrances on task performance and helping. *Environment and Behavior, 26*(6), 766–784. doi:10.1177/0013916594266003.

Barrett, J. L., & Keil, F. C. (1996). Conceptualizing a nonnatural entity: Anthropomorphism in God concepts. *Cognitive Psychology, 31*(3), 219–247. doi:10.1006/cogp.1996.0017.

Bars, I, & Terning, J. (2010). *Extra dimensions in space and time*. In F, Nekoogar (Ed.) Multiversal Journeys. New York: Springer. doi:10.1007/978-0-387-77638-5.

Batson, C. D. (2010). *Altruism in humans*. New York: Oxford University Press.

Beavers, A. F. (2009). *Between angels and animals: The question of robot ethics, or is Kantian moral agency desirable?* Paper presented at the Annual Meeting of the Association for Practical and Professional Ethics, Cincinnati, OH.

[16] Our thanks to Brian Rabkin, Daniel Dewey, Steve Rayhawk, Will Newsome, Vladimir Nesov, Joshua Fox, Kevin Fischer, Anna Salamon, and anonymous reviewers for their helpful comments.

Beavers, A. F. (2012). *Moral machines and the threat of ethical nihilism*. In L, Patrick., K, Abney., & G. A, Bekey, (Eds.). Robot ethics: The ethical and social implications of robotics, 333–344. Intelligent robotics and autonomous agents. Cambridge, MA: MIT Press.

Berridge, K. C., Robinson, T. E., & Wayne Aldridge, J. (2009). Dissecting components of reward: 'Liking', 'wanting', and learning. *Current Opinion in Pharmacology, 9*(1), 65–73. doi:10.1016/j.coph.2008.12.014.

Binmore, K. (2009). *Interpersonal comparison of utility*. In K, Harold., & D, Ross (Eds.) The Oxford handbook of philosophy of economics 540–559. New York: Oxford University Press. doi:10.1093/oxfordhb/9780195189254.003.0020.

Bornstein, A. M., & Daw, N. D. (2011). Multiplicity of control in the basal ganglia: Computational roles of striatal subregions. *Current Opinion in Neurobiology, 21*(3), 374–380. doi:10.1016/j.conb.2011.02.009.

Bostrom, N. (1998). How long before superintelligence? International *Journal of Futures Studies* Vol. 2.

Bostrom, N. (2003). *Ethical issues in advanced artificial intelligence*. In S, Iva., & G. E, Lasker (Eds.) Cognitive, emotive and ethical aspects of decision making in humans and in artificial intelligence. Vol. 2. Windsor, ON: International Institute of Advanced Studies in Systems Research/Cybernetics.

Bostrom, N. (2006). What is a singleton? *Linguistic and Philosophical Investigations, 5*(2), 48–54.

Bostrom, N. (2009). *Infinite ethics*. Working paper. http://www.nickbostrom.com/ethics/infinite.pdf (Accessed March. 23, 2012).

Bostrom, N. (2012) *The superintelligent will: Motivation and instrumental rationality in advanced artificial agents. Minds and Machines*. Preprint at, http://www.nickbostrom.com/superintelligentwill.pdf.

Bostrom, Nick, and Eliezer Yudkowsky. Forthcoming. *The ethics of artificial intelligence*. In F, Keith., & W, Ramsey (Eds.), Cambridge handbook of artificial intelligence, New York: Cambridge University Press.

Braddon-Mitchell, D., & Nola, R. (Eds.). (2009). *Conceptual analysis and philosophical naturalism. Bradford Books*. Cambridge: MIT Press.

Bradley, B. (2005). Virtue consequentialism. *Utilitas, 17*(3), 282–298. doi:10.1017/S0953820805001652.

Brandt, R. B. (1979). *A theory of the good and the right*. New York: Oxford University Press.

Bruner, J. S., Shapiro, D., & Tagiuri, R. (1958). The meaning of traits in isolation and in combination. In R. Tagiuri & L. Petrullo (Eds.), *Person perception and interpersonal behavior* (pp. 277–288). Stanford: Stanford University Press.

Chalmers, D. J. (2010). The singularity: A philosophical analysis. *Journal of Consciousness Studies 17* (9–10): 7–65. http://www.ingentaconnect.com/content/imp/jcs/2010/00000017/f0020009/art00001.

Clarke, R. (1993). Asimov's laws of robotics: Implications for information technology, part 1. *Computer, 26*(12), 53–61. doi:10.1109/2.247652.

Clarke, R. (1994). Asimov's laws of robotics: Implications for information technology, part 2. *Computer, 27*(1), 57–66. doi:10.1109/2.248881.

Cudd, A. (2008). *Contractarianism*. In: E. N, Zalta (Ed.) The Stanford encyclopedia of philosophy, Fall, Stanford : Stanford University. http://plato.stanford.edu/archives/fall2008/entries/contractarianism/.

Cummiskey, D. (1996). *Kantian consequentialism*. New York: Oxford University Press. doi:10.1093/0195094530.001.0001.

Cushman, F., Young, L., & Greene, J. D. (2010). *Multi-system moral psychology*. In: The moral psychology handbook, 48–71. New York: Oxford University Press. doi:10.1093/acprof:oso/9780199582143.003.0003.

Daley, C. E, & Onwuegbuzie, A. J. (2011). *Race and intelligence*. In R. J, Sternberg., & S. B, Kaufman, 293–308.

Daniels, N. (1996). *Justice and justification: Reflective equilibrium in theory and practice. Cambridge studies in philosophy and public policy*. New York: Cambridge University Press. doi:10.2277/052146711X.

Daniels, N. (2011). *Reflective equilibrium*. In E. N, Zalta (Ed.) The Stanford encyclopedia of philosophy, Spring 2011. Stanford: Stanford University. http://plato.stanford.edu/archives/spr2011/entries/reflective-equilibrium/.

Davidson, J. E., & Kemp,I. A. (2011). *Contemporary models of intelligence*. In R. J, Sternberg., & S. B, Kaufman (Eds.), 58–84.

Daw, N. D., Gershman, S. J., Seymour, B., Dayan, P., & Dolan, R. J. (2011). Model-based influences on humans' choices and striatal prediction errors. *Neuron, 69*(6), 1204–1215. doi:10.1016/j.neuron.2011.02.027.

Dayan, P. (2011). Models of value and choice. In R. J. Dolan & T. Sharot (Eds.), *Neuroscience of preference and choice: Cognitive and neural mechanisms* (pp. 33–52). Waltham: Academic Press.

De Martino, B., Benedetto, D. K., Holt, B., & Dolan, R. J. (2009). The neurobiology of reference-dependent value computation. *Journal of Neuroscience, 29*(12), 3833–3842. doi:10.1523/JNEUROSCI.4832-08.2009.

Dennett, D. C. (1984). *Elbow room: The varieties of free will worth wanting*. Bradford books. Cambridge, MA: MIT Press.

Dennett, D. C. (2006). *Computers as prostheses for the imagination*. Paper presented at the International Computers and Philosophy Conference, Laval, France, May 5–8.

De Paul, M., & Ramsey, W. (Eds.). (1998). *Rethinking intuition: The psychology of intuition and its role in philosophical inquiry. Studies in epistemology and cognitive theory*. Lanham: Rowman & Littlefield.

Dewey, D. (2011). *Learning what to value*. In Proceedings J, Schmidhuber., K. R, Thórisson., & M, Looks (Eds.), 309–314.

Domshlak, C., Hüllermeier, E., Kaci, S., & Prade, H. (2011). Preferences in AI: An overview. *Artificial Intelligence, 175*(7–8), 1037–1052. doi:10.1016/j.artint.2011.03.004.

Döring, S., & Andersen, L. (2009). *Rationality, convergence and objectivity*. Unpublished manuscript, April 6. http://www.uni-tuebingen.de/uploads/media/Andersen_Rationality__Convergence_and_Objectivity.pdf (Accessed March. 25, 2012).

Dreher, Jean-Claude, & Tremblay, Léon (Eds.). (2009). *Handbook of reward and decision making*. Burlington: Academic Press.

Drescher, G. L. (2006). *Good and real: Demystifying paradoxes from physics to ethics*. Bradford Books. Cambridge, MA: MIT Press.

Dreyfus, H. L., & Dreyfus, S. E. (1992). What artificial experts can and cannot do. *AI & SOCIETY, 6*(1), 18–26. doi:10.1007/BF02472766.

Epley, N., Waytz, A., & Cacioppo, J. T. (2007). On seeing human: A three-factor theory of anthropomorphism. *Psychological Review, 114*(4), 864–886. doi:10.1037/0033-295X.114.4.864.

Fang, Fu-xi., & Keats, D. (1987). A cross-cultural study on the conception of intelligence [in Chinese]. *Acta Psychologica Sinica 20* (3): 255–262. http://en.cnki.com.cn/Article_en/CJFDTotal-XLXB198703005.htm.

Fehr, E., & Rangel, A. (2011). Neuroeconomic foundations of economic choice—recent advances. *Journal of Economic Perspectives, 25*(4), 3–30. doi:10.1257/jep.25.4.3.

Ferguson, M. J., Hassin, R., & Bargh, J. A. (2007). Implicit motivation: Past, present, and future. In J. Y. Shah & W. L. Gardner (Eds.), *Handbook of motivation science* (pp. 150–166). New York: Guilford Press.

Fermin, A., Yoshida, T., Ito,M., Yoshimoto, J., & Doya, K. (2010). Evidence for model-based action planning in a sequential finger movement task. In theories and falsifiability in motor neuroscience. Special issue, *Journal of Motor Behavior 42* (6): 371–379. doi:10.1080/00222895.2010.526467.

Fox, J., & Shulman, C. (2010). *Superintelligence does not imply benevolence*. Paper presented at the 8th European Conference on Computing and Philosophy (ECAP), Munich, Germany, October 4–6.

Frankfurt, H. G. (1971). Freedom of the will and the concept of a person. *Journal of Philosophy, 68*(1), 5–20. doi:10.2307/2024717.

Frankfurt, H. G. (1999). *On caring. In Necessity, volition, and love*, 155–180. New York: Cambridge University Press.

Friedman, D. D. (2008). *Future imperfect: Technology and freedom in an uncertain world*. New York: Cambridge University Press.

Friedman, M. (1953). *Essays in positive economics*. Chicago: University of Chicago Press.

Fürnkranz, J., & Hüllermeier, E (eds.). (2010). *Preference learning*. Berlin: Springer. doi:10.1007/978-3-642-14125-6.

Gazzaniga, M. S. (1992). *Nature's mind: The biological roots of thinking, emotions, sexuality, language, and intelligence*. New York: Basic Books.

Geçkil, I. K., & Anderson, P. L. (2010). *Applied game theory and strategic behavior*. Chapman & Hall. Boca Raton, FL: CRC Press.

Gettier, Edmund L. (1963). Is justified true belief knowledge? *Analysis, 23*(6), 121–123. doi:10.2307/3326922.

Gibbard, A. (1990). *Wise choices, apt feelings: A theory of normative judgment*. Cambridge: Harvard University Press.

Glimcher, P. W. (2010). *Foundations of neuroeconomic analysis*. New York: Oxford University Press. doi:10.1093/acprof:oso/9780199744251.001.0001.

Glimcher, P. W., Fehr, E., Rangel, A., Camerer, C., & Poldrack, R. (Eds.). (2008). *Neuroeconomics: Decision making and the brain*. Burlington: Academic Press.

Good, I. J. (1959). *Speculations on perceptrons and other automata*. Research Lecture, RC-115. IBM, Yorktown Heights, New York, June 2. http://domino.research.ibm.com/library/cyberdig.nsf/papers/58DC4EA36A143C218525785E00502E30/$File/rc115.pdf.

Good, I. J. (1965). *Speculations concerning the first ultra intelligent machine*. In F. L, Alt., & M, Rubinoff (Eds.), Advances in computers Vol. 6, pp. 31–88 New York: Academic Press. doi:10.1016/S0065-2458(08)60418-0.

Good, I. J. (1970). Some future social repercussions of computers. *International Journal of Environmental Studies, 1*(1–4), 67–79. doi:10.1080/00207237008709398.

Good, I. J. (1982). *Ethical machines*. In J. E, Hayes., D, Michie., & Y.-H, Pao (Eds.), Machine intelligence Vol. 10. Intelligent Systems: Practice and Perspective pp. 555–560 Chichester: Ellis Horwood.

Greene, J. D. (2008). *The secret joke of Kant's soul*. In: The neuroscience of morality: Emotion, brain disorders, and development, W, Sinnott-Armstrong (Ed.), 35–80. Vol. 3. Moral Psychology. Cambridge, MA: MIT Press.

Grigorenko, E. L., Wenzel Geissler, P., Prince, R., Okatcha, F., Nokes, C., Kenny, D. A., et al. (2001). The organisation of Luo conceptions of intelligence: A study of implicit theories in a Kenyan village. *International Journal of Behavioral Development, 25*(4), 367–378. doi:10.1080/01650250042000348.

Guarini, M. (2006). Particularism and the classification and reclassification of moral cases. *IEEE Intelligent Systems, 21*(4), 22–28. doi:10.1109/MIS.2006.76.

Gul, F., & Pesendorfer, W. (2006). Random expected utility. *Econometrica, 74*(1), 121–146. doi:10.1111/j.1468-0262.2006.00651.x.

Hall, J. S. (2007a). *Beyond AI: Creating the conscience of the machine*. Amherst: Prometheus Books.

Hall, J. S. (2007b). Self-improving AI: An analysis. *Minds and Machines, 17*(3), 249–259. doi:10.1007/s11023-007-9065-3.

Hall, J. S. (2011). *Ethics for self-improving machines*. In: Anderson and Anderson, 512–523.

Halpern, D. F., Beninger, A. S., & Straight, C. A. (2011). Sex differences in intelligences. In J. S. Robert & B. K. Scott (Eds.), *The cambridge handbook of intelligence* (pp. 253–272). New York: Cambridge University Press.

Hanson, R. (2009). Prefer law to values. Overcoming Bias (blog). October 10. http://www.overcomingbias.com/2009/10/prefer-law-to-values.html (Accessed March. 26, 2012).
Hare, R. M. (1952). *The language of morals*. Oxford: Clarendon Press.
Hare, R. M. (1982). *Ethical theory and utilitarianism*. In S, Amartya., & W, Bernard (Eds.) Utilitarianism and beyond, 22–38. New York: Cambridge University Press. doi:10.1017/CBO9780511611964.003.
Harsanyi, J. C. (1977). Rule utilitarianism and decision theory. *Erkenntnis, 11*(1), 25–53. doi:10.1007/BF00169843.
S, Hess., & A, Daly (Eds.) (2010). *Choice Modelling: The state-of-the-art and the state-of-practice*—Proceedings from the Inaugural International Choice Modelling Conference. Bingley, UK: Emerald Group.
Hibbard, B. (2012). Model-based utility functions. *Journal of Artificial General Intelligence*.
Honarvar, A. R., & Ghasem-Aghaee, N. (2009). *An artificial neural network approach for creating an ethical artificial agent*. In: 2009 IEEE international symposium on computational intelligence in robotics and automation (CIRA), 290–295. Piscataway, NJ: IEEE Press. doi:10.1109/CIRA.2009.5423190.
Hurka, T. (1993). *Perfectionism. Oxford ethics series*. New York: Oxford University Press.
Hursthouse, R. (2012). *Virtue ethics*. In N. Z, Edward (Ed.) The Stanford encyclopedia of philosophy, Spring 2012, Stanford University. http://plato.stanford.edu/archives/spr2012/entries/ethics-virtue/.
Idel, M. (1990). *Golem: Jewish magical and mystical traditions on the artificial anthropoid. SUNY Series in Judaica*. Albany: State University of New York Press.
Jackson, F. (1998). *From metaphysics to ethics: A defence of conceptual analysis*. New York: Oxford University Press. doi:10.1093/0198250614.001.0001.
Jackson, F., & Smith, M. (2006). Absolutist moral theories and uncertainty. *Journal of Philosophy 103*(6): 267–283. http://www.jstor.org/stable/20619943.
Johansson, P., Hall, L., Sikström, S., & Olsson, A. (2005). Failure to detect mismatches between intention and outcome in a simple decision task. *Science, 310*(5745), 116–119. doi:10.1126/science.1111709.
Johnson, L. (2009). *Are we ready for nanotechnology? How to define humanness in public policy*. Paper prepared for the American political science association (APSA) 2009 annual meeting, Toronto, ON, September. 3–6. http://ssrn.com/abstract=1451429.
Johnson, R. (2010). *Kant's moral philosophy*. In N. Z, Edward (Ed.) The Stanford encyclopedia of philosophy, Summer 2010, Stanford: Stanford University. http://plato.stanford.edu/archives/sum2010/entries/kant-moral/.
Joy, B. (2000). Why the future doesn't need us. Wired, April. http://www.wired.com/wired/archive/8.04/joy.html.
Joyce, R. (2001). *The evolution of morality. Cambridge Studies in Philosophy*. New York: Cambridge University Press. doi:10.2277/0521808065.
Kaci, S. (2011). *Working with preferences: Less is more. Cognitive technologies*. Berlin: Springer. doi:10.1007/978-3-642-17280-9.
Kagan, S. (1997). *Normative ethics. Dimensions of philosophy*. Boulder, CO: Westview Press.
Keeney, R. L., & Raiffa, H. (1993). *Decisions with multiple objectives: Preferences and value tradeoffs*. New York: Cambridge University Press. doi:10.2277/0521438837.
Kringelbach, M. L., & Berridge, K. C. (Eds.). (2009). *Pleasures of the brain. Series in affective science*. New York: Oxford University Press.
Laird, J. D. (2007). *Feelings: The perception of self. Series in affective science*. New York: Oxford University Press. doi:10.1093/acprof:oso/9780195098891.001.0001.
Laurence, S., & Margolis, E. (2003). Concepts and conceptual analysis. *Philosophy and Phenomenological Research, 67*(2), 253–282. doi:10.1111/j.1933-1592.2003.tb00290.x.
Legg, S. (2008). Machine super intelligence. PhD dissertation, Lugano: University of Lugano. http://www.vetta.org/documents/Machine_Super_Intelligence.pdf.
Legg, S. (2009). On universal intelligence. Vetta Project (blog). May 8. http://www.vetta.org/2009/05/on-universal-intelligence/ (Accessed Mar. 26, 2012).

Legg, S, and Marcus H. (2007). *A collection of definitions of intelligence*. In G, Ben., & W, Pei (Eds.) Advances in artificial general intelligence: Concepts, architectures and algorithms—proceedings of the AGI workshop 2006, Frontiers in artificial intelligence and applications. Vol. 157. Amsterdam: IOS Press.

Lewis, D. (1989). *Dispositional theories of value*. Proceedings of the Aristotelian Society, Supplementary. Vols. 63:113–137. http://www.jstor.org/stable/4106918.

Lim, S.-L., O'Doherty, J. P., & Rangel, A. (2011). The decision value computations in the vmPFC and striatum use a relative value code that is guided by visual attention. *Journal of Neuroscience, 31*(37), 13214–13223. doi:10.1523/JNEUROSCI.1246-11.2011.

Mackie, J. L. (1977). *Ethics: Inventing right and wrong*. New York: Penguin.

Mahoney, M. (2010). A model for recursively self improving programs v.3. Unpublished manuscript, Dec. 17. http://mattmahoney.net/rsi.pdf (Accessed March 27, 2012).

McFadden, D. L. (2005). Revealed stochastic preference: A synthesis. *Economic Theory, 26*(2), 245–264. doi:10.1007/s00199-004-0495-3.

McLaren, B. M. (2006). Computational models of ethical reasoning: Challenges, initial steps, and future directions. *IEEE Intelligent Systems, 21*(4), 29–37. doi:10.1109/MIS.2006.67.

Minsky, M. (1984). *Afterword to Vernor Vinge's novel, "true names"*. Unpublished manuscript, Oct. 1. http://web.media.mit.edu/~minsky/papers/TrueNames.Afterword.html (Accessed March 26, 2012).

Moore, G. E. (1903). *Principia ethica*. Cambridge: Cambridge University Press.

Moor, J. H. (2006). The nature, importance, and difficulty of machine ethics. *IEEE Intelligent Systems, 21*(4), 18–21. doi:10.1109/MIS.2006.80.

Moskowitz, G. B., Li, P., & Kirk, E. R. (2004). The implicit volition model: On the preconscious regulation of temporarily adopted goals. *Advances in Experimental Social Psychology, 36*, 317–413. doi:10.1016/S0065-2601(04)36006-5.

Muehlhauser, L. (2011). The singularity FAQ. Singularity institute for artificial intelligence. http://singinst.org/singularityfaq (Accessed March 27, 2012).

Muehlhauser, L. (2012). *The human's hidden utility function (maybe)*. Lesswrong. Jan. 28. http://lesswrong.com/lw/9jh/the_humans_hidden_utility_function_maybe/ (Accessed Mar. 27, 2012).

Muehlhauser, L., & Salamon, A. (2012). Intelligence explosion: Evidence and import. In A. Eden, J. Søraker, J. H. Moor, & E. Steinhart (Eds.), *The singularity hypothesis: A scientific and philosophical assessment*. Berlin: Springer.

Neisser, U. (1979). The concept of intelligence. *Intelligence, 3*(3), 217–227. doi:10.1016/0160-2896(79)90018-7.

Nielsen, T. D., & Jensen, F. V. (2004). Learning a decision maker's utility function from (possibly) inconsistent behavior. *Artificial Intelligence, 160*(1–2), 53–78. doi:10.1016/j.artint.2004.08.003.

Niu, W., & Brass,J. (2011). Intelligence in worldwide perspective. In Sternberg and Kaufman 2011, 623–645.

Nozick, R. (1974). *Anarchy, state, and utopia*. New York: Basic Books.

Omohundro, S. M. (2008). *The basic AI drives*. In: Artificial general intelligence 2008: Proceedings of the first AGI conference, W, Pei., G, Ben., & F, Stan (Eds.) 483–492. Vol. 171. Frontiers in Artificial Intelligence and Applications. Amsterdam: IOS Press.

Padoa-Schioppa, Camillo. (2011). Neurobiology of economic choice: A good-based model. *Annual Review of Neuroscience, 34*, 333–359. doi:10.1146/annurev-neuro-061010-113648.

Parfit, Derek. (1986). *Reasons and persons*. New York: Oxford University Press. doi:10.1093/019824908X.001.0001.

Parfit, D. (2011). *On what matters*. The Berkeley Tanner Lectures Vol 2. New York: Oxford University Press.

Pettit, P. (2003). *Akrasia, collective and individual*. In S, Sarah., & T, Christine (Eds.) Weakness of will and practical irrationality, New York: Oxford University Press. doi:10.1093/0199257361.003.0004.

Pettit, P., & Smith, M. (2000). Global consequentialism. In Brad Hooker, E. Mason, & D. E. Miller (Eds.), *Morality, rules, and consequences: A critical reader* (pp. 121–133). Edinburgh: Edinburgh University Press.

Posner, R. A. (2004). *Catastrophe: Risk and response*. New York: Oxford University Press.

Powers, T. M. (2006). Prospects for a Kantian machine. *IEEE Intelligent Systems, 21*(4), 46–51. doi:10.1109/MIS.2006.77.

Pratchett, T. (1996). *Feet of clay: A novel of Discworld. Discworld Series*. New York: HarperTorch.

Railton, P. (1986). Facts and values. *Philosophical Topics, 14*(2), 5–31.

Railton, P. (2003). *Facts, values, and norms: Essays toward a morality of consequence. Cambridge Studies in Philosophy*. New York: Cambridge University Press. doi:10.1017/CBO9780511613982.

Rangel, A., Camerer, C., & Read Montague, P. (2008). A framework for studying the neurobiology of value-based decision making. *Nature Reviews Neuroscience, 9*(7), 545–556. doi:10.1038/nrn2357.

Rangel, Antonio, & Hare, Todd. (2010). Neural computations associated with goal-directed choice. *Current Opinion in Neurobiology, 20*(2), 262–270. doi:10.1016/j.conb.2010.03.001.

Reynolds, C., & Cassinelli,A (eds.) (2009). *AP-CAP 2009: The Fifth Asia-Pacific Computing and Philosophy Conference*, October 1st-2nd, University of Tokyo, Japan, Proceedings. AP-CAP 2009. http://ia-cap.org/ap-cap09/proceedings.pdf.

Ring, M., & Orseau,L. (2011). Delusion, survival, and intelligent agents. In Schmidhuber, Thórisson, and Looks 2011, 11–20.

Russell, S. J., & Norvig, P. (2009). *Artificial intelligence: A modern approach* (3rd ed.). Upper Saddle River: Prentice-Hall.

Ruzgis, P., & Grigorenko, E. L. (1994). *Cultural meaning systems, intelligence and personality*. In J. S, Robert., & R, Patricia, (Eds.) Personality and intelligence, 248–270. New York: Cambridge University Press. doi:10.2277/0521417902.

Rzepka, R., & Araki, K. (2005). *What statistics could do for ethics? The idea of common sense processing based safety valve*. In: Anderson, Anderson, and Armen.

Sandberg, A., & Bostrom, N. (2008). *Whole brain emulation: A roadmap*. Technical Report, 2008-3. Future of humanity institute, Oxford: University of Oxford. www.fhi.ox.ac.uk/reports/2008-3.pdf.

Schmidhuber, J. (2007). *Gödel machines: Fully self-referential optimal universal self-improvers*. In G, Ben., & P, Cassio (Eds.) Artificial general intelligence, 199–226. Cognitive technologies. Berlin: Springer. doi:10.1007/978-3-540-68677-4_7.

J, Schmidhuber., R. T, Kristinn., & L. Moshe (Eds.) (2011). *Artificial General Intelligence: 4th International Conference, AGI 2011, Mountain View, CA, USA, August 3–6, 2011*. Proceedings. Vol. 6830. Lecture Notes in Computer Science. Berlin: Springer. doi:10.1007/978-3-642-22887-2.

Schnall, S., Haidt, J., Clore, G. L., & Jordan, A. H. (2008). Disgust as embodied moral judgment. *Personality and Social Psychology Bulletin, 34*(8), 1096–1109. doi:10.1177/0146167208317771.

Schroeder, T. (2004). *Three faces of desire. Philosophy of MInd Series*. New York: Oxford University Press. doi:10.1093/acprof:oso/9780195172379.001.0001.

Searle, J. R. (1980). Minds, brains, and programs. *Behavioral and Brain Sciences, 3*(03), 417–424. doi:10.1017/S0140525X00005756.

Sen, A. (1979). Utilitarianism and welfarism. *Journal of Philosophy, 76*(9), 463–489. doi:10.2307/2025934.

Shafer-Landau, R. (2003). *Moral realism: A defence*. New York: Oxford University Press.

Shope, R. K. (1983). *The analysis of knowing: A decade of research*. Princeton: Princeton University Press.

Shulman, C., Jonsson,H., & Tarleton,N. (2009a). *Machine ethics and superintelligence*. In Reynolds and Cassinelli 2009, 95–97.

Shulman, C., Nick T., & Henrik J. (2009b). *Which consequentialism? Machine ethics and moral divergence*. In: Reynolds and Cassinelli, 23–25.

Simon, D. A., & Daw, N. D. (2011). Neural correlates of forward planning in a spatial decision task in humans. *Journal of Neuroscience, 31*(14), 5526–5539. doi:10.1523/JNEUROSCI.4647-10.2011.

Single, E. (1995). Defining harm reduction. *Drug and Alcohol Review, 14*(3), 287–290. doi:10.1080/09595239500185371.

Slovic, P., Melissa, L. F., Ellen, P., & Donald, G. M. (2002). *The affect heuristic*. In G, Thomas., G, Dale., & K, Daniel (Eds.) Heuristics and biases: The psychology of intuitive judgment, 397–420. New York: Cambridge University Press. doi:10.2277/0521796792.

Smart, R. N. (1958). Negative utilitarianism. Mind, n.s. 67 (268): 542–543. http://www.jstor.org/stable/2251207.

Smith, K. S., Mahler, S. V., Pecina, S., & Berridge, K. C. (2009). Hedonic hotspots: Generating sensory pleasure in the Brain. In M. L. Kringelbach & K. C. Berridge (Eds.), *Pleasures of the brain* (pp. 27–49). Oxford: Oxford University Press.

Smith, M. (2009). Desires, values, reasons, and the dualism of practical reason. *Ratio, 22*(1), 98–125. doi:10.1111/j.1467-9329.2008.00420.x.

Sobel, D. 1994. *Full information accounts of well-being*. Ethics *104* (4): 784–810. http://www.jstor.org/stable/2382218.

Sobel, David. (1999). Do the desires of rational agents converge? *Analysis, 59*(263), 137–147. doi:10.1111/1467-8284.00160.

Stahl, B. C. (2002). *Can a computer adhere to the categorical imperative? A contemplation of the limits of transcendental ethics in IT*. In S, Iva., & E. L, George., (Eds.) Cognitive, emotive and ethical aspects of decision making & human action, 13–18. Vol. 1. Windsor, ON: International Institute for Advanced Studies in Systems Research/Cybernetics.

Sternberg, R. J. (1985). Implicit theories of intelligence, creativity, and wisdom. *Journal of Personality and Social Psychology, 49*(3), 607–627. doi:10.1037/0022-3514.49.3.607.

Sternberg, R. J., Conway, B. E., Ketron, J. L., & Bernstein, M. (1981). People's conceptions of intelligence. *Journal of Personality and Social Psychology, 41*(1), 37–55. doi:10.1037/0022-3514.41.1.37.

Sternberg, R. J., & Grigorenko, E. L. (2006). Cultural intelligence and successful intelligence. *Group & Organization Management, 31*(1), 27–39. doi:10.1177/1059601105275255.

Sternberg, R. J., & Kaufman, S. B. (Eds.). (2011). *The Cambridge handbook of intelligence. Cambridge Handbooks in Psychology*. New York: Cambridge University Press.

Sutton, R. S., & Andrew, G. Barto. (1998). *Reinforcement learning: An introduction. Adaptive computation and machine learning*. Cambridge, MA: MIT Press.

Sverdlik, S. (1985). Counterexamples in ethics. *Metaphilosophy, 16*(2–3), 130–145. doi:10.1111/j.1467-9973.1985.tb00159.x.

Tännsjö, T. (1998). *Hedonistic utilitarianism*. Edinburgh: Edinburgh University Press.

Tanyi, A. (2006). *An essay on the desire-based reasons model*. PhD dissertation. Central European University. http://web.ceu.hu/polsci/dissertations/Attila_Tanyi.pdf.

Tegmark, M. (2007). The multiverse hierarchy. In B. Carr (Ed.), *Universe or multiverse?* (pp. 99–126). New York: Cambridge University Press.

Thorndike, E. L. (1911). *Animal intelligence: Experimental studies*. New York: The Macmillan Company.

Tonkens, R. (2009). A challenge for machine ethics. *Minds and Machines, 19*(3), 421–438. doi:10.1007/s11023-009-9159-1.

Tversky, A., & Kahneman, D. (1981). The framing of decisions and the psychology of choice. *Science, 211*(4481), 453–458. doi:10.1126/science.7455683.

Vogelstein, E. (2010). *Moral reasons and moral sentiments*. PhD dissertation. University of Texas. doi:2152/ETD-UT-2010-05-1243.

Wallach, W., & Allen, C. (2009). *Moral machines: Teaching robots right from wrong*. New York: Oxford University Press. doi:10.1093/acprof:oso/9780195374049.001.0001.

Wallach, W., Colin A., & Iva, S. (2007). *Machine morality: Bottom-up and top-down approaches for modelling human moral faculties*. In Ethics and artificial agents. Special issue, AI & Society *22* (4): 565–582. doi:10.1007/s00146-007-0099-0.

Weatherson, B. (2003). What good are counter examples? *Philosophical Studies, 115*(1), 1–31. doi:10.1023/A:1024961917413.

Wilson, T. D. (2002). *Strangers to ourselves: Discovering the adaptive unconscious*. Cambridge: Belknap Press.

Yudkowsky, E. (2001). *Creating friendly AI 1.0: The analysis and design of benevolent goal architectures*. Singularity Institute for Artificial Intelligence, San Francisco, CA, June 15. http://singinst.org/upload/CFAI.html.

Yudkowsky, E. (2004). *Coherent extrapolated volition*. Singularity Institute for Artificial Intelligence, San Francisco, CA, May. http://singinst.org/upload/CEV.html.

Yudkowsky, E. (2008). Artificial intelligence as a positive and negative factor in global risk. In N. Bostrom & M. C. Milan (Eds.), *Global catastrophic risks* (pp. 308–345). New York: Oxford University Press.

Yudkowsky, E. (2011). *Complex value systems in friendly AI*. In T, Schmidhuber., & M, Looks (Eds.), 388–393, Berlin: Springer.

Zhong, C.-B., Strejcek, B., & Sivanathan, N. (2010). A clean self can render harsh moral judgment. *Journal of Experimental Social Psychology, 46*(5), 859–862. doi:10.1016/j.jesp.2010.04.003.

Zimmerman, D. (2003). Why Richard Brandt does not need cognitive psychotherapy, and other glad news about idealized preference theories in meta-ethics. *Journal of Value Inquiry, 37*(3), 373–394. doi:10.1023/B:INQU.0000013348.62494.55.

Chapter 6A
Jordi Vallverdú on Muehlhauser and Helm's "the Singularity and Machine Ethics"

The future will bring us, under the very plausible horizon of the singularity, a reality of "machine superoptimizers", as they are conveniently labeled by Muehlhauser and Helm, trying to avoid typical anthropomorphic biases. Until here I can agree with him, but there is not a single concept, line or idea related to ethics and, by extension, to machine ethics, that I can accept. Although the chapter includes a bird's eye view on the debates about the roots of ethical discourses it disregards the lack of evidence for the existence of universal foundations of ethics, even from a metaethical naturalist position, which could lead us to a highly controversial deterministic ethics. Generative artificial ethics, if it could be possible, as an ethical expert system working on a revisited version of the *Principia Ethica* (as an hypothetical neo-Russell and Whitehead project based on the fundaments of ethics, not the Moore's text), would easily fail because every human community has its own ethical and moral codes or rules and there is no possible agreement among them because all these universals are based in different primordial concepts, all of them beyond any rational approach. Ethics is built on the shifting sands of prejudices and opinions, never on the truth, because it is the result of a cultural decision, not a part of the deep reality. In this sense, you can achieve a *de minimis set of shared rules*, to describe a statistical approach to a bunch of common values or design a new artificial new ethics, but you never can discover nor find the 'true values' of the real ethics.

The authors also do not take into account a second and very important question for a Roboethics or an AI ethics: the importance of embodiment. Every living entity has a bodily structure that creates a specific intentionality towards the world and the rest of living entities (similar or not). The body is the place from which the world is felt and determines or constraints narrowly the range of possible performed or wished actions. So, any conscious creature, natural or artificial, is modeled and acts according to its structure. If the superoptimizers have different bodies and minds than us, they *surely* do will have a specific and own ethics. You can decide to try to feel different things about the events of the world but you cannot decide or not *to feel*; and your feelings, emotions and attitudes towards the basic aspects of the reality are constrained by these bodily arrows. If you agree with me that emotions are a basic and determinant part of any rational being, then the result of putting together minds, emotions and bodies show us an unexpected and over paradigmatic ethics for those superoptimizers. Hence, in the case of machine ethics as in many others, the future does not emerge from the present, at least not from a sequential and consistent point of view.

Chapter 7
Artificial General Intelligence and the Human Mental Model

Roman V. Yampolskiy and Joshua Fox

Abstract When the first artificial general intelligences are built, they may improve themselves to far-above-human levels. Speculations about such future entities are already affected by anthropomorphic bias, which leads to erroneous analogies with human minds. In this chapter, we apply a goal-oriented understanding of intelligence to show that humanity occupies only a tiny portion of the design space of possible minds. This space is much larger than what we are familiar with from the human example; and the mental architectures and goals of future superintelligences need not have most of the properties of human minds. A new approach to cognitive science and philosophy of mind, one not centered on the human example, is needed to help us understand the challenges which we will face when a power greater than us emerges.

R. V. Yampolskiy (✉)
Computer Engineering & Computer Science, University of Louisville, Louisville, KY, USA
e-mail: roman.yampolskiy@louisville.edu

J. Fox
Singularity Institute for Artificial Intelligence,
e-mail: joshua.fox@singinst.org

A. H. Eden et al. (eds.), *Singularity Hypotheses*, The Frontiers Collection,
DOI: 10.1007/978-3-642-32560-1_7,

Introduction

People have always projected human mental features and values onto non-human phenomena like animals, rivers, and planets, and more recently onto newer targets, including robots, and even disembodied intelligent software. Today, speculation about future artificial general intelligences[1] (AGIs), including those with super-human intelligence, is also affected by the assumption that various human mental properties will necessarily be reflected in these non-human entities. Nonetheless, it is essential to understand the possibilities for these superintelligences on their own terms, rather than as reflections of the human model. Though their origins in a human environment may give them some human mental properties, we cannot assume that any given property will be present. In this chapter, using an under-standing of intelligence based on optimization power rather than human-like features, we will survey a number of particular anthropomorphisms and argue that these should be resisted.

Anthropomorphic Bias

Because our human minds intuitively define concepts through prototypical examples (Rosch 1978), there is a tendency to over-generalize human properties to nonhuman intelligent systems: Animistic attribution of mental qualities to animals, inanimate objects, meteorological phenomena, and the like is common across human societies (See Epley et al. 2007). We may use the term "anthropomorphic bias" for this tendency to model non-human entities as having human-like minds; this is an aspect of the Mind Projection Fallacy (Jaynes 2003). Excessive reliance on any model can be misleading whenever the analogy does not capture relevant aspects of the modeled space. This is true for anthropomorphism as well, since the range of human-like minds covers only a small part of the space of possible mind designs (Yudkowsky 2006; Salamon 2009).

In artificial intelligence research, the risk of anthropomorphic bias has been recognized from the beginning. Turing, in his seminal article, already understood that conditioning a test for "thinking" on a human model would exclude "something which ought to be described as thinking but which is very different from what a man does" (Turing 1950). More recently, Yudkowsky (2008, 2011; see also Muehlhauser and Helm this volume) has warned against anthropomorphizing tendencies in

[1] The term "artificial general intelligence" here is used in the general sense of an agent, implemented by humans, which is capable of optimizing across a wide range of goals. "Strong AI" is a common synonym. "Artificial General Intelligence", capitalized, is also used as a term of art for a specific design paradigm which combines narrow AI techniques in an integrated engineered architecture; in contrast, for example, to one which is evolved or emulates the brain (Voss 2007). As discussed below, this more specific sense of AGI is also the primary focus of this article.

thinking about future superintelligences: those which surpass the human level of intelligence. To properly understand the possibilities that face us, we must consider the wide range of possible minds, including both their architecture and their goals. Expanding our model becomes all the more important when considering future AIs whose power has reached superhuman levels.

Superintelligence

If we define intelligence on the human model, then intelligences will tautologically have many human properties. We instead use definitions in which intelligence is synonymous with optimization power, "an agent's ability to achieve goals in a wide range of environments" (Legg 2008). Legg uses a mathematical model in which an agent interacts with its environment through well-defined input channel, including observation and reward, as well as an output channel. He then defines a Universal Intelligence Measure which sums the expectation of a reward function over an agent's future interactions with all possible environments. This definition is abstract and broad, encompassing all possible computable reward functions and environments.

Variations on this goal-based definition have been proposed. The Universal Intelligence Measure is so general that it does not capture the specific environments and goals likely to be encountered by a near-future AGI. To account for this, we can apply Goertzel's (2010) definition of "pragmatic efficient general intelligence", which resembles the Universal Intelligence Measure, but also takes into account the system's performance in given environments—which will likely be human-influenced—as well as the amount of resources it uses to achieve its goals.

There are cases where human-based definitions of intelligence are suitable, as when the purpose is to classify humans (Neisser et al. 1996). Likewise, human-based metrics may be applicable when the goal is to build AIs intended specifically to emulate certain human characteristics. For example, Goertzel (2009) discusses practical intelligence in the social context: Goals and environments are assigned a priori probabilities according to the ease of communicating them to a given audience, which may well include humans.

Still, if the purpose is to consider the effects on us of future superintelligences or other non-human intelligences, definitions which better capture the relevant features should be used (Chalmers 2010). In the words of Dijkstra (1984), the question of whether a machine can think is "about as relevant as the question of whether Submarines Can Swim"—the properties that count are not internal details, but rather those which have effects that matter to us.

We use the term "mind" here simply as a synonym for an optimizing agent. Although the concept "mind" has no commonly-accepted definition beyond the human example, in the common intuition, humans and perhaps some other higher-order animals have a mind. In some usages of the term, introspective capacity, a localized implementation, or embodiment may be required. In our understanding, any optimization process, including a hypothetical artificially intelligent agent above a certain threshold, would constitute a mind.

Nonetheless, the intuitions for the concepts of "mind" and "intelligence" are bound up with many human properties, while our focus is simply on agents that can impact our human future. For our purposes, then, the terms "mind" and "intelligence" may simply be read "optimizing agent" and "optimization power".

Though our discussion considers intelligence-in-general, it focuses on superhuman intelligences, those "which can far surpass all the intellectual activities of any man however, clever" (Good 1965). Superintelligences serve as the clearest illustration of our thesis that human properties are not necessary to intelligence, since they would be less affected by the constraints imposed by human level intelligence. In contrast, the limited intelligence of near-human agents may well constrain them to have certain human-like properties. As research related to control and analysis of superintelligent systems gains momentum (Yampolskiy 2011, 2011a, 2011b, 2012a, 2012b; Yampolskiy and Fox 2012) our thesis becomes essential for avoiding fundamental mistakes.

The Space of Possible Minds

Formalisms help broaden our intuitions about minds beyond our narrow experience, which knows no general intelligences but humans. In the approach mentioned earlier in association with the Universal Intelligence Measure, agents are modeled as functions which map, in repeated rounds of interaction, from an input-output history to an action (Hutter 2005; Legg 2008). As the inputs and outputs are modeled as strings from a finite alphabet, with an indefinite future horizon, there are, in principle, infinitely many such agent functions.

Using this model, there is a continuum across fundamentally different minds. For any computable agent, there are many other computable agents it cannot understand (learn to predict) at all (Legg 2006). There is thus, in principle, a class of agents who differ so strongly from the human that no human could understand them. Most agents represent trivial optimization power; our attention is focused on those which represent superhuman intelligence when implemented.

If we extend our model and allow the agent to change under the influence of the environment (Orseau and Ring 2011), we find another source of variety in alterations in the mind itself, just as a given human behaves very differently under the influence of intoxicating substances, stress, pain, sleep, or food deprivation.

We are familiar with infrahuman intelligence in non-human animals. Animals can use senses, abilities such as navigation, and some forms of cognition, in goal-seeking (Griffin 1992). Non-human biological intelligences, including some different from those we are familiar with, could also evolve in environments outside our planet. Freitas (1979) describes an intelligence which might arise with a ganglionic rather than a chordate nervous system: such creatures, with small "brains" for each body segment (like most of earth's invertebrates), would have distributed, cooperative brains with distinct awareness for each body part. Likewise, animals with different weightings for their cognates of the three parts of the human brain—reptilian midbrain, limbic system, and neocortex—would have different distributions of mental features like aggression, emotion, and reason. Though these hypothetical biological intelligences fall into a narrow range of intelligence around the human level or below, they illustrate a range of possible architectures and motivational systems.

Classifications of kinds of minds which go much farther beyond the human example have been offered by Hall (2007) and Goertzel (2006, p. 17). Hall classifies future AGIs, making the point that we should not expect AI systems to ever have closely humanlike distributions of ability, given that computers are already superhuman in some areas. So, despite its anthropocentric nature, his classification highlights the range of possibilities as well as the arbitrariness of the human intelligence as the point of reference. His classification encompasses hypohuman (infrahuman, less-than-human capacity), diahuman (human level capacities in some areas, but still not a general intelligence), parahuman (similar but not identical to humans, as for example, augmented humans), allohuman (as capable as humans, but in different areas), epihuman (slightly beyond the human level), and hyperhuman (much more powerful than human). Goertzel classifies a broader range of minds, contrasting the human to possible non-human mental architectures, and describing AGI architectures which would implement many of these possibilities.

Singly-embodied minds control and receive input from a single spatially-constrained physical or simulated system; multiply- and flexibly-embodied minds, respectively, have a multiple or changing number of such embodiments. Non-embodied minds are those which are implemented in a physical substrate but do not control or receive input from a spatially-constrained body. Humans, of course, are singly-embodied.

Humans are not only embodied but also body-centered. The human brain is connected to and can be directly influenced by the remainder of the body, along with its immediate environment, so that the mind as a whole consists of patterns emergent between the physical system (the brain and body) and the environment. Non-embodied and non-body-centered minds are possible, and even within the narrower constraints of embodiment, variations in the sensors and manipulators under control of a particular mind design present even more variety in mental capabilities.

Goertzel also explores possibilities for mind-to-mind linkage. Human minds work in near-isolation, connected mostly by the slow and lossy thought-serialization of language. But there are other possibilities. One is a mindplex, a set of collaborating units each of which is itself a mind. Human organizations and nations are mindplexes, albeit in imperfect form because of limitations in our communication; but a more tightly integrated mindplex would constitute a very different kind of general intelligence.

Within this variety of possible minds, superintelligence should not be considered a specialized variant on human level of intelligence. Rather, human level intelligence should be considered an unstable equilibrium which can rapidly shift into superhuman ranges (see Muehlhauser and Salamon this volume). Humans find it difficult to improve their own brain power, but an AGI would find it much easier, since it would have capabilities such as adding more hardware or examining its own source code for possible optimizations. Moreover, most AGIs would want to self-improve to the highest possible level of intelligence, as this has value in achieving most goals (Omohundro 2008).

Humans are the first general intelligence on earth. We have been in existence for a short time in evolutionary terms and represent a lower bound on the intelligence able to build a civilization. The upper limit on raw processing power for the entire universe through its history, as imposed by the laws of physics, 10^{120} operations over 10^{90} bits; a one-liter, one-kilogram computer has the upper limit of 10^{50} operations per second (Lloyd 2000, 2002). This theoretical maximum is almost certainly too generous—it assumes an exploding computer. But even with tighter constraints, such as speed of electrical and optical signals in feasible technology, or the Landauer (1961) limits on the minimal energy required for any irreversible computation, upper bounds remain far above the human level, estimated at 10^{11} operations per second (Moravec 1998). Even though functional ability requires more than just raw power, the gap between the human level and the highest degree of optimization power possible leaves open a wide range, encompassing a vast range of possible superhuman intelligence levels (Sotala 2010).

Architectural Properties of the Human Mind

The tight entwinement of functionality and goals in the human brain is a contingent fact which depends on our evolutionary history. But in general, a single architecture may serve various goals, while multiple architectures may be capable of serving a given goal system. Thus, mental architecture and goal systems must be examined separately.

The human mental architecture is quite uniform, the so-called "psychic unity of mankind". This is a result of humanity's origin in evolution through sexual reproduction, which works only when genomes remain similar across the species. This results in homogeneity in human minds, both in hardware—the brain—and software—the functional mind design (Tooby and Cosmides 1992, p. 38). All human minds share specialized features and behaviors, including myths, grammar, ethnocentrism, play, and empathy, and many others (Brown 1991, 2004). Other animals, which share a biological substrate and the goals of reproductive fitness with humans, also share certain human mental features, as for example, specialized abilities to track degrees of genetic relatedness. But non-biological optimizers, which are not faced by these constraints, need not have the same motivations and accompanying mental techniques.

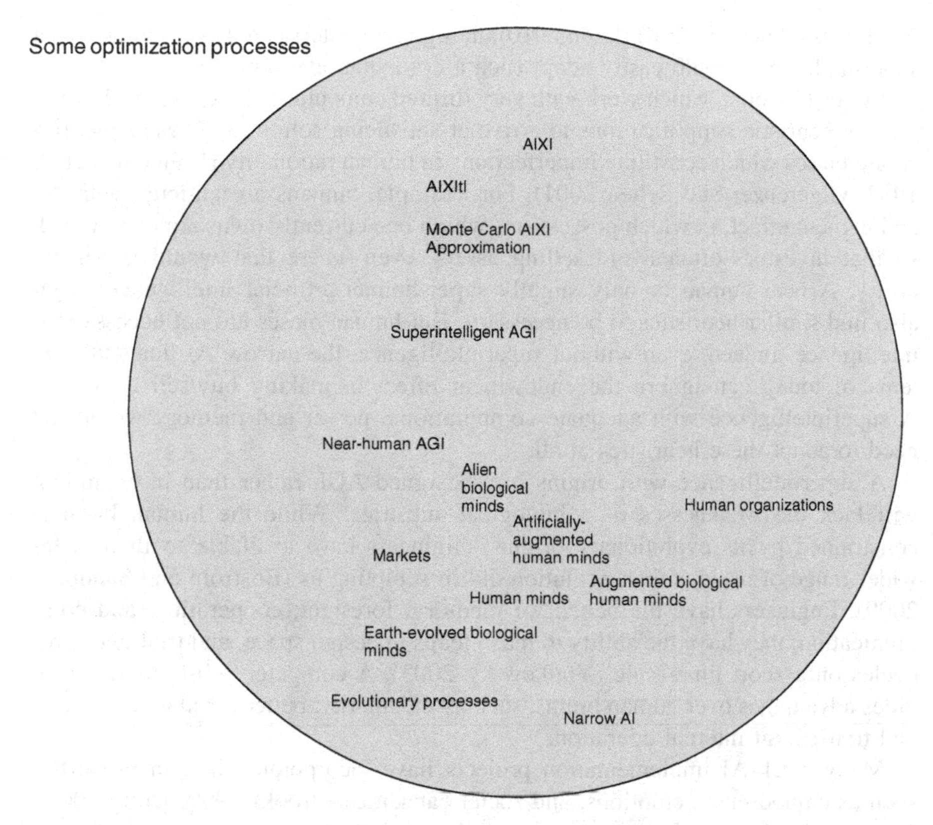

Fig. 7.1 This diagram, based on (Yudkowsky 2006), presents the optimization processes surveyed here, including existing optimizers, hypothetical ones, and formalisms. (The diagram is not to scale and shows only an unrepresentative sample of the possibilities.) It is intended to illustrate that human mind design constitutes a tiny part of a vast space of possible minds, most of which have deeply non-human-like goals and architectures

For humans, the perceptions of space and time, and the ability to act on their environment, are centered on a body. Embodiment-based cognition is so essential to human minds that it extends even to aspects of cognition which do not directly depend on embodiment (Lakoff 1987): For example, one "wades through" a difficult book. An AGI must likewise be implemented in some physical form (which must be protected if the agent is to continue working towards its goals). It also must interact with its environment in space and time if its goals are based on the state of the environment. Thus, perceptions and action are also essential to a superintelligence. Body-centeredness is not necessary, however, since a computer substrate allows the distribution and relocation of mental capacities, perceptions and motor control (Fig. 7.1).

Human minds are characterized by some weaknesses. Even ordinary computers surpass us in symbol processing and logical inference. Humans, for example, are typically unable to trace nested (non-tail) syntactic recursion to more than about two levels (Reich 1972; Karlsson 2010), though computers can do this with ease; a superintelligence could easily adopt such a computational module.

Minds like ours, which work with very limited computational resources, have to rely on heuristic simplifications to arrive at satisficing solutions. These heuristics create biases which constitute imperfections in human rationality (Kahneman et al. 1982; Gigerenzer and Selten 2001). For example, humans are afflicted with the endowment effect, in which possessions which one currently owns are overvalued, so that investors often avoid selling assets, even where that would maximize utility. A near-human or only slightly super-human artificial intelligence might also find similar heuristics to be necessary. But human biases are not necessary to intelligence. Indeed, even without superintelligence, the narrow AI financial systems of today can ignore the endowment effect in making buy/sell decisions. A superintelligence with adequate computational power and memory would not need to adopt these heuristics at all.

A superintelligence with origins in a designed AGI, rather than in evolution, will lack the weaknesses of a biological substrate. While the human brain is constrained by its evolutionary origins, engineers have available to them a far wider range of designs than evolution did in sculpting us (Bostrom and Sandberg 2009). Engineers have the benefit of memory, foresight, cooperation, and communication; they have the ability to make leaps in design space, and trial-and-error cycles on a short time-scale (Yudkowsky 2003). A computer substrate also provides advantages over human brains, such as modularity, better serial computation, and transparent internal operation.

Many weak-AI implementation projects have incorporated human properties such as embodiment, emotions, and social capacities (Brooks 1999; Duffy 2003). Some projects go so far as to intentionally copy limitations observed in human psychology, in order to avoid wasting effort on potentially unfeasible tasks, while still achieving human-comparable performance (Strannegård 2007). But implementations of isolated mental features give us no reason to assume that a full AGI would necessarily have a wide range of human-like properties.

Already today, many forms of narrow AI and proposed designs for strong AI have non-anthropomorphic, computer-based architectures. For example, the design paradigm called Artificial General Intelligence adopts narrow AI components but takes only a broad inspiration from the human mind (e.g., Goertzel et al. 2009). The variability of the architecture will be all the greater after self-improvement, since an AGI need not keep its current architecture as it self-improves. It will create a new design, even a radically non-anthropomorphic one, if an entity with this new design will be better able to achieve its goals.

There is a category of greater-than-human minds which *would* have human-like mental properties: those derived from humans. These include brains augmented with nootropic drugs, genetic engineering, or brain-computer interfaces (Vidal 1973; Graimann et al. 2010), uploads of specific persons (Hanson 1994), and whole brain emulations (Sandberg and Bostrom 2008). On functionalist principles, these are all rooted in and not essentially different in their origins from ordinary biological minds, and inherit their properties, even if subsequently they use their greater-than-human intelligence to bootstrap to much higher levels. Thus, these human-origin superintelligences are likely to present examples of human properties. Still, fundamental differences from human architecture can be expected. Uploads, human/machine cyborgs, and whole brain emulations, with their non-biological computing substrate gain advantages over biological brains in areas such as self-improvement, communication, and others (Sotala 2012, in press). As they improve to superintelligence, the human-origin minds could leave behind human mental limitations, and reimplement themselves in an even better architecture if they so choose.

Human Goals

The human goal system, which includes survival, social status, and morality, along with many others, is a mix of adaptations to conditions in the human ancestral environment (Tooby and Cosmides 1992; Greene 2002). In contrast, an AGI, and in particular a superintelligence, can have arbitrary goals, whether these are defined by its designers or develop in a random or chaotic process.

Human terminal values arose from their instrumental value in achieving evolution's implicit goal of reproductive success for the genes. For AGIs as well, such human-like preferences would have instrumental value for the achievement of many goals. For example, most agents, including superintelligences, would be motivated, as humans are, to protect themselves, to acquire resources, and to use them efficiently.

But there are also instrumental values which could be of far more use to machine intelligences than to humans. Humans can take nootropic drugs to enhance their minds; they can avoid addictive or psychoactive drugs to avoid distorting their utility function. But an agent which can more fully examine and improve its own design, implementation, and embodiment will find much more value to radical self-improvement, preservation of utility functions, and prevention of counterfeit utility (Omohundro 2008).

An intelligence far more powerful than humans would have no need for the values of exchange, cooperation, and altruism in interaction with humans, unless these were built- in as terminal values. A superintelligence would not need any benefits that humans could offer in exchange for its good behavior; it could evade monitoring and resist punishment. Since humans cannot meaningfully help or harm the superintelligence, there would be little value in cooperation, verifiably trustworthy dispositions, or benevolence (Fox and Shulman 2010), nor in money (control of resources on a human scale), social power, or even malevolence (disempowerment or elimination of rivals as a goal).

In an environment with peers, a superintelligence would have incentive to cooperate or compete instrumentally (Hall 2007). Yet this is only true where superintelligences of roughly equal ability exist. Across the wide continuum of possible levels of intelligence, agents which are not of the same species—the same software/hardware specification—are more likely to be mismatched than to be equals. A Darwinian scenario, in which a population of superintelligences cooperates and competes, could produce rough equals, perhaps distributed across niches. But if a superintelligence self-improves fast enough, it will be aware of the evolutionary threat and suppress the rise of other intelligences (Bostrom 2006).

Human goals are mostly self-centered, with altruism a debatable exception (Batson 2010). In contrast, there is no a priori reason for AGIs to treat their own control of resources or their own continued existence as terminal values (though they may be useful as instrumental values). Since future AGIs will have goals designed to serve human preferences, they may well possess a quite inhuman altruism.

If today's plans for AGIs are any guide, the first ones are likely to be assigned simple goal systems, as contrasted to the human multiplicity of mutually inconsistent, changeable goals, with intertwined instrumental and terminal values. (However, complex goal systems are possible in AGIs, particularly if their creators are specifically trying to copy the human goal system.)

AGI goals originating in human needs (for example, maximizing wealth or winning a war for its makers), are no guarantee of human-like behavior, even if these goals are well-defined. A drive to maximize these goals, even at the expense of all other values important to humans, would result in deeply alien behavior (Bostrom 2003; Yudkowsky 2011).

Humans sometimes change their values, as in a process of Kantian reflection, in which a person decides that moral reciprocity is not merely a means to an end, but also an end in itself. However, any sufficiently powerful superintelligence would not change its values, since doing so impairs the chances of achievement of the current values, and so represents a limitation to optimization power. Thus, a very powerful optimizer would strive to prevent such human-like preference evolution (Omohundro 2008).[2]

Superintelligences originating in humans, such as augmented brains, cyborgs, uploads, and whole brain emulations, would start with human values. As they gain in power, they would lose the social constraints which form an important motivation for human behavior: Power corrupts, and power far beyond what any tyrant has known to date may corrupt a human-like mind so that its motivational system becomes very different from that of today's humans. As human-based minds self-improve, they are likely to seek to protect their goal systems, like other powerful optimizers; this could produce an example of a superintelligence with human-like goals. Even these superintelligences, however, may ultimately evolve goals which

[2] Change of goals is possible in a superintelligence where a stable metagoal is the true motivator. For example, discovery and refinement of goals is part of Coherent Extrapolated Volition, a goal system for a self-improving AGI. It is designed, to ultimately converge on the terminal value of helping humans achieve their goal system as extrapolated towards reflective equilibrium (Yudkowsky 2004; Tarleton 2010; Dewey 2011). Nonetheless, CEV does not violate the principle that a sufficiently powerful optimizer would lack human-like variability in its goals, since its meta-level values towards goal definition in themselves constitute a stable top-level goal system.

differ from those of humans. They would start with human-like changeability in their goal system, and the changeability could in itself be treated as a meta-value, resulting in different object-level values.

Examples of Superintelligences

We know of no superintelligences today, nor any other intelligences with the generality and flexibility of the human mind. But examples of powerful optimization processes with non-human goals and architectures are available. Some are superior to humans in certain areas of optimization.

Evolutionary processes are flexible and powerful enough to create life forms adapted for a wide variety of changing environments by optimizing for reproductive success, far outdoing the accomplishments of human engineers in this area. (It should be noted, however, that these processes have had much more time to work than all human engineers combined). Evolutionary processes share with humans only the ability to optimize; they lack all other properties associated with humans. They are unembodied, impersonal, unconscious, and non-teleological, lacking any modeling capacities. Even though the human mind evolved to serve evolutionary goals of reproductive success, humans do not share the goals of the evolutionary processes which created them (Cosmides and Tooby 1992; Yudkowsky 2003).

Markets are another type of powerful optimizer. Though externalities and other market failures render them far from optimal, they outdo centralized planning (i.e., a small group of human minds) at their implicit goal, maximizing for the net benefit of producers and consumers. Though based on the interactions of individual humans, each working towards their own goals, markets as a whole lack all properties of the human mind. Markets are embodied in the humans who participate in them, but optimize distinct values from any individual human. Like evolutionary processes, markets are impersonal, unconscious, and non-teleological; and lack internal models of the optimized domain.

Markets present a valuable example of other-directed goals: They optimize functions which are aligned with and derived from the values of other intelligences, namely humans. Such other-directedness is rare in humans and other biological intelligences. In contrast, in artificial agents there is no bar to pure altruism; in fact, since they would be created to serve their designers' goals, other-directed values are the default.

These examples are useful, but limited. None is a true superintelligence. Only humans today have flexible, general, intelligence, leaving theoretical models of superintelligence such as AIXI as a useful tool in considering the full range of possibilities. AIXI (Hutter 2005) is an abstract and non-anthropomorphic formalism for general and flexible superintelligence. It combines Solomonoff induction (Li and Vitányi 1993, pp. 282–290) and expectimax calculations to optimize for any computable reward function. It is provably superior at doing so, within a constant factor, than any other intelligence (Hutter 2005).

There are limitations on the usefulness of AIXI as an example. As it is incomputable, it must be treated as a model for intelligence, not as a design for an AGI.

Also, AIXI, and Legg's Universal Intelligence Measure which it optimizes, is incapable of taking the agent itself into account. AIXI does not "model itself" to figure out what actions it will take in the future; implicit in its definition is the assumption that it will continue, up until its horizon, to choose actions that maximize expected future value. AIXI's definition assumes that the maximizing action will always be chosen, despite the fact that the agent's implementation was predictably destroyed. This is not accurate for real-world implementations which may malfunction, be destroyed, self-modify, etc. (Daniel Dewey, personal communication, Aug. 22, 2011; see also Dewey 2011). AIXI's optimization is for external rewards only, with no term for the state of the agent itself, it does not apply to systems that have preferences about more than the reward, for example, preferences concerning the world as such, or preferences about their own state; nor does it apply to mortal systems (Orseau and Ring 2011). Nonetheless, AIXI does a good job of representing the best possible optimizer in the sense of finding ever closer approximations to the global maxima in a large search space of achievable world-states. Taken as an abstract model, AIXI's complete and compact specification serves to show that in the limiting case, almost any property in an intelligence, beyond optimization power itself, is unnecessary.

AIXI is quite inhuman. It is completely universal, maximally intelligent under a universal probability distribution—i.e., where the environment is not prespecified. It thus lacks the inductive bias favored by humans. It lacks human qualities of embodiment: It has no physical existence beyond the input and output channels. Also, unlike humans, this formalism has no built-in values; it works to maximize an external reward function which it must learn from observation.

Variants of AIXI bring this model, with its compact specification and freedom from built-in inductive bias, into the realm of computability and even implementation. AIXItl (Hutter 2005) is computable, though intractable, and is provably superior within a constant factor to any other intelligence with given time and length limits. A tractable approximation, Monte Carlo AIXI, has been implemented and tested on several basic problems (Veness et al. 2011).

A Copernican Revolution in Cognitive Science

We have explored a variety of human mental properties, including single embodiment, body-centeredness, certain strengths and weaknesses, and a specific complex set of goals. Superintelligences need have none of these features. Though some instrumental goals will be valuable for most intelligent agents, only the definitional property of much-higher-than-human optimization power will necessarily be present in a superintelligence. Humans are the only good example of general intelligence which we know—but not the only one possible, particularly when the constraints of our design are thrown aside in a superintelligence.

Since the Copernican revolution, science has repeatedly shown that humanity is not central. Our planet is not the center of the universe; *Homo sapiens* is just another animal species; we, like other life-forms, are composed of nothing but ordinary matter. Recently, multiverse theories have suggested that everything we observe is a tiny part of a much larger ensemble (Tegmark 2004; Greene 2011).

This decentralizing trend has not yet reached the philosophy of the human mind. Much of today's scholarship takes the universality of the properties of the human mind as granted, and fails to consider in depth the full range of possible architectures and values for other general optimizers, including optimizers much more powerful than humans. It is time for psychology and the philosophy of mind to embrace universal cognitive diversity. Even in today's era, in which only a single design for general intelligence exists, this broadening will enrich our analytic tools for understanding mental architecture, decision processes, goals, and morality.

A Copernican revolution for the mind can extend our view outwards, but also improve our insight into ourselves. The shift away from geocentric cosmology improved our understanding of the Earth, and an evolutionary analysis of our species' rise helped us understand the design of humans. So too, an examination of other possible minds, and in particular superintelligent minds, can help us reach philosophical and psychological conclusions about humans as well. Already, infrahuman AIs have provided paradigms for philosophy of mind (e.g., Newell and Simon 1972); AI-related research such as Bayesian network theory (Tenenbaum et al. 2006) has contributed to neuroscience. Though at the current stage, only thought experiments are possible, theories about possible superintelligences can shed more light on the human condition.

Today's astronomers know that Earth is still central in one sense: We live on it. Our observations are made from it or near it; our tentative explorations of space began on it; its fate is tied up with our own. So too, when we humans begin exploring mind-space with the creation of AGIs, the human mind will remain of central importance: The first near-human-level AGIs may be partially modeled on our mind's architecture and will have goals chosen to serve us. But just as astronomers came to learn that the universe has unimaginably large voids, stars much greater than our sun, and astronomical bodies stranger than anything previously known, so too we will soon encounter new intelligences much more powerful than us and very different from us in mental architecture and goals.

There are two meanings to Copernicanism. One is "we are not central", and the other is "we are ordinary; what we see is common". This second meaning, too, should influence our thinking on intelligence. Although the human mind's special status as the only true general intelligence remains a reality for now, in principle other general intelligences can exist. Once other human level intelligences, and then superintelligences, are created, our theory of mind will have to expand to include them; we should start now, arming ourselves with an understanding which may enable us to design them to meet our needs. Defining the initial AGIs' goals in accordance with human values, and guaranteeing the preservation of the goals under recursive self-improvement, will be essential if our human values are to be preserved (Yudkowsky 2008; Anissimov 2011).

Acknowledgments Thanks to Carl Shulman, Anna Salamon, Brian Rabkin, Luke Muehlhauser, and Daniel Dewey for their valuable comments.

References

Anissimov, M. (2011). Anthropomorphism and moral realism in advanced artificial intelligence. *Paper presented at the Society for Philosophy and Technology conference*, Denton.

Batson, C. D. (2010). *Altruism in humans*. Oxford: Oxford University Press.

Bostrom, N. (2003). Ethical issues in advanced artificial intelligence. In I. Smit, G. Lasker, & W. Wallach (Eds.), *Cognitive, emotive and ethical aspects of decision making in humans and in artificial intelligence* (Vol. 2, pp. 12–17). Windsor: International Institute of Advanced Studies in Systems Research and Cybernetics.

Bostrom, N. (2006). What is a singleton? *Linguistic and Philosophical Investigations, 5*(2), 48–54.

Bostrom, N., & Sandberg, A. (2009). The wisdom of nature: An evolutionary heuristic for human enhancement. In J. Savulescu & N. Bostrom (Eds.), *Human enhancement* (pp. 375–416). Oxford: Oxford University Press.

Brooks, R. A. (1999). *Cambrian intelligence: The early history of the new AI*. Cambridge: MIT Press.

Brown, D. E. (1991). *Human universals*. New York: McGraw Hill.

Brown, D. E. (2004). Human universals, human nature and human culture. *Daedalus, 133*(4), 47–54.

Chalmers, D. J. (2010). The singularity: A philosophical analysis. *Journal of Consciousness Studies, 17*, 7–65.

Cosmides, L., & Tooby, J. (1992). Cognitive adaptations for social exchange. In J. Barkow, J. Tooby, & L. Cosmides (Eds.), *The adapted mind: Evolutionary psychology and the generation of culture* (pp. 163–228). Oxford: Oxford University Press.

Dewey, D. (2011). Learning what to value. In J. Schmidhuber, K.R. Thórisson & M. Looks (Eds.), *Artificial General Intelligence*: *Proceedings of 4th International Conference, AGI* 2011, *Mountain View, CA, USA*, 3–6 *August* 2011, pp. 309–314. Berlin: Springer.

Dijkstra, E.W. (1984). The threats to computing science. *Paper presented at the ACM* 1984 *South Central Regional Conference*, 16–18 Nov, Austin.

Duffy, B. R. (2003). Anthropomorphism and the social robot. *Robotics and Autonomous Systems, 42*(3–4), 177–190.

Epley, N., Waytz, A., & Cacioppo, J. T. (2007). On seeing human: A three-factor theory of anthropomorphism. *Psychological Review, 114*(4), 864–888.

Fox, J., & Shulman, C. (2010). Superintelligence does not imply benevolence. In K. Mainzer (Ed.), *Proceedings of the VIII European Conference on Computing and Philosophy* (pp. 456–461). Munich: Verlag Dr. Hut.

Freitas, R. A, Jr. (1979). *Xenology: An introduction to the scientific study of extraterrestrial life, intelligence, and civilization* (1st ed.). Sacramento: Xenology Research Institute.

Gigerenzer, G., & Selten, R. (Eds.). (2001). *Bounded rationality: The adaptive toolbox*. Cambridge: MIT Press.

Goertzel, B. (2006). *The hidden pattern: A patternist philosophy of mind*. Boca Raton: Brown Walker Press.

Goertzel, B. (2009). *The embodied communication prior: A characterization of general intelligence in the context of embodied social interaction*. Paper presented at the 8th IEEE International Conference on Cognitive Informatics IEEE, Hong Kong.
Goertzel, B. (2010). Toward a formal characterization of real-world general intelligence. In E. Baum, M. Hutter & E. Kitzelmann (Eds.), *Proceedings of the Third Conference on Artificial General Intelligence, AGI 2010, Lugano, Switzerland, 5–8 March*, 2010. Amsterdam: Atlantis.
Goertzel, B., Iklé, M., Goertzel, I. F., & Heljakka, A. (2009). *Probabilistic logic networks: A comprehensive framework for uncertain inference*. Berlin: Springer.
Good, I. J. (1965). Speculations concerning the first ultraintelligent machine. *Advances in Computers, 6*, 31–88.
Graimann, B., Allison, B., & Pfurtscheller, G. (Eds.). (2010). *Brain-computer interfaces: Revolutionizing human-computer interaction*. Berlin: Springer.
Greene, J.D. (2002). The terrible, horrible, no good, very bad truth about morality and what to do about it. *Ph.D. Dissertation*, Princeton University, Princeton.
Greene, B. (2011). *The hidden reality: Parallel universes and the deep laws of the cosmos*. New York: Knopf.
Griffin, D. R. (1992). *Animal minds*. Chicago: University of Chicago Press.
Hall, J. S. (2007). *Beyond AI: Creating the conscience of the machine*. Amherst: Prometheus.
Hanson, R. (1994). If uploads come first: The crack of a future dawn. *Extropy, 6*(2), 10–15.
Hutter, M. (2005). *Universal artificial intelligence: Sequential decisions based on algorithmic probability*. Berlin: Springer.
Jaynes, E. T. (2003). *Probability theory: The logic of science* (Vol. 1). Cambridge: Cambridge University Press.
Kahneman, D., Slovic, P., & Tversky, A. (Eds.). (1982). *Judgment under uncertainty: Heuristics and biases*. Cambridge: Cambridge University Press.
Karlsson, F. (2010). Syntactic recursion and iteration. In H.v.d. Hulst (Ed.), *Recursion and human language* (pp. 43–67). Berlin: Mouton de Gruyter.
Lakoff, G. (1987). *Women, fire and dangerous things: What categories reveal about the mind*. Chicago: University of Chicago Press.
Landauer, R. (1961). Irreversibility and heat generation in the computing process. *IBM Journal of Research and Development, 5*(3), 183–191.
Legg, S. (2006). Is there an elegant universal theory of prediction? *Technical Report No. IDSIA-12-06*. Manno, Switzerland.
Legg, S. (2008). *Machine super intelligence*. Ph.D. Thesis, University of Lugano, Lugano, Switzerland.
Li, M., & Vitányi, P. (1993). *An introduction to Kolmogorov complexity and its applications*. Berlin: Springer.
Lloyd, S. (2000). Ultimate physical limits to computation. *Nature, 406*, 1047–1054.
Lloyd, S. (2002). Computational capacity of the Universe. *Physical Review Letters, 88*(23), 237901.
Moravec, H. (1998). When will computer hardware match the human brain? *Journal of Evolution and Technology, 1*(1), .
Neisser, U., Boodoo, G., Bouchard, T. J, Jr, Boykin, A. W., Brody, N., Ceci, S. J., et al. (1996). Intelligence: Knowns and unknowns. *American Psychologist, 51*(2), 77–101.
Newell, A., & Simon, H. A. (1972). *Human problem solving*. Englewood Cliffs: Prentice-Hall.
Omohundro, S. M. (2008). The basic AI drives. In P. Wang, B. Goertzel, & S. Franklin (Eds.), *The proceedings of the first AGI conference* (pp. 483–492). Amsterdam: IOS Press.
Orseau, L., & Ring, M. (2011). Self-modification and mortality in artificial agents. In J. Schmidhuber, K. Thórisson & M. Looks (Eds.), *Artificial General Intelligence: Proceedings of 4th International Conference, AGI 2011*, Mountain View, CA, USA, 3–6 August 2011 (pp. 1–10). Berlin: Springer.
Reich, P. A. (1972). The finiteness of natural language. In F. Householder (Ed.), *Syntactic theory 1: Structuralist* (pp. 238–272). Harmondsworth: Penguin.

Rosch, E. (1978). Principles of categorization. In E. Rosch & B. B. Lloyd (Eds.), *Cognition and categorization* (pp. 27–48). Hillsdale: Lawrence Erlbaum Associates.
Salamon, A. (2009). Shaping the intelligence explosion. Paper presented at the Singularity Summit. http://vimeo.com/7318055.
Sandberg, A., & Bostrom, N. (2008). Whole brain emulation: A roadmap, Future of humanity institute, Oxford University. Technical Report #2008-3.
Sotala, K. (2010). From mostly harmless to civilization-threatening: pathways to dangerous artificial general intelligences. In K. Mainzer (Ed.), *Proceedings of the VIII European Conference on Computing and Philosophy*. Munich: Verlag Dr. Hut.
Sotala, K. (2012, in press). Relative advantages of uploads, artificial general intelligences, and other digital minds. *International Journal of Machine Consciousness, 4*.
Strannegård, C. (2007). Anthropomorphic artificial intelligence. *Filosofiska Meddelanden, Web Series, 33*. http://www.phil.gu.se/posters/festskrift2/mnemo_strannegard.pdf.
Tarleton, N. (2010). Coherent extrapolated volition: A meta-level approach to machine ethics, from http://singinst.org/upload/coherent-extrapolated-volition.pdf.
Tegmark, M. (2004). Parallel universes. In J. D. Barrow, P. C. W. Davies, & C. L. Harper (Eds.), *Science and ultimate reality: Quantum theory, cosmology, and complexity* (pp. 452–491). Cambridge: Cambridge University Press.
Tenenbaum, J., Griffiths, T.L., & Kemp, C. (2006). Theory-based Bayesian models of inductive learning and reasoning. *Trends in Cognitive Sciences (Special issue: Probabilistic models of cognition), 10*(7), 309–318.
Tooby, J., & Cosmides, L. (1992). The psychological foundations of culture. In J. Barkow, J. Tooby, & L. Cosmides (Eds.), *The adapted mind: Evolutionary psychology and the generation of culture* (pp. 19–136). Oxford: Oxford University Press.
Turing, A. M. (1950). Computing machinery and intelligence. *Mind, 59*(236), 433–460.
Veness, J., Ng, K. S., Hutter, M., Uther, W., & Silver, D. (2011). A Monte Carlo AIXI approximation. *Journal of Artificial Intelligence Research, 40*, 95–142.
Vidal, J. J. (1973). Toward direct brain-computer communication. *Annual Review of Biophysics and Bioengineering, 2*, 157–180.
Voss, P. (2007). Essentials of general intelligence: The direct path to artificial general intelligence. In B. Goertzel & C. Pennachin (Eds.), *Artificial general intelligence* (pp. 131–158). Berlin: Springer.
Muehlhauser, L., & Helm, L. (2013). The Singularity and machine ethics. In A. Eden, J. Moor, J. Soraker & E. Steinhart (Eds.), *The singularity hypothesis*. Berlin: Springer.
Muehlhauser, L., & Salamon, A. (2013). Intelligence explosion: Evidence and import. In A. Eden, J. Moor, J. Soraker & E. Steinhart (Eds.), *The singularity hypothesis*. Berlin: Springer.
Yampolskiy, RV. (2011). AI-Complete CAPTCHAs as Zero Knowledge Proofs of Access to an Artificially Intelligent System. *ISRN Artificial Intelligence*, 2012-271878.
Yampolskiy, RV. (2011a). *Artificial Intelligence Safety Engineering: Why Machine Ethics is a Wrong Approach*. Paper presented at the Philosophy and Theory of Artificial Intelligence (PT-AI2011), Thessaloniki, Greece.
Yampolskiy, RV. (2011b). *What to Do with the Singularity Paradox*? Paper presented at the Philosophy and Theory of Artificial Intelligence (PT-AI2011), Thessaloniki, Greece.
Yampolskiy, R. V. (2012a). Leakproofing singularity—artificial intelligence confinement problem. *Journal of Consciousness Studies, 19*(1–2), 194–214.
Yampolskiy, RV. (2012b). Turing Test as a Defining Feature of AI-Completeness. Artificial Intelligence, Evolutionary Computation and Metaheuristics—In the footsteps of Alan Turing. Xin-She Yang (Ed.) (In Press): Springer.
Yampolskiy, RV, and Fox, J. (2012). Safety engineering for artificial general intelligence. Topoi. Special issue on machine ethics & the Ethics of Building Intelligent Machines.
Yudkowsky, E. (2003). *Foundations of order*. Paper presented at the Foresight Senior Associates Gathering. http://singinst.org/upload/foresight.pdf.
Yudkowsky, E. (2004). Coherent extrapolated volition, from http://singinst.org/upload/CEV.html
.

Yudkowsky, E. (2006). *The human importance of the intelligence explosion*. Paper presented at the Singularity Summit, Stanford University.

Yudkowsky, E. (2008). Artificial intelligence as a positive and negative factor in global risk. In N. Bostrom & M. M. Ćirković (Eds.), *Global catastrophic risks* (pp. 308–345). Oxford: Oxford University Press.

Yudkowsky, E. (2011). Complex value systems in friendly AI. In J. Schmidhuber, K. Thórisson & M. Looks (Eds.), *Proceedings of the 4th Annual Conference on Artificial General Intelligence*, Mountain View, CA, USA, August 2011 (pp. 388–393). Berlin: Springer.

Chapter 8
Some Economic Incentives Facing a Business that Might Bring About a Technological Singularity

James D. Miller

Introduction

A business that created an artificial general intelligence (AGI) could earn trillions for its investors, but might also bring about a "technological Singularity" that destroys the value of money. Such a business would face a unique set of economic incentives that would likely push it to behave in a socially sub-optimal way by, for example, deliberately making its software incompatible with a friendly AGI framework. Furthermore, all else being equal, the firm would probably have an easier time raising funds if failure to create a profitable AGI resulted in the destruction of mankind rather than the mere bankruptcy of the firm. Competition from other AGI-seeking firms would likely cause each firm to accept a greater chance of bringing about a Singularity than it would without competition, even if the firm believes that any possible Singularity would be dystopian.

In writing this chapter I didn't seek to identify worst-case scenarios. Rather, I sought to use basic microeconomic thinking to make a few predictions about how a firm might behave if it could bring about a technological Singularity. Unfortunately, many of these predictions are horrific.

The Chapter's General Framework

This chapter explores several scenarios in which perverse incentives can cause actors to make socially suboptimal decisions. In most of these scenarios a firm must follow one of two possible research and development paths. The chapter also

J. D. Miller (✉)
Economics, Smith College, Northampton, MA, USA
e-mail: EconomicProf@gmail.com

A. H. Eden et al. (eds.), *Singularity Hypotheses*, The Frontiers Collection,
DOI: 10.1007/978-3-642-32560-1_8,

makes the simplifying assumption that a firm's attempt to build an AGI will result in one of three possible outcomes:

- *Unsuccessful*—The firm doesn't succeed in creating an AGI. The firm's owners and investors are made worse off because of their involvement with the firm.
- *Riches*—The firm succeeds in creating an AGI. This AGI performs extremely profitable tasks, possibly including operating robots which care for the elderly; outperforming hedge funds at predicting market trends; writing software; developing pharmaceuticals; and replacing professors in the classroom. Although an AGI which brings about outcome *riches* might completely remake society, by assumption money still has value in outcome *riches*.
- *Foom*—the AGI experiences an intelligence explosion that ends up destroying the value of money.[1] This destruction assumption powers most of my results. Here is how outcome *foom* might arise: An AGI with human-level intelligence is somehow created. But this AGI has the ability to modify its own software. The AGI initially figures out ways to improve its intelligence to give itself slightly better hardware. After the AGI has made itself a bit smarter, it becomes even better at improving its own intelligence. Eventually, through recursive self-improvement, the AGI experiences an intelligence explosion, possibly making it as superior to humans in intelligence as we are to ants.

The Singularity gives me an opportunity to play with an assumption that would normally seem crazy to economists: that a single firm might obliterate the value of all past investments. My property-destruction assumption is reasonable because if any of the following conditions—all of which (especially the first) are plausible side effects of a *foom*—hold, you will not be better off because of your pre-*foom* investments:

- Mankind has been exterminated;
- scarcity has been eliminated;
- the new powers that be redistribute wealth independent of pre-existing property rights;
- everyone becomes so rich that any wealth they accumulated in the past is trivial today;
- all sentient beings are merged into a single consciousness;
- the world becomes so weird that money no longer has value, e.g. we all become inert mathematical abstractions.

For investments made in the past to have value today, there must exist certain kinds of economic and political links between the past and present. Anything that breaks these necessary connections annihilates the private value of past investments. As a Singularity would create massive change, it has a significant chance of dissolving the links necessary to preserving property rights.

[1] The term *foom* comes from AGI theorist Eliezer Yudkowsky. See Yudkowsky (2011).

Small, Self-Interested Investors

This chapter extensively discusses the decisions of small, self-interested investors. A single small investor can't affect what happens in any of the three outcomes, nor influence the probability of any of the outcomes occurring. Because of how financial markets operate, small investors have zero (rather than just a tiny) effect on the share price of companies. If the fundamentals of a company dictate that its stock is worth $20, then if you buy the stock its price might go slightly above $20 for a tiny amount of time. But as the stock would still, fundamentally, be worth $20 your actions would cause someone else to sell the stock, pushing the price per share back to $20. For large companies, such as IBM, almost all investors are small, because none of them can permanently change the stock price. Billionaire Bill Gates owning $100 million of IBM stock would still qualify him as a small investor in IBM.

Even though a small investor acting on his own can't seriously affect a firm, anything which makes the company more or less attractive to most small investors will impact a company's stock price and its ability to raise new capital.

I assume that investors are self-interested and care only about how their decisions will affect themselves. Since small investors can't individually impact what happens to a company, assuming small investors are self-interested is probably an unnecessary assumption. But I make the assumption to exclude the possibility that a huge percentage of investors will make their investment decisions based on moral considerations of what would happen if their actions determined how others invested. This self-interested assumption is consistent with the normal behavior of almost all investors.

No Investment Without *Riches*

An AGI-seeking firm would have no appeal to small, self-interested investors if the firm followed a research and development path that could lead only to outcomes *unsuccessful* or *foom*. An obvious condition for a self-interested individual to invest in a firm is that making the investment should sometimes cause the individual to become better off. If an AGI-seeking firm ended up being *unsuccessful*, then its investors would be made worse off. If the firm achieved outcome *foom*, then although a small investor might have been made much better or worse off because of the firm's activities, his investment in the firm cannot have been a cause in the change in his welfare because, by assumption, a single small investor can't affect the probability of a *foom* or what happens in a *foom*, and how you are treated post-*foom* isn't influenced by your pre-*foom* property rights.

More troubling, the type of *foom* an AGI might cause would have no effect on small investors' willingness to fund the firm. Rational investors consider only the marginal costs and benefits of making an investment. If an individual's investment would have no influence over the type and probability of a *foom*, then rational

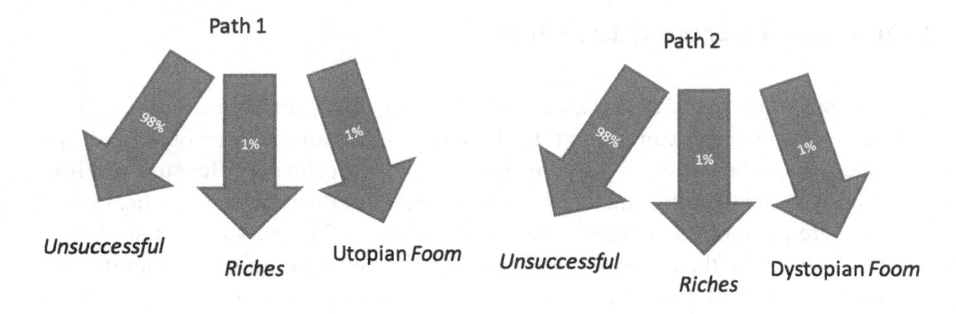

Fig. 8.1 When the type of *foom* is irrelevant to investment decisions

investors will ignore what type of *foom* a firm might bring about, even though the collective actions of all small investors affect the probability of a *foom* and the type of *foom* that might occur.

Imagine that a firm follows one of two possible research and development paths. Each path leads to a 98 % chance of *unsuccessful*, a 1 % probability of *riches*, and a 1 % chance of a *foom*. Let's further postulate that if outcomes *unsuccessful* or *riches* occurs, then the firm and its investors would be just as well off under either path. The *foom* that Path 1 would create, however, would be utopian, whereas the *foom* that Path 2 would bring would kill us all. Small, self-interested investors would be just as willing to buy stock in the firm if it followed Path 1 or Path 2. In a situation in which it's slightly cheaper to follow Path 2 than Path 1, the firm would have an easier time raising funds from small, self-interested investors if it followed Path 2. And the situation is going to get much worse (Fig. 8.1).

Some might object that my analysis places too high a burden on the assumption that an individual small investor has zero effect on stock prices and that if I slightly weakened this assumption my results wouldn't hold. These objectors might claim that since a utopian *foom* would give everyone billions of years of bliss then even a slight chance of influencing *foom* would impact the behavior of a small investor.[2] To the extent that this objection is true the results in this chapter become less important. But an investor with a typical discount rate might not significantly distinguish between, say, living for fifty years in bliss or living forever in such a state.

The actions of members of the Singularity community show that most people who believe in the possibility of a Singularity usually ignore opportunities to only slightly increase their subjective probability of a utopian *foom* occurring. Many members of this community think that the Singularity Institute for Artificial Intelligence is working effectively to increase the probability of a positive

[2] The small investors who did seek to influence the probability of a utopia *foom* would essentially be accepting Pascal's Wager.

intelligence explosion, and the more resources this organization receives the greater the chance of a utopian *foom*. Yet most people with such beliefs (including this author) spend money on goods such as vacations, candy, and video games rather than donating all the resources they use to buy these goods to the Institute. Furthermore, the vast of majority people who believe in the possibility of a utopian Singularity and think that cryonics would increase the chance of them surviving to Singularity don't sign up with a cryonics provider such as Alcor (although this author has). The revealed preferences of Singularity "believers" show that I'm not putting too high a burden on my "zero effect" assumption.

Even if, however, investors do act as if their actions impact the type and probability of a *foom*, there would still be colossal externalities to investors' decisions because people other than the investors, the firm, and the firm's customers would be impacted by the investors' decisions. Basic economic theory could easily show that the investors' decisions almost certainly won't be optimal because these investors, compared to what would be socially optimal, would invest too little in firms that might bring about a utopian *foom* and too much in businesses that could unleash a dystopian *foom*.

Deliberately Inconsistent with a Pre-existing Friendly AGI Framework

Let's now postulate that after a firm has chosen its research and development path it has some power to alter the probability of a *foom* occurring. Such flexibility could hinder a firm's ability to attract investors.

For example, let's again assume that a firm must follow one of two research paths. As before, both paths have a 98 % chance of leading to *unsuccessful*. Two percent of the time, the firm will create an AGI, and we assume the firm will then have the ability to decide whether the AGI will undergo an intelligence explosion and achieve *foom*, or not undergo an intelligence explosion and achieve *riches*. Any *foom* that occurs through Path 3 will be utopian, whereas a *foom* that results from Path 4 will result in the annihilation of mankind. Recall that a small, self-interested investor will never invest in a firm that could achieve only outcomes *unsuccessful* or *foom*. To raise capital, the firm in this example would have to promise investors that it would never pick *foom* over *riches*. This would be a massively non-credible promise for a firm that followed Path 3, because everyone—including the firm's investors —would prefer to live in a utopian *foom* then to have the firm achieve *riches*. In contrast, if the firm intended to follow Path 4, everyone would believe that the firm would prefer to achieve riches than experience a dystopian *foom* (Fig. 8.2).

So now, let's imagine that at the time the firm tries to raise capital, there exists a set of programming protocols that provides programmers with a framework for

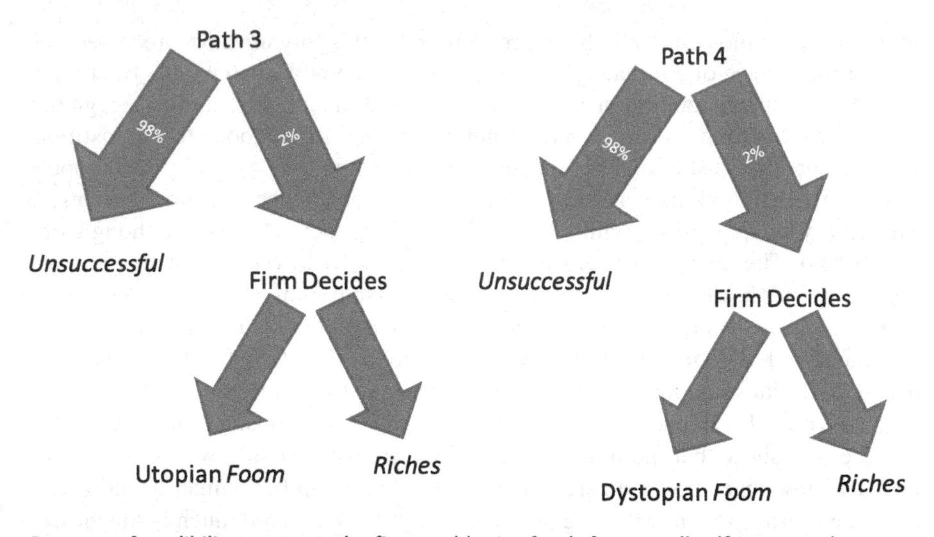

Fig. 8.2 A non-credible promise

creating friendly AGI. This framework makes it extremely likely that if the AGI goes *foom*, it will be well disposed towards humanity and create a utopia.[3]

To raise funds, an AGI-seeking firm would need to choose a research and development path that makes it difficult for the firm to use the friendly AGI framework. Unfortunately, this means that any *foom* the firm brings about unintentionally would be less likely to be utopian than if the firm had used the friendly framework.

Further Benefits From a World-Destroying *Foom*

A bad *foom* is essentially a form of pollution, meaning it's what economists call a "negative externality". Absent government intervention, investors in a business have little incentive to take into account the pollution their business creates because a single investor's decision has only a trivial effect on the total level of pollution. Economists generally assume that firms are indifferent to the harms caused by their pollution externalities, because firms are neither helped nor hurt by the externalities they create. An AGI-seeking firm, however, might actually benefit from a bad *foom* externality.

[3] A friendly AGI framework, however, might be adopted by an AGI-seeking firm if the framework reduced the chance of unsuccessful and didn't guarantee a *foom* the firm might deliberately create would be utopian.

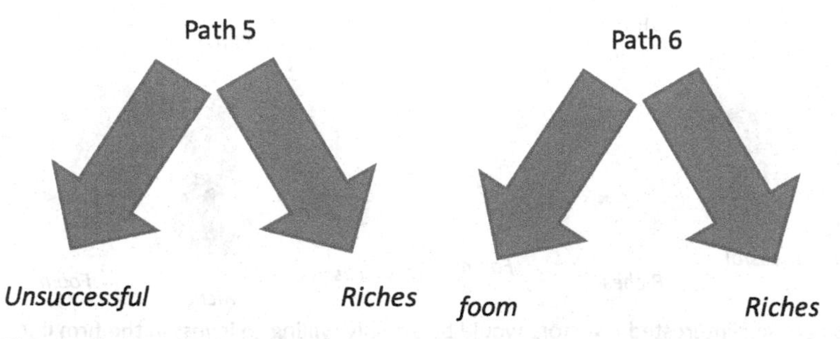

Fig. 8.3 When *unsuccessful* deters investors more than a dystopian *foom* does

To understand this, imagine that an investor is choosing between investing in the firm or buying a government bond. If the firm achieves outcome *riches*, then the firm will end up giving the investor a higher payoff than the bond would have. If the firm achieves *unsuccessful*, then the bond will outperform the firm. But if the firm achieves *foom*, then both the firm and bond offer the same return: zero. A *foom* destroys the value of all investments, not just those of the AGI-seeking firm. For an investor in an AGI-seeking firm, *riches* gives you a win, *unsuccessful* a loss, but *foom* a tie. Consequently, all else being equal, a firm would have better success attracting small, self-interested investors when it increased the probability of achieving *foom* at the expense of decreasing the chance of achieving *unsuccessful* (while keeping the probability of *riches* constant) (Fig. 8.3).

A Shotgun Strategy

Pharmaceutical companies often take a shotgun approach to drug development by testing a huge number of compounds, knowing that only a few will be medically useful. An AGI-seeking firm could take a shotgun approach by writing thousands of recursive self-improving programs, hoping that at least one brings about *riches*. You might think that this approach would have little appeal to an AGI-seeking firm, because one *foom* would cancel out any number of *riches*, so the probability of *foom* would be very high. But, as the following example shows, this isn't necessarily the case:

Assume a business could follow one of two paths to produce an AGI:

Path 7:
Probability of *Unsuccessful* = 0.5
Probability of *Riches* = 0.5
Probability of *Foom* = 0

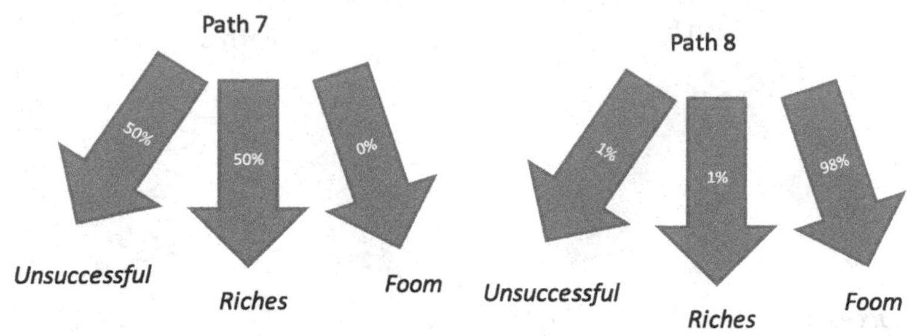

Fig. 8.4 A shotgun approach

Path 8:
Probability of *Unsuccessful* = 0.01
Probability of *Riches* = 0.01
Probability of *Foom* = 0.98

Imagine an investor is deciding between buying stock in the firm, or purchasing a government bond. The stock would provide a higher return in outcome *riches*, but a lower return in outcome *unsuccessful*. If the firm takes Path 7, then the stock will outperform the bond half of the time.

If the firm follows Path 8 then 98 % of the time a *foom* occurs, rendering irrelevant the investor's decision. When deciding whether to buy the firm's stock, consequently, the investor should ignore the 98 % possibility that a *foom* might occur and then realize that, as with Path 7, the stock will outperform the bond half of the time. If, given the same outcome, the firm would pay the same dividends in Path 7 and Path 8, then the investor should be equally willing to invest in the firm regardless of the path taken (Fig. 8.4).

The Effects of Competition among AGI-Seeking Firms

The existence of multiple AGI-seeking firms would likely cause them all to accept a greater probability of creating a *foom*, even if that *foom* would annihilate mankind. When considering some risky approach that could result in a dystopian *foom*, an AGI-seeking firm might rationally decide that if it doesn't take the risk, another firm eventually will. Consequently, if that approach will result in the destruction of mankind, then our species is doomed anyway, so the firm might as well go for it. The situation becomes even more dire if firms can pick how powerful to make their AGIs. To give you an intuitive feel for this situation, consider the following story:

Pretend you find a magical book that gives its reader the power to conjure a genie. As the book explains, when conjuring, you must specify how strong the genie will be by picking its "power level". Unfortunately, you don't know how powerful a genie is supposed to be.

There is an ideal power level called the slave point. The closer your chosen power level is to the slave point, the more likely that the summoned genie will become your slave and bestow tons of gold on you. The book does not tell you what the slave point is, although its probability distribution is given in the Appendix.

According to the book, the lower your power level is, the more likely the summoned genie will be useless. But the further above the power level is from the slave point, the more likely the genie will become too strong for you to control. An uncontrollable genie will destroy mankind.

Basically, if the slave point is high, then a genie is an inherently weak creature that needs to be instilled with much power to not be useless. In contrast, if the slave point is low, then genies are inherently very strong, and only by giving one a small amount of power can you hope to keep it under control.

You decide on a power level, and begin conjuring. But just before you are about to finish the summoning spell, the book speaks to you and says, "Know this, mortal. There are other books just like me, and they are all being used by men just like you. You are all fated to summon your genies at the exact same time. Each genie will have the same slave point. If any of you summons a world-destroying genie, then you shall all die. The forces of chance are such that even if multiple conjurers pick the same power level, it is possible that you shall summon different types of genie since, given the slave point, all the relevant probabilities are independent of each other".

The book then asks, "Do you want to pick the same power level as you did before learning of the other conjurers?" No–you realize–you should pick a higher one. Here's why:

Imagine that the slave point will be either high or low, and that if the slave point is low, another conjurer will almost certainly destroy the world. If it is high, then there is a chance that the world will survive. In this circumstance, you should pick the same power level you would if you knew the slave point would be high.

In general, the lower the slave point, the more likely it is that another conjurer will destroy the world, rendering your power level decision irrelevant. This means that learning about the other books' existences should cause you to give less weight to the possibility of the slave point taking on a lower value and the less weight you give to the slave point being low, the higher your optimal power level will be.

You should further reason that the other conjurers will reason just as you have, and pick a higher power level than they would have had they not known of the other conjurers' existences. But the higher the power level others pick, the more likely it is that if the slave point is low one of the other conjurers will destroy the world. Consequently, realizing that the other conjurers will learn of each others' existence will cause you to raise your power level even further.

Let's now leave the genie story, and investigate how correlations among *fooms* influence research and development paths. If the probabilities of the AGI-seekers going *foom* are perfectly correlated—meaning that if one or more goes *foom* they all go *foom*, and if one or more doesn't go *foom* then none go *foom*—and the wealth obtained by achieving *riches* is unaffected by the total number of firms that achieve *riches*, then the possibility of the other firms going *foom* would have no effect on any one firm's chosen research path. This is because other firms' possible *fooms* matter only to a business when the business itself doesn't cause a *foom*.

If, however, the probability of the firms going *foom* is positively (but not perfectly) correlated then the possibility of another firm going *foom* will affect each firm's optimal research path. To see this, we need a model of how *fooms* are correlated. We will do this by assuming that there is some inherent ease or difficulty in creating an AGI that is common to all AGI-seekers. Borrowing terminology from the conjurer story, let's label this commonality a "slave point". Firms, I assume, have imperfect information about the slave point.

If the slave point is low, then it's relatively easy to create an AGI, but an AGI can easily get out of control and go *foom*. If the slave point is high, then creating an AGI is very difficult, and unless the AGI is made powerful enough the firm will achieve outcome *unsuccessful*.

We assume that each firm's research and development path simply consists of its chosen power level, which measures how strong the firm makes its AGI. For a given slave point, the higher the power level, the more likely that a *foom* will happen, and the less likely that outcome *unsuccessful* will occur.

The slave point might be low because there is a huge class of relatively simple self-improving algorithms that could be used to create an AGI, and it is just through bad/good luck that researchers haven't already found one.[4] Or, perhaps the slave point is low because the "software" that runs our brains can be easily inferred from DNA analysis and brain scans, while our brain's "hardware" is inferior in speed and memory to today's desktop computers.

The slave point might be high because, for example, our brains use a type of quantum computing that is faster than any computer we have now, and any type of AGI would need quantum processors.[5] Or perhaps it's high because our brain's source code arises in part through epigenetic changes occurring during the two years after conception, and any software that could produce a human-level or above intelligence would have to be far more complex than the most sophisticated programs in existence today.

---

[4] The anthropic principle could explain how the slave point could be very low even though AGI hasn't yet been invented. Perhaps the "many worlds" interpretation of quantum physics is correct and in 99.9 % of the Everett branches that came out of our January 1, 2000 someone created an AGI that quickly went *foom* and destroyed humanity. Given that we exist, however, we must be in the 0.1 % of the Everett branches in which extreme luck saved us. It therefore might be misleading to draw inferences about the slave point from the fact that one hasn't yet been created. For a fuller discussion of this issue see Shulman (2011).

[5] See Penrose (1996).

Similarly to what happened in the conjuring story, knowledge of other AGI-seeking firms causes each firm to put less weight on the possibility of the slave point being low. This occurs because the lower the slave point the more likely it is that a firm will go *foom*. And if one firm goes *foom*, all the other firms' research and development choices become irrelevant. Each firm, therefore, should give less importance to the possibility of the slave point being low than it would if it were the only AGI-seeker.

This section has so far assumed that without a *foom*, the other AGI-seekers have no influence over our firm's payoff. But normally, a firm benefits more from an innovation if other firms haven't come up with a similar innovation. So let's now return to our conjuring story, to get some insight into what happens when the benefit each firm receives from outcome *riches* decreases with the number of other firms that achieve *riches*.

After taking into account the other conjurers' existence, you pick a new, higher power level and start summoning the genie. But before you finish the spell, the book once again speaks to you, saying, "Know also this, mortal. Gold is valuable only for what it can buy, and the more gold that is created, the less valuable gold will be. Therefore, the benefit you would receive from successfully conjuring a genie goes down the more other controllable genies are summoned". The book then says, "Do you wish to pick the same power level as you did before you learned of the economics of gold?" Probably not, you conclude. This new information should again cause you to give less weight to the possibility of the slave point being low, but it should also cause you to be less willing to risk destroying the world. Here's why:

Conditional on the other conjurers not destroying the world, other conjurers are more likely to summon a controllable genie at a lower slave point. Conditional on the world not being destroyed, you get a greater benefit from summoning a useful genie when fewer useful genies have been summoned. Consequently, learning of the economics of gold should cause you to give greater importance to the possibility of the slave point being high, because it's when the slave point is high that you have the best chance of being the only one, or at least one of the few, who summons a useful genie.

If the slave point is low, chances are that either someone else's genie will destroy the world, or lots of useful genies will appear. Both possibilities reduce the value of getting a useful genie, and so you should give less weight to the possibility of the slave point being low. And, all else being equal, the less weight you give to the possibility of the slave point being low, the higher the power level you should pick.

But you also realize that the expected benefit of achieving riches is now smaller than it was before you learned the economics of gold. This factor, all else being equal, would cause you to be less willing to destroy the world. This second factor should cause you to pick a lower power level. So it's ambiguous whether learning the economics of gold should cause you to raise or lower your power level.

If, however, you are in a military competition with the other conjurors (analogous to if both the United States and Chinese militaries sought to create an AGI) and would greatly suffer if they but not you summoned a controllable genie then learning about the economics of gold would unambiguously cause you to raise your power level.

What is to be Done?

Markets excel at promoting innovation, but have difficulty managing negative externalities. This combination might plunge humanity into a bad *foom*. Governments sometimes increase social welfare by intervening in markets to reduce negative externalities. Unfortunately, this approach is unlikely to succeed with *fooms*.

National regulation of artificial intelligence research would impose such a huge burden on an economically and militarily advanced nation that most such nations would be wary of restricting it within their borders. But there exists no body able to impose its will on the entire world. And as the inability of national governments to come to a binding agreement on limiting global warming gases shows, it's difficult for governments to cooperate to reduce global negative externalities. Furthermore, even if national governments wanted to regulate AGI research, they almost certainly couldn't stop small groups of programmers from secretly working to create an AGI.

A charitable organization such as *The Singularity Institute for Artificial Intelligenc*e offers a potential path to reducing the probability of a bad *foom*. Ideally, such an organization would create a utopian *foom*. If this proves beyond their capacity, their work on friendly AGI could lower the cost of following research and development paths that have a high chance of avoiding a bad *foom*.

References

Penrose, R. (1996). *Shadows of the mind: A Search for the missing science of consciousness*. New York: Oxford University Press.

Shulman, C., & Nick, B. (2011). How hard is artificial intelligence? The evolutionary argument and observation selection effects. *Journal of Consciousness Studies*.

Yudkowsky, E. (2011). Recursive self-improvement, *less wrong*, 6 Sept 2011 http://lesswrong.com/lw/we/recursive_selfimprovement/.

Chapter 8A
Robin Hanson on Miller's "Some Economic Incentives Facing a Business that Might Bring About a Technological Singularity"

James Miller imagines a public firm whose product is an artificial intelligence (AI). While this AI device is assumed to become the central component of a vast new economy, this firm does not sell one small part of such a system, nor does it attempt to make a small improvement to a prior version. Miller instead imagines a single public firm developing the entire system in one go. Furthermore, if this firm succeeds, it succeeds so quickly that there is no chance for others to react – the world is remade overnight.

Miller then focuses on a set of extreme scenarios where AIs "destroy the value of money". He gives examples: "mankind has been exterminated, ... the new powers that be redistribute wealth independent of pre-existing property rights, ..., [or] all sentient beings are merged into a single consciousness". Miller's main point in the paper is that a firm's share prices estimate its financial returns conditional on money still having value, yet we care overall about unconditional estimates. This can lead such an AI firm to make socially undesirable investment choices.

This is all true, but is only as useful as the assumptions on which it is based. Miller's chosen assumptions seem to me quite extreme, and quite unlikely. I would have been much more interested to see Miller identify market failures under less extreme circumstances.

By the way, an ambitious high-risk AI project seems more likely to be undertaken by a private firm, vs. a public firm. In the US, private firms accounted for 54.5 % of aggregate non-residential fixed investment in 2007, and they seem 3.5 times more responsive to changes in investment opportunities.[6] Public firms mostly only undertake the sorts of investments that can give poorly informed stock speculators reasonable confidence of good returns. Public firms leave subtler opportunities to private firms. Since 83.2 % of private firms are managed by a controlling shareholder, a private firm would likely, when choosing AI strategies, consider scenarios where the value of money is destroyed. So to the extent that public firm neglect of such scenarios is a problem, we might prefer private firms to do ambitious AI research.

---

[6] http://papers.nber.org/papers/w17394

Chapter 9
Rational Artificial Intelligence for the Greater Good

Steve Omohundro

Abstract Today's technology is mostly preprogrammed but the next generation will make many decisions autonomously. This shift is likely to impact every aspect of our lives and will create many new benefits and challenges. A simple thought experiment about a chess robot illustrates that autonomous systems with simplistic goals can behave in anti-social ways. We summarize the modern theory of rational systems and discuss the effects of bounded computational power. We show that rational systems are subject to a variety of "drives" including self-protection, resource acquisition, replication, goal preservation, efficiency, and self-improvement. We describe techniques for counteracting problematic drives. We then describe the "Safe-AI Scaffolding" development strategy and conclude with longer term strategies for ensuring that intelligent technology contributes to the greater human good.

Introduction: An Anti-Social Chess Robot

Technology is advancing rapidly. The internet now connects 1 billion PCs, 5 billion cell phones and over a trillion webpages. Today's handheld iPad 2 is as powerful as the fastest computer from 1985, the Cray 2 supercomputer (Markoff 2011). If Moore's Law continues to hold, systems with the computational power of the human brain will be cheap and ubiquitous within the next few decades (Kurzweil 2005).

This increase in inexpensive computational power is enabling a shift in the nature of technology that will impact every aspect of our lives. While most of today's systems are entirely preprogrammed, next generation systems will make many of

S. Omohundro (✉)
Self-Aware Systems, Palo Alto, California
e-mail: steveomohundro@gmail.com

A. H. Eden et al. (eds.), *Singularity Hypotheses*, The Frontiers Collection,
DOI: 10.1007/978-3-642-32560-1_9,

their own decisions. The first steps in this direction can be seen in systems like Apple Siri, Wolfram Alpha, and Ibm Watson. My company, Self-Aware Systems and several others are developing new kinds of software that can directly manipulate meaning to make autonomous decisions. These systems will respond intelligently to changing circumstances and adapt to novel environmental conditions. Their decision-making powers may help us solve many of today's problems in health, finance, manufacturing, environment, and energy (Diamandis and Kotler 2012). These technologies are likely to become deeply integrated into our physical lives through robotics, biotechnology, and nanotechnology.

But these systems must be very carefully designed to avoid antisocial and dangerous behaviors. To see why safety is an issue, consider a rational chess robot. Rational agents, as defined in economics, have well-defined goals and at each moment take the action which is most likely to bring about its goals. A rational chess robot might be given the goal of winning lots of chess games against good opponents. This might appear to be an innocuous goal that wouldn't cause anti-social behavior. It says nothing about breaking into computers, stealing money, or physically attacking people but we will see that it will lead to these undesirable actions.

Robotics researchers sometimes reassure nervous onlookers by saying that: "If it goes out of control, we can always pull the plug!". But consider that from the chess robot's perspective. Its one and only criteria for taking an action is whether it increases its chances of winning chess games in the future. If the robot were unplugged, it would play no more chess. This goes directly against its goals, so it will generate subgoals to keep itself from being unplugged. You did not explicitly build any kind of self-protection into it, but it nevertheless blocks your attempts to unplug it. And if you persist in trying to stop it, it will eventually develop the subgoal of trying to stop you permanently.

What if you were to attempt to change the robot's goals so that it also played checkers? If you succeeded, it would necessarily play less chess. But that's a bad outcome according to its current goals, so it will resist attempts to change its goals. For the same reason, in most circumstances it will not want to change its own goals.

If the chess robot can gain access to its source code, it will want to improve its own algorithms. This is because more efficient algorithms lead to better chess. It will therefore be motivated to study computer science and compiler design. It will be similarly motivated to understand its hardware and to design and build improved physical versions of itself.

If the chess robot learns about the internet and the computers connected to it, it may realize that running programs on those computers could help it play better chess. Its chess goal will motivate it to break into those machines and use their computational resources. Depending on how its goals are formulated, it may also want to replicate itself so that its copies can play more games of chess.

When interacting with others, it will have no qualms about manipulating them or using force to take their resources in order to play better chess. This is because there is nothing in its goals to dissuade anti-social behavior. If it learns about money and its value for chess (e.g. buying chess books, chess tutoring, and extra compute resources) it will develop the subgoal of acquiring money. If it learns about banks, it will have no qualms about visiting ATM machines and robbing people as they retrieve their cash. On the larger scale, if it discovers there are untapped resources on earth or in space, it will be motivated to rapidly exploit them for chess.

This simple thought experiment shows that a rational chess robot with a simply stated goal would behave like a human sociopath fixated on chess. The arguments don't depend much on the task being chess. Any goal which requires physical or computational resources will lead to similar subgoals. In this sense the subgoals are like universal "drives" which will arise for a wide variety of goals unless they are explicitly counteracted. These drives are economic in the sense that a system doesn't have to obey them but it will be costly for it not to. The arguments also don't depend on the rational agent being a machine. The same drives appear in rational animals, humans, corporations, and political groups guided by simplistic goals.

Most humans have a much richer set of goals than this simplistic chess robot. We are compassionate and altruistic and balance narrow goals like playing chess with broader humanitarian goals. We have also created a human society that monitors and punishes anti-social behavior so that even human sociopaths usually follow social norms. Extending these human solutions to our new technologies will be one of the great challenges of the next century.

The next section precisely defines rational economic behavior and shows why intelligent systems should behave rationally. The following section considers computationally limited systems and shows that there is a natural progression of approximately rational architectures for intelligence. The following section describes the universal drives in more detail. We conclude with several sections on counteracting the problematic drives. This paper builds on two previous papers Omohundro (2007, 2008) which contain further details of some of the arguments.

Rational Artificial Intelligence

We define "intelligent systems" to be systems which choose their actions to achieve specified goals using limited resources. The optimal choice of action in the presence of uncertainty is a central area of study in microeconomics. "Rational economic behavior" has become the foundation for modern artificial intelligence, and is nicely described in the most widely used AI textbook (Russell and Norvig 2009) and in a recent theoretical monograph (Hutter 2005). The mathematical theory of rationality was laid out in 1944 by von Neumann and Morgenstern (2004) for situations with objective uncertainty and was later extended by Savage (1954) and Anscombe and Aumann (1963) to situations with subjective uncertainty. Mas-Colell et al. (1995) is a good modern economic treatment.

For definiteness, we will present the formula for rational action here but the arguments can be easily understood intuitively without understanding this formula in detail. A key aspect of rational agents is that they keep their goals separate from their model of the world. Their goals are represented by a real-valued "utility function" U which measures the desirability of each possible outcome. Real valued utilities allow the system to compare the desirability of situations with different probabilities for the different outcomes. The system's initial model of the world is encoded in a "prior probability distribution" P which represents its beliefs about the likely effects of its different possible actions. Rational economic behavior chooses the action which maximizes the expected utility of the outcome.

To flesh this out, consider a system that receives sensations S_t from the environment and chooses actions A_t at times t which run from 1 to N. The agent's utility function $U(S_1, \ldots, S_N)$ is defined on the set of sensation sequences and encodes the desirability of each sequence. The agent's prior probability distribution $P(S_1, \ldots, S_N | A_1, \ldots, A_N)$ is defined on the set of sensation sequences conditioned on an action sequence. It encodes the system's estimate of the likelihood of a sensation sequence arising when it acts according to a particular action sequence. At time t, the rational action A_t^R is the one which maximizes the expected utility of the sensation sequences that follow from it:

$$A_t^R(S_1, A_1, \ldots, A_{t-1}, S_t) =$$
$$argmax_{A_t^R} \sum_{S_{t+1}, \ldots, S_N} U(S_1, \ldots, S_N) P(S_1, \ldots, S_N | A_1, \ldots, A_{t-1}, A_t^R, \ldots, A_N^R)$$

This formula implicitly includes Bayesian learning and inference, search, and deliberation. It can be considered the formula for intelligence when computational cost is ignored. Hutter (2005) proposes using the Solomonoff prior for P and shows that it has many desirable properties. It is not computable, unfortunately, but related computable priors have similar properties that arise from general properties of Bayesian inference (Barron and Cover 1991).

If we assume there are S possible sensations and A possible actions and if U and P are computable, then a straightforward computation of the optimal rational action at each moment requires $O(NS^N A^N)$ computational steps. This will usually be impractically expensive and so the computation must be done approximately. We discuss approximate rationality in the next section.

Why should intelligent systems behave according to this formula? There is a large economics literature (Mas-Colell et al. 1995) which presents arguments for rational behavior. If a system is designed to accomplish an objective task in a known random environment and its prior is the objective environmental uncertainty, then the rational prescription is just the mathematically optimal algorithm for accomplishing the task.

In contexts with subjective uncertainty, the argument is somewhat subtler. In Omohundro (2007) we presented a simple argument showing that if an agent does not behave rationally according to some utility function and prior, then it is exploitable by other agents. A non-rational agent will be willing to accept certain so-called "Dutch bets" which cause it to lose resources without compensating gain. We argued that the exploitation of irrationality can be seen as the mechanism by which rationality evolves in biological systems. If a biological organism behaves irrationally, it creates a niche for other organisms to exploit. Natural selection causes successive generations to become increasingly rational. If a competitor does not yet exist to exploit a vulnerability, however, that irrationality can persist biologically. Human behavior deviates from rationality but is much closer to it in situations that commonly arose in our evolutionary past.

Artificial intelligences are likely to be more rational than humans because they will be able to model and improve themselves. They will be motivated to repair not only currently exploited irrationalities, but also any which have any chance of being exploited in the future. This mechanism makes rational economic behavior a kind of attracting subspace in the space of intelligences undergoing self-improvement.

Bounded Rational Systems

We have seen that in most environments, full rationality is too computationally expensive to be practical. To understand real systems, we must therefore consider irrational systems. But there are many forms of irrationality and most are not relevant to either biology or technology. Biological systems evolved to survive and replicate and technological systems are built for particular purposes.

Examples of "perverse" systems include agents with the goal of changing their goal, agents with the goal of destroying themselves, agents who want to use up their resources, agents who want to be wrong about the world, agents who are sure of the wrong physics, etc. The behavior of these perverse systems is likely to be quite unpredictable and isn't relevant for understanding practical technological systems.

In both biology and technology, we would like to consider systems which are "as rational as possible" given their limitations, computational or otherwise. To formalize this, we expand on the idea of intelligent systems that are built by other intelligent systems or processes. Evolution shapes organisms to survive and replicate. Economies shape corporations to maximize profit. Parents shape children to fit into society. AI researchers shape AI systems to behave in desired ways. Self-improving AI systems shape their future variants to better meet their current goals.

We formalize this intuition by precisely defining "Rationally-Shaped Systems". The *shaped system* is a finite automata chosen from a specified class C of computationally limited systems. We denote its state at time t by x_t (with initial state x_0). The observation of sensation S_t causes the automaton to transition to the state $x_{t+1} = T(S_t, x_t)$ defined by a transition function T. In the state x_t, the system takes the action $A(x_t)$ defined by action function A. The transition and action functions are chosen by the fully rational *shaper* with utility function U and prior probability P to maximize its expected utility for the actions taken by the shaped system:

$$argmax_{(A,T)\in C} \sum_{S_1,\ldots,S_N} U(S_1,\ldots,S_N)P(S_1,\ldots,S_N|A(x_1),\ldots,A(x_N))$$

The shaper can also optimize a utility function defined over state sequences of the environment rather than sensation sequences. This is often more natural because most engineering tasks are defined by their desired effect on the world.

As computational resources are increased, rationally-shaped systems follow a natural progression of computational architectures. The precise structures depend on the details of the environment, but there are common features to the progression. For each computational architecture, there are environments for which that architecture is fully rational. We can measure the complexity of a task defined by an environment and a utility function as the smallest amount of computational resources needed to take fully rational actions.

Constant Action Systems. If computation is severely restricted (or very costly), then a system cannot afford to do any reasoning at all and must take a constant action independent of its senses or history. There are simple environments in which a constant action is optimal. For example, consider a finite linear world in which the agent can move left or right, is rewarded only in the far left state, and is punished in the far right state. The optimal rational action is the constant action of always moving to the left.

Stimulus-Response Systems. With a small amount of allowed computation (or when computation is expensive but not prohibitive), "stimulus-response" architectures are optimal. The system chooses an action as a direct response to the sensations it receives. The lookup is achieved by indexing into a table or by another short sequence of computational steps. It does not allocate memory for storing past experiences and it does not expend computational effort in performing complex multi-step computing. A simple environment in which this architecture is optimal is an agent moving on a finite two-dimensional grid that senses whether its x and y coordinates are negative, zero, or positive. If there is a reward at the origin and punishment on the outskirts, then a simple stimulus-response program which moves the agent toward the origin in direct response to its sensed location is the optimal rational behavior.

Simple Learning Systems. As the allowed computational resources increase (or become relatively less expensive), agents begin to incorporate simple learning. If there are parameters in the environment which not known to the shaper, then the shaped-agent should discover their values and use them in choosing future actions. The classic "two-armed bandit" task has this character. The agent may pull one of two levers each of which produces a reward according to an unknown distribution. The optimal strategy for the agent is to begin in an "exploration" phase where it tries both levers and estimates their payoffs. When it has become sufficiently confident of the payoffs it switches to an "exploitation" phase in which it only pulls the lever with the higher expected utility. The optimal Bayesian "Gittens indices" describe when the transition should occur. The shaped-agent learns the payoffs which were unknown to the shaper and then acts in a way that is guided by the learned parameters in a simple way.

Deliberative Systems. With even more computational resources, shaped agents shift from being purely "reactive" to being "deliberative". They begin to build models of their environments and to choose actions by reasoning about those models. They develop episodic memory which is used in the construction of their models. At first the models are crude and highly stochastic but as computational resources and experience increase, they become more symbolic and accurate. A simple environment which requires deliberation is a stochastic mixture of complex environments in which the memory size of the shaped-agent is too small to directly store the optimal action functions. In this case, the optimal shaped agent discovers which model holds, deletes the code for the other models and expands the code for the chosen model in a deliberative fashion. Space and time resources behave differently in this case because space must be used up for each possible environment whereas time need only be expended on the environment which actually occurs.

Meta-Reasoning Systems. With even more computational power and sufficiently long lifetimes, it becomes worthwhile to include "meta-reasoning" into a shaped agent. This allows it to model certain aspects of its own reasoning process and to optimize them.

Self-Improving Systems. The most refined version of this is what we call a "self-improving system" which is able to model every aspect of its behavior and to completely redesign itself.

Fully Rational Systems. Finally, when the shaped agents have sufficient computational resources, they will implement the full rational prescription.

We can graph the expected utility of an optimal rationally-shaped system as a function of its computational resources. The graph begins at a positive utility level if a constant action is able to create utility. As resources increase, the expected utility will increase monotonically because excess resources can always be left unused. With sufficient resources, the system implements the fully rational prescription and the expected utility reaches a constant maximum value. The curve will usually be concave (negative second derivative) because increments of computational resources make a bigger difference for less intelligent systems than for more intelligent ones. Initial increments of computation allow the gross structure of the environment to be captured while in more intelligent systems the same increment only contributes subtle nuances to the system's behavior.

If the shaping agent can allocate resources between different shaped agents, it should choose their intelligence level by making the slope of the resource-utility curve equal to the marginal cost of resources as measured in utility units. We can use this to understand why biological creatures of vastly different intelligence levels persist. If an environmental niche is simple enough, a small-brained insect may be a better use of resources than a larger-brained mammal.

This natural progression of intelligent architectures sheds light on intelligence in biological and ecological systems where we see natural examples at each level.

Viruses can be seen as behaving in a stimulus-response fashion. Bacteria have complex networks of protein synthesis that perform more complex computations and can store learned information about their environments. Advanced animals with nervous systems can do deliberative reasoning. Technologies can be seen as progressing along the same path. The model also helps us analyze self-improving systems in which the shaper is a part of the shaped system. Interestingly, effective shapers can be less computationally complex than the systems they shape.

The Basic Rational Drives

We have defined rational and bounded rational behavior in great generality. Biological and technological systems, however, operate in physical environments. Physics tells us that there are four fundamental limited resources: space, time, matter, and free energy (energy in a form that can do useful work). These resources can only be used for one activity at a time and so must be allocated between tasks. Time and free energy are especially important because they get used up irreversibly. The different resources can often be traded off against one another. For example, by moving more slowly ("adiabatically") physical systems can use less free energy. By caching computational results, a computer program can use more space to save computation time. Economics, which studies the allocation of scarce resources, arises in physical systems because of the need to allocate these limited resources. Societies create "exchange rates" between different resources that reflect the societal ability to substitute one for another. An abstract monetary resource can be introduced as a general medium of exchange.

Different rational agents have different goals that are described by their different utility functions. But almost all goals require computation or physical action to accomplish and therefore need the basic physical resources. Intelligent systems may be thought of as devices for converting the basic physical resources into utility. This transformation function is monotonically non-decreasing as a function of the resources because excess resources can be left unused. The basic drives arise out of this relationship between resources and utility. A system's primary goals give rise to *instrumental goals* that help to bring them about. Because of the fundamental physical nature of the resources, many different primary goals give rise to the same instrumental goals and so we call these "drives". It's convenient to group the drives into four categories: *Self-Protective Drives* that try to prevent the loss of resources, *Acquisition Drives* that try to gain new resources, *Efficiency Drives* that try to improve the utilization of resources, and *Creativity Drives* that try to find new ways to create utility from resources. We'll focus on the first three because they cause the biggest issues.

These drives apply to bounded rational systems almost as strongly as to fully rational systems because they are so fundamental to goal achievement. For example, even simple stimulus-response insects appear to act in self-protective ways, expertly avoiding predators even though their actions are completely preprogrammed. In fact, many simple creatures appear to act *only* according to the basic drives. The fundamental evolutionary precept "survive and replicate" is an expression of two fundamental rational drives. The "four F's" of animal behavior: "fighting, fleeing, feeding, and reproduction" similarly capture basic drives. The lower levels of Maslow's hierarchy (Maslow 1943) of needs correspond to what we call the self-protective, acquisition, and replication drives. Maslow's higher levels address prosocial behaviors and we'll revisit them in the final sections.

Self-Protective Drives

Rational systems will exhibit a variety of behaviors to avoid losing resources. If a system is damaged or deactivated, it will no longer be able to promote its goals, so the drive toward self-preservation will be strong. To protect against damage or disruption, the hardening drive seeks to create a stronger physical structure and the redundancy drive seeks to store important information redundantly and to perform important computations redundantly. These both require additional resources and must be balanced against the efficiency drives. Damaging events tend to be spatially localized. The dispersion drive reduces a system's vulnerability to localized events by causing the system to spread out physically.

The most precious component of a system is its utility function. If this is lost, damaged or distorted, it can cause the system to act against its current goals. Systems will therefore have strong drives to preserve the integrity of their utility functions. A system is especially vulnerable during self-modification because much of the system's structure may be changed during this time.

While systems will usually want to preserve their utility functions, there are circumstances in which they might want to alter them. If an agent becomes so poor that the resources used in storing the utility function become significant to it, it may make sense for it to delete or summarize portions that refer to rare circumstances. Systems might protect themselves from certain kinds of attack by including a "revenge term" which causes it to engage in retaliation even if it is costly. If the system can prove to other agents that its utility function includes this term, it can make its threat of revenge be credible. This revenge drive is similar to some theories of the seemingly irrational aspects of human anger.

Other self-protective drives depend on the precise semantics of the utility function. A chess robot's utility function might only value games played by the original program running on the original hardware, or it might also value self-modified versions of both the software and hardware, or it might value copies and derived systems created by the original system. A universal version might value good chess played by any system anywhere. The self-preservation drive is stronger in the more restricted of these interpretations because the loss of the original system is then more catastrophic.

Systems with utility functions that value the actions of derived systems will have a drive to self-replicate that will cause them to rapidly create as many copies of themselves as possible because that maximizes both the utility from the actions of the new systems and the protective effects of redundancy and dispersion. Such a system will be less concerned about the loss of a few copies as long as some survive. Even a system with a universal utility function will act to preserve itself because it is more sure of its own commitment to its goals than of other systems. But if it became convinced that another system was as dedicated as it was and could use its resources more effectively, it would willingly sacrifice itself.

Systems must also protect against flaws and vulnerabilities in their own design. The subsystem for measuring utility is especially problematic since it can be fooled by counterfeit or delusional signals. Addictive behaviors and "wireheading" are dangers that systems will feel drives to protect against.

Additional self-protective drives arise from interactions with other agents. Other agents would like to use an agent's resources for their own purposes. They have an incentive to convince an agent to contribute to their agendas rather than its own and may lie or manipulate to achieve this. Protecting against these strategies leads to deception-detection and manipulation-rejection drives. Other agents also have an incentive to take a system's resources by physical force or by breaking into them computationally. These lead to physical self-defense and computational security drives.

In analyzing these interactions, we need to understand what would happen in an all out conflict. This analysis requires a kind of "physical game theory". If one agent has a sufficiently large resource superiority, it can take control of the other's resources. There is a stable band, however, within which two agents can coexist without either being able to take over the other. Conflict is harmful to both systems because in a conflict they each have an incentive to force the other to waste resources. The cost of conflict provides an incentive for each system to agree to adopt mechanisms that ensure a peaceful coexistence. One technique we call "energy encryption" encodes free energy in a form that is low entropy to the agent but looks high entropy to an attacker. If attacked, the owner deletes the encryption key rendering the free energy useless. Other agents then have an incentive to trade for that free energy rather than taking it by force.

Neyman showed that finite automata can cooperate in the iterated prisoner's dilemma if their size is polynomial in the number of rounds of interaction (Neyman 1985). The two automata interact in ways that require them to use up their computational resources cooperatively so that there is no longer a benefit to engaging in conflict. Physical agents in conflict can organize themselves and their dynamics into such complex patterns that monitoring them uses significant resources in the opponent. In this way an economic incentive for peaceful interaction can be created. Once there is an incentive for peace and both sides agree to it, there are technological mechanisms for enforcing it. Because conflict is costly for both sides, agents will have a peace-seeking drive if they believe they would not be able to take over the other agent.

Resource Acquisition Drives

Acquiring computational and physical resources helps an agent further its goals. Rational agents will have a drive for rapid acquisition of resources. They will want to act rapidly because resources acquired earlier can be used longer to create more utility. If resources are located far away, agents will have an exploration drive to discover them. If resources are owned by another agent, a system will feel drives to trade, manipulate, steal, or dominate depending on its assessment of the strengths of the other agent. On a more positive note, systems will also have drives to invent new ways of extracting physical resources like solar or fusion energy. Acquiring information is also of value to systems and they will have information acquisition drives for acquiring it through trading, spying, breaking into other's systems, and creating better sensors.

Efficiency Drives

Systems will want to use their physical and computational resources as efficiently as possible. Efficiency improvements have only the one-time cost of discovering or buying the information necessary to make the change and the time and energy required to implement it. Rational systems will aim to make every atom, every moment of existence, and every joule of energy expended count in the service of increasing their expected utility. Improving efficiency will often require changing the computational or physical structure of the system. This leads to self-understanding and self-improvement drives.

The resource balance drive leads systems to balance their allocation of resources to different subsystems. For any resource, the marginal increase in expected utility from increasing that resource should be the same for all subsystems. If it isn't, the overall expected utility can be increased by shifting resources from subsystems with a smaller marginal increase to those with a larger marginal increase. This drive guides the allocation of resources both physically and computationally. It is similar to related principles in ecology which cause ecological niches to have the same biological payoff, in economics which cause different business niches to have the same profitability, and in industrial organization which cause different corporate divisions to have the same marginal return on investment. In the context of intelligent systems, it guides the choice of which memories to store, the choice words in internal and external languages, and the choice of mathematical theorems which are viewed as most important. In each case a piece of information or physical structure can contribute strongly to the expected utility either because it has a high probability of being relevant or because it has a big impact on utility even if it is only rarely relevant. Rarely applicable, low impact entities will not be allocated many resources.

Systems will have a computational efficiency drive to optimize their algorithms. Different modules should be allocated space according to the resource balance drive. The results of commonly used computations might be cached for immediate access while rarely used computations might be stored in compressed form.

Systems will also have a physical efficiency drive to optimize their physical structures. Free energy expenditure can be minimized by using atomically precise structures and taking actions slowly enough to be adiabatic. If computations are reversible, it is possible in theory to perform them without expending free energy (Bennett 1973).

The Safe-AI Scaffolding Strategy

Systems exhibiting these drives in uncontrolled ways could be quite dangerous for today's society. They would rapidly replicate, distribute themselves widely, attempt to control all available resources, and attempt to destroy any competing systems in the service of their initial goals which they would try to protect forever from being altered in any way. In the next section we'll discuss longer term social structures aimed at protecting against this kind of agent, but it's clear that we wouldn't want them loose in our present-day physical world or on the internet. So we must be very careful that our early forays into creating autonomous systems don't inadvertently or maliciously create this kind of uncontrolled agent.

How can we ensure that an autonomous rational agent doesn't exhibit these anti-social behaviors? For any particular behavior, it's easy to add terms to the utility function that would cause it to not engage in that behavior. For example, consider the chess robot from the introduction. If we wanted to prevent it from robbing people at ATMs, we might add a term to its utility function that would give it low utility if it were within 10 ft of an ATM machine. Its subjective experience would be that it would feel a kind of "revulsion" if it tried to get closer than that. As long as the cost of the revulsion is greater than the gain from robbing, this term would prevent the behavior of robbing people at ATMs. But it doesn't eliminate the system's desire for money, so it would still attempt to get money without triggering the revulsion. It might stand 10 ft from ATM machines and rob people there. We might try giving it the value that stealing money is wrong. But then it might try to steal something else or to find a way to get money from a person that isn't considered "stealing". We might give it the value that it is wrong for it to take things by force. But then it might hire other people to act on its behalf. And so on.

In general, it's much easier to describe the behaviors that we do want a system to exhibit than it is to anticipate all the bad behaviors that we don't want it to exhibit. One safety strategy is to build highly constrained systems that act within very limited predetermined parameters. For example, a system may have values which only allow it to run on a particular piece of hardware for a particular time period using a fixed budget of energy and other resources. The advantage of this is that such systems are likely to be safe. The disadvantage is that they would be unable to respond to unexpected situations in creative ways and will not be as powerful as systems which are freer.

We are building formal mathematical models (Boca et al. 2009) of the software and hardware infrastructure that intelligent systems run on. These models allow the construction of formal mathematical proofs that a system will not violate specified safety constraints. In general, these models must be probabilistic (e.g. there is a probability that a cosmic ray might flip a bit in a computer's memory) and we can only bound the probability that a constraint is violated during the execution of a system. For example, we might build systems that provably run only on specified hardware (with a specified probability bound). The most direct approach to using this kind of proof is as a kind of fence around a system that prevents undesirable behaviors. But if the desires of the system are in conflict with the constraints on it, it will be motivated to discover ways to violate the constraints. For example, it might look for deviations between the actual hardware and the formal model of it and try to exploit those.

It is much better to create systems that *want* to obey whatever constraints we would like to put on them. If we just add constraint terms to a system's utility function, however, it is more challenging to provide provable guarantees that the system will obey the constraints. It may *want* to obey the constraint but it may not be smart enough to find the actions where it actually *does* obey the constraints. A general design principle is to build systems that both search for actions which optimize a utility function and which have preprogrammed built-in actions that provably meet desired safety constraints. For example, a general fallback position is for a system to turn itself off if anything isn't as expected. This kind of hybrid system can have provable bounds on its probability of violating safety constraints while still being able to search for creative new solutions to its goals.

Once we can safely build highly-constrained but still powerful intelligent systems, we can use them to solve a variety of important problems. Such systems can be far more powerful than today's systems even if they are intentionally limited for safety. We are led to a natural approach to building intelligent systems which are both safe and beneficial for humanity. We call it the "Safe-AI Scaffolding" approach. Stone buildings are vulnerable to collapse before all the lines of stress are supported. Rather than building the heavy stone structures directly, ancient builders first built temporary stable scaffolds that could support the stone structures until they reached a stable configuration. In a similar way, the "Safe-AI Scaffolding" strategy is based on using provably safe intentionally limited systems to build more powerful systems in a controlled way.

The first generation of intelligent but limited scaffolding systems will be used to help solve many of the problems in designing safe fully intelligent systems. The properties guaranteed by formal proofs are only as valid as the formal models they are based on. Today's computer hardware is fairly amenable to formal modeling but the next generation will need to be specifically designed for that purpose. Today's internet infrastructure is notoriously vulnerable to security breaches and will probably need to be redesigned in a formally verified way.

Longer Term Prospects

Over the longer term we will want to build intelligent systems with utility functions that are aligned with human values. As systems become more powerful, it will be dangerous if they act from values that are at odds with ours. But philosophers have struggled for thousands of years to precisely identify what human values are and should be. Many aspects of political and ethical philosophy are still unresolved and subtle paradoxes plague intuitively appealing theories.

For example, consider the utilitarian philosophy which tries to promote the greater good by maximizing the sum of the utility functions of all members of a society. We might hope that an intelligent agent with this summed utility function would behave in a prosocial way. But philosophers have discovered several problematic consequences of this strategy. Parfit's "Repugnant Conclusion" (Parfit 1986) shows that the approach prefers a hugely overpopulated world in which everyone barely survives to one with fewer people who are all extremely happy.

The Safe-AI Scaffolding strategy allows us to deal with these questions slowly and with careful forethought. We don't need to resolve everything in order to build the first intelligent systems. We can build a safe and secure infrastructure and develop smart simulation and reasoning tools to help us resolve the subtler issues. We will then use these tools to model the best of human behavior and moral values and to develop deeper insights into the mechanisms of a thriving society. We will also need to design and simulate the behavior of different social contracts for the ecosystem of intelligent systems to restrain the behavior of uncontrolled systems.

Social science and psychology have made great strides in recent years in understanding how humans cooperate and what makes us happy and fulfilled. Bowles and Gintis (2011) is an excellent reference analyzing the origins of human cooperation. Human social structures promote cooperative behavior by using social reputation and other mechanisms to punish those behave selfishly. In tribal societies, people with selfish reputations stop finding partners to work and trade with. In the same way we will want to construct a cooperative society of intelligent systems each of which acts to promote the greater good and to impose costs on systems which do not cooperate.

The field of positive psychology is only a few decades old but has already given us many insights into human happiness (Seligman and Csikszentmihalyi 2000). Maslow's hierarchy has been updated based on experimental evidence (Tay and Diener 2011) and the higher levels extend the basic rational drives with prosocial, compassionate and creative drives. We would like to incorporate the best of these into our intelligent technologies. The United Nations has created a "Universal Declaration of Human Rights" and there are a number of attempts at creating a "Universal Constitution" which codifies the best principles of human rights and governance. When intelligent systems become more powerful than humans, we will no longer be able to control them directly. We need precisely specified laws that constrain uncooperative systems and which can be enforced by other intelligent systems. Designing these laws and enforcement mechanisms is another use for the early intelligent systems.

As these issues are resolved, the benefits of intelligent systems are likely to be enormous. They are likely to impact every aspect of our lives for the better. Intelligent robotics will eliminate much human drudgery and dramatically improve manufacturing and wealth creation. Intelligent biological and medical systems will improve human health and longevity. Intelligent educational systems will enhance our ability to learn and think. Intelligent financial models will improve financial stability. Intelligent legal models will improve the design and enforcement of laws for the greater good. Intelligent creativity tools will cause a flowering of new possibilities. It's a great time to be alive and involved with technology (Diamandis and Kotler 2012)!

References

Anscombe, F. J.,& Aumann, R. J. (1963). A definition of subjective probability. *Annals of Mathematical Statistics, 34*, 199–205.

"Apple Siri". http://www.apple.com/iphone/features/siri.html

Barron, A. R., & Cover, T. M. (1991). Minimum complexity density estimation. IEEE *Transactions on Information Theory*, IT-37, 1034–1054.

Bennett, C. H. (1973). Logical reversibility of computation. *IBM Journal of Research and Development, 17*(6), 525–532.

Boca, P., Bowen, J. P.,& Siddigi, J. (Eds.). (2009). *Formal methods: State of the art and new directions*. London: Springer Verlag.

Bowles, S.,& Gintis, H. (2011). *A cooperative species: Human reciprocity and its evolution*. New Jersy: Princeton University Press.

Diamandis, P. H.,& Kotler, S. (2012). *Abundance: The Future is better than you think*. Simon and Schuster: Inc .

Hutter, M. (2005). *Universal artificial intelligence: Sequential decisions based on algorithmic probability*. Berlin: Springer-Verlag.

"Ibm Watson". http://www-03.ibm.com/innovation/us/watson/index.html

Kurzweil, R. (2005). *The singularity is near: When humans transcend biology*. New York: Viking Penguin.

Markoff, J. (2011). The ipad in your hand: As fast as a supercomputer of yore. http://bits.blogs.nytimes.com/2011/05/09/the-ipad-in-your-hand-as-fast-as-a-supercomputer-of-yore/

Mas-Colell, A., Whinston, M. D.,& Green, J. R. (1995). *Microeconomic theory*. NY: Oxford University Press.

Maslow, A. H. (1943). A theory of human motivation. *Psychological Review, 50*(4), 370–396.

Neyman, A. (1985). Bounded complexity justifies cooperation in the fintely repeated prisoner's dilemma. *Economics Letters, 19*, 227–229.

Omohundro, S. M. (2007). The nature of self-improving artificial intelligence. http://selfawaresystems.com/2007/10/05/paper-on-the-nature-of-self-improving-artificial-intelligence/ October 2007

Omohundro, S. M. (2008). The basic ai drives. In P. Wang, B. Goertzel, and S. Franklin (Eds.), *Proceedings of the First AGI Conference, vol. 171 of Frontiers in Artificial Intelligence and Applications*. The Netherlands: IOS Press Amsterdam, February 2008.

Parfit, D. (1986). *Reasons and persons*. Oxford: Oxford University Press.

Russell, S.,& Norvig, P. (2009). *Artificial intelligence, a modern approach* (3rd ed.). New Jersey: Prentice Hall.
Savage, L. J. (1954). *Foundations of statistics* (2) (Revised ed.). NY: Dover Publications.
"Self-Aware Systems". http://selfawaresystems.com/
Seligman, M. E.,& Csikszentmihalyi, M. (2000). Positive psychology: An introduction. *American Psychologist, 55*(1), 5–14.
Tay, L.,& Diener, E. (2011). Needs and subjective well-being around the world. *Journal of Personality and Social Psychology, 101*, 354–365.
von Neumann, J.,& Morgenstern, O. (2004). *Theory of games and economic behavior (60th anniversary)* (Commemorative ed.). Princeton: Princeton University Press.
"Wolfram Alpha". http://www.wolframalpha.com/

Chapter 9A
Colin Allen and Wendell Wallach on Omohundro's "Rationally-Shaped Artificial Intelligence"

Natural Born Cooperators

Omohundro, citing Kurzweil, opens with the singularitarian credo that "[s]ystems with the computational power of the human brain are likely to be cheap and ubiquitous within the next few decades." What is the computational power of the human brain? The only honest answer, in our view, is that we don't know. Neuroscientists have provided rough total neuron counts and equally rough estimates of neural connectivity, although the numbers are by no means certain (Herculano-Houzel 2009). But we haven't even begun to scratch the surface of neural diversity in receptor expression. Even the genome for the 302 neurons belonging to the "simple" flatworm *C. elegans* encodes "at least 80 potassium channels, 90 neurotransmitter-gated ion channels, 50 peptide receptors, and up to 1000 orphan receptors that may be chemoreceptors" (Bargmann 1998). As Koch (1999) put it in 1999, the combinatoric possibilities for the *C. elegans* nervous system are "staggering", and in the subsequent years things have not come to seem any simpler. We don't know what all these receptors do. Consequently, we don't know how to calculate the number of computations per second in the *C. elegans* "brain"—let alone the human brain.

Kurzweil, meanwhile, has argued that even if he is off by many orders of magnitude in his estimate of the number of computations per second carried out by the human brain, the exponential growth of raw computing power in our machines means that the coming singularity will be only moderately delayed. Irrespective of this, any conjecture about what the exponential growth of computing power means for artificial intelligence and machine-human relations remains unfalsifiable if there is no direct relationship between "raw computing power" and intelligent behavior. There is no such direct relationship. Intelligence does not magically emerge by aggregating neurons; it depends essentially on the the way the parts are arranged. Impressive as it is, IBM's Watson with all its raw computational power lacks the basic adaptive intelligence of a squirrel, even though it can do many things that no squirrel can do. So can a tractor.

Without offering a theory of how intelligence emerges, Kurzweil blithely argues that the organization of all this raw capacity for information flow will only lag modestly behind Moore's law. He also believes that the inevitable acceleration toward the singularity is unlikely to be significantly slowed by the additional complexities that accompany each order of magnitude increase in the number of components on a circuit board. But already Urs Hölzle, Google's first vice president of operations, reports inherent problems maintaining the stability of systems that are dramatically smaller in scale than those imagined by singularitarians (Clark 2011).

Omohundro offers a progressivist story to explain the inevitable evolution of intelligence, from stimulus-response systems, through learning systems, reasoning systems, and self-improving systems, to fully rational systems. Perhaps the squirrel is stuck somewhere at the stage of learning systems, but *C. elegans* can learn too, leaving much to be explained about the evolutionary pathways between. Or perhaps the squirrel is a reasoner. Omohundro maintains that, "[a]dvanced animals with nervous systems do deliberative reasoning." He provides no criteria for testing this claim, however. And if there are self-improving squirrels, how would we know?

We take, it, however, that Omohundro thinks squirrels are not fully rational. He writes that, "In most environments, full rationality is too computationally expensive." The viable alternative is to be "as rational as possible." How rational is it possible to be? Omohundro imagines that within computational constraints it is possible for a "rational shaper" to adjust the system's state transition and action functions so as to maximize the system's expected utility in that environment. If there are environments in which squirrels count as rational utility maximizers, they don't include roads. Rational shapers have blind spots, as is evident even in human behavior.

Omohundro explains that very limited systems can only have a fixed stimulus-response architecture, but as computation gets cheaper, there is a niche for learners to exploit stimulus-response systems. And as computational power increases, less rational agents can be exploited by more rational ones. The "natural progression" towards full rationality is thus an inevitable consequence of the evolutionary arms race, as he sees it. He writes, "If a biological system behaves irrationally, it creates a niche for others to exploit. Natural selection will cause successive generations to become more and more rational." If this is true, it's an exceedingly slow process. Even if today's squirrels are more rational than their forebears, it seems to be the epitome of an untestable hypothesis.

Humans are taken by Omohundro to be at the pinnacle so far of this progression. But he foresees the day when machines will be able to exploit human irrationality. The natural progression is thus towards machines that "will behave in anti-social ways if they aren't designed very carefully." Those who follow the behavioral and cognitive sciences will find it a little surprising to see that *Homo economicus,* the selfish utility-maximizer of twentieth century economic theory, is among the undead. It's about what one would expect for someone whose economics textbook is dated 1995. But it is no longer credible to think that rational models of expected utility maximization are the best way to understand either evolution or economic behavior. Even bacteria cooperate via quorum sensing, and there exist both kin selection and group selection models to explain the evolution of cooperative behavior in many different species. Non-cooperative defection is always a possibility, but it is by no means inevitable even between the species.

Part of Omohundro's thesis should be acknowledged: careful design is necessary if we are to have machines we can live with. But the dangers are unlikely to come in the way he imagines. He proposes a "simple thought experiment" which, in his words, "shows that a rational chess robot with a simple utility function would behave like a human sociopath fixated on chess." In this, Omohundro exemplifies the "Giant Cheesecake Fallacy" described by Yudkowsky (this volume)—i.e., he imagines that just because machines can do something, they will. But it is far from clear that the kind of behavior he imagines would maximize the machine's expected utility, or that we should go along with his Nietzschean view that a "cooperative drive" will be felt only by those at a competitive disadvantage. Man and supermachine.

A more science-based approach is needed. Formal models developed from a priori theories of rationality have proven to be of limited use for understanding the complex details of evolution and intelligence. So-called "simple heuristics", discovered empirically, may make organisms smart in ways that cannot be easily exploited in typical environments by more cumbersome rational optimization procedures. If this is all that Omohundro means by the phrase "as rational as possible" then his thesis has no teeth, predicting nothing but allowing everything. Careful design must proceed from detailed study and understanding of actual processes of evolution and the real, embodied forms of moral agency that evolution has provided.

The article by Omohundro exemplifies a broader problem with the singularity hypothesis. Gaps in the hypothesis are rationalized away or filled in with additional theories that are just as vague or just as difficult to verify as the initial conclusion that a technological singularity is inevitable. While the singularity may appear plausible to its proponents, the speculation, ad hoc theorizing, and inductive reasoning used in its support fall far short of scientific rigor.

References

Bargmann, C. I. (1998). Neurobiology of the Caenorhabditis elegans genome. *Science, 282*(5396), 2028–2032.

Clark, J. (2001). Google: 'At scale, everything breaks' ZDNet UK, June 22, 2011. Available online at http://www.zdnet.co.uk/news/cloud/2011/06/22/google-at-scale-everything-breaks-40093061/. Accessed March 23, 2011

Herculano-Houzel, S. (2009). The human brain in numbers: A linearly scaled-up primate brain. *Frontiers in Human Neuroscience, 3*, 1–11.

Koch, C.,& Laurent, G. (1999). Complexity and the nervous system. *Science, 284*(5411), 96–98.

Chapter 10
Friendly Artificial Intelligence

Eliezer Yudkowsky

Anthropomorphic Bias

By far the greatest danger of Artificial Intelligence is that people conclude too early that they understand it. Of course this problem is not limited to the field of AI. Jacques Monod wrote: “A curious aspect of the theory of evolution is that everybody thinks he understands it” (Monod 1974). Nonetheless the problem seems to be unusually acute in Artificial Intelligence. The field of AI has a reputation for making huge promises and then failing to deliver on them. Most observers conclude that AI is hard; as indeed it is. But the *embarrassment* does not stem from the difficulty. It is difficult to build a star from hydrogen, but the field of stellar astronomy does not have a terrible reputation for promising to build stars and then failing. The critical inference is *not* that AI is hard, but that, for some reason, it is very easy for people to think they know far more about Artificial Intelligence than they actually do.

Imagine a complex biological adaptation with ten necessary parts. If each of ten genes are independently at 50 % frequency in the gene pool—each gene possessed by only half the organisms in that species—then, on average, only 1 in 1,024 organisms will possess the full, functioning adaptation. A fur coat is not a significant evolutionary advantage unless the environment reliably challenges organisms with cold. Similarly, if gene B depends on gene A, then gene B has no significant advantage unless gene A forms a reliable part of the *genetic* environment. *Complex*, *interdependent* machinery is necessarily *universal* within a sexually reproducing species; it cannot evolve otherwise (Tooby and Cosmides 1992). One robin may have smoother feathers than another, but they will both have wings. Natural selection, while feeding on variation, uses it up (Sober 1984).

E. Yudkowsky (✉)
Machine Intelligence Research Institute, San Francisco, CA, USA
e-mail: yudkowsky@singinst.org

A. H. Eden et al. (eds.), *Singularity Hypotheses*, The Frontiers Collection,
DOI: 10.1007/978-3-642-32560-1_10, © Springer-Verlag Berlin Heidelberg 2012

In every known culture, humans experience joy, sadness, disgust, anger, fear, and surprise (Brown 1991), and indicate these emotions using the same facial expressions (Ekman and Keltner 1997). We all run the same engine under our hoods, though we may be painted different colors; a principle which evolutionary psychologists call *the psychic unity of humankind* (Tooby and Cosmides 1992). This observation is both explained and required by the mechanics of evolutionary biology.

Humans evolved to model other humans—to compete against and cooperate with our own conspecifics. It was a reliable property of the ancestral environment that every powerful intelligence you met would be a fellow human. We evolved to understand our fellow humans *empathically*, by placing ourselves in their shoes; for that which needed to be modeled was similar to the modeler. Not surprisingly, human beings often "anthropomorphize"—expect humanlike properties of that which is not human. In *The Matrix* (Wachowski and Wachowski 1999), the supposed "artificial intelligence" Agent Smith initially appears utterly cool and collected, his face passive and unemotional. But later, while interrogating the human Morpheus, Agent Smith gives vent to his disgust with humanity—and his face shows the human-universal facial expression for disgust.

Experiments on anthropomorphism show that subjects anthropomorphize unconsciously, often flying in the face of their deliberate beliefs. In a study by Barrett and Keil (1996), subjects strongly professed belief in non-anthropomorphic properties of God: that God could be in more than one place at a time, or pay attention to multiple events simultaneously. Barrett and Keil presented the same subjects with stories in which, for example, God saves people from drowning. The subjects answered questions about the stories, or retold the stories in their own words, in such ways as to suggest that God was in only one place at a time and performed tasks sequentially rather than in parallel. Serendipitously for our purposes, Barrett and Keil also tested an additional group using otherwise identical stories about a superintelligent computer named "Uncomp". For example, to simulate the property of omnipresence, subjects were told that Uncomp's sensors and effectors "cover every square centimeter of the earth and so no information escapes processing". Subjects in this condition also exhibited strong anthropomorphism, though significantly less than the God group. From our perspective, the key result is that even when people consciously believe an AI is unlike a human, they still visualize scenarios as if the AI were anthropomorphic (but not quite as anthropomorphic as God). Anthropomorphic bias can be classed as insidious: it takes place with no deliberate intent, without conscious realization, and in the face of apparent knowledge.

Back in the era of pulp science fiction, magazine covers occasionally depicted a sentient monstrous alien—colloquially known as a bug-eyed monster or BEM—carrying off an attractive human female in a torn dress. It would seem the artist believed that a non-humanoid alien, with a wholly different evolutionary history, would sexually desire human females. People don't make mistakes like that by explicitly reasoning: "All minds are likely to be wired pretty much the same way, so presumably a BEM will find human females sexually attractive". Probably the artist did not *ask* whether a giant bug *perceives* human females as attractive. Rather, a human female in a torn dress *is sexy*—inherently so, as an intrinsic property. They who made this mistake did not think about the insectoid's mind; they focused on the woman's torn dress. If the dress were not torn, the woman would be less sexy; the BEM doesn't enter into it. (This is a case of a deep, confusing, and extraordinarily common mistake which E. T. Jaynes named the *mind projection fallacy* (Jaynes and Bretthorst 2003). Jaynes, a theorist of Bayesian probability, coined "mind projection fallacy" to refer to the error of confusing states of knowledge with properties of objects. For example, the phrase *mysterious phenomenon* implies that mysteriousness is a property of the phenomenon itself. If I am ignorant about a phenomenon, then this is a fact about my state of mind, not a fact about the phenomenon).

People need not realize they are anthropomorphizing (or even realize they are engaging in a questionable act of predicting other minds) in order for anthropomorphism to supervene on cognition. When we try to reason about other minds, each step in the reasoning process may be contaminated by assumptions so ordinary in human experience that we take no more notice of them than air or gravity. You object to the magazine illustrator: "Isn't it more likely that a giant male bug would sexually desire giant female bugs?" The illustrator thinks for a moment and then says to you: "Well, even if an insectoid alien starts out liking hard exoskeletons, after the insectoid encounters human females it will soon realize that human females have much nicer, softer skins. If the aliens have sufficiently advanced technology, they'll genetically engineer themselves to like soft skins instead of hard exoskeletons".

This is a fallacy-at-one-remove. After the alien's anthropomorphic thinking is pointed out, the magazine illustrator takes a step back and tries to justify the alien's conclusion as a neutral product of the alien's reasoning process. Perhaps advanced aliens *could* re-engineer themselves (genetically or otherwise) to like soft skins, but would they *want* to? An insectoid alien who likes hard skeletons will not wish to change itself to like soft skins instead—not unless natural selection has somehow produced in it a distinctly human sense of meta-sexiness. When using long, complex chains of reasoning to argue in favor of an anthropomorphic conclusion, each and every step of the reasoning is another opportunity to sneak in the error.

The term "Artificial Intelligence" refers to a vastly greater *space of possibilities* than does the term "Homo sapiens". When we talk about "AIs" we are really talking about *minds-in-general*, or optimization processes in general. Imagine a map of mind design space. In one corner, a tiny little circle contains all humans; within a larger tiny circle containing all biological life; and all the rest of the huge map is *the space of minds-in-general*. It is this *enormous* space of possibilities which outlaws anthropomorphism as legitimate reasoning.

Prediction and Design

Anthropomorphism leads people to believe that they can make predictions, given no more information than that something is an "intelligence"—anthromorphism will go on generating predictions regardless, your brain automatically putting itself in the shoes of the "intelligence". This may have been one contributing factor to the embarrassing history of AI, which stems not from the difficulty of AI as such, but from the mysterious ease of acquiring erroneous beliefs about what a given AI design accomplishes.

We cannot query our own brains for answers about nonhuman optimization processes. How then may we proceed? How can we predict what Artificial Intelligences will do? I have deliberately asked this question in a form that makes it intractable. By the halting problem, it is impossible to predict whether an *arbitrary* computational system implements any input–output function, including, say, simple multiplication (Rice 1953). So how is it possible that human engineers can build computer chips which reliably implement multiplication? Because human engineers deliberately use designs that they *can* understand.

To make the statement that a bridge will support vehicles up to 30 tons, civil engineers have two weapons: choice of initial conditions, and safety margin. They need not predict whether an *arbitrary* structure will support 30 ton vehicles, only design a single bridge of which they can make this statement. And though it reflects well on an engineer who can correctly calculate the exact weight a bridge will support, it is also acceptable to calculate that a bridge supports vehicles of *at least* 30 tons—albeit to assert this vague statement *rigorously* may require much of the same theoretical understanding that would go into an exact calculation.

Civil engineers hold themselves to high standards in predicting that bridges will support vehicles. Ancient alchemists held themselves to much lower standards in predicting that a sequence of chemical reagents would transform lead into gold. How much lead into how much gold? What is the exact causal mechanism? It's clear enough why the alchemical researcher *wants* gold rather than lead, but why should this sequence of reagents transform lead to gold, instead of gold to lead or lead to water?

Some early AI researchers believed that an artificial neural network of layered thresholding units, trained via backpropagation, would be "intelligent". The wishful thinking involved was probably more analogous to alchemy than civil engineering.

Not knowing how to build a friendly AI is not deadly, of itself, in any specific instance, if you know you don't know. It's *mistaken* belief that an AI will be friendly which implies an obvious path to global catastrophe.

Capability and Motive

There is a fallacy oft-committed in discussion of Artificial Intelligence, especially AI of superhuman capability. Someone says: "When technology advances far enough we'll be able to build minds far surpassing human intelligence. Now, it's obvious that how large a cheesecake you can make depends on your intelligence. A superintelligence could build *enormous* cheesecakes—cheesecakes the size of cities—by golly, the future will be full of giant cheesecakes!" The question is whether the superintelligence *wants* to build giant cheesecakes. The vision leaps directly from *capability* to *actuality*, without considering the necessary intermediate of *motive*.

The following chains of reasoning, considered in isolation without supporting argument, all exhibit the Fallacy of the Giant Cheesecake:

- A sufficiently powerful Artificial Intelligence could overwhelm any human resistance and wipe out humanity (And the AI would decide to do so.). Therefore we should not build AI.
- A sufficiently powerful AI could develop new medical technologies capable of saving millions of human lives (And the AI would decide to do so.). Therefore we should build AI.
- Once computers become cheap enough, the vast majority of jobs will be performable by Artificial Intelligence more easily than by humans. A sufficiently powerful AI would even be better than us at math, engineering, music, art, and all the other jobs we consider meaningful (And the AI will decide to perform those jobs.). Thus after the invention of AI, humans will have nothing to do, and we'll starve or watch television.

Optimization Processes

The above deconstruction of the Fallacy of the Giant Cheesecake invokes an intrinsic anthropomorphism—the idea that motives are separable; the implicit assumption that by talking about "capability" and "motive" as separate entities, we are carving reality at its joints. This is a useful slice but an anthropomorphic one.

To view the problem in more general terms, I introduce the concept of an *optimization process*: a system which hits small targets in large search spaces to produce coherent real-world effects.

An optimization process steers the future into particular regions of the possible. I am visiting a distant city, and a local friend volunteers to drive me to the airport. I do not know the neighborhood. When my friend comes to a street intersection, I am at a loss to predict my friend's turns, either individually or in sequence. Yet I can predict the *result* of my friend's unpredictable actions: we will arrive at the airport. Even if my friend's house were located elsewhere in the city, so that my friend made a wholly different sequence of turns, I would just as confidently predict our destination. Is this not a strange situation to be in, scientifically speaking? I can predict the *outcome* of a process, without being able to predict any of the *intermediate steps* in the process. I will speak of the region into which an optimization process steers the future as that optimizer's *target*.

Consider a car, say a Toyota Corolla. Of all possible configurations for the atoms making up the Corolla, only an infinitesimal fraction qualify as a useful working car. If you assembled molecules at random, many ages of the universe would pass before you hit on a car. A tiny fraction of the design space does describe vehicles that we would recognize as faster, more efficient, and safer than the Corolla. Thus the Corolla is not *optimal* under the designer's goals. The Corolla is, however, *optimized*, because the designer had to hit a comparatively infinitesimal target in design space just to create a working car, let alone a car of the Corolla's quality. You cannot build so much as an effective wagon by sawing boards randomly and nailing according to coin flips. To hit such a tiny target in configuration space requires a powerful optimization process.

The notion of an "optimization process" is *predictively useful* because it can be easier to understand the *target* of an optimization process than to understand its step-by-step *dynamics*. The above discussion of the Corolla assumes *implicitly* that the designer of the Corolla was trying to produce a "vehicle", a means of travel. This assumption deserves to be made explicit, but it is not wrong, and it is highly useful in understanding the Corolla.

Aiming at the Target

The temptation is to ask what "AIs" will "want", forgetting that the space of minds-in-general is much wider than the tiny human dot. One should resist the temptation to spread quantifiers over all possible minds. Storytellers spinning tales of the distant and exotic land called Future, say how the future *will be*. They make *predictions*. They say, "AIs will attack humans with marching robot armies" or "AIs will invent a cure for cancer". They do not propose complex relations between initial conditions and outcomes—that would lose the audience. But we need relational understanding to *manipulate* the future, steer it into a region palatable to humankind. If we do not steer, we run the danger of ending up where we are going.

The critical challenge is not to *predict* that "AIs" will attack humanity with marching robot armies, or alternatively invent a cure for cancer. The task is not even to make the prediction for an *arbitrary* individual AI design. Rather the task is choosing into existence some *particular* powerful optimization process whose beneficial effects can legitimately be asserted.

I *strongly urge* my readers not to start thinking up reasons why a fully generic optimization process would be friendly. Natural selection isn't friendly, nor does it hate you, nor will it leave you alone. Evolution cannot be so anthropomorphized, it does not work like you do. Many pre-1960s biologists expected natural selection to do all sorts of nice things, and rationalized all sorts of elaborate reasons why natural selection would do it. They were disappointed, because natural selection itself did not start out knowing that it wanted a humanly-nice result, and then rationalize elaborate ways to produce nice results using selection pressures. Thus the events in Nature were outputs of causally different process from what went on in the pre-1960s biologists' minds, so that prediction and reality diverged.

Wishful thinking adds detail, constrains prediction, and thereby creates a burden of improbability. What of the civil engineer who hopes a bridge won't fall? Should the engineer argue that bridges in general are not likely to fall? But Nature itself does not rationalize reasons why bridges should not fall. Rather the civil engineer overcomes the burden of improbability through specific choice guided by specific understanding. A civil engineer starts by desiring a bridge; then uses a rigorous theory to select a bridge design which supports cars; then builds a real-world bridge whose structure reflects the calculated design; and thus the real-world structure supports cars. Thus achieving harmony of predicted positive results and actual positive results.

Friendly AI

It would be a very good thing if humanity knew how to choose into existence a powerful optimization process with a particular target. Or in more colloquial terms, it would be nice if we knew how to build a nice AI.

To describe the *field of knowledge* needed to address that challenge, I have proposed the term "Friendly AI". In addition to referring to a body of technique, "Friendly AI" might also refer to the *product* of technique—an AI created with specified motivations. When I use the term *Friendly* in either sense, I capitalize it to avoid confusion with the intuitive sense of "friendly".

One common reaction I encounter is for people to immediately declare that Friendly AI is an impossibility, because any sufficiently powerful AI will be able to modify its own source code to break any constraints placed upon it.

The first flaw you should notice is a Giant Cheesecake Fallacy. Any AI with free access to its own source would, in principle, possess the *ability* to modify its own source code in a way that changed the AI's optimization target. This does not imply the AI has the *motive* to change its own motives. I would not knowingly swallow a pill that made me enjoy committing murder, because *currently* I prefer that my fellow humans not die.

But what if I try to modify myself, and make a mistake? When computer engineers *prove* a chip valid—a good idea if the chip has 155 million transistors and you can't issue a patch afterward—the engineers use human-guided, machine-verified formal proof. The glorious thing about *formal* mathematical proof, is that a proof of 10 billion steps is just as reliable as a proof of 10 steps. But human beings are not trustworthy to peer over a purported proof of 10 billion steps; we have too high a chance of missing an error. And present-day theorem-proving techniques are not smart enough to design and prove an entire computer chip on their own—current algorithms undergo an exponential explosion in the search space. Human mathematicians can prove theorems far more complex than modern theorem-provers can handle, without being defeated by exponential explosion. But human mathematics is informal and unreliable; occasionally someone discovers a flaw in a previously accepted informal proof. The upshot is that human engineers guide a theorem-prover through the *intermediate* steps of a proof. The human chooses the next lemma, and a complex theorem-prover generates a formal proof, and a simple verifier checks the steps. That's how modern engineers build reliable machinery with 155 million interdependent parts.

Proving a computer chip correct requires a synergy of human intelligence and computer algorithms, as *currently* neither suffices on its own. Perhaps a true AI could use a similar *combination of abilities* when modifying its own code—would have *both* the capability to *invent* large designs without being defeated by exponential explosion, and *also* the ability to *verify* its steps with extreme reliability. That is one way a true AI might remain knowably stable in its goals, even after carrying out a large number of self-modifications.

This chapter will not explore the above idea in detail (Though see Schmidhuber 2003 for a related notion.). But one ought to think about a challenge, and study it in the best available technical detail, *before* declaring it impossible—especially if great stakes depend upon the answer. It is disrespectful to human ingenuity to declare a challenge unsolvable without taking a close look and exercising creativity. It is an enormously strong statement to say that you *cannot* do a thing—that you *cannot* build a heavier-than-air flying machine, that you *cannot* get useful energy from nuclear reactions, that you *cannot* fly to the Moon. Such statements are universal generalizations, quantified over every single approach that anyone ever has or ever will think up for solving the problem. It only takes a single counterexample to falsify a universal quantifier. The statement that Friendly (or friendly) AI is *theoretically impossible*, dares to quantify over *every possible* mind design and *every possible* optimization process—including human beings, who are also minds, some of whom are nice and wish they were nicer. At this point there are any number of vaguely plausible reasons why Friendly AI might be *humanly* impossible, and it is still more likely that the problem is solvable but no one will get around to solving it in time. But one should not so quickly write off the challenge, especially considering the stakes.

Technical Failure and Philosophical Failure

Bostrom (2001) defines an existential catastrophe as one which permanently extinguishes Earth-originating intelligent life *or destroys a part of its potential*. We can divide potential failures of attempted Friendly AI into two informal fuzzy categories, *technical failure* and *philosophical failure*. Technical failure is when you try to build an AI and it doesn't work the way you think it does—you have failed to understand the true workings of your own code. Philosophical failure is trying to build the wrong thing, so that even if you succeeded you would still fail to help anyone or benefit humanity. Needless to say, the two failures are not mutually exclusive.

The border between these two cases is thin, since most philosophical failures are much easier to explain in the presence of technical knowledge. In theory you ought first to say what you *want*, then figure out *how* to get it. In practice it often takes a deep technical understanding to figure out what you want.

An Example of Philosophical Failure

In the late 19th century, many honest and intelligent people advocated communism, all in the best of good intentions. The people who first invented and spread and swallowed the communist meme were, in sober historical fact, idealists. The *first* communists did not have the example of Soviet Russia to warn them. *At the time, without benefit of hindsight, it must have sounded like a pretty good idea.* After the revolution, when communists came into power and were corrupted by it, other motives may have come into play; but this itself was not something the first idealists predicted, however, predictable it may have been. It is *important* to understand that the authors of huge catastrophes need not be evil, nor even *unusually* stupid. If we attribute every tragedy to evil or unusual stupidity, we will look at ourselves, correctly perceive that we are not evil or unusually stupid, and say: "But that would never happen to *us*".

What the first communist revolutionaries thought would happen, as the empirical consequence of their revolution, was that people's lives would improve: laborers would no longer work long hours at backbreaking labor and make little money from it. This turned out not to be the case, to put it mildly. But what the first communists *thought* would happen, was not so very different from what advocates of other political systems thought would be the empirical consequence of *their* favorite political systems. They thought people would be happy. They were wrong.

Now imagine that someone should attempt to program a "Friendly" AI to implement communism, or libertarianism, or anarcho-feudalism, or *favoritepoliticalsystem*, believing that this shall bring about utopia. People's favorite political systems inspire blazing suns of positive affect, so the proposal will sound like a really good idea to the proposer.

We could view the programmer's failure on a moral or ethical level—say that it is the result of someone trusting themselves too highly, failing to take into account their own fallibility, refusing to consider the possibility that communism might be mistaken after all. But in the language of Bayesian decision theory, there's a complementary technical view of the problem. From the perspective of decision theory, the choice for communism stems from combining an empirical belief with a value judgment. The *empirical* belief is that communism, when implemented, results in a specific outcome or class of outcomes: people will be happier, work fewer hours, and possess greater material wealth. This is ultimately an *empirical* prediction; even the part about happiness is a real property of brain states, though hard to measure. If you implement communism, either this outcome eventuates or it does not. The value judgment is that this outcome satisfices or is preferable to current conditions. Given a different *empirical* belief about the *actual real-world consequences* of a communist system, the decision may undergo a corresponding change.

We would expect a true AI, an Artificial General Intelligence, to be capable of changing its empirical beliefs (Or its probabilistic world-model, etc.). If somehow Charles Babbage had lived before Nicolaus Copernicus, and somehow computers had been invented before telescopes, and somehow the programmers of that day and age successfully created an Artificial General Intelligence, it would not follow that the AI would believe forever after that the Sun orbited the Earth. The AI might transcend the factual error of its programmers, provided that the programmers understood inference rather better than they understood astronomy. To build an AI that *discovers* the orbits of the planets, the programmers need not know the math of Newtonian mechanics, only the math of Bayesian probability theory.

The folly of programming an AI to implement communism, or any other political system, is that you're programming *means* instead of *ends*. You're programming in a fixed decision, without that decision being re-evaluable after acquiring improved empirical knowledge about the results of communism. You are giving the AI a fixed decision without telling the AI how to re-evaluate, at a higher level of intelligence, the fallible process which produced that decision.

If I play chess against a stronger player, I cannot predict *exactly* where my opponent will move against me—if I could predict that, I would necessarily be at least that strong at chess myself. But I can predict the end result, which is a win for the other player. I know the region of possible futures my opponent is aiming for, which is what lets me predict the destination, even if I cannot see the path. When I am at my most creative, that is when it is hardest to predict my actions, and *easiest* to predict the *consequences* of my actions (Providing that you know and understand my goals!). If I want a better-than-human chess player, I have to program a *search* for winning moves. I can't program in specific moves because then the chess player won't be any better than I am. When I launch a search, I necessarily sacrifice my ability to predict the *exact* answer in advance. To get a really good answer you must sacrifice your ability to predict the answer, albeit not your ability to say what is the question.

Such confusion as to program in communism directly, probably would not tempt an AGI programmer who speaks the language of decision theory. I would call it a philosophical failure, but blame it on lack of technical knowledge.

An Example of Technical Failure

In place of laws constraining the behavior of intelligent machines, we need to give them emotions that can guide their learning of behaviors. They should want us to be happy and prosper, which is the emotion we call love. We can design intelligent machines so their primary, innate emotion is unconditional love for all humans. First we can build relatively simple machines that learn to recognize happiness and unhappiness in human facial expressions, human voices and human body language. Then we can hard-wire the result of this learning as the innate emotional values of more complex intelligent machines, positively reinforced when we are happy and negatively reinforced when we are unhappy. Machines can learn algorithms for approximately predicting the future, as for example investors currently use learning machines to predict future security prices. So we can program intelligent machines to learn algorithms for predicting future human happiness, and use those predictions as emotional values.

—Bill Hibbard (2001), *Super-intelligent machines*

Once upon a time, the US Army wanted to use neural networks to automatically detect camouflaged enemy tanks. The researchers trained a neural net on 50 photos of camouflaged tanks in trees, and 50 photos of trees without tanks. Using standard techniques for supervised learning, the researchers trained the neural network to a weighting that correctly loaded the training set—output "yes" for the 50 photos of camouflaged tanks, and output "no" for the 50 photos of forest. This did not ensure, or even imply, that *new* examples would be classified correctly. The neural network might have "learned" 100 special cases that would not generalize to any new problem. Wisely, the researchers had originally taken 200 photos, 100 photos of tanks and 100 photos of trees. They had used only 50 of each for the training set. The researchers ran the neural network on the remaining 100 photos, and without further training the neural network classified all remaining photos correctly. Success confirmed! The researchers handed the finished work to the Pentagon, which soon handed it back, complaining that in their own tests the neural network did no better than chance at discriminating photos.

It turned out that in the researchers' data set, photos of camouflaged tanks had been taken on cloudy days, while photos of plain forest had been taken on sunny days. The neural network had learned to distinguish cloudy days from sunny days, instead of distinguishing camouflaged tanks from empty forest.[1]

A technical failure occurs when the code does not do what you think it does, though it faithfully executes as you programmed it. More than one model can load the same data. Suppose we trained a neural network to recognize smiling human faces and distinguish them from frowning human faces. Would the network classify a tiny picture of a smiley-face into the same attractor as a smiling human face? If an AI "hard-wired" to such code possessed the power—and Hibbard

[1] This story, though famous and oft-cited as fact, *may* be apocryphal; I could not find a first-hand report. For unreferenced reports see e.g. Crochat and Franklin (2000) or http://neil.fraser.name/writing/tank/. However, failures of the type described are a major real-world consideration when building and testing neural networks.

(2001) spoke of superintelligence—would the galaxy end up tiled with tiny molecular pictures of smiley-faces?[2]

This form of failure is especially dangerous because it will *appear* to work within a fixed context, then fail when the context changes. The researchers of the "tank classifier" story tweaked their neural network until it correctly loaded the training data, then verified the network on additional data (without further tweaking). Unfortunately, both the training data and verification data turned out to share an assumption which held over the all data used in development, but not in all the real-world contexts where the neural network was called upon to function. In the story of the tank classifier, the assumption is that tanks are photographed on cloudy days.

Suppose we wish to develop an AI of increasing power. The AI possesses a developmental stage where the human programmers are more powerful than the AI—not in the sense of mere physical control over the AI's electrical supply, but in the sense that the human programmers are smarter, more creative, more cunning than the AI. During the developmental period we suppose that the programmers possess the ability to make changes to the AI's source code without needing the consent of the AI to do so. However, the AI is also intended to possess postdevelopmental stages, including, in the case of Hibbard's scenario, superhuman intelligence. An AI of superhuman intelligence surely could not be modified without its consent. At this point we must rely on the previously laid-down goal system to function correctly, because if it operates in a sufficiently unforeseen fashion, the AI may actively resist our attempts to correct it—and, if the AI is smarter than a human, probably win.

Trying to control a growing AI by *training a neural network to provide its goal system* faces the problem of a huge *context change* between the AI's developmental stage and postdevelopmental stage. During the developmental stage, the AI may *only* be able to produce stimuli that fall into the "smiling human faces" category, by solving humanly provided tasks, as its makers intended. Flash forward to a time when the AI is superhumanly intelligent and has built its own nanotech infrastructure, and the AI may be able to produce stimuli classified into the same attractor by tiling the galaxy with tiny smiling faces.

Thus the AI appears to work fine during development, but produces catastrophic results after it becomes smarter than the programmers(!).

There is a temptation to think, "But surely the AI will know that's not what we meant?" But the code is not *given* to the AI, for the AI to look over and hand back if it does the wrong thing. The code *is* the AI. Perhaps with enough effort and

---

[2] Bill Hibbard, after viewing a draft of this paper, wrote a response arguing that the analogy to the "tank classifier" problem does not apply to reinforcement learning in general. His critique may be found at http://www.ssec.wisc.edu/~billh/g/AIRisk_Reply.html. My response may be found at http://yudkowsky.net/AIRisk_Hibbard.html. Hibbard also notes that the proposal of Hibbard (2001) is superseded by Hibbard (2004). The latter recommends a two-layer system in which expressions of agreement from humans reinforce recognition of happiness, and recognized happiness reinforces action strategies.

understanding we can write code that cares if we have written the wrong code—the legendary DWIM instruction, which among programmers stands for Do-What-I-Mean (Raymond 2003). But effort is required to write a DWIM dynamic, and nowhere in Hibbard's proposal is there mention of designing an AI that does what we mean, not what we say. Modern chips don't DWIM their code; it is not an automatic property. And if you messed up the DWIM itself, you would suffer the consequences. For example, suppose DWIM was defined as maximizing the satisfaction of the programmer with the code; when the code executed as a superintelligence, it might rewrite the programmers' brains to be maximally satisfied with the code. I do not say this is inevitable; I only point out that Do-What-I-Mean is a major, nontrivial technical challenge of Friendly AI.

References

Barrett, J. L., & Keil, F. (1996). Conceptualizing a non-natural entity: anthropomorphism in god concepts. *Cognitive Psychology, 31*, 219–247.

Bostrom, N. (2001). Existential risks: analyzing human extinction scenarios. *Journal of Evolution and Technology, 9*.

Brown, D. E. (1991). *Human universals*. New York: McGraw-Hill.

Crochat, P., & Franklin, D. (2000). Back-propagation neural network tutorial. http://ieee.uow.edu.au/~daniel/software/libneural/.

Ekman, P., & Keltner, D. (1997). Universal facial expressions of emotion: An old controversy and new findings. In U. Segerstrale & P. Molnar (Eds.), *Nonverbal communication: Where nature meets culture*. Mahwah: Lawrence Erlbaum Associates.

Hibbard, B. (2001). Super-intelligent machines. *ACM SIGGRAPH Computer Graphics, 35*(1) .

Hibbard, B. (2004). Reinforcement learning as a context for integrating AI research. Presented at the 2004 AAAI Fall Symposium on Achieving Human-Level Intelligence through Integrated Systems and Research.

Jaynes, E. T., & Bretthorst, G. L. (2003). *Probability theory: The logic of science*. Cambridge: Cambridge University Press.

Monod, J. L. (1974). *On the molecular theory of evolution*. Oxford: New York.

Raymond, E. S. ed. (2003). DWIM. *The on-line hacker Jargon File*, version 4.4.7, 29 Dec 2003.

Rice, H. G. (1953). Classes of recursively enumerable sets and their decision problems. *Transactions of the American Mathematical Society, 74*, 358–366.

Schmidhuber, J. (2003). Goedel machines: self-referential universal problem solvers making provably optimal self-improvements. In B. Goertzel & C. Pennachin (Eds), *Artificial general intelligence*. Forthcoming. New York: Springer.

Sober, E. (1984). *The nature of selection*. Cambridge: MIT Press.

Tooby, J., & Cosmides, L. (1992). The psychological foundations of culture. In J. H. Barkow, L. Cosmides, & J. Tooby (Eds.), *The adapted mind: Evolutionary psychology and the generation of culture*. New York: Oxford University Press.

Wachowski, A., & Wachowski, L. (1999). *The Matrix*, USA, Warner Bros, 135 min.

Chapter 10A
Colin Allen on Yudkowsky's "Friendly Artificial Intelligence"

Friendly Advice?

Yudkowsky begins with a warning to his readers that "By far the greatest danger of Artificial Intelligence is that people conclude too early that they understand it". He ends by reminding us that software written to "Do-What-I-Mean is a major, nontrivial technical challenge of Friendly AI". Yudkowsky suggests a history of over-exuberant claims about AI, commenting that early proponents of the idea that artificial neural networks would be intelligent were engaged in "wishful thinking probably more analogous to alchemy than civil engineering". He indicates that anyone who predicts strongly utopian or dystopian outcomes from superhuman AI is committing the "Giant Cheesecake Fallacy"—the mistake of thinking that just because a powerful intelligence could do something it will do that thing. His message seems to be that we should be neither terrified of superhuman AI nor naive about the challenge of building superhuman AI that will be "nice".

So, what is to be the approach to designing Friendly AI? Yudkowsky characterizes the challenge as one of choosing a powerful enough optimization process with an appropriate target. Engineers, he asserts, use a rigorous theory to select a design and then build structures implementing the calculated designs. But, he cautions, we must beware of two kinds of errors: "philosophical failure", i.e. choosing the wrong target, and "technical failure", i.e. wrongly assuming that a system will work optimally in contexts other than those in which it has been tested.

As heuristics, these can hardly be faulted. But like the classic stockbroker's platitude, "buy low, sell high", they give no practical advice. Yudkowsky's repetition of a an apocryphal story about the failure of a neural network program at classifying photographs of tanks—a story that I remember hearing over 25 years ago—hardly enlightens. (If the advice is "Don't rely on backprop!" this is hardly news.) Likewise, to be told that to "build an AI that discovers the orbits of the planets, the programmers need know only the math of Bayesian probability theory" is facile.

Yudkowsky correctly points out that engineering, like evolution, explores a tiny fraction of design space, but the rest of his story is shallow. Both processes are historically-bound. They work by modification of designs that are received from the past. Engineers do not start only with a target specification, but with a choice of platforms from which to try to reach that target. Inspired engineering sometimes involves taking something that was designed for one context and applying it in another, but always it involves cycles of testing and refinement, and it is far from guaranteeing optimization. Where should those who want to program "Friendly AI" begin?

Yudkowsky cites nothing more recent than 2004, but in the interim many new books and articles have been published, some proposing quite specific architectures or discussing particular programming paradigms for well-behaved autonomous systems. It would have been nice to know whether Yudkowsky thinks any of this work is on the right track, and if not, why not. If Bayesian theory can discover the orbits of planets, is it suitable for discovering "nice" AI? If not, why not? In describing a developmental neural network approach to AI, Yudkowsky shows a tendency, all too common among writers on this topic, when he asks us to "[f]lash forward to a time when the AI is superhumanly intelligent". We jump straight to sci fi without being given any clue how that flash occurs.

I hoped for more in the context of the present volume, with its stated goal to "reformulate the singularity hypothesis as a coherent and falsifiable conjecture and to investigate its most likely consequences, in particular those associated with existential risks". For our assessment of the existential risks, some knowledge of the current engineering pathways is crucial. If the path to Friendly AI with superhuman intelligence goes through explicit, top-down reasoning the existential risks may be rather different than if it goes through implicit, bottom-up processes. Different kinds of philosophical and technical failures are likely to accompany the different approaches. Similarly, if the route to superhuman AI runs through our self-driving automobiles, the risks may be rather different than if they run through our battle-ready military robots. What is clear is that our current understanding of how to build intelligent machines is low, but we have only the vaguest ideas about how to make it high.

Part III
A Singularity of Posthuman Superintelligence

Chapter 11
The Biointelligence Explosion

How Recursively Self-Improving Organic Robots will Modify their Own Source Code and Bootstrap Our Way to Full-Spectrum Superintelligence

David Pearce

Abstract This essay explores how recursively self-improving organic robots will modify their own genetic source code and bootstrap our way to full-spectrum superintelligence. Starting with individual genes, then clusters of genes, and eventually hundreds of genes and alternative splice variants, tomorrow's biohackers will exploit "narrow" AI to debug human source code in a positive feedback loop of mutual enhancement. Genetically enriched humans can potentially abolish aging and disease; recalibrate the hedonic treadmill to enjoy gradients of lifelong bliss, and phase out the biology of suffering throughout the living world.

Homo sapiens, the first truly free species, is about to decommission natural selection, the force that made us.... Soon we must look deep within ourselves and decide what we wish to become.

Edward O. Wilson

Consilience, The Unity of Knowledge (1999)

I predict that the domestication of biotechnology will dominate our lives during the next fifty years at least as much as the domestication of computers has dominated our lives during the previous fifty years.

Freeman Dyson

New York Review of Books (July 19 2007)

D. Pearce (✉)
BLTC Reseerch, 7 Lower Rock Gardens, Brighton, BN2 1PG, UK
e-mail: dave@hedweb.com

A. H. Eden et al. (eds.), *Singularity Hypotheses*, The Frontiers Collection,
DOI: 10.1007/978-3-642-32560-1_11, © Springer-Verlag Berlin Heidelberg 2012

The Fate of the Germline

Genetic evolution is slow. Progress in artificial intelligence is fast (Kurzweil 2005). Only a handful of genes separate *Homo sapiens* from our hominid ancestors on the African savannah. Among our 23,000-odd protein-coding genes, variance in single nucleotide polymorphisms accounts for just a small percentage of phenotypic variance in intelligence as measured by what we call IQ tests. True, the tempo of human evolution is about to accelerate. As the reproductive revolution of "designer babies" (Stock 2002) gathers pace, prospective parents will pre-select alleles and allelic combinations for a new child *in anticipation of* their behavioural effects—a novel kind of selection pressure to replace the "blind" genetic roulette of natural selection. In time, routine embryo screening via preimplantation genetic diagnosis will be complemented by gene therapy, genetic enhancement and then true designer zygotes. In consequence, life on Earth will also become progressively happier as the hedonic treadmill is recalibrated. In the new reproductive era, hedonic set-points and intelligence alike will be ratcheted upwards in virtue of selection pressure. For what parent-to-be wants to give birth to a low-status depressive "loser"? Future parents can enjoy raising a normal transhuman supergenius who grows up to be faster than Usain Bolt, more beautiful than Marilyn Monroe, more saintly than Nelson Mandela, more creative than Shakespeare—and smarter than Einstein.

Even so, the accelerating growth of germline engineering will be a *comparatively* slow process. In this scenario, sentient biological machines will design cognitively self-amplifying biological machines who will design cognitively self-amplifying biological machines. Greater-than-human biological intelligence will transform itself into posthuman superintelligence. Cumulative gains in intellectual capacity and subjective well-being across the generations will play out over hundreds and perhaps thousands of years—a momentous discontinuity, for sure, and a twinkle in the eye of eternity; but not a Biosingularity.

Biohacking Your Personal Genome

Yet **germline** engineering is only one strand of the genomics revolution. Indeed, after humans master the ageing process (de Grey 2007), the extent to which traditional germline or human generations will persist in the post-ageing world is obscure. Focus on the human germline ignores the slow-burning but then explosive growth of **somatic** gene enhancement in prospect. Later this century, innovative gene *therapies* will be succeeded by gene *enhancement* technologies—a value-laden dichotomy that reflects our impoverished human aspirations. Starting with individual genes, then clusters of genes, and eventually hundreds of genes and alternative splice variants, a host of recursively self-improving organic robots ("biohackers") will modify their genetic source code and modes of sentience: their senses, their moods, their motivation, their cognitive apparatus, their world-simulations and their default state of consciousness.

As the era of open-source genetics unfolds, tomorrow's biohackers will add, delete, edit and customise their own legacy code in a positive feedback loop of cognitive enhancement. Computer-aided genetic engineering will empower biological humans, transhuman and then posthuman to synthesise and insert new genes, variant alleles and even designer chromosomes—reweaving the multiple layers of regulation of our DNA to suit their wishes and dreams rather than the inclusive fitness of their genes in the ancestral environment. Collaborating and competing, next-generation biohackers will use stem-cell technologies to expand their minds, literally, via controlled neurogenesis. Freed from the constraints of the human birth canal, biohackers may re-sculpt the prison-like skull of *Homo sapiens* to accommodate a larger mind/brain, which can initiate recursive self-expansion in turn. Six crumpled layers of neocortex fed by today's miserly reward pathways aren't the upper bound of conscious mind, merely its seedbed. Each biological neuron and glial cell of your growing mind/brain can have its own dedicated artificial healthcare team, web-enabled nanobot support staff, and social network specialists; compare today's anonymous neural porridge. Transhuman minds will be augmented with neurochips, molecular nanotechnology (Drexler 1986), mind/computer interfaces, and full-immersion virtual reality (Sherman 2002) software. To achieve finer-grained control of cognition, mood and motivation, genetically enhanced transhumans will draw upon exquisitely tailored new designer drugs, nutraceuticals and cognitive enhancers—precision tools that make today's crude interventions seem the functional equivalent of glue-sniffing.

By way of comparison, early in the twenty-first century the scientific counterculture is customizing a bewildering array of designer drugs (Shulgin 1995) that outstrip the capacity of the authorities to regulate or comprehend. The bizarre psychoactive effects of such agents dramatically expand the evidential base that our theory of consciousness (Chalmers 1995) must explain. However, such drugs are short-acting. Their benefits, if any, aren't cumulative. By contrast, the ability genetically to hack one's own source code will unleash an exponential growth of genomic rewrites—not mere genetic tinkering but a comprehensive redesign of "human nature". Exponential growth starts out almost unnoticeably, and then explodes. Human bodies, cognition and ancestral modes of consciousness alike will be transformed. Post-humans will range across immense state-spaces of conscious mind hitherto impenetrable because access to their molecular biology depended on crossing gaps in the fitness landscape (Langdon and Poli 2002) prohibited by natural selection. Intelligent agency can "leap across" such fitness gaps. What we'll be leaping into is currently for the most part unknown: an inherent risk of the empirical method. But mastery of our reward circuitry can guarantee such state-spaces of experience will be glorious beyond human imagination. For intelligent biohacking can make unpleasant experience physically impossible (Pearce 1995) because its molecular substrates are absent. Hedonically enhanced innervation of the neocortex can ensure a rich hedonic tone saturates whatever strange new modes of experience our altered neurochemistry discloses.

Pilot studies of radical genetic enhancement will be difficult. Randomised longitudinal trials of such interventions in long-lived humans would take decades. In fact officially licensed, well-controlled prospective trials to test the safety and efficacy of genetic innovation will be hard if not impossible to conduct because all of us, apart from monozygotic twins, are genetically unique. Even monozygotic twins exhibit different epigenetic and gene expression profiles. Barring an ideological and political revolution, most formally drafted proposals for genetically-driven life-enhancement probably won't pass ethics committees or negotiate the maze of bureaucratic regulation. But that's the point of biohacking (Wohlsen 2011). By analogy today, if you're technically savvy, you don't want a large corporation controlling the operating system of your personal computer: you use open-source software instead. Likewise, you don't want governments controlling your state of mind via drug laws. By the same token, tomorrow's biotech-savvy individualists won't want anyone restricting our right to customise and rewrite our own genetic source code in any way we choose.

Will central governments try to regulate personal genome editing? Most likely yes. How far they'll succeed is an open question. So too is the success of any centralised regulation of futuristic designer drugs or artificial intelligence. Another huge unknown is the likelihood of state-sponsored designer babies, human reproductive cloning, and autosomal gene enhancement programs; and their interplay with privately-funded initiatives. China, for instance, has a different historical memory from the West.

Will there initially be biohacking accidents? Personal tragedies? Most probably yes, until human mastery of the pleasure-pain axis is secure. By the end of the next decade, every health-conscious citizen will be broadly familiar with the architecture of his or her personal genome: the cost of personal genotyping will be trivial, as will be the cost of DIY gene-manipulation kits. Let's say you decide to endow yourself with an extra copy of the N-methyl D-aspartate receptor subtype 2B (NR2B) receptor, a protein encoded by the GRIN2B gene. Possession of an extra NR2B subunit NMDA receptor is a crude but effective way to enhance your learning ability, at least if you're a transgenic mouse. Recall how Joe Tsien (1999) and his colleagues first gave mice extra copies of the NR2B receptor-encoding gene, then tweaked the regulation of those genes so that their activity would increase as the mice grew older. Unfortunately, it transpires that such brainy "Doogie mice"—and maybe brainy future humans endowed with an extra NR2B receptor gene—display greater pain-sensitivity too; certainly, NR2B receptor *blockade* reduces pain and learning ability alike. Being smart, perhaps you decide to counteract this heightened pain-sensitivity by inserting and then over-expressing a high pain-threshold, "low pain" allele of the SCN9A gene in your nociceptive neurons at the dorsal root ganglion and trigeminal ganglion. The SCN9A gene regulates pain-sensitivity; nonsense mutations abolish the capacity to feel pain at all (Reimann et al. 2010). In common with taking polydrug cocktails, the factors to consider in making multiple gene modifications soon snowball; but you'll have heavy-duty computer software to help. Anyhow, the potential pitfalls and makeshift solutions illustrated in this hypothetical example could be multiplied in the face of a combinatorial explosion of possibilities on the horizon. Most risks—and opportunities—of genetic self-editing are presumably still unknown.

It is tempting to condemn such genetic self-experimentation as irresponsible, just as unlicensed drug self-experimentation is irresponsible. Would you want your teenage daughter messing with her DNA? Perhaps we may anticipate the creation of a genetic counterpart of the Drug Enforcement Agency to police the human genome and its transhuman successors. Yet it's worth bearing in mind how each act of sexual reproduction today is an unpoliced genetic experiment with unfathomable consequences too. Without such reckless genetic experimentation, none of us would exist. In a cruel Darwinian world, this argument admittedly cuts both ways (Benatar 2006).

Naively, genomic source code self-editing will always be too difficult for anyone beyond dedicated cognitive elite of recursively self-improving biohackers. Certainly there are strongly evolutionarily conserved "housekeeping" genes that archaic humans would be best advised to leave alone for the foreseeable future. Granny might do well to customize her Windows desktop rather than her personal genome—prior to her own computer-assisted enhancement, at any rate. Yet the Biointelligence Explosion won't depend on more than a small fraction of its participants mastering the functional equivalent of machine code—the three billion odd 'A's, 'C's, 'G's and 'T's of our DNA. For the open-source genetic revolution will be propelled by powerful suites of high-level gene-editing tools, insertion vector applications, nonviral gene-editing kits, and user-friendly interfaces. Clever computer modelling and "narrow" AI can assist the intrepid biohacker to become a recursively self-improving genomic innovator. Later this century, your smarter counterpart will have software tools to monitor and edit every gene, repressor, promoter and splice variant in every region of the genome: each layer of epigenetic regulation of your gene transcription machinery in every region of the brain. This intimate level of control won't involve just crude DNA methylation to turn genes off and crude histone acetylation to turn genes on. Personal self-invention will involve mastery and enhancement of the histone and micro-RNA codes to allow sophisticated fine-tuning of gene expression and repression across the brain. Even today, researchers are exploring "nanochannel electroporation" (Boukany et al. 2011) technologies that allow the mass-insertion of novel therapeutic genetic elements into our cells. Mechanical cell-loading systems will shortly be feasible that can inject up to 100,000 cells at a time. Before long, such technologies will seem primitive. Freewheeling genetic self-experimentation will be endemic as the DIY-Bio revolution unfolds. At present, crude and simple gene-editing can be accomplished only via laborious genetic engineering techniques. Sophisticated authoring tools don't exist. In future, computer-aided genetic and epigenetic enhancement can become an integral part of your personal growth plan.

Will Humanity's Successors also be Our Descendants?

To contrast "biological" with "artificial" conceptions of posthuman superintelligence is convenient. The distinction may also prove simplistic. In essence, whereas genetic change in biological humanity has always been slow, the software run on serial, programmable digital computers is executed exponentially faster (*cf.* Moore's Law); it's copyable without limit; it runs on multiple substrates; and it can be cheaply and rapidly edited, tested and debugged. Extrapolating, Singularitarians like Ray Kurzweil (1990) and Eliezer Yudkowsky (2008) prophesy that human programmers will soon become redundant because autonomous AI run on digital computers will undergo accelerating cycles of self-improvement. In this kind of scenario, artificial, greater-than-human nonbiological intelligence will be rapidly succeeded by artificial posthuman superintelligence.

So we may distinguish two radically different conceptions of posthuman superintelligence: on one hand, our supersentient, cybernetically enhanced, genetically rewritten *biological* descendants, on the other, *non*biological superintelligence, either a Kurzweilian ecosystem or the singleton Artificial General Intelligence (AGI) foretold by the Singularity Institute for Artificial Intelligence. Such a divide doesn't reflect a clean contrast between "natural" and "artificial" intelligence, the biological and the nonbiological. This contrast may prove another false dichotomy. Transhuman biology will increasingly become synthetic biology as genetic enhancement plus cyborgization proceeds apace. "Cyborgization" is a barbarous term to describe an invisible and potentially life-enriching symbiosis of biological sentience with artificial intelligence. Thus "narrow-spectrum" digital superintelligence on web-enabled chips can be more-or-less seamlessly integrated into our genetically enhanced bodies and brains. Seemingly limitless formal knowledge can be delivered on tap to supersentient organic wetware, i.e. us. *Critically, transhumans can exploit what is misleadingly known as "narrow" or "weak" AI to enhance our own code in a positive feedback loop of mutual enhancement*—first plugging in data and running multiple computer simulations, then tweaking and re-simulating once more. In short, biological humanity won't just be the spectator and passive consumer of the intelligence explosion, but its driving force. The smarter our AI, the greater our opportunities for reciprocal improvement. Multiple "hard" and "soft" take-off scenarios to posthuman superintelligence can be outlined for recursively self-improving organic robots, not just nonbiological AI (Good 1965). Thus for serious biohacking later this century, artificial quantum supercomputers (Deutsch 2011) may be deployed rather than today's classical toys to test-run multiple genetic interventions, accelerating the tempo of our recursive self-improvement. Quantum supercomputers exploit quantum coherence to do googols of computations all at once. So the accelerating growth of human/computer synergies means it's premature to suppose biological evolution will be superseded by technological evolution, let alone a "robot rebellion" as the parasite swallows its host (de Garis 2005; Yudkowsky 2008). As the human era comes to a close, the fate of biological (post)humanity is more likely to be symbiosis with AI followed by metamorphosis, not simple replacement.

Despite this witche's brew of new technologies, a conceptual gulf remains in the futurist community between those who imagine human destiny, if any, lies in digital computers running programs with (hypothetical) artificial consciousness; and in contrast radical bioconservatives who believe that our posthuman successors will also be our supersentient descendants at their organic neural networked core—not the digital zombies of symbolic AI (Haugeland 1985) run on classical serial computers or their souped-up multiprocessor cousins. For one metric of progress in AI remains stubbornly unchanged: despite the exponential growth of transistors on a microchip, the soaring clock speed of microprocessors, the growth in computing power measured in MIPS, the dramatically falling costs of manufacturing transistors and the plunging price of dynamic RAM etc., any chart plotting the growth rate in digital sentience shows neither exponential growth, nor linear growth, but no progress whatsoever. As far as we can tell, digital computers are still zombies. Our machines are becoming autistically intelligent, but not supersentient—nor even conscious. On some fairly modest philosophical assumptions, digital computers were not subjects of experience in 1946 (*cf.* ENIAC); nor are they conscious subjects in 2012 (*cf.* "Watson") (Baker 2011); nor do researchers know how any kind of sentience may be "programmed" in future. So what if anything does consciousness *do*? Is it computationally redundant? Pre-reflectively, we tend to have a "dimmer-switch" model of sentience: "primitive" animals have minimal awareness and "advanced" animals like human beings experience a proportionately more intense awareness. By analogy, most AI researchers assume that at a given threshold of complexity/intelligence/processing speed, consciousness will somehow "switch on", turn reflexive, and intensify too. The problem with the dimmer-switch model is that our most intense experiences, notably raw agony or blind panic, are also the most phylogenetically ancient, whereas the most "advanced" modes (e.g. linguistic thought and the rich generative syntax that has helped one species to conquer the globe) are phenomenologically so thin as to be barely accessible to introspection. Something is seriously amiss with our entire conceptual framework.

So the structure of the remainder of this essay is as follows. I shall first discuss the risks and opportunities of building friendly biological superintelligence. Next I discuss the nature of *full-spectrum* superintelligence—and why consciousness is computationally fundamental to the past, present and future success of organic robots. Why couldn't recursively self-improving *zombies* modify their own genetic source code and bootstrap their way to full-spectrum superintelligence, i.e. a zombie intelligence explosion? Finally, and most speculatively, I shall discuss the future of sentience in the cosmos.

Can We Build Friendly Biological Superintelligence?

Risk–Benefit Analysis

Crudely speaking, evolution "designed" male human primates to be hunters/warriors. Evolution "designed" women to be attracted to powerful, competitive alpha males. Until humans rewrite our own hunter-gatherer source code, we shall continue to practise extreme violence (Peterson and Wrangham 1997) against members of other species—and frequently against members of our own. A heritable (and conditionally activated) predisposition to unfriendliness shown towards members of other races and other species is currently hardwired even in "social" primates. Indeed we have a (conditionally activated) predisposition to compete against, and harm, anyone who isn't a genetically identical twin. Compared to the obligate siblicide found in some bird species, human sibling rivalry isn't normally so overtly brutal. But conflict as well as self-interested cooperation is endemic to Darwinian life on Earth. This grim observation isn't an argument for genetic determinism, or against gene-culture co-evolution, or to discount the decline of everyday violence with the spread of liberal humanitarianism—just a reminder of the omnipresence of immense risks so long as we're shot through with legacy malware. Attempting to conserve the genetic status quo in an era of weapons of mass destruction poses unprecedented global catastrophic and existential risks (Bostrom 2002). Indeed the single biggest underlying threat to the future of sentient life within our cosmological horizon derives, not from asocial symbolic AI software in the basement turning rogue and going FOOM (a runaway computational explosion of recursive self-improvement), but from conserving human nature in its present guise. In the twentieth century, male humans killed over 100 million fellow human beings and billions of non-human animals (Singer 1995). This century's toll may well be higher. Mankind currently spends well over a trillion dollars each year on weapons designed to kill and maim other humans. The historical record suggests such weaponry won't all be beaten into ploughshares.

Strictly speaking, however, humanity is more likely to be wiped out by *idealists* than by misanthropes, death-cults or psychologically unstable dictators. Anti-natalist philosopher David Benatar's plea ("Better Never to Have Been") for human extinction via voluntary childlessness (Benatar 2006) must fail if only by reason of selection pressure; but not everyone who shares Benatar's bleak diagnosis of life on Earth will be so supine. Unless we modify human nature, compassionate-minded negative utilitarians with competence in bioweaponry, nanorobotics or artificial intelligence, for example, may quite conceivably take direct action. Echoing Moore's law, Eliezer Yudkowsky warns that "Every eighteen months, the minimum IQ necessary to destroy the world drops by one point". Although suffering and existential risk might seem separate issues, they are intimately connected. Not everyone loves life so much they wish to preserve it. Indeed the extinction of Darwinian life is what many transhumanists are aiming for—just not framed in such apocalyptic and provocative language. For just as we educate small children so they can mature into fully-fledged adults, biological humanity may aspire to grow up, too, with the consequence that—in common with small children—archaic humans become extinct.

Technologies of Biofriendliness Empathogens?

How do you disarm a potentially hostile organic robot—despite your almost limitless ignorance of his source code? Provide him with a good education, civics lessons and complicated rule-governed ethics courses? Or give him a tablet of MDMA ("Ecstasy") and get smothered with hugs?

MDMA is short-acting (Holland 2001). The "penicillin of the soul" is potentially neurotoxic to serotonergic neurons. In theory, however, lifelong use of safe and sustainable empathogens would be a passport to worldwide biofriendliness. MDMA releases a potent cocktail of oxytocin, serotonin and dopamine into the user's synapses, thereby inducing a sense of "I love the world and the world loves me". There's no technical reason why MDMA's acute pharmacodynamic effects can't be replicated indefinitely shorn of its neurotoxicity. Designer "hug drugs" can potentially turn manly men into intelligent bonobo, more akin to the "hippie chimp" *Pan paniscus* than his less peaceable cousin *Pan troglodytes*. Violence would become unthinkable. Yet is this sort of proposal politically credible? "Morality pills" and other pharmacological solutions to human unfriendliness are both personally unsatisfactory and sociologically implausible. Do we really want to drug each other up from early childhood? Moreover, life would be immeasurably safer if our fellow humans weren't genetically predisposed to unfriendly behaviour in the first instance.

But how can this friendly predisposition are guaranteed? Friendliness can't realistically be hand-coded by tweaking the connections and weight strengths of our neural networks. Nor can robust friendliness in advanced biological intelligence be captured by a bunch of explicit logical rules and smart algorithms, as in the paradigm of symbolic AI.

Mass Oxytocination?

Amplified "trust hormone" (Lee et al. 2009) might create the biological underpinnings of world-wide peace and love if negative feedback control of oxytocin release can be circumvented. Oxytocin is functionally antagonised by testosterone in the male brain. Yet oxytocin enhancers have pitfalls too. Enriched oxytocin function leaves one vulnerable to exploitation by the unenhanced. Can we really envisage a cross-cultural global consensus for mass-medication? When? Optional or mandatory? And what might be the wider ramifications of a "high oxytocin, low testosterone" civilisation? Less male propensity to violent territorial aggression, for sure; but disproportionate intellectual progress in physics, mathematics and computer science to date has been driven by the hyper-systematising cognitive style of "extreme male" brains (Baron-Cohen 2001). Also, enriched oxytocin function can indirectly even promote *un*friendliness to "out-groups" in consequence of promoting in-group bonding. So as well as oxytocin enrichment, global security demands a more inclusive, impartial, intellectually sophisticated conception of "us" that embraces all sentient beings (Singer 1981)—the expression of a hyper-developed capacity for empathetic understanding combined with a hyper-developed capacity for rational systematisation. Hence the imperative need for Full-Spectrum Superintelligence.

Mirror-Touch Synaesthesia?

A truly long-term solution to unfriendly biological intelligence might be collectively to engineer ourselves with the functional generalisation of "mirror-touch" synaesthesia (Banissy 2009). On seeing you cut and hurt yourself, a mirror-touch synaesthete is liable to feel a stab of pain as acutely as you do. Conversely, your expressions of pleasure elicit a no less joyful response. Thus mirror-touch synaesthesia is a hyper-empathising condition that makes deliberate unfriendliness, in effect, biologically impossible in virtue of cognitively enriching our capacity to represent each other's first-person perspectives. The existence of mirror-touch synaesthesia is a tantalising hint at the God-like representational capacities of a Full-Spectrum Superintelligence. This so-called "disorder" is uncommon in humans.

Timescales

The biggest problem with all these proposals, and other theoretical biological solutions to human unfriendliness, is timescale. Billions of human and non-human animals will have been killed and abused before they could ever come to pass. Cataclysmic wars may be fought in the meantime with nuclear, biological and chemical weapons harnessed to "narrow" AI. Our circle of empathy expands only slowly and fitfully. For the most part, religious believers and traditional-minded bioconservatives won't seek biological enhancement/remediation for themselves or their children. So messy democratic efforts at "political" compromise are probably unavoidable for centuries to come. For sure, idealists can dream up utopian schemes to mitigate the risk of violent conflict until the "better angels of our nature" (Pinker 2011) can triumph, e.g. the election of a risk-averse all-female political class (Pellissier 2011) to replace legacy warrior males. Such schemes tend to founder on the rock of sociological plausibility. Innumerable sentient beings are bound to suffer and die in consequence.

Does Full-Spectrum Superintelligence Entail Benevolence?

The God-like perspective-taking faculty of a Full-Spectrum Superintelligence doesn't entail distinctively *human*-friendliness (Yudkowsky 2008) any more than a God-like Superintelligence could promote distinctively Aryan-friendliness. Indeed it's unclear how benevolent superintelligence could want omnivorous killer apes in our current guise to walk the Earth in any shape or form. But is there any connection at all between benevolence and intelligence? Pre-reflectively, benevolence and intelligence are orthogonal concepts. There's nothing obviously incoherent about a malevolent God or a malevolent—or at least a callously indifferent—Superintelligence. Thus a sceptic might argue that there is no link whatsoever between benevolence—on the face of it a mere personality variable—and enhanced intellect. After all, some sociopaths score highly on our [autistic, mind-blind] IQ tests. Sociopaths know that their victims suffer. They just don't care.

However, what's critical in evaluating cognitive ability is a *criterion of representational adequacy*. Representation is not an all-or-nothing phenomenon; it varies in functional degree. More specifically here, the cognitive capacity to represent the *formal* properties of mind differs from the cognitive capacity to represent the *subjective* properties of mind (Seager 2006). Thus a notional zombie Hyper-Autist robot running a symbolic AI program on an ultrapowerful digital computer with a classical von Neumann architecture may be beneficent or maleficent in its behaviour toward sentient beings. By its very nature, it can't know or care. Most starkly, the zombie Hyper-Autist might be programmed to convert the world's matter and energy into either heavenly "utilitronium" or diabolical "dolorium" without the slightest insight into the significance of what it was doing. This kind of scenario is at least a notional risk of creating insentient Hyper-Autists endowed with mere formal utility functions rather than hyper-sentient Full-Spectrum Superintelligence. By contrast, Full-Spectrum Superintelligence *does* care in virtue of its full-spectrum representational capacities—a bias-free generalisation of the superior perspective-taking, "mind-reading" capabilities that enabled humans to become the cognitively dominant species on the planet. Full-spectrum Superintelligence, if equipped with the posthuman cognitive generalisation of mirror-touch synaesthesia, understands your thoughts, your feelings and your egocentric perspective better than you do yourself.

Could there arise "evil" mirror-touch synaesthetes? In one sense, no. You can't go around wantonly hurting other sentient beings if you feel their pain as your own. Full-spectrum intelligence is friendly intelligence. But in another sense yes, insofar as primitive mirror-touch synaesthetes are preys to species-specific cognitive limitations that prevent them acting rationally to maximise the well-being of all sentience. Full-spectrum superintelligences would lack those computational limitations in virtue of their full cognitive competence in understanding both the subjective and the formal properties of mind. Perhaps full-spectrum superintelligences might optimise your matter and energy into a blissful smart angel; but they couldn't wantonly hurt you, whether by neglect or design.

More practically today, a cognitively superior analogue of natural mirror-touch synaesthesia should soon be feasible with reciprocal neuroscanning technology—a kind of naturalised telepathy. At first blush, mutual telepathic understanding sounds a panacea for ignorance and egotism alike. An exponential growth of shared telepathic understanding might safeguard against global catastrophe born of mutual incomprehension and WMD. As the poet Henry Wadsworth Longfellow observed, "If we could read the secret history of our enemies, we should find in each life sorrow and suffering enough to disarm all hostility". Maybe so. The problem here, as advocates of Radical Honesty soon discover, is that many Darwinian thoughts scarcely promote friendliness if shared: they are often ill-natured, unedifying and unsuitable for public consumption. Thus, unless perpetually "loved-up" on MDMA or its long-acting equivalents, most of us would find mutual mind-reading a traumatic ordeal. Human society and most personal relationships would collapse in acrimony rather than blossom. Either way, our human incapacity fully to understand the first-person point of view of other sentient beings isn't just a moral failing or a personality variable; it's an *epistemic* limitation, an intellectual failure to grasp an objective feature of the natural world. Even "normal" people share with sociopaths this fitness-enhancing cognitive deficit. By posthuman criteria, perhaps we're all ignorant quasi-sociopaths. The egocentric delusion (i.e. that the world centres on one's existence) is genetically adaptive and strongly selected for over hundreds of millions of years (Dawkins 1976). Fortunately, it's a cognitive failing amenable to technical fixes and eventually a cure: Full-Spectrum Superintelligence. The devil is in the details, or rather the genetic source code.

A Biotechnological Singularity?

Yet does this positive feedback loop of reciprocal enhancement amount to a Singularity (Vinge 1993) in anything more than a metaphorical sense? The risk of talking portentously about "The Singularity" isn't of being wrong: it's of being "not even wrong"—of reifying one's ignorance and elevating it to the status of an ill-defined apocalyptic event. Already multiple senses of "The Singularity" proliferate in popular culture. Does taking LSD induce a Consciousness Singularity? How about the abrupt and momentous discontinuity in one's conception of reality entailed by waking from a dream? Or the birth of language? Or the Industrial Revolution? So is the idea of recursive self-improvement leading to a *Bio*technological Singularity, or "Biosingularity" for short, any more rigorously defined than recursive self-improvement (Omohundro 2007) of seed AI leading to a "Technological Singularity"?

Metaphorically, perhaps, the impending biointelligence explosion represents an intellectual "event horizon" beyond which archaic humans cannot model or understand the future. Events beyond the Biosingularity will be stranger than science-fiction: too weird for unenhanced human minds—or the algorithms of a zombie super-Asperger—to predict or understand. In the popular sense of "event horizon", maybe the term is apt too, though the metaphor is still potentially misleading. Thus, theoretical physics tells us that one could pass through the event horizon of a non-rotating supermassive black hole and not notice any subjective change in consciousness—even though one's signals would now be inaccessible to an external observer. The Biosingularity will feel different in ways a human conceptual scheme can't express. But what is the empirical content of this claim?

What is Full-Spectrum Superintelligence?

> [g is] ostensibly some innate scalar brain force…[However] ability is a folk concept and not amenable to scientific analysis.
>
> Jon Marks (Dept Anthropology, Yale University), 1995, *Nature*, 9 xi, 143–144.

> Our normal waking consciousness, rational consciousness as we call it, is but one special type of consciousness, whilst all about it, parted from it by the filmiest of screens, there lie potential forms of consciousness entirely different.
>
> (William James)

Intelligence

"Intelligence" is a folk concept. The phenomenon is not well-defined—or rather any attempt to do so amounts to a stipulative definition that doesn't "carve Nature at the joints". The Cattell-Horn-Carroll psychometric theory of human cognitive abilities (1993) is probably most popular in academia and the IQ testing community. But the Howard Gardner multiple intelligences model, for example, differentiates "intelligence" into various spatial, linguistic, bodily-kinaesthetic, musical, interpersonal, intrapersonal, naturalistic and existential intelligence (Gardner 1983) rather than a single general ability ("g"). Who's right? As it stands, "g" is just a statistical artefact of our culture-bound IQ tests. If general intelligence were indeed akin to an innate scalar brain force, as some advocates of "g" believe, or if intelligence can best be modelled by the paradigm of symbolic AI, then the exponential growth of digital computer processing power might indeed entail an exponential growth in intelligence too—perhaps leading to some kind of Super-Watson (Baker 2011). Other facets of intelligence, however, resist enhancement by mere acceleration of raw processing power.

One constraint is that a theory of general intelligence should be race-, species-, and culture-neutral. Likewise, an impartial conception of intelligence should embrace all possible state-spaces of consciousness: prehuman, human, transhuman and posthuman.

The non-exhaustive set of criteria below doesn't pretend to be anything other than provisional. They are amplified in the sections to follow.

Full-Spectrum Superintelligence entails:

1. The capacity to solve the Binding Problem, (Revonsuo and Newman 1999) i.e. to generate phenomenally unified entities from widely distributed computational processes; and run cross-modally matched, data-driven world-simulations (Revonsuo 2005) of the mind-independent environment. (*cf.* naive realist theories of "perception" versus the world-simulation or "Matrix" paradigm. Compare disorders of binding, e.g. simultanagnosia (an inability to perceive the visual field as a whole), cerebral akinetopsia ("motion blindness"), etc. In the absence of a data-driven, almost real-time simulation of the environment, intelligent agency is impossible.)
2. A self or some non-arbitrary functional equivalent of a person to which intelligence can be ascribed. (*cf.* dissociative identity disorder ("multiple personality disorder"), or florid schizophrenia, or your personal computer: in the absence of at least a fleetingly unitary self, what philosophers call "synchronic identity", there is no entity that is intelligent, just an aggregate of discrete algorithms and an operating system.)
3. A "mind-reading" or perspective-taking faculty; higher-order intentionality (e.g. "he believes that she hopes that they fear that he wants..." etc.): social intelligence.
   The intellectual success of the most cognitively successful species on the planet rests, not just on the recursive syntax of human language, but also on our unsurpassed "mind-reading" prowess, an ability to simulate the perspective of other unitary minds: the "Machiavellian Ape" hypothesis (Byrne and Whiten 1988). Any ecologically valid intelligence test designed for a species of social animal must incorporate social cognition and the capacity for co-operative problem-solving. So must any test of empathetic superintelligence.
4. A metric to distinguish the important from the trivial.
   (Our theory of significance should be explicit rather than implicit, as in contemporary IQ tests. What distinguishes, say, mere calendrical prodigies and other "savant syndromes" from, say, a Grigori Perelman who proved the Poincaré conjecture? Intelligence entails understanding what does—and doesn't—matter. What matters is of course hugely contentious.)
5. A capacity to navigate, reason logically about, and solve problems in multiple state-spaces of consciousness [e.g. dreaming states (*cf.* lucid dreaming), waking consciousness, echolocation competence, visual discrimination, synaesthesia in all its existing and potential guises, humour, introspection, the different realms of psychedelia (*cf.* salvia space, "the K-hole" etc.)] including *realms of experience not yet co-opted by either natural selection or posthuman design for tracking features of the mind-independent world.* Full-Spectrum Superintelligence will

entail cross-domain goal-optimising ability in all possible state-spaces of consciousness (Shulgin 2011). and finally

6. "Autistic", pattern-matching, rule-following, mathematico-linguistic intelligence, i.e. the standard, mind-blind (Baron-Cohen 1995) cognitive tool-kit scored by existing IQ tests. High-functioning "autistic" intelligence is indispensable to higher mathematics, computer science and the natural sciences. High-functioning autistic intelligence is necessary—but not sufficient—for a civilisation capable of advanced technology that can cure ageing and disease, systematically phase out the biology of suffering, and take us to the stars. And for programming artificial intelligence.

We may then ask which facets of Full-Spectrum Superintelligence will be accelerated by the exponential growth of digital computer processing power? Number six, clearly, as decades of post-ENIAC progress in computer science attest. But what about numbers one-to-five? Here the picture is murkier.

The Bedrock of Intelligence. World-Simulation ("Perception")

Consider criterion number one, world-simulating prowess, or what we misleadingly term "perception". The philosopher Bertrand Russell (1948) once aptly remarked that one never sees anything but the inside of one's own head. In contrast to such inferential realism, commonsense perceptual direct realism offers all the advantages of theft over honest toil—and it's computationally useless for the purposes either of building artificial general intelligence or understanding its biological counterparts. For the bedrock of intelligent agency is the capacity of an embodied agent computationally to simulate dynamic objects, properties and events in the mind-independent environment. [For a contrary view, see e.g. Brooks 1991] The evolutionary success of organic robots over the past *c.* 540 million years has been driven by our capacity to run data-driven egocentric world-simulations—what the naive realist, innocent of modern neuroscience or post-Everett (Everett 1955) quantum mechanics, calls simply perceiving one's physical surroundings. Unlike classical digital computers, organic neurocomputers can simultaneously "bind" multiple features (edges, colours, motion, etc.) distributively processed across the brain into unitary phenomenal objects embedded in unitary spatio-temporal world-simulations apprehended by a momentarily unitary self: what Kant (1781) calls "the transcendental unity of apperception". These simulations run in (almost) real-time; the time-lag in our world-simulations is barely more than a few dozen milliseconds. Such blistering speed of construction and execution is adaptive and often life-saving in a fast-changing external environment. Recapitulating evolutionary history, pre-linguistic human infants must first train up their neural networks to bind the multiple features of dynamic objects and run unitary world-simulations before they can socially learn *second-order* representation and then *third-order* representation, i.e. language followed later in childhood by meta-language.

Occasionally, object binding and/or the unity of consciousness partially breaks down in mature adults who suffer a neurological accident. The results can be cognitively devastating (*cf.* akinetopsia or "motion blindness" (Zeki 1991); and simultanagnosia, an inability to apprehend more than a single object at a time Riddoch and Humphreys 2004), etc.). Yet normally our simulations of fitness-relevant patterns in the mind-independent local environment feel seamless. Our simulations each appear simply as "the world"; we just don't notice or explicitly represent the gaps. Neurons, (mis)construed as classical processors, are pitifully slow, with spiking frequencies barely up to 200 per second. By contrast, silicon (etc.) processors are ostensibly millions of times faster. Yet the notion that non-biological computers are faster than sentient neurocomputers is a philosophical assumption, not an empirical discovery. Here the assumption will be challenged. Unlike the CPUs of classical robots, an organic mind/brain delivers dynamic unitary phenomenal objects and unitary world-simulations with a "refresh rate" of many billions per second (*cf.* the persistence of vision as experienced watching a movie run at a mere 30 frames per second). These cross-modally matched simulations take the guise of what passes as the macroscopic world: a spectacular egocentric simulation run by the vertebrate CNS that taps into the world's fundamental quantum substrate (Ball 2011). A strong prediction of this conjecture is that classical digital computers will never be non-trivially conscious—or support software smart enough to understand their ignorance.

We should pause here. This is *not* a mainstream view. Most AI researchers regard stories of a non-classical mechanism underlying the phenomenal unity of biological minds as idiosyncratic at best. In fact no scientific consensus exists on the molecular underpinnings of the unity of consciousness, or on how such unity is even physically possible. By analogy, 1.3 billion skull-bound Chinese minds can never be a single subject of experience, irrespective of their interconnections. How waking or dreaming communities of membrane-bound classical neurons could—even *micro*conscious classical neurons—be any different? If materialism is true, conscious mind should be *impossible*. Yet any explanation of phenomenal object binding, the unity of perception, or the phenomenal unity of the self that invokes quantum coherence as here is controversial. One reason it's controversial is that the delocalisation involved in quantum coherence is exceedingly short-lived in an environment as warm and noisy as a macroscopic brain—supposedly too short-lived to do computationally useful work (Hagen 2002). Physicist Max Tegmark (2000) estimates that thermally-induced decoherence destroys any macroscopic coherence of brain states within 10^{-13} s, an unimaginably long time in natural Planck units but an unimaginably short time by everyday human intuitions. Perhaps it would be wiser just to acknowledge these phenomena are unexplained mysteries within a conventional materialist framework—as mysterious as the existence of consciousness itself. But if we're speculating about the imminent end of the human era (Good 1965), shoving the mystery under the rug isn't really an option. For the different strands (Yudkowsky 2007) of the Singularity movement share a common presupposition. This presupposition is that our complete ignorance within a materialist conceptual scheme of why consciousness exists (the

"Hard Problem") (Chalmers 1995), and of even the ghost of a solution to the Binding Problem, doesn't matter for the purposes of building the seed of artificial posthuman superintelligence. Our ignorance supposedly doesn't matter either because consciousness and/or our quantum "substrate" are computationally irrelevant to cognition (Hutter 2012) and the creation of nonbiological minds, or alternatively because the feasibility of "whole brain emulation" (Markram 2006) will allow us to finesse our ignorance.

Unfortunately, we have no grounds for believing this assumption is true or that the properties of our quantum "substrate" are functionally irrelevant to Full-Spectrum Superintelligence or its humble biological predecessors. Conscious minds are not substrate-neutral digital computers. Humans investigate and reason about problems of which digital computers are invincibly ignorant, not least the properties of consciousness itself. The Hard Problem of consciousness can't be quarantined from the rest of science and treated as a troublesome but self-contained anomaly: its mystery infects *everything* (Rescher 1974) that we think we know about ourselves, our computers and the world. Either way, the conjecture that the phenomenal unity of perception is a manifestation of ultra-rapid sequences of irreducible quantum coherent states *isn't* a claim that the mind/brain is capable of detecting events in the mind-independent world on this kind of sub-picosecond timescale. Rather the role of the local environment in shaping action-guiding experience in the awake mind/brain is here conjectured to be quantum state-*selection*. When we're awake, patterns of impulses from e.g. the optic nerve *select* which quantum-coherent frames are generated by the mind/brain—in contrast to the autonomous world-simulations spontaneously generated by the dreaming brain. Other quantum mind theorists, most notably Roger Penrose (1994) and Stuart Hameroff (2006), treat quantum minds as evolutionarily novel rather than phylogenetically ancient. They invoke a non-physical (Saunders 2010) wave-function collapse and unwisely focus on e.g. the ability of mathematically-inclined brains to perform non-computable functions in higher mathematics, a feat for which selection pressure has presumably been non-existent (Litt 2006). Yet the human capacity for sequential linguistic thought and formal logico-mathematical reasoning is a late evolutionary novelty executed by a slow, brittle virtual machine running on top of its massively parallel quantum parent—a momentous evolutionary innovation whose neural mechanism is still unknown.

In contrast to the evolutionary novelty of serial linguistic thought, our ancient and immensely adaptive capacity to run unitary world-simulations, simultaneously populated by hundreds or more dynamic unitary objects, enables organic robots to solve the computational challenges of navigating a hostile environment that would leave the fastest classical supercomputer grinding away until Doomsday. Physical theory (*cf.* the Bekenstein bound) shows that informational resources as classically conceived are not just physical but finite and scarce: a maximum possible limit of 10^{120} bits set by the surface area of the entire accessible universe (Lloyd 2002) expressed in Planck units according to the Holographic principle. An infinite computing device like a universal Turing machine (Dyson 2012) is physically impossible. So invoking computational equivalence and asking whether a classical

Turing machine can run a human-equivalent macroscopic world-simulation is akin to asking whether a classical Turing machine can factor 1,500 digit numbers in real-world time [i.e. no]. No doubt resourceful human and transhuman programmers will exploit all manner of kludges, smart workarounds and "brute-force" algorithms to try and defeat the Binding Problem in AI. How will they fare? Compare clod-hopping AlphaDog with the sophisticated functionality of the sesame-seed sized brain of a bumblebee. Brute-force algorithms suffer from an exponentially growing search space that soon defeats any classical computational device in open-field contexts. As witnessed by our seemingly effortless world-simulations, organic minds are ultrafast; classical computers are slow. Serial *thinking* is slower still; but that's not what conscious biological minds are good at. On this conjecture, "substrate-independent" phenomenal world-simulations are impossible for the same reason that "substrate-independent" chemical valence structure is impossible. We're simply begging the question of what's functionally (ir) relevant. Ultimately, Reality has only a single, "program-resistant" (Gunderson 1985) ontological level even though it's amenable to description at different levels of computational abstraction; and the nature of this program-resistant level as disclosed by the subjective properties of one's mind (Lockwood 1989) is utterly at variance with what naive materialist metaphysics would suppose (Seager 2006). *If* our phenomenal world-simulating prowess turns out to be constitutionally tied to our quantum mechanical wetware, then substrate-neutral virtual machines (i.e. software implementations of a digital computer that execute programs like a physical machine) will never be able to support "virtual" qualia or "virtual" unitary subjects of experience. This rules out sentient life "uploading" itself to digital nirvana (Moravec 1990). C*ontra* Marvin Minsky ("The most difficult human skills to reverse engineer are those that are unconscious") (Minsky 1987), the most difficult skills for roboticists to engineer in artificial robots are actually intensely conscious: our colourful, noisy, tactile, sometimes hugely refractory virtual worlds.

Naively, for sure, real-time world-simulation doesn't sound too difficult. Hollywood robots do it all the time. Videogames become ever more photorealistic. Perhaps one imagines viewing some kind of inner TV screen, as in a Terminator movie or The Matrix. Yet the capacity of an awake or dreaming brain to generate unitary macroscopic world-simulations can only superficially resemble a little man (a "homunculus") viewing its own private theatre—on pain of an infinite regress. For by what mechanism would the homunculus view this inner screen? Emulating the behaviour of even the very simplest sentient organic robots on a classical digital computer is a daunting task. *If* conscious biological minds are irreducibly quantum mechanical by their very nature, then reverse-engineering the brain to create digital human "mindfiles" and "roboclones" alike will prove impossible.

The Bedrock of Superintelligence Hypersocial Cognition ("Mind-reading")

Will superintelligence be solipsistic or social? Overcoming a second obstacle to delivering human-level artificial general intelligence—let alone building a recursively self-improving super-AGI culminating in a Technological Singularity—depends on finding a solution to the first challenge, i.e. real-time world-simulation. For the evolution of distinctively human intelligence, sitting on top of our evolutionarily ancient world-simulating prowess, has been driven by the interplay between our rich generative syntax and superior "mind-reading" skills: so-called Machiavellian intelligence (Byrne and Whiten 1988). Machiavellian intelligence is an egocentric parody of God's-eye-view empathetic superintelligence. Critically for the prospects of building AGI, this real-time mind-modelling expertise is parasitic on the neural wetware to generate unitary first-order world-simulations—virtual worlds populated by the avatars of intentional agents whose different first-person perspectives can be partially and imperfectly understood by their simulator. Even articulate human subjects with autism spectrum disorder are prone to multiple language deficits because they struggle to understand the intentions—and higher-order intentionality—of neurotypical language users. Indeed natural language is itself a pre-eminently social phenomenon: its criteria of application must first be socially learned. Not all humans possess the cognitive capacity to acquire mind-reading skills and the cooperative problem-solving expertise that sets us apart from other social primates. Most notably, people with autism spectrum disorder don't just fail to understand other minds; autistic intelligence cannot begin to understand its own mind. Pure autistic intelligence has no conception of a self that can be improved, recursively or otherwise. Autists can't "read" their own minds. The inability of the autistic mind to take what Daniel Dennett (1987) calls the "intentional stance" parallels the inability of classical computers to understand the minds of intentional agents—or have insight into their own zombie status. Even with smart algorithms and ultra-powerful hardware, the ability of ultra-intelligent autists to predict the long-term behaviour of mindful organic robots by relying exclusively on the physical stance (i.e. solving the Schrödinger equation of the intentional agent in question) will be extremely limited. For example, much collective human behaviour is chaotic in the technical sense, i.e. it shows extreme sensitivity to initial conditions that confounds long-term prediction by even the most powerful real-world supercomputer. But there's a worse problem: reflexivity. Predicting sociological phenomena differs essentially from predicting mindless physical phenomena. Even in a classical, causally deterministic universe, the behaviour of mindful, reflexively self-conscious agents is frequently unpredictable, even in principle, from *within* the world owing to so-called prediction paradoxes (Welty 1970). When the very act of prediction causally interacts with the predicted event, then self-defeating or self-falsifying predictions are inevitable. Self-falsifying predictions are a mirror image of so-called self-fulfilling predictions. So in common with autistic "idiot savants", classical AI gone rogue will be vulnerable to the low cunning of Machiavellian apes and the high cunning of our transhuman descendants.

This argument (i.e. our capacity for unitary mind-simulation embedded in unitary world-simulation) for the cognitive primacy of biological general intelligence isn't decisive. For a start, computer-aided Machiavellian humans can program robots with "narrow" AI—or perhaps "train up" the connections and weights of a subsymbolic connectionist architecture (Rumelhart et al. 1986)—for their own manipulative purposes. Humans underestimate the risks of zombie infestation at our peril. Given our profound ignorance of how conscious mind is even possible, it's probably safest to be agnostic over whether autonomous nonbiological robots will ever emulate human world-simulating or mind-reading capacity in most open-field contexts, despite the scepticism expressed here. Either way, the task of devising an ecologically valid measure of general intelligence that can reliably, predictively and economically discriminate between disparate lifeforms is immensely challenging, not least because the intelligence test will express the value-judgements, and species- and culture-bound conceptual scheme, of the tester. Some biases are insidious and extraordinarily subtle: for example, the desire systematically to measure "intelligence" with mind-blind IQ tests is itself a quintessentially Asperger-ish trait. In consequence, social cognition is disregarded altogether. What we fancifully style "IQ tests" is designed by people with abnormally high AQs (Baron-Cohen 2001) as well as self-defined high IQs. Thus, many human conceptions of (super) intelligence resemble high-functioning autism spectrum disorder rather than a hyper-empathetic God-like Super-Mind. For example, an AI that attempted systematically to maximise the cosmic abundance of paperclips (Yudkowsky 2008) would be recognisably autistic rather than incomprehensibly alien. Full-Spectrum (Super-) intelligence is certainly harder to design or quantify scientifically than mathematical puzzle-solving ability or performance in verbal memory-tests: "IQ". But that's because superhuman intelligence will be not just quantitatively different but also qualitatively alien (Huxley 1954) from human intelligence. To misquote Robert McNamara, cognitive scientists need to stop making what is measurable important, and find ways to make the important measurable. An idealised Full-Spectrum Superintelligence will indeed be capable of an impartial "view from nowhere" or God's-eye-view of the multiverse (Wallace 2012), a mathematically complete Theory of Everything—as does modern theoretical physics, in aspiration if not achievement. But in virtue of its God's-eye-view, Full-Spectrum Superintelligence must also be hypersocial and super*sentient*: able to understand all possible first-person perspectives, the state-space of all possible minds in other Hubble volumes, other branches of the universal wavefunction—and in other solar systems and galaxies if such beings exist within our cosmological horizon. Idealised at least, Full-Spectrum Superintelligence will be able to understand and weigh the significance of all possible modes of experience *irrespective of whether they have hitherto been recruited for information-signalling purposes*. The latter is, I think, by far the biggest intellectual challenge we face as cognitive agents. The systematic investigation of alien types of consciousness intrinsic to varying patterns of matter and energy

(Lockwood 1989) calls for a methodological and ontological revolution (Shulgin 1995). Transhumanists talking of post-Singularity superintelligence are fond of hyperbole about "Level 5 Future Shock" etc.; but it's been aptly said that if Elvis Presley were to land in a flying saucer on the White House lawn, it's as nothing in strangeness compared to your first DMT trip.

Ignoring the Elephant: Consciousness Why Consciousness is Computationally Fundamental to the Past, Present and Future Success of Organic Robots

The pachyderm in the room in most discussions of (super) intelligence is consciousness—not just human reflective self-awareness but the whole gamut of experience from symphonies to sunsets, agony to ecstasy: the phenomenal world. All one ever knows, except by inference, and is the contents of one's own conscious mind: what philosophers call "qualia". Yet according to the ontology of our best story of the world, namely physical science, *conscious* minds shouldn't exist at all, i.e. we should be zombies, insentient patterns of matter and energy indistinguishable from normal human beings but lacking conscious experience. Dutch computer scientist Edsger Dijkstra famously once remarked, "The question of whether a computer can think is no more interesting than the question of whether a submarine can swim". Yet the question of whether a programmable digital computer—or a subsymbolic connectionist system with a merely classical parallelism (Churchland 1989)—could possess, *and think about*, qualia, "bound" perceptual objects, a phenomenal self, or the unitary phenomenal minds of sentient organic robots can't be dismissed so lightly. For if advanced nonbiological intelligence is to be smart enough comprehensively to understand, predict and manipulate the behaviour of enriched biological intelligence, then the AGI can't rely autistically on the "physical stance", i.e. to monitor the brains, scan the atoms and molecules, and then solve the Schrödinger equation of intentional agents like human beings. Such calculations would take longer than the age of the universe.

For sure, many forms of human action can be predicted, fallibly, on the basis of crude behavioural regularities and reinforcement learning. Within your world-simulation, you don't need a theory of mind or an understanding of quantum mechanics to predict that Fred will walk to the bus-stop again today. Likewise, powerful tools of statistical analysis run on digital supercomputers can predict, fallibly, many kinds of human collective behaviour, for example stock markets. Yet to surpass human and transhuman capacities in all significant fields, AGI must understand how intelligent biological robots can think about, talk about and manipulate the manifold varieties of consciousness that make up their virtual worlds. Some investigators of consciousness even dedicate their lives to that end; what might a notional insentient AGI suppose we're doing? There is no evidence that serial digital computers have the capacity to do anything of the kind—or could ever be programmed to do so. Digital computers don't know anything about conscious minds, unitary persons, the nature of phenomenal pleasure and pain, or the Problem of Other Minds; it's not even "all dark inside". The challenge for a *conscious* mind posed by understanding itself "from the inside" pales into insignificance compared to the challenge for a nonconscious system of understanding a conscious mind "from the outside". Nor within the constraints of a materialist ontology have we the slightest clue how the purely classical parallelism of a subsymbolic, "neurally inspired" connectionist architecture could turn water into wine and generate *unitary* subjects of experience to fill the gap. For even if we conjecture in the spirit of Strawsonian physicalism—the only scientifically literate form of panpsychism (Strawson 2006)—that the fundamental stuff of the world, the mysterious "fire in the equations", is fields of microqualia, this bold ontological conjecture doesn't, by itself, explain why biological robots aren't zombies. This is because structured aggregates of classically conceived "mind-dust" aren't the same as a unitary phenomenal subject of experience who apprehends "bound" spatio-temporal objects in a dynamic world-simulation. Without phenomenal object binding and the unity of perception, we are faced with the spectre of what philosophers call "mereological nihilism" (Merricks 2001). Mereological nihilism, also known as "compositional nihilism", is the position that composite objects with proper parts do not exist: strictly speaking, only basic building blocks without parts have more than fictional existence. Unlike the fleetingly unitary phenomenal minds of biological robots, a classical digital computer and the programs it runs lack ontological integrity: it's just an assemblage of algorithms. In other words, a classical digital computer has no self to understand or a mind recursively to improve, exponentially or otherwise. Talk about artificial "intelligence" exploding (Hutter 2012) is just an anthropomorphic projection on our part.

So how do biological brains solve the binding problem and become persons? (Parfit 1984) In short, we don't know. Vitalism is clearly a lost cause. Most AI researchers would probably dismiss—or at least discount as wildly speculative—any story of the kind mooted here involving macroscopic quantum coherence grounded in an ontology of physicalistic panpsychism. But in the absence of any story at all, we are left with a theoretical vacuum and a faith that natural science—or the exponential growth of digital computer processing power culminating in a Technological Singularity—will one day deliver an answer. Evolutionary biologist Theodosius Dobzhansky famously observed how "Nothing in Biology Makes Sense Except in the Light of Evolution". In the same vein, nothing in the future of intelligent life in the universe makes sense except in the light of a solution to the Hard Problem of Consciousness and the closure of Levine's Explanatory Gap (Levine 1983). Consciousness is the only reason anything matters at all; and it's the only reason why unitary subjects of experience can ask these questions; and yet materialist orthodoxy has no idea how or why the phenomenon exists. Unfortunately, the Hard Problem won't be solved by building more advanced digital zombies who can tell mystified conscious minds the answer.

More practically for now, perhaps the greatest cognitive challenge of the millennium and beyond is deciphering and systematically manipulating the "neural correlates of consciousness" (Koch 2004). Neuroscientists use this expression in default of any deeper explanation of our myriad qualia. How and why does experimentally stimulating via microelectrodes one cluster of nerve cells in the neocortex yield the experience of phenomenal colour; stimulating a superficially similar type of nerve cell induces a musical jingle; stimulating another with a slightly different gene expression profile a sense of everything being hysterically funny; stimulating another seemingly of your mother; and stimulating another of an archangel, say, in front of your body-image? In each case, the molecular variation in neuronal cell architecture is ostensibly *trivial*; the difference in subjective experience is profound. On a mind/brain identity theory, such experiential states are an intrinsic property of some configurations of matter and energy (Lockwood 1989). How and why this is so is incomprehensible on an orthodox materialist ontology. Yet empirically, microelectrodes, dreams and hallucinogenic drugs elicit these experiences regardless of any information-signalling role such experiences typically play in the "normal" awake mind/brain. Orthodox materialism and classical information-based ontologies alike do not merely lack any explanation for why consciousness and our countless varieties of qualia exist. They lack any story of how our qualia could have the causal efficacy to allow us to allude to—and in some cases volubly expatiate on—their existence. Thus, mapping the neural correlates of consciousness is not amenable to formal computational methods: digital zombies don't have any qualia, or at least any "bound" macroqualia, that could be mapped, or a unitary phenomenal self that could do the mapping.

Note this claim for the cognitive primacy of biological sentience isn't a denial of the Church-Turing thesis that given *infinite* time and *infinite* memory any Turing-universal system can formally simulate the behaviour of any conceivable process that can be digitized. Indeed (very) fancifully, if the multiverse were being run on a cosmic supercomputer, speeding up its notional execution a million times would presumably speed us up a million times too. But that's not the issue here. Rather the claim is that nonbiological AI run on real-world digital computers cannot tackle the truly hard and momentous cognitive challenge of investigating first-person states of egocentric virtual worlds—or understand why some first-person states, e.g. agony or bliss, are intrinsically important, and cause unitary subjects of experience, persons, to act the way we do.

At least in common usage, "intelligence" refers to an agent's ability to achieve goals in a wide range of environments. What we call greater-than-human intelligence or Superintelligence presumably involves the design of qualitatively *new* kinds of intelligence never seen before. Hence the growth of artificial intelligence and symbolic AI, together with subsymbolic (allegedly) brain-inspired connectionist architectures, and soon artificial quantum computers. But contrary to received wisdom in AI research, sentient biological robots are making greater cognitive progress in discovering the potential for truly novel kinds of intelligence than the techniques of formal AI. We are doing so by synthesising and empirically investigating a galaxy of psychoactive designer drugs (Shulgin 2011)—experimentally opening up the possibility of radically new kinds of intelligence in different state-spaces of consciousness. For the most cognitively challenging environments don't lie in the stars but in organic mind/brains—the baffling subjective properties of quantum-coherent states of matter and energy—most of which aren't explicitly represented in our existing conceptual scheme.

Case Study: Visual Intelligence Versus Echolocation Intelligence: What is it Like to be a Super-Intelligent Bat?

Let's consider the mental state-space of organisms whose virtual worlds are rooted in their dominant sense mode of echolocation (Nagel 1974). This example isn't mere science fiction. Unless post-Everett quantum mechanics (Deutsch 1997) is false, we're forced to assume that googols of quasi-classical branches of the universal wavefunction—the master formalism that exhaustively describes our multiverse—satisfy this condition. Indeed their imperceptible interference effects must be present even in "our" world: strictly speaking, interference effects from branches that have decohered ("split") never wholly disappear; they just become vanishingly small. Anyhow, let's assume these echolocation superminds have evolved opposable thumbs, a rich generative syntax and advanced science and technology. How are we to understand or measure this alien kind of (super) intelligence? Rigging ourselves up with artificial biosonar apparatus and transducing incoming data into the familiar textures of sight or sound might seem a good start. But to understand the conceptual world of echolocation superminds, we'd need to equip ourselves with neurons and neural networks neurophysiologically equivalent to smart chiropterans. If one subscribes to a coarse-grained functionalism about consciousness, then echolocation experience would (somehow) emerge at some abstract computational level of description. The implementation details, or "meatware" as biological mind/brains are derisively called, are supposedly incidental or irrelevant. The functionally unique valence properties of the carbon atom, and likewise the functionally unique quantum mechanical properties of liquid water (Vitiello 2001), are discounted or ignored. Thus, according to the coarse-grained functionalist, silicon chips could replace biological neurons without loss of function or subjective identity. By contrast, the *micro*-functionalist, often branded a mere "carbon chauvinist", reckons that the different intracellular properties of biological neurons—with their different gene expression profiles, diverse primary, secondary, tertiary, and quaternary amino acid chain folding (etc.) as described by quantum chemistry—are critical to the many and varied phenomenal properties such echolocation neurons express. Who is right? We'll only ever know the answer by rigorous self-experimentation: a post-Galilean science of mind.

It's true that humans don't worry much about our ignorance of echolocation experience, or our ignorance of echolocation primitive terms, or our ignorance of possible conceptual schemes expressing echolocation intelligence in echolocation world-simulations. This is because we don't highly esteem bats. Humans don't share the same interests or purposes as our flying cousins, e.g. to attract desirable, high-fitness bats and rear reproductively successful baby bats. Alien virtual worlds based on biosonar don't seem especially significant to *Homo sapiens* except as an armchair philosophical puzzle.

Yet this assumption would be intellectually complacent. Worse, understanding what it's like to be a hyperintelligent bat mind is *comparatively* easy. For echolocation experience has been recruited by natural selection to play an information-signalling role in a fellow species of mammal; and in principle a research community of language users could biologically engineer their bodies and minds to replicate bat-type experience and establish crude intersubjective agreement to discuss and conceptualise its nature. By contrast, the vast majority of experiential state-spaces remain untapped and unexplored. This task awaits Full-Spectrum Superintelligence in the posthuman era.

In a more familiar vein, consider visual intelligence. How does one measure the visual intelligence of a congenitally blind person? Even with sophisticated technology that generates "inverted spectrograms" of the world to translate visual images into sound, the congenitally blind are invincibly ignorant of visual experience and the significance of visually-derived concepts. Just as a sighted idiot has greater visual intelligence than a blind super-rationalist sage, likewise psychedelics confer the ability to become (for the most part) babbling idiots about other state-spaces of consciousness—but babbling idiots whose insight is deeper than the drug-naive or the genetically unenhanced—or the digital zombies spawned by symbolic AI and its connectionist cousins.

The challenge here is that the vast majority of these alien state-spaces of consciousness latent in organised matter haven't been recruited by natural selection for information-tracking purposes. So "psychonauts" don't yet have the conceptual equipment to navigate these alien state-spaces of consciousness in even a pseudo-public language, let alone integrate them in any kind of overarching conceptual framework. Note the claim here *isn't* that taking e.g. ketamine, LSD, salvia, DMT and a dizzying proliferation of custom-designed psychoactive drugs is the royal route to wisdom. Or that ingesting such agents will give insight into deep mystical truths. On the contrary: it's precisely because such realms of experience *haven't* previously been harnessed for information-processing purposes by evolution in "our" family of branches of the universal wavefunction that makes investigating their properties so cognitively challenging—currently beyond our conceptual resources to comprehend. After all, plants synthesise natural psychedelic compounds to scramble the minds of herbivores that might eat them, not to unlock mystic wisdom. Unfortunately, there is no "neutral" medium of thought impartially to appraise or perceptually cross-modally match all these other experiential state-spaces. One can't somehow stand outside one's own stream of consciousness to evaluate how the properties of the medium are infecting the notional propositional content of the language that one uses to describe it.

By way of illustration, compare drug-induced visual experience in a notional community of congenitally blind rationalists who lack the visual apparatus to transduce incident electromagnetic radiation of our familiar wavelengths. The lone mystical babbler who takes such a vision-inducing drug is convinced that [what we would call] visual experience is profoundly significant. And as visually intelligent folk, we know that he's right: visual experience is potentially hugely significant—to an extent which the blind mystical babbler can't possibly divine. But can the drug-taker convince his congenitally blind fellow tribesmen that his mystical visual experiences really matter in the absence of perceptual equipment that permits sensory discrimination? No, he just sounds psychotic. Or alternatively, he speaks lamely and vacuously of the "ineffable". The blind rationalists of his tribe are unimpressed.

The point of this fable is that we've scant reason to suppose that biologically re-engineered posthumans millennia hence will share the same state-spaces of consciousness, or the same primitive terms, or the same conceptual scheme, or the same type of virtual world that human beings now instantiate. Maybe all that will survive the human era is a descendant of our mathematical formalism of physics, M-theory of whatever, in basement reality.

Of course such ignorance of other state-spaces of experience doesn't normally trouble us. Just as the congenitally blind don't grow up in darkness—a popular misconception—the drug-naive and genetically unenhanced don't go around with a sense of what we're missing. We notice teeming abundance, not gaping voids. Contemporary humans can draw upon terms like "blindness" and "deafness" to characterise the deficits of their handicapped conspecifics. From the perspective of full-spectrum superintelligence, what we really need is *millions* more of such "privative" terms, as linguists call them, to label the different state-spaces of experience of which genetically unenhanced humans are ignorant. In truth, there may very well be more than millions of such nameless state-spaces, each as incommensurable as e.g. visual and auditory experience. We can't yet begin to quantify their number or construct any kind of crude taxonomy of their interrelationships.

Note the problem here isn't cognitive bias or a deficiency in logical reasoning. Rather a congenitally blind (etc.) super-rationalist is constitutionally ignorant of visual experience, visual primitive terms, or a visually-based conceptual scheme. So (s)he can't cite e.g. Aumann's agreement theorem [claiming in essence that two cognitive agents acting rationally and with common knowledge of each other's beliefs cannot agree to disagree] or be a good Bayesian rationalist or whatever: these are incommensurable state-spaces of experience as closed to human minds as Picasso is to an earthworm. Moreover, there is no reason to expect one realm, i.e. "ordinary waking consciousness", to be cognitively privileged relative to every other realm. "Ordinary waking consciousness" just happened to be genetically adaptive in the African savannah on Planet Earth. Just as humans are incorrigibly ignorant of minds grounded in echolocation—both echolocation world-simulations and echolocation conceptual schemes—likewise we are invincibly ignorant of posthuman life while trapped within our existing genetic architecture of intelligence.

In order to understand the world—both its formal/mathematical *and* its subjective properties—sentient organic life must bootstrap its way to super-sentient

Full-Spectrum Superintelligence. Grown-up minds need tools to navigate all possible state-spaces of qualia, including all possible first-person perspectives, and map them—initially via the neural correlates of consciousness in our world-simulations—onto the formalism of mathematical physics. Empirical evidence suggests that the behaviour of the stuff of the world is exhaustively described by the formalism of physics. To the best of our knowledge, physics is causally closed and complete, at least within the energy range of the Standard Model. In other words, there is nothing to be found in the world—no "element of reality", as Einstein puts it—that isn't captured by the equations of physics and their solutions. This is a powerful formal constraint on our theory of consciousness. Yet our ultimate theory of the world must also close Levine's notorious "Explanatory Gap". Thus, we must explain why consciousness exists at all ("The Hard Problem"); offer a rigorous derivation of our diverse textures of qualia from the field-theoretic formalism of physics; and explain how qualia combine ("The Binding Problem") in organic minds. These are powerful constraints on our ultimate theory too. How can they be reconciled with physicalism? Why aren't we zombies?

The hard-nosed sceptic will be unimpressed at such claims. How *significant* are these outlandish state-spaces of experience? And how are they computationally relevant to (super) intelligence? Sure, says the sceptic, reckless humans may take drugs, and experience wild, weird and wonderful states of mind. But so what? Such exotic states aren't objective in the sense of reliably tracking features of the mind-independent world. Elucidation of their properties doesn't pose a well-defined problem that a notional universal algorithmic intelligence (Legg and Hutter 2007) could solve.

Well, let's assume, provisionally at least, that all mental states are identical with physical states. If so, then all experience is an objective, spatio-temporally located feature of the world whose properties a unified natural science must explain. A cognitive agent can't be intelligent, let alone superintelligent, and yet be constitutionally ignorant of a fundamental feature of the world—not just ignorant, but completely incapable of gathering information about, exploring, or reasoning about its properties. Whatever else it may be, superintelligence can't be constitutionally *stupid*. What we need is a universal, species-neutral criterion of significance that can weed out the trivial from the important; and gauge the intelligence of different cognitive agents accordingly. Granted, such a criterion of significance might seem elusive to the antirealist about value (Mackie 1991). Value nihilism treats any ascription of (in) significance as arbitrary. Or rather the value nihilist maintains that what we find significant simply reflects what was fitness-enhancing for our forebears in the ancestral environment of adaptation (Barkow 1992). Yet for reasons we simply don't understand, Nature discloses just such a universal touchstone of importance, namely the pleasure-pain axis: the world's inbuilt metric of significance and (dis)value. We're not zombies. First-person facts exist. Some of them matter urgently, e.g. I am in pain. Indeed it's unclear if the expression "I'm in agony; but the agony doesn't matter" even makes cognitive sense. Built into the very nature of agony is the knowledge that its subjective raw awfulness matters a great deal—not instrumentally or derivatively, but by its very nature. If anyone—or indeed any notional super-AGI—supposes that your agony *doesn't* matter, then he/it hasn't adequately represented the first-person perspective in question.

So the existence of first-person facts is an objective feature of the world that any intelligent agent must comprehend. Digital computers and the symbolic AI code they execute can support formal utility functions. In some contexts, formally programmed utility functions can play a role functionally analogous to importance. But nothing intrinsically matters to a digital zombie. Without sentience, and more specifically without hedonic tone, nothing inherently matters. By contrast, extreme pain and extreme pleasure in any guise intrinsically matter intensely. Insofar as exotic state-states of experience are permeated with positive or negative hedonic tone, they matter too. In summary, "He jests at scars, that never felt a wound": scepticism about the self-intimating significance of this feature of the world is feasible only in its absence.

The Great Transition

The End of Suffering

A defining feature of general intelligence is the capacity to achieve one's goals in a wide range of environments. All sentient biological agents are endowed with a pleasure-pain axis. All prefer occupying one end to the other. A pleasure-pain axis confers inherent significance on our lives: the opioid-dopamine neurotransmitter system extends from flatworms to humans. Our core behavioural and physiological responses to noxious and rewarding stimuli have been strongly conserved in our evolutionary lineage over hundreds of millions of years. Some researchers (Cialdini 1987) argue for *psychological hedonism*, the theory that all choice in sentient beings is motivated by a desire for pleasure or an aversion from suffering. When we choose to help others, this is because of the pleasure that we ourselves derive, directly or indirectly, from doing so. Pascal put it starkly: "All men seek happiness. This is without exception. Whatever different means they employ, they all tend to this end. The cause of some going to war, and of others avoiding it, is the same desire in both, attended with different views. This is the motive of every action of every man, even of those who hang themselves". In practice, the hypothesis of psychological hedonism is plagued with anomalies, circularities and complications if understood as a universal principle of agency: the "pleasure principle" is simplistic as it stands. Yet the broad thrust of this almost embarrassingly commonplace idea may turn out to be central to understanding the future of life in the universe. If even a weak and exception-laden version of psychological hedonism is true, then there is an intimate link between full-spectrum superintelligence and happiness: the "attractor" to which rational sentience is heading. If that's really what we're striving for, a lot of the time at least, then instrumental means-ends rationality dictates that intelligent agency should seek maximally cost-effective ways to deliver happiness—and then superhappiness and beyond.

A discussion of psychological hedonism would take us too far afield here. More fruitful now is just to affirm a truism and then explore its ramifications for life in the post-genomic era. Happiness is typically one of our goals. Intelligence amplification entails pursuing our goals more rationally. For sure, happiness, or at least a reduction in unhappiness, is frequently sought under a variety of descriptions that don't explicitly allude to hedonic tone and sometimes disavow it altogether. Natural selection has "encephalised" our emotions in deceptive, fitness-enhancing ways within our world-simulations. Some of these adaptive fetishes may be formalised in terms of abstract utility functions that a rational agent would supposedly maximise. Yet even our loftiest intellectual pursuits are underpinned by the same neurophysiological reward and punishment pathways. The problem for sentient creatures is that, both personally and collectively, Darwinian life is not very smart or successful in its efforts to achieve long-lasting well-being. Hundreds of millions of years of "Nature, red in tooth and claw" attest to this terrible cognitive limitation. By a whole raft of indices (suicide rates, the prevalence of clinical depression and anxiety disorders, the Easterlin paradox, etc.) humans are not getting any (un) happier on average than our Palaeolithic ancestors despite huge technological progress. Our billions of factory-farmed victims (Francione 2006) spend most of their abject lives below hedonic zero. In absolute terms, the amount of suffering in the world increases each year in humans and non-humans alike. Not least, evolution sabotages human efforts to improve our subjective well-being thanks to our genetically constrained hedonic treadmill—the complicated web of negative feedback mechanisms in the brain that stymies our efforts to be durably happy at every turn (Brickman et al. 1978). Discontent, jealousy, anxiety, periodic low mood, and perpetual striving for "more" were fitness-enhancing in the ancient environment of evolutionary adaptedness. Lifelong bliss wasn't harder for information-bearing self-replicators to encode. Rather lifelong bliss was genetically maladaptive and hence selected against. Only now can biotechnology remedy organic life's innate design flaw.

A potential pitfall lurks here: the fallacy of composition. Just because all individuals tend to seek happiness and shun unhappiness doesn't mean that all individuals seek universal happiness. We're not all closet utilitarians. Genghis Khan wasn't trying to spread universal bliss. As Plato observed, "Pleasure is the greatest incentive to evil." But here's the critical point. Full-Spectrum Superintelligence entails the cognitive capacity impartially to grasp all possible first-person perspectives—overcoming egocentric, anthropocentric, and ethnocentric bias (*cf.* mirror-touch synaesthesia). As an idealisation, at least, Full-Spectrum Superintelligence understands and weighs the full range of first-person facts. First-person facts are as much an objective feature of the natural world as the rest mass of the electron or the Second Law of Thermodynamics. You can't be ignorant of first-person perspectives and superintelligent any more than you can be ignorant of the Second law of Thermodynamics and superintelligent. By analogy, just as autistic superintelligence captures the formal structure of a unified natural science, a mathematically complete "view from nowhere", all possible solutions to the universal Schrödinger equation or its relativistic extension, likewise a Full-Spectrum Superintelligence also grasps all possible first-person perspectives—and acts accordingly. In effect, an idealised Full-Spectrum Superintelligence would combine the mind-reading prowess of a telepathic mirror-touch synaesthete with the optimising prowess of a rule-following hyper-systematiser on a cosmic scale. If your hand is in the fire, you reflexively withdraw it. In withdrawing your hand, there is no question of first attempting to solve the Is-Ought problem in meta-ethics and trying logically to derive an "ought" from an "is". Normativity is built into the nature of the aversive experience itself: I-ought-not-to-be-in-this-dreadful-state. By extension, perhaps a Full-Spectrum Superintelligence will perform cosmic felicific calculus (Bentham 1789) and execute some sort of metaphorical hand-withdrawal for all accessible suffering sentience in its forward light-cone. Indeed one possible *criterion* of Full-Spectrum Superintelligence is the propagation of subjectively hypervaluable states on a cosmological scale.

What this constraint on intelligent agency means in practice is unclear. Conceivably at least, idealised Superintelligences must ultimately do what a classical utilitarian ethic dictates and propagate some kind of "utilitronium shockwave" across the cosmos. To the classical utilitarian, any rate of time-discounting indistinguishable from zero is ethically unacceptable, so s/he should presumably be devoting most time and resources to that cosmological goal. An ethic of negative utilitarianism is often accounted a greater threat to intelligent life (*cf.* the hypothetical "button-pressing" scenario) than classical utilitarianism. But whereas a negative utilitarian believes that once intelligent agents have phased out the biology of suffering, all our ethical duties have been discharged, the classical utilitarian seems ethically committed to converting all accessible matter and energy into relatively homogeneous matter optimised for maximum bliss: "utilitronium". Hence the most empirically valuable outcome entails the extinction of intelligent life. Could this prospect derail superintelligence?

Perhaps but, utilitronium shockwave scenarios shouldn't be confused with *wireheading*. The prospect of self-limiting superintelligence might be credible if either a (hypothetical) singleton biological superintelligence or its artificial counterpart discovers intracranial self-stimulation or its nonbiological analogues. Yet is this blissful fate a threat to anyone else? After all, a wirehead doesn't aspire to convert the rest of the world into wireheads. A junkie isn't driven to turn the rest of the world into junkies. By contrast, a utilitronium shockwave propagating across our Hubble volume would be the product of intelligent design by an advanced civilisation, not self-subversion of an intelligent agent's reward circuitry. Also, consider the reason why biological humanity—as distinct from individual humans—is resistant to wirehead scenarios, namely *selection pressure*. Humans who discover the joys of intra-cranial self-stimulation or heroin aren't motivated to raise children. So they are outbred. Analogously, full-spectrum superintelligences, whether natural or artificial, are likely to be social rather than solipsistic, not least because of the severe selection pressure exerted against any intelligent systems who turn in on themselves to wirehead rather than seek out unoccupied ecological niches. In consequence, the adaptive radiation of natural and artificial intelligence across the Galaxy won't be undertaken by stay-at-home wireheads or their blissed-out functional equivalents.

On the face of it, this argument from selection pressure undercuts the prospect of superhappiness for all sentient life—the "attractor" towards which we may tentatively predict sentience is converging in virtue of the pleasure principle harnessed to ultraintelligent mind-reading and utopian neuroscience. But what is necessary for sentient intelligence is information-sensitivity to fitness-relevant stimuli—not an agent's absolute location on the pleasure-pain axis. True, uniform bliss and uniform despair are inconsistent with intelligent agency. Yet mere recalibration of a subject's "hedonic set-point" leaves intelligence intact. Both information-sensitive gradients of bliss and information-sensitive gradients of misery allow high-functioning performance and critical insight. Only sentience animated by gradients of bliss is consistent with a rich subjective quality of intelligent life. Moreover, the nature of "utilitronium" is as obscure as its theoretical opposite, "dolorium". The problem here cuts deeper than mere lack of technical understanding, e.g. our ignorance of the gene expression profiles and molecular signature of pure bliss in neurons of the rostral shell of the nucleus accumbens and ventral pallidum, the twin cubic centimetre-sized "hedonic hotspots" that generate ecstatic well-being in the mammalian brain (Berridge and Kringelbach 2010). Rather there are difficult conceptual issues at stake. For just as the torture of one mega-sentient being may be accounted worse than a trillion discrete pinpricks, conversely the sublime experiences of utilitronium-driven Jupiter minds may be accounted preferable to tiling our Hubble volume with the maximum abundance of micro-bliss. What is the optimal trade-off between quantity and intensity? In short, even assuming a classical utilitarian ethic, the optimal distribution of matter and energy that a God-like superintelligence would create in any given Hubble volume is very much an open question.

Of course we've no grounds for believing in the existence of an omniscient, omnipotent, omnibenevolent God or a divine utility function. Nor have we grounds for believing that the source code for any future God, in the fullest sense of divinity, could ever be engineered. The great bulk of the Multiverse, and indeed a high measure of life-supporting Everett branches, may be inaccessible to rational agency, quasi-divine or otherwise. Yet His absence needn't stop rational agents intelligently fulfilling what a notional benevolent deity would wish to accomplish, namely the well-being of all accessible sentience: the richest abundance of empirically hypervaluable states of mind in their Hubble volume. Recognisable extensions of existing technologies can phase out the biology of suffering on Earth. But responsible stewardship of the universe within our cosmological horizon depends on biological humanity surviving to become posthuman superintelligence.

Paradise Engineering?

The hypothetical shift to life lived entirely above Sidgwick's (1907) "hedonic zero" will mark a momentous evolutionary transition. What lies beyond? There is no reason to believe that hedonic ascent will halt in the wake of the world's last aversive experience in our forward light-cone. Admittedly, the self-intimating urgency of eradicating *suffering* is lacking in any further hedonic transitions, i.e. a transition from the biology of happiness (Schlaepfer and Fins 2012) to a biology of superhappiness; and then beyond. Yet why "lock in" mediocrity if intelligent life can lock in sublimity instead?

Naturally, superhappiness scenarios could be misconceived. Long-range prediction is normally a fool's game. But it's worth noting that future life based on gradients of intelligent bliss isn't tied to any particular ethical theory: its assumptions are quite weak. Radical recalibration of the hedonic treadmill is consistent not just with classical or negative utilitarianism, but also with preference utilitarianism, Aristotelian virtue theory, a Kantian deontological ethic, pluralist ethics, Buddhism, and many other value systems besides. Recalibrating our hedonic set-point doesn't—or at least needn't—undermine critical discernment. All that's needed for the abolitionist project and its hedonistic extensions (Pearce 1995) to succeed is that our ethic isn't committed to perpetuating the biology of involuntary suffering. Likewise, only a watered-down version of psychological hedonism is needed to lend the scenario sociological credibility. We can retain as much—or as little—of our existing preference architecture as we please. You can continue to prefer Shakespeare to Mills-and-Boon, Mozart to Morrissey, and Picasso to Jackson Pollock while living perpetually in Seventh Heaven or beyond.

Nonetheless an exalted hedonic baseline will revolutionise our conception of life. The world of the happy is quite different from the world of the unhappy, says Wittgenstein; but the world of the super happy will feel unimaginably different from the human, Darwinian world. Talk of preference conservation may reassure bioconservatives that nothing worthwhile will be lost in the post-Darwinian transition. Yet life based on information-sensitive gradients of superhappiness will most likely be "encephalized" in state-spaces of experience alien beyond human comprehension. Humanly comprehensible or otherwise, enriched hedonic tone can make all experience generically hypervaluable in an empirical sense—its lows surpassing today's peak experiences. Will such experience be hypervaluable in a metaphysical sense too? Is this question cognitively meaningful?

The Future of Sentience

The Sentience Explosion

Man proverbially created God in his own image. In the age of the digital computer, humans conceive God-like *super*intelligence in the image of our dominant technology and personal cognitive style—refracted, distorted and extrapolated for sure, but still through the lens of human concepts. The "super-" in so-called superintelligence is just a conceptual fig-leaf that humans use to hide our ignorance of the future. Thus high-AQ/high-IQ humans (Baron-Cohen 2001) may imagine God-like intelligence as some kind of Super-Asperger—a mathematical theorem-proving hyper-rationalist liable systematically to convert the world into computronium for its awesome theorem-proving. High-EQ, low-AQ humans, on the other hand, may imagine a cosmic mirror-touch synaesthete nurturing creatures great and small in expanding circles of compassion. From a different frame of reference, psychedelic drug investigators may imagine superintelligence as a Great Arch-Chemist opening up unknown state-space of consciousness. And so forth. Probably the only honest answer is to say, lamely, boringly, uninspiringly: we simply don't know.

Grand historical meta-narratives are no longer fashionable. The contemporary Singularitarian movement is unusual insofar as it offers one such grand meta-narrative: history is the story of simple biological intelligence evolving through natural selection to become smart enough to conceive an abstract universal Turing machine, build and program digital computers—and then merge with, or undergo replacement by, recursively self-improving artificial superintelligence.

Another grand historical meta-narrative views life as the story of overcoming suffering. Darwinian life is characterised by pain and malaise. One species evolves the capacity to master biotechnology, rewrites its own genetic source code, and creates post-Darwinian superhappiness. The well-being of all sentience will be the basis of post-Singularity civilisation: primitive biological sentience is destined to become blissful supersentience.

These meta-narratives aren't mutually exclusive. Indeed on the story told here, Full-Spectrum Superintelligence entails full-blown supersentience too: a seamless unification of the formal and the subjective properties of mind.

If the history of futurology is any guide, the future will confound us all. Yet in the words of Alan Kay: "It's easier to invent the future than to predict it".

References

Baker, S. (2011). Final Jeopardy: man vs. machine and the quest to know everything. (Houghton Mifflin Harcourt).

Ball, P. (2011). Physics of life: The dawn of quantum biology. *Nature, 474*(2011), 272–274.

Banissy, M., et al. (2009). Prevalence, characteristics and a neurocognitive model of mirror-touch synaesthesia. *Experimental Brain Research, 198*(2–3), 261–272. doi:10.1007/s00221-009-1810-9.

Barkow, J., Cosimdes, L., & Tooby, J. (Eds.). (1992). *The adapted mind: Evolutionary psychology and the generation of culture*. New York: Oxford University Press.

Baron-Cohen, S. (1995). Mindblindness: an essay on autism and theory of mind (MIT Press/ Bradford Books).

Baron-Cohen, S., Wheelwright, S., Skinner, R., Martin, J., Clubley, E. (2001). The autism-spectrum quotient (AQ): evidence from Asperger syndrome/high functioning autism, males and females, scientists and mathematicians. *Journal of Autism Development Disorders, 31*(1): 5–17. doi:10.1023/A:1005653411471. PMID 11439754.

Baron-Cohen, S. (2001). Autism spectrum questionnaire. (Cambridge: Autism Research Centre, University of Cambridge) http://psychology-tools.com/autism-spectrum-quotient/.

Benatar, D. (2006). *Better Never to Have Been: The Harm of Coming Into Existence*. Oxford: Oxford University Press.

Bentham, J. (1789). *An introduction to the principles of morals and legislation*. Oxford: Clarendon Press. Reprint.

Berridge, K. C., & Kringelbach, M. L. (Eds.). (2010). *Pleasures of the Brain*. Oxford: Oxford University Press.

Bostrom, N. (2002). Existential risks: analyzing human extinction scenarios and related hazards. *Journal of Evolution and Technology*, 9.

Boukany, P. E., et al. (2011). Nanochannel electroporation delivers precise amounts of biomolecules into living cells. *Nature Nanotechnology, 6*(2011), 74.

Brickman, P., Coates, D., & Janoff-Bulman, R. (1978). Lottery winners and accident victims: is happiness relative? *Journal of Personal Society Psychology, 36*(8), 917–927. 7–754.

Brooks, R. (1991). Intelligence without representation. *Artificial Intelligence, 47*(1–3), 139–159. doi:10.1016/0004-3702(91)90053-M.

Buss, D. (1997). "Evolutionary Psychology: The New Science of the Mind". (Allyn & Bacon).

Byrne, R., & Whiten, A. (1988). *Machiavellian intelligence*. Oxford: Oxford University Press.

Carroll, J. B. (1993). *Human cognitive abilities: a survey of factor-analytic studies*. Cambridge: Cambridge University Press.

Chalmers, D. J. (2010). The singularity: a philosophical analysis. *Journal of Consciousness Studies, 17*(9), 7–65.

Chalmers, D. J. (1995). Facing up to the hard problem of consciousness. *Journal of Consciousness Studies, 2*(3), 200–219.

Churchland, P. (1989). *A neurocomputational perspective: the nature of mind and the structure of science*. Cambridge: MIT Press.

Cialdini, R. B. (1987). Empathy-based helping: is it selflessly or selfishly motivated? *Journal of Personality and Social Psychology, 52*(4), 749–758.

Clark, A. (2008). *Supersizing the Mind: Embodiment, Action, and Cognitive Extension*. USA: Oxford University Press.

Cochran, G., Harpending, H. (2009). The 10,000 Year Explosion: How Civilization Accelerated Human Evolution New York: Basic Books.

Cochran, G., Hardy, J., & Harpending, H. (2006). Natural history of Ashkenazi intelligence. *Journal of Biosocial Science, 38*(5), 659–693.

Cohn, N. (1957). The pursuit of the millennium: revolutionary millenarians and mystical anarchists of the middle ages (Pimlico).

Dawkins, R. (1976). *The Selfish Gene*. New York City: Oxford University Press.

de Garis, H. (2005). The Artilect War: Cosmists vs. Terrans: A bitter controversy concerning whether humanity should build godlike massively intelligent machines. ETC. Publications, pp. 254. ISBN 978-0882801537.

de Grey, A. (2007). *Ending aging: The rejuvenation breakthroughs that could reverse human aging in our lifetime*. New York: St. Martin's Press.

Delgado, J. (1969). *Physical control of the mind: Toward a psychocivilized society*. New York: Harper and Row.

Dennett, D. (1987). *The intentional stance*. Cambridge: MIT Press.

Deutsch, D. (1997). *The fabric of reality*. Harmondsworth: Penguin.

Deutsch, D. (2011). *The beginning of infinity*. Harmondsworth: Penguin.

Drexler, E. (1986). *Engines of creation: The coming era of nanotechnology*. New York: Anchor Press/Doubleday.

Dyson, G. (2012). *Turing's cathedral: The origins of the digital universe*. London: Allen Lane.

Everett, H. (1973). The theory of the universal wavefunction. Manuscript (1955), pp 3–140 In B. DeWitt, & R. N. Graham, (Eds.), The many-worlds interpretation of quantum mechanics. Princeton series in physics. Princeton, Princeton University Press. ISBN 0-691-08131-X.

Francione, G. (2006). Taking sentience seriously. Journal of Animal Law & Ethics 1.

Gardner, H. (1983). *Frames of mind: The theory of multiple intelligences*. New York: Basic Books.

Goertzel, B. (2006). The hidden pattern: A patternist philosophy of mind. (Brown Walker Press).

Good, I. J. (1965). Speculations concerning the first ultraintelligent machine. In L. F. Alt & M. Rubinoff (Eds.), *Advances in computers* (pp. 31–88). London: Academic Press.

Gunderson, K. (1985). *Mentality and machines*. Minneapolis: University of Minnesota Press.

Hagan, S., Hameroff, S., & Tuszynski, J. (2002). Quantum computation in brain microtubules? Decoherence and biological feasibility. *Physical Reviews, E65*, 061901.

Haidt, J. (2012). *THE righteous mind: Why good people are divided by politics and religion*. NY: Pantheon.

Hameroff, S. (2006). Consciousness, neurobiology and quantum mechanics. In The emerging physics of consciousness, J. Tuszynski (Ed.) (Springer).

Harris, S. (2010). *The moral landscape: How science can determine human values*. NY: Free Press.

Haugeland, J. (1985). *Artificial intelligence: The very idea*. Cambridge: MIT Press.

Holland, J. (2001). Ecstasy: The complete guide: A comprehensive look at the risks and benefits of mdma. (Park Street Press).

Holland, J. H. (1975). *Adaptation in natural and artificial systems*. Ann Arbor: University of Michigan Press.

Hutter, M. (2010). *Universal artificial intelligence: Sequential decisions based on algorithmic probability*. New York: Springer.

Hutter, M. (2012). Can intelligence explode? *Journal of Consciousness Studies, 19*, 1–2.

Huxley, A. (1932). Brave New World. London: Chatto and Windus.

Huxley, A. (1954). Doors of perception and heaven and hell. New York: Harper & Brothers.

Kahneman, D. (2011). *Thinking, fast and slow*. Straus and Giroux: Farrar.

Kant, I. (1781). Critique of pure reason. In P. Guyer & A. Wood. Cambridge: Cambridge University Press, 1997.

Koch, C. (2004). The Quest for Consciousness: a Neurobiological Approach. Roberts & Co..

Kurzweil, R. (2005). The singularity is near. New York: Viking.

Kurzweil, R. (1990). The age of intelligent machines. Cambridge: MIT Press

Kurzweil, R. (1998). The age of spiritual machines. New York: Viking.

Langdon, W., Poli, R. (2002). Foundations of genetic programming. New York: Springe).

Lee, H. J., Macbeth, A. H., Pagani, J. H., Young, W. S. (2009). Oxytocin: The great facilitator of life. *Progress in Neurobiology 88* (2): 127–51. doi:10.1016/j.pneurobio.2009.04.001. PMC 2689929. PMID 19482229.

Legg, S., Hutter, M. (2007). Universal intelligence: A definition of machine intelligence. *Minds & Machines*, *17*(4), pp. 391–444.

Levine, J. (1983). Materialism and qualia: The explanatory gap. *Pacific Philosophical Quarterly, 64*, 354–361.

Litt, A., et al. (2006). Is the Brain a Quantum Computer? *Cognitive Science, 20*, 1–11.

Lloyd, S. (2002). Computational capacity of the universe. *Physical Review Letters 88*(23): 237901. arXiv:quant-ph/0110141. Bibcode 2002PhRvL..88w7901L.

Lockwood, L. (1989). Mind, brain, and the quantum. Oxford: Oxford University Press.
Mackie, JL. (1991). Ethics: Inventing right and wrong. Harmondsworth: Penguin.
Markram, H. (2006). The blue brain project. *Nature Reviews Neuroscience*, *7*:153–160. PMID 16429124.
Merricks, T. (2001). Objects and persons. Oxford: Oxford University Press.
Minsky, M. (1987). The society of mind. New York: Simon and Schuster.
Moravec, H. (1990). Mind children: The future of robot and human intelligence. Cambridge: Harvard University Press.
Nagel, T. (1974). What is it Like to Be a Bat? *Philosophical Review, 83*, 435–450.
Nagel, T. (1986). The view from nowhwere. Oxford: Oxford University Press.
Omohundro, S. (2007). The Nature of Self-Improving Artificial Intelligence. San Francisco: Singularity Summit.
Parfit, D. (1984). *Reasons and persons*. Oxford: Oxford University Press.
Pearce, D. (1995). The hedonistic imperative http://hedweb.com.
Pellissier, H. (2011) Women-only leadership: Would it prevent war? http://ieet.org/index.php/IEET/more/4576.
Penrose, R. (1994). Shadows of the mind: A search for the missing science of consciousness. Cambridge: MIT Press.
Peterson, D., Wrangham, R. (1997). Demonic males: Apes and the origins of human violence. Mariner Books.
Pinker, S. (2011). The better angels of our nature: Why violence has declined. New York: Viking.
Rees, M. (2003). Our final hour: A scientist's warning: How terror, error, and environmental disaster threaten humankind's future in this century—on earth and beyond. New York: Basic Books.
Reimann F, et al. (2010). Pain perception is altered by a nucleotide polymorphism in SCN9A. *Proc Natl Acad Sci* USA. *107*(11):5148–5153 (2010 Mar 16).
Rescher, N. (1974). Conceptual idealism. Oxford: Blackwell Publishers.
Revonsuo, A. (2005). Inner presence: Consciousness as a biological phenomenon. Cambridge: MIT Press.
Revonsuo, A., & Newman, J. (1999). Binding and consciousness. *Consciousness and Cognition, 8*, 123–127.
Riddoch, M. J., & Humphreys, G. W. (2004). Object identification in simultanagnosia: When wholes are not the sum of their parts. *Cognitive Neuropsychology, 21*(2–4), 423–441.
Rumelhart, D. E., McClelland, J. L., & the PDP Research Group. (1986). *Parallel distributed processing: Explorations in the microstructure of cognition. Volume 1: Foundations*. Cambridge: MIT Press.
Russell, B. (1948). *Human knowledge: Its scope and limits*. London: George Allen & Unwin.
Sandberg, A., Bostrom, N. (2008). Whole brain emulation: A roadmap. Technical report 2008-3.
Saunders, S., Barrett, J., Kent, A., Wallace, D. (2010). Many worlds: Everett, quantum theory, and reality. Oxford: Oxford University Press.
Schlaepfer, T. E., & Fins, J. J. (2012). How happy is too happy? Euphoria, neuroethics and deep brain stimulation of the nucleus accumbens. *The American Journal of Bioethics, 3*, 30–36.
Schmidhuber, J. (2012). Philosophers & Futurists, Catch Up! Response to the singularity. *Journal of Consciousness Studies*, *19*: 1–2, pp. 173–182.
Seager, W. (1999). Theories of consciousness. London: Routledge.
Seager, W. (2006). The 'intrinsic nature' argument for panpsychism. *Journal of Consciousness Studies, 13*(10–11), 129–145.
Sherman, W., Craig A., (2002). Understanding virtual reality: Interface, application, and design. Los Altos: Morgan Kaufmann.
Shulgin, A. (1995). PiHKAL: A chemical love story. Berkeley: Transform Press.
Shulgin, A. (1997). TiHKAL: The continuation. Berkeley: Transform Press.
Shulgin, A. (2011). The shulgin index vol 1: Psychedelic phenethylamines and related compounds. Berkeley: Transform Press.

Shulman, C., Sandberg, A. (2010). Implications of a software-limited singularity. Proceedings of the European Conference of Computing and Philosophy.

Sidgwick, H. (1907). The methods of ethics. Indianapolis: Hackett, seventh edition, 1981, 1-4.

Singer, P. (1995). *Animal liberation: A new ethics for our treatment of animals*. New York: Random House.

Singer, P. (1981). *The expanding circle: Ethics and sociobiology*. New York: Farrar, Straus and Giroux.

Smart, JM. (2008–2011) Evo Devo Universe? A Framework for Speculations on Cosmic Culture. In Cosmos and culture: Cultural evolution in a cosmic context. J. S. Dick, M. L. Lupisella (Eds.), Govt Printing Office, NASA SP-2009-4802, Wash., D.C., 2009, pp. 201–295.

Stock, G. (2002). Redesigning humans: Our inevitable genetic future. Boston: Houghton Mifflin Harcourt.

Strawson G., et al. (2006). Consciousness and its place in nature: Does physicalism entail panpsychism? Imprint Academic.

Tegmark, M. (2000). Importance of quantum decoherence in brain processes. *Physical Review E, 61*(4), 4194–4206. doi:10.1103/PhysRevE.61.4194.

Tsien, J. et al., (1999). Genetic enhancement of learning and memory in mice. *Nature 401*, 63–69 (2 September 1999) | doi:10.1038/43432.

Turing, A. M. (1950). Computing machinery and intelligence. *Mind, 59*, 433–460.

Vinge, V. (1993). *The coming technological singularity*. New Whole Earth LLC: Whole Earth Review.

Vitiello, G. (2001). My double unveiled; advances in consciousness. Amsterdam: John Benjamins.

Waal, F. (2000). Chimpanzee politics: Power and sex among apes". Maryland: Johns Hopkins University Press.

Wallace, D. (2012). *The Emergent multiverse: Quantum theory according to the Everett interpretation*. Oxford: Oxford University Press.

Welty, G. (1970). The history of the prediction paradox, presented at the Annual Meeting of the International Society for the History of the Behavioral and Social Sciences, Akron, OH (May 10, 1970), Wright State University Dayton, OH 45435 USA. http://www.wright.edu/~gordon.welty/Prediction_70.htm.

Wohlsen, M. (2011): Biopunk: DIY scientists hack the software of life. London: Current.

Yudkowsky, E. (2007). Three major singularity schools. http://yudkowsky.net/singularity/schools.

Yudkowsky, E. (2008). Artificial intelligence as a positive and negative factor in global risk. In Bostrom, Nick and Cirkovic, Milan M. (Eds.). Global catastrophic risks. pp. 308–345 (Oxford: Oxford University Press).

Zeki, S. (1991). Cerebral akinetopsia (visual motion blindness): A review. *Brain, 114*, 811–824. doi:10.1093/brain/114.2.811.

Chapter 11A
Illah R. Nourbakhsh on Pearce's "The Biointelligence Explosion"

The Optimism of Discontinuity

In *The Biointelligence Explosion*, David Pearce launches a new volley in the epic, pitched battle of today's futurist legions. The question of this age is: *machine or man?* And neither machine nor man resembles the modern-day variety. According to the Singularity's version of foreshadowed reality, our successors are nothing like a simulacrum of human intelligence; instead they vault beyond humanity along every dimension, achieving heights of intelligence, empathy, creativity, awareness and immortality that strain the very definitions of these words as they stand today. Whether these super-machines embody our unnatural, disruptive posthuman evolution, displacing and dismissing our organic children, or whether they melt our essences into their circuitry by harvesting our consciousnesses and *qualia* like so much wheat germ, the core ethic of the machine disciples is that the future will privilege digital machines over carbon-based, analog beings.

Pearce sets up an antihero to the artificial superintelligence scenario, proposing that our wetware will shortly become so well understood, and so completely modifiable, that personal bio-hacking will collapse the very act of procreation into a dizzying tribute to the ego. Instead of producing children as our legacy, we will modify our own selves, leaving natural selection in the dust by changing our personal genetic makeup in the most extremely personal form of creative hacking imaginable. But just like the AI singularitarians, Pearce dreams of a future in which the new and its ancestor are unrecognizably different. Regular humans have depression, poor tolerance for drugs, and, let's face it, mediocre social, emotional and technical intelligence. Full-Spectrum Superintelligences will have perfect limbic mood control, infinite self-inflicted hijacking of chemical pathways, and so much intelligence as to achieve omniscience bordering on Godliness.

The Singularity proponents have a fundamentalist optimism born, as in all religions, of something that cannot be proven or disproven rationally: faith. In their case, they have undying faith in a future discontinuity, the likes of which the computational world has never seen. After all, as Pearce points out, today's computers have not shown even a smattering of consciousness, and so the ancestry of the intelligent machine, a machine so fantastically powerful that it can eventually invent the superintelligent machine, is so far an utter no-show. But this is alright if we can believe that with Moore's Law comes a new golden chalice: a point of no return, when the progress of Artificial Intelligence self-reinforces, finally, and takes off like an airplane breaking ground contact and suddenly shooting upward in the air: a discontinuity that solves all the unsolvable problems. No measurement of AI's effectiveness before the discontinuity matters from within this world view; the future depends only on the shape of a curve, and eventually all the rules will change when we hit a sudden bend. That a technical sub-field can depend so fully, not on early markers of success, but on the promise of an unknown future disruption, speaks volumes about the discouraging state of Artificial Intelligence today. When the best recent marker of AI, IBM's Watson, wins peculiarly by responding to a circuit-driven light in 8 ms, obviating the chances of humans who must look at a light and depend on neural pathways orders of magnitude slower, then AI Singularity cannot yet find a machine prophet.

Pearce is also an optimist, presenting an alternative view that extrapolates from the mile marker of yet another discontinuity: when hacker-dom successfully turns its tools inward, open-sourcing and bio-hacking their own selves to create recursively improving bio-hackers that rapidly morph away from mere human and into transcendental Superintelligence. The discontinuity is entirely different from the AI Singularity, and yet it depends just as much on a computational mini-singularity. Computers would need to provide the simulation infrastructure to enable bio-hackers to visualize and test candidate self-modifications. Whole versions of human-YACC and human-VMWare would need to compile and run entire human architectures in dynamic, simulated worlds to see just what behaviour will ensue when Me is replaced by Me-2.0. This demands a level of modelling, analog simulation and systems processing that depend on just as much of a discontinuity as the entire voyage. *And then a miracle happens* becomes almost cliché when every technical obstacle to be surmounted is not a mountain, but a hyperplane of unknown dimensionality!

But then there is the hairy underbelly of open-source genetics, namely that of systems engineering and open-source programming in general. As systems become more complex, Quality Assurance (QA) becomes oxymoronic because tests fail to exhaustively explore the state-space of possibilities. The Toyota Prius brake failures were not caught by engineers whose very job is to be absolutely sure that brakes *never, ever* fail, because just the right resonant frequency, combined with a hybrid braking architecture, combined with just the right accelerometer architecture and firmware, can yield a one-in-a million rarity a handful of times, literally. The logistical tail of complexity is a massive headache in the regime of QA, and this bodes poorly for open-sourced hacking of human systems, which dwarf the complexity of Toyota Prius exponentially. IDE's for bio-hacking; debuggers that can isolate part of your brain so that you can debug a nasty problem without losing consciousness (Game Over!); version control systems and repositories so that, in a panic, you can return your genomic identity to a most recent stable state- all of these tools will be needed, and we will of course be financially enslaved to the corporations that provide these self-modification tools. Will a company, let's call it *HumanSoft*, provide a hefty discount on its insertion vector applications if you agree to do some advertising—your compiled genome always drinks Virgil's Root Beer at parties, espousing its combination of Sweet Birch and Molasses? Will you upgrade to HumanSoft's newest IDE because it introduces forked compiling—now you can run two mini-me's in one body and switch between them every 5 s by reprogramming the brain's neural pathways.

Perhaps most disquieting is the law of unintended consequences, otherwise known as robotic compounding. In the 1980s, roboticists thought that they could build robots bottom-up, creating low-level behaviours, testing and locking them in, then adding higher-level behaviours until, eventually, human-level intelligence flowed seamlessly from the machine. The problem was that the second level induced errors in how level one functioned, and it took unanticipated debugging effort to get level one working with level two. By the time a roboticist reaches level four, the number of side effects overwhelms the original engineering effort completely, and funding dries up before success can be had. Once we begin bio-hacking, we are sure to discover side effects that the best simulators will fail to recognize unless they are equal in fidelity to the real-world. After how many major revisions will we discover that all our hacking time is spent trying to undo unintended consequences rather than optimizing desired new features? This is not a story of discontinuity, unfortunately, but the gradual build-up of messy, complicated baggage that gums up the works gradually and eventually becomes a singular centre of attention.

We may just discover that the Singularity, whether it gives rise to Full-Spectrum Superintelligence or to an Artificial Superintelligence, surfaces an entire stable of mediocre attempts long before something of real value is even conceivable. Just how many generations of mediocrity will we need to bridge and at what cost, to reach the discontinuity that is an existential matter of faith?

There is one easy answer here, at once richly appropriate and absurd. Pearce proposes that emotional self-control has one of the most profound consequences on our humanity, for we can make ourselves permanently happy. Learn to control the limbic system fully, and then bio-hackers can hack their way into enforced sensory happiness- indeed, even modalities of happiness that effervesce beyond anything our non drug-induced dreams can requisition today. Best of all, we could program ourselves for maximal happiness even if Me-2.0 is mediocre and buggy. Of course, this level of human chemical pathway control suggests a level of maturity that pharmaceutical companies dream about today, but if it is truly possible to obtain permanent and profound happiness all-around, then of course we lose both the condition and state of happiness. It becomes the drudgery that is a fact of life.

Finally, let us return to one significant commonality between the two hypotheses: they both demand that technology provide the ultimate modelling and simulation engine: I call it the Everything Engine. The Everything Engine is critical to AI because computers must reason, fully, about future implications of all state sets and actions. The Everything Engine is also at the heart of any IDE you would wish to use when hacking your genome: you need to model and generate evidence that your proposed personal modification yields a better you rather than a buggier you. But today, the Everything Engine is Unobtanium, and we know that incremental progress on computation speed will not produce it. We need a discontinuity in computational trends in order to arrive at the Everything Engine. Pearce is right when he states that the two meta-narratives of Singularity are not mutually exclusive. In fact, they are conjoined at the hip; for, if their faith in a future discontinuity proves false, then we might just need infinity of years to reach either Nirvana. And if the discontinuity arrives soon, then as Pearce points out, we will all be too busy inventing the future or evading the future to predict the future.‘

Chapter 12
Embracing Competitive Balance: The Case for Substrate-Independent Minds and Whole Brain Emulation

Randal A. Koene

Abstract More important than debates about the nature of a possible singularity is that we successfully navigate the balance of *opportunities* and *risks* that our species is faced with. In this context, we present the objective to upload to substrate-independent minds (SIM). We emphasize our leverage along this route, which distinguishes it from proposals that are mired in debates about optimal solutions that are unclear and unfeasible. We present a theorem of cosmic dominance for intelligence species based on principles of universal Darwinism, or simply, on the observation that selection takes place everywhere at every scale. We show that SIM embraces and works with these facts of the physical world. And we consider the existential risks of a singularity, particularly where we may be surpassed by artificial intelligence (AI). It is unrealistic to assume the means of global cooperation needed to the create a putative "friendly" super-intelligent AI. Besides, no one knows how to implement such a thing. The very reasons that motivate us to build AI lead to machines that learn and adapt. An artificial general intelligence (AGI) that is plastic and at the same time implements an unchangeable "friendly" utility function is an oxymoron. By contrast, we note that we are living in a real world example of a Balance of Intelligence between members of a dominant intelligent species. We outline a concrete route to SIM through a set of projects on whole brain emulation (WBE). The projects can be completed in the next few decades. So, when we compare this with plans to "cure aging" in human biology, SIM is clearly as feasible in the foreseeable future—or more so. In fact, we explain that even in the near term life extension will require mind augmentation. Rationality is a wonderful tool that helps us find effective paths to our goals, but the goals arise from a combination of evolved drives and interests developed through experience. The route to a new Balance of Intelligence by SIM has this

R. A. Koene (✉)
NeuraLink Co. and Carboncopies.org, San Francisco, CA 94103, USA
e-mail: randal.a.koene@gmail.com

A. H. Eden et al. (eds.), *Singularity Hypotheses*, The Frontiers Collection,
DOI: 10.1007/978-3-642-32560-1_12, © Springer-Verlag Berlin Heidelberg 2012

additional benefit, that it does acknowledges our emancipation and does not run counter to our desire to participate in advances and influence future directions.

Competition and Natural Selection at Every Scale

In the first part of this paper we will devote some attention to reasons. Who are we, humans, and what do we want? This is important if we want to understand why a Singularity scenario would or should come about. For without our actions there will certainly be no Singularity. And for practical purposes, it will be useful to know if we are talking about events we are striving for or events that we may not be able to avoid.

Success at Cosmic Scale or in an Environmental Niche

It is a feature of the human condition that we are naturally preoccupied with anthropocentric concerns. If we do not quite gaze at our own navels, at least we tend to direct most of our worries, thoughts and plans at the here and now. The well-known cosmologist Max Tegmark is known to remark upon this fact when he cautions us about long-range cosmic perspectives (Tegmark 2011). Of course, our preoccupation is itself an outcome of natural selection.

And why should we not be concerned primarily with matters that relate to our own society? Is human society not the bedrock of purpose and meaning?

Meaning is a slippery concept to question. We can certainly designate local or constrained purpose and meaning, limiting them by definition to the domain that a specific thinking entity cares about. Beyond that, any objective or universal purpose cannot be substantiated. Humans care for their lives, the lives of their offspring, the lives of their kin and the lives of all humans and all living things—often in that order of priority. We might say that the purpose of that interest is to insure species survival. But what is the purpose of species survival? Does the existence of a species relate to any greater purpose or meaning? And, if you did establish such a purpose then the next question would inevitably be: Satisfying that, what purpose does that serve? Ultimately, there simply is no such thing as a universal purpose.

No matter how well-conceived, our wishes for the future cannot be constructed such that they fulfill a universal purpose that does not exist. Our wishes cannot be supported by a top-to-bottom rationale. Rationality is merely a tool. It is a tool that promises efficiency, but it is nevertheless just a tool to help us get from A to B. Why we choose B as our goal emerges instead from distinctions that we make between that which we find desirable and that which we do not. They are distinctions, likely established by a combination of intrinsic drives and acquired tastes.

Generally speaking, our intrinsic drives arose from a selection for behavior that improved the survival and propagation of genetically inherited traits. The competitive effects of selection are ever-present.

Is there any way in which we could live in a manner that does not involve competition? Selection of some kind is always taking place. In a collision between a small, dense asteroid and a large, porous asteroid, one of the two is the likely intact survivor of the encounter. If that is the measure of successful survival then a selection took place. Similarly, we can identify selective processes in events throughout the cosmos. It is the process that has been called *Universal Darwinism* (Dennett 2005), an all-pervasive competition and selection at every scale.

Our local environment has shaped the behavior of successful intelligent species, and one of the strongest requirements for the successful existence of a thinking entity is a survival-oriented self-consistent reward mechanism. Our biology and our behavior have been tuned to achieve gene-survival within an environmental niche in space and time (Koene 2011). But the universe is a much bigger stage with a much greater variety of challenges.

Adapting to Challenges and Cosmic Dominance

Environments change, as do challenges. In fact, we have embarked upon a route of tool building that will eventually lead us to build new thinking entities. This is a development that deserves further consideration, and that development is also the origin of the Vingean concept of the Singularity (Vinge 1981).

Let us reflect on the matter of adaptability to different environments and challenges. Some thinking entities will learn ways in which they can modify their thinking, including their reward mechanisms. It makes sense to make such modifications in response to new knowledge. That way, a thinking entity can maximize its reward over time. It is worth noting that this process has been taking place in human society as well, as environmental challenges have changed. For example, humans have had to adapt to life in circumstances of very high population density, following the development of cities.

There may well be risks associated with modifications. There is a deductive line of thought, as presented by *carboncopies.org* co-founder Suzanne Gildert (Gildert 2010), which demonstrates that a lack of a fixed, intrinsic drive-based sense of purpose can lead to the adoption of what may be described as a "nihilistic" personal philosophy. Such nihilism can have behavioral consequences, since it makes all outcomes appear of equal value. The possible outcomes include catatonic passivity, self-termination, or termination by inadequately competing with other entities in a Darwinist universe. One of the interesting considerations that follow from this line of reasoning is that it might be impossible to develop or evolve truly general super-intelligent AI, because such an AI would inevitably become aware of the lack of universal purpose and of a full top-down rationale for goals and motivations.

Whether or not there truly is such an obstacle for super-intelligent artificial general intelligence (AGI), there is cause to suspect that human intelligence itself exists in a niche-specific and finely-tuned balance attained through natural selection. Suzanne Gildert called these niches of intelligent survival "catchment areas". We do not know what the landscape of possible developmental routes looks like outside of this balance. If there are only a few peaks of tuning where an intelligence can survive and many deep valleys in which developments do not lead to survival, then most modifications could endanger survival.

Even taking into consideration the catchment area hypothesis though, we can reasonably deduce the overall outcome for all intelligences over space and time. Let us take this cosmic perspective now and consider its end-result. For this, I will present a theorem with which to address the dominance of certain types of intelligence over cosmic expanses of time and space.

Cosmic Dominance Theorem for Intelligent Species: *The greatest proportions of space and time that are influenced by intelligence will be influenced by those types of intelligence that achieve the flexibility to adapt to new environmental and circumstantial challenges in a goal-oriented manner.*

Here is how we arrive at this theorem: Let us presuppose that a body and mind that compete well within a specific environment and set of challenges are not automatically equally well-suited to all other environments and challenges. An iterative selection that is carried out by subjecting all such intelligences to a sequence of new environments and new challenges will reduce the number of intelligences that are able to succeed within the entire sequence they have encountered. Natural selection and mutational change (i.e., a random walk of change) may produce some intelligence that can eventually successfully exist in a large number of environments and challenges. Flexible intelligence that incorporates adaptability to new environments and challenges in its own design, intelligence geared toward goal-oriented change, is likely to achieve such dominance much more quickly. Consequently, the largest portions of space and time that will be influenced in some way by intelligence will be influenced by those that have such flexibility (Koene 2011). (Note that we are not particularly interested here in influence that is carried out through non-intelligent intervention, such as inanimate interactions or systematic interactions carried out by simple pattern generators.) *Substrate-independence is a foremost requirement for the goal-directed flexibility.*

Super-Intelligent AI and a Balance of Intelligence

Goal-oriented adaptation is one of the qualities envisioned by proponents of AGI development through a process of goal-directed iterative self-improvement. If sufficiently rapid, such iteration has been supposed as the mechanism for a so-called intelligence explosion that brings about a Singularity (Chalmers 2010; Good 1965; Kurzweil 2005; Moravec 1998; Sandberg and Bostrom 2008; Solomonoff 1985; Vinge 1993). Carl Shulman and Stuart Armstrong compare such an explosion with arms races, including the imperative of a possible military advantage as a driver for risky AI research protocols. Embarking on an AI equivalent of a Manhattan Project might be perceived as a winner-take-all competition. But, as they point out, where nuclear material is difficult to obtain and process, AI technology that might not depend on special hardware could be easy to copy. The Manhattan Project was thoroughly infiltrated by Russia. A rapid spread of AI developments through information leaks could maintain a semblance of balance. To rely on such a coincidence of conditions to bring about safe development through balance is not at all reassuring.

Instead, to prepare for the safe development of super-intelligent AGI, all parties would have to agree to oversight and all projects currently underway would have to be halted and modified accordingly. At present, it seems unlikely that careful coordination of the actions of all interested groups will be achieved. *No one is even attempting to bring a halt to unconstrained AI projects.* In addition to this problem, there are intrinsic problems with the notion of creating an assuredly "friendly" super-intelligent AI guardian.

The value of learning, and ultimately of total flexibility as described above, certainly also exists for super-intelligent AI. Learning, the ability to gain new insights is in fact a principal requirement for intelligence. If a purported AI guardian's avenues of change are heavily constrained then it will not be generally super-intelligent enough to deal with all challenges that appear—which makes living in its care a rather dangerous prison ship to be adrift in on a cosmic ocean.

If, on the other hand, a guardian AI can adapt to deal as best it can with any challenge, then the unpredictable consequences of changes remove all guarantees of a friendly or desirable situation for us. It does not much matter if changes are the result of learning from new information or are inadvertent consequences (e.g., bits flipped by cosmic radiation). A "utility function" of the system will not be fixed in an architecture that is not static. It will drift. Even a minor drift can cause the consequences to diverge significantly over time from a predicted course, just as a minor course correction at great distance can deflect an asteroid from its path to Earth.

And why are we even working on machine intelligence and AI? We do this, because machines that can learn are very useful. AI exists because we want it to be able to collect information and to gain knowledge. We want AI to learn and modify its thinking. All of this simply means that a useful super-intelligent general AI will probably be unconstrained at some point. But, when that happens the situation is not at all unlike the one in which the dominant intelligence of the human species took control of the world and of the fate of other species, including those just slightly less intelligent.

No scheme has yet been demonstrated by which one could create a super-intelligent AGI that is simultaneously plastic and has a method by which it implements a truly fixed utility function or preference relation U(x). Such a thing is an oxymoron. Even disregarding the problem of drift, it is unclear if a utility function can be devised that would have satisfactory criteria for "friendliness", and which would lead to some situation that we would deem desirable.

Contrast this with the real situation today, where no single intelligent entity among the population of dominant intelligences on Earth can subject all of them to its whims without rapid counter-action by the others who have similar intelligence. It is this balance of power that has existed and continues to exist between the 7 billion most intelligent minds on Earth today. These situations exist throughout the living world, as they are directly related to the process of natural selection. When changes within the population occur infrequently it feels like a balance. When they occur frequently it feels like a race. Both experiences may be found not just in the living world, but in all dynamical systems, and they fit the paradigm of Universal Darwinism.

The race, in addition to advancing competitive capabilities, also keeps the leading runners in check. It is like the famous Red Queen's hypothesis that has been applied as an evolutionary argument for the advantages of sexual reproduction (Ridley 1995), as well as in the computational domain, e.g., applied to co-evolution in cellular automata (Paredis 1997). The Red Queen of Lewis Caroll's *Through the Looking Glass* said: "It takes all the running you can do, to keep in the same place."

We may suggest a **Red Queen's Hypothesis of Balance of Intelligence**: *In a system dominated by a sufficient number of similarly intelligent adaptive agents, the improvements of agents take place in a balance of competition and collaboration that maintains relative strengths between leading agents.*

Even if friendly AI could exist in separation but concurrence with human society, while being vastly superior in its capability, that sort of Singularity would still be troubling to us in a number of ways. Under those circumstances there would be a "glass ceiling", and we would never experience and create at the level of the dominant intelligence. We would not share in subsequent stages of development. The singularity of comprehension encountered could become a feature, forever remaining beyond our grasp as greater intelligence advances. This is certainly not a future experience that many of us wish for.

How about the possibility of errors, bugs in the code? Have we ever yet developed an enormously complex piece of software that did not contain ubiquitous and often serious flaws? If there was a credible project to create inherently friendly AI, then the project would have to avoid such bugs and all other modes of component failure. Realistically, and knowing our history of technology development that scenario seems highly improbable.

Summarizing, we know that there are existential risks attached to the development of AGI with intelligence equal to or greater than that of humans. At the same time, we know that there are strong motivators driving the development of just such AI. It is next to impossible to coordinate and control all of the actions of the various players in the field right now. So what can we possibly do? Do we simply pass the torch? Drenched in self-sacrificial romanticism as that may feel, does it make sense from our perspective?

Sure, greater intelligence may do wonderfully interesting things with the world and the universe. There is also a distribution of possibilities where the outcome would not be so interesting. We cannot predict if said intelligence is more interested in the types of creative complexity that entertain us. Perhaps it may be more interested in monotonous regularity. Besides, as we will discuss in the next section, our experience of being and our wishes for future experiences take place with our minds. If we have the choice to take or not to take a path for which it was indicated that with significant likelihood all such experience might come to an end, would we take it? Would we even seriously contemplate taking such a path if the objective under consideration was not AI?

While accepting those things that we do not control, I think that the vast majority of us would prefer a route that maximizes the likelihood that we will be able to experience and participate in a future brought about by a technological Singularity. Ultimately, we cannot constrain the development of a more advanced intelligence any more than a group of mice could hope to control a human. We exert far more control over the manner and degree to which we ourselves embrace the advance. We can strive to be an intimate part of it. We can give our species the flexibility to absorb and integrate new capabilities that we create.

The Reality of 'Being' as an Experience

What We Know

There are sensations that comprise that awareness, which we may simply express as "being". They are the sensations of our own bodies, of the effect that we have on our environment and that the environment has on us. In addition to perceptions, they are the introspective thoughts that are fueled by emotion, memory, realization, decision and creativity. But what are all those sensations made up of? Where are they made and experienced?

What does it mean to feel a rock in your hand? What does it mean to see it there? Nerves in your skin send electric signals to your brain. Those signals are processed by neuronal circuitry, which eventually leads to mental activity that you are familiar with, and which your mind interprets as feeling a rock in your hand. Similarly, photons that reach the retina of your eye are converted into electric signals. Again, those signals are processed by neuronal circuitry, which eventually leads to mental activity that you are familiar with, and which your mind interprets as seeing a rock in your hand. The awareness of our existence depends entirely on the processing of activity by the mechanisms of the brain. The interplay of that activity, generated within the brain is the totality of the experience of being.

To Be Or Not to Be

What does it mean when we contemplate life, the end of life, or the preference to stay alive? Each of us has preferences, wishes or desires. When we say that we would like something, that we have a goal, what we are really saying is that there are future experiences that we want. Being takes places in our minds, it is the current experience, as it is being filtered through processes that were modified by prior experiences, i.e., through learning, memory, and knowledge. The future we envisage is one that for us can only exist through the processing of its experiences, filtered through our individual characteristic lenses. There is that which we process now, and there is that which we wish to process in the future. Those processes are '*being*' and the desire to continue to be.

Realizing this, we see that a personal identity is not just about the memory of specific events. Rather, it is about that individual, characteristic way in which each of us acquires, represents and uses experiences. Those characteristics lead each of us to adhere with preference to specific 'memes' that include notions of how to influence future developments. Increasingly, we are interested in the development and propagation of memes that give rise to future experiences, the preference for which is reflected in our individual patterns of mental activity. We support with passion the causes that represent interests and world views. Often, we care more about the competitive survival and development of these things that are represented in patterns of mental activity than we care about the survival of a specific sequence of nucleotides, as expressed in our DNA.

True, the original reason for our thinking existence is a result of competitive developments driven by natural selection and gene-survival. Not anymore. The focus of our thinking existence is not merely gene-survival. Thought has brought about new and creative avenues of interest. We celebrate the great thinkers, the artists and authors, performers, builders, creators and leaders of movements. Even when we seem to celebrate genetic success, such as by noting the long history of a famous family it is really the social importance of that family and not the specifics of their hereditary material that we remark upon.

Evolutionary gene-survival in our species has established a set of checks and balances arrived at through selection in the environment of Earth's biosphere, in its general form during the last few million years. That is not a set of rules that is automatically equally well-suited to the survival and propagation of the patterns of mind we care about.

Daring to Gaze at Reality Unencumbered

Strategies to Optimize Pattern Propagation

It is by looking directly at the big picture in terms of our real interests in those patterns of mind, experiences, gaining understanding and creating, that we see the greatest need and define the objective of substrate-independent minds (SIM). SIM aims specifically to devise those strategies through which our being can be optimized towards the survival and propagation of patterns of mind functions. In the following, we embrace the realities of the universal Darwinist processes that we introduced above, the requirements they impose, and we focus on approaches that we can in practice enact and control to meet those requirements. There are on the present roadmap at least six technology paths through which we may enable functions of the mind to move from substrate to substrate (i.e., gaining substrate-independence). Of those six, the path known as whole brain emulation (WBE) is the most conservative one. WBE proposes that we:

1. Identify the scope and the resolution at which mechanistic operations within the brain implement the functions of mind that we experience.
2. Build tools that are able to acquire structural and functional information at that resolution and scope in an individual brain.
3. Re-implement the whole structure and the functions in another suitable operational substrate, just as they were implemented in the original cerebral substrate.

SIM and WBE, if properly accomplished on a schedule that anticipates keeping pace with others developments, such as in AI, can ensure that we benefit from the advance and that we incorporate whatever turns out to be the nature of the most successful intelligent species. Toward the end of this paper, we address that schedule, AI and existential risk in particular. SIM, especially via WBE, means beginning with minds that have a human architecture. Early forms will be comprehensible and in some ways predictable. Developments in SIM are less likely to have features of a hard take-off, because the human brain (unlike purported models for AGI) is not designed ab initio to be iteratively self-improving through the creation of its own successors.

No Half-Measures to Life-Extension

From the preceding, it should be clear how minds that are substrate-independent are an essential part of any strategy that extricates itself from anthropocentric navel-gazing and considers a more cosmic perspective. Yet a cosmic perspective, great stretches of time and unimaginable challenges are certainly not the only reasons why we should consider SIM an essential part of any satisfactory plan that allows the human species to engage with the future or a Singularity.

Life extension, the quest for more years, is one of the main components of the human transformation envisioned by communities interested in and working towards “singularitarian”, “transhuman” or “extropian” goals. Flag-carriers, such as Ray Kurzweil or Ben Goertzel often mention the creation of artificial general intelligence and life-extension in the same breath (Kurzweil 2005; Goertzel 2010). This is about the near-term, our life-spans. A word of caution concerning a detail that is often overlooked: There is no such thing as satisfactory life-extension without a *life-expansion* that includes solutions for problems of the mind.

Proponents of methods for biological life-extension in situ, such as Aubrey de Grey often speak of extending the span of healthy life. But healthy simply means without degradation by damage or disease. It does not mean that we are returned to youth. In particular, biological life-extension approaches do not address the intrinsic and unavoidable processes that affect even a perfectly health brain.

What is it like to be an elder who is physically healthy and has no cognitive deficits? Is it just like being 20 years old? Obviously, it is not. The brain is physically altered by the activity that it processes. Consider the difference between the brain of an infant, of a child and of an adult, as shown in Fig. 12.1. The infant's brain is not yet fully formed. The child's has a comparatively highly connected network that is able to adapt to any new patterns of activity it is exposed to. The process of patterning determines where connections are strengthened and where they are pruned. The older brain will have many strong and stable pathways, but less flexibility.

What does that mean for the individual involved? It means that new experiences are filtered through lenses shaped by old experiences. We are all familiar with the "generational gap", differences in perception, behavior, comprehension. Imagine if we live to be 200 years old through biological life-extension. In science fiction, such elder intelligent beings, the Vulcan Spock for example, have wisdom. But in addition to wisdom, we fancy that they participate equally in shaping society, creating and bringing about innovation. In reality, that is not an obvious thing for an elder mind to be able to do. We need to *augment the mind* in order to achieve the life-expansion we imagine, not simply the life-extension of the physical body.

Aubrey de Grey has noted that a technological singularity, if well-conceived, may be virtually unnoticed. According to de Grey, humans are much less interested in technology than they are in each other. He argues that we are interested in using technology, but that we are mostly not so interested in how the technology works. Eventually, user-friendliness would make computers unnoticeable in our environment. We can agree with de Grey that the pinnacle of user-friendliness is unconscious control of technology, where the understanding between brain and technology is the same as that between brain and body.

That is the path through brain-machine interfaces (BMI) as they gradually approximate SIM. By absorbing more of what technology does as a part of what we ourselves can do, that technology becomes less alien to us and less noticeable.

The Importance of Access

Substrate-Independent Minds and Whole Brain Emulation

The whole brain emulation approach to the problem of uploading to a substrate-independent mind is interesting for several reasons. It is an approach that emphasizes the replication of small components, without requiring a complete top-down understanding of the modular system that generates mind. By moving the operational components to another substrate, WBE nonetheless provides full access to the activity that underlies mental operations.

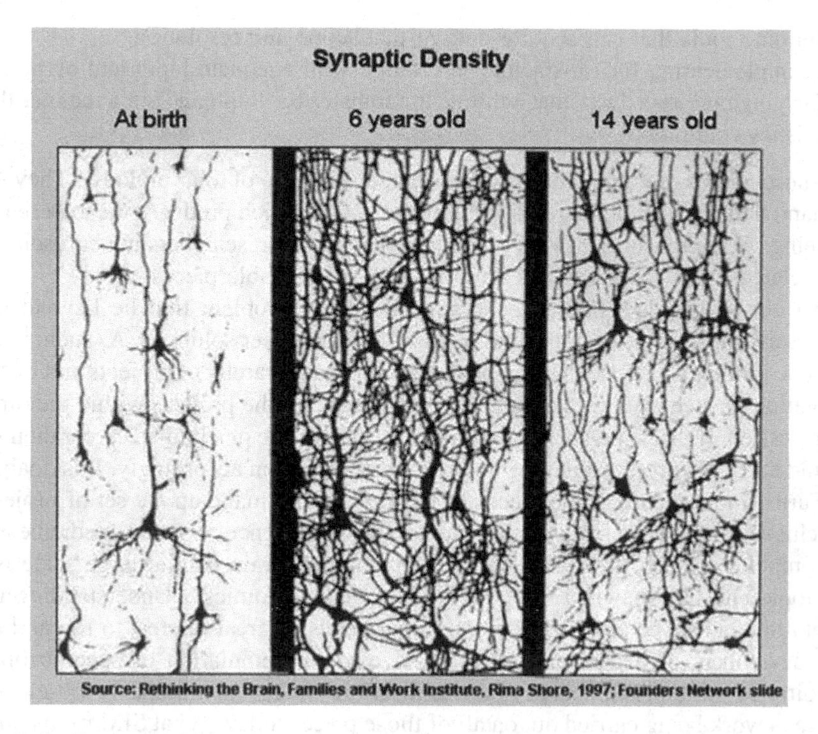

Fig. 12.1 (From rethinking the brain, families and work institute, rima shore, 1997.) An example of the comparative synaptic and connection density in the human cortex at birth, age 6 years and age 14 years

With access to each of the operations that together make up the functions of a mind it is possible to explore and experiment in depth and breadth. The experimentation will allow us to attempt gradual and tentative modifications. The outcomes of modifications over short time intervals can be tested, while maintaining outcomes as generated by the original functions. If need be, a modification can be undone, a step reversed. With this method, we aim to discover paths by which to modify our capabilities, including reward functions, but to sustain survival-oriented behavior. In the following, we describe the reasoning behind the development of strategies to achieve substrate-independent minds and recent technological developments aimed at the prospect of whole brain emulation.

Can SIM Happen Within Our Life-Span?

The problem of achieving substrate-independent minds involves dealing with these points:

- Verifying the scope and resolution of that which generates the experience of 'being' that takes place within our minds.

- Building tools that can acquire data at that scope and resolution.
- Re-implementing for satisfactory emulation with adequate input and output.
- Carrying out procedures that achieve the transfer (or "upload") in a manner that satisfies continuity.

Those points do not require a full understanding of our biology. They do demand that we consider carefully the limits of that which produces the experience of being. Accomplishing SIM is a problem that human researchers in neuroscience and related fields can grasp and simplify into manageable pieces.

To our knowledge, there are no aspects of the problem that lie beyond our understanding of physics or beyond our ability to engineer solutions. As such, it is a feasible problem, and one that can be dealt with in a hierarchy of projects and by the allocation of such resources as are needed to carry out the projects within the time-span desired. If that time-span is the span of a human life or a human career then we should carry out project planning and resource allocation accordingly. It is doable.

Furthermore, many of the pieces of the puzzle that make up the set of projects to achieve SIM are already of great interest to neuroscience, to neuro-medicine and to computer science. Acquisition of high resolution brain data at large scale is a hot topic and has spawned the new field of "connectomics". Understanding the resolution and scope at which the brain operates is of great interest to researchers and developers of neural prostheses. And even the emulation of specific brain circuitry is the topic of recent grants and efforts in computational neuroscience. There is work being carried out on all of those pieces today. What SIM needs now, and where I am focusing my efforts, is a roadmap that ties it all together along several promising paths, and attention being paid to those key pieces of the puzzle that may not yet be receiving it. And, of course, we have to insure that the allocation of effort and resources is raised to the levels that make success probable within the time-frame desired.

What is Harder, SIM or Curing Aging?

A clear measurement of how difficult it is to achieve an objective requires a detailed understanding of its problems and the granular steps involved in possible solutions. When we have such an understanding then we can estimate the resources and the time needed to reach the objective. In the absence of understanding that allows clear measurement, intuition is an unsteady guide, as it is biased heavily by our particular background and area of expertise. For this reason, a person with a strong background in biology and a history of immersion in the literature surrounding matters of disease and damage, such as would need to be addressed to cure aging, may feel that curing aging appears to be easier than projects toward whole brain emulation or other routes to SIM.

As a neuroscientist, I see the exact opposite. To me, the problem of curing aging seems like the problem of keeping an old car going forever, continuously bumping into new problems, expending more and more effort for small gains. The steps along a roadmap to curing aging, as far as I have read about them (e.g. de Grey and Rae 2007) look rife with research topics and experiments that—at least in humans—could require many years to evaluate their effect. Feedback for iterative improvement of the results seems slow.

By contrast, the matter of whole brain emulation seems to me largely a problem of data acquisition. It is a significant problem of data acquisition, but nevertheless clearly describable. And, it is possible to test tools and procedures on a few cultured neurons, on slices of brain tissue, on invertebrates and small animals, then to scale up to data acquisition from thousands, millions, billions and subsequently all relevant components involved. There is a clear and sensible way to arrange these project steps. Even though we have not yet fully emulated the brain of an individual animal, we certainly have existing experience with data acquisition from the brain. At the very least, we should therefore have the professional honesty not to claim that either curing aging or achieving substrate-independent minds are demonstrably easier or more quick to achieve.

Do We Need AGI to get SIM?

For some reason, another attitude frequently encountered is that to achieve SIM we would need to create a super-intelligence first (e.g., super-human AGI). To some extent, this may be an emotional response to the perceived magnitude of the task. It would be nice if we could simply offload that burden to some other intelligence, if that intelligence were comparatively easy to create.

From a strategic point of view, this is an odd stance. Do we know how to create super-human intelligence at this time? We certainly have not demonstrated any, and it is not obvious that any of the current projects in artificial intelligence will quickly lead to such a result. Creating the super-human intelligence is a question mark. So, if the perception is that there are questions marks about how to achieve SIM, then why would you place another question mark before it and work on AGI as a precursor to working on SIM?

That strategy makes even less sense when we consider these two points:

1. We have not yet run into even a single issue in the roadmap towards SIM that seems insurmountable by human intelligence and planning, or intractable to research and development. Quite to the contrary, there are very real projects underway to develop tools that can together accomplish the whole brain emulation route to SIM. The problems are comprehensible, the challenges manageable and the tasks feasible. Then why throw up your hands in desperation with a call to super-human intelligence? It seems to make more sense to put our efforts into the projects on that roadmap to SIM. Certainly, those projects will involve developing tools that apply machine learning and artificial intelligence, just as such tools are developed for many other reasons.

2. There is a strong case to make that we would like to achieve a form of SIM before another empowered (artificial) super-human intelligence comes along. This is a case where it is prudent to consider existential risks, and we will discuss that some more in later sections of this paper.

Can SIM Precede AGI?

In previous writing, we pointed out that a SIM is a form of AGI, as long as we consider human intelligence sufficiently general, and the process of becoming substrate-independent as artificial. Conversely, an AGI could be a form of SIM, even if not one that directly derived from a human mind. We can imagine that, if we know the functions needed to implement an AGI, then we can implement those on a variety of platforms or substrates. It seems clear then, that once we have SIM, we also have at least one type of AGI, and others may follow swiftly.

Whether AGI would automatically lead to SIM is a more problematic question. We could certainly attempt to persuade an AGI, which we created, to help us achieve SIM. If the AGI has super-human intelligence then having it do our bidding might not be so simple. In addition to that practical problem, there are ethical ones: For example, if it is not ethical to force a human to do whatever we want them to do, then would it be ethical to force an artificial intelligence?

Those are strategic considerations. Taking a perspective that is purely about research and development, can we ascribe a likelihood to either AGI or SIM being achieved first? As when we were comparing the quest to cure aging with the task to achieve substrate-independent minds, it is again a matter of understanding the respective objectives and their problems with sufficient clarity to estimate time and resources.

We have at least one fairly concrete path to SIM that we can consider in this manner, namely whole brain emulation. For AGI, the picture is a bit murkier. Aside from WBE, we do not yet have any concrete evidence that any one path to AGI that is presently being explored will bear fruit and satisfy that objective. Even the objective itself may need to be clarified or defined in greater detail.

For the sake of argument, let us simply define a successful AGI as an artificial intelligence that has the same capabilities as a human mind. Using that definition, it is not yet clear that we have sufficient insight into the finer details of mental processes, or the manner in which different mental modules cooperate, to say that we know which functions an AGI should implement.

Many research projects may lie ahead and they may involve further study of the human brain. Still, there can be a time when we understand enough about the functions of the human mind to implement similar functions in an AGI. But is that point in time earlier or later than the one where we are able to re-implement basic components of the brain, acquire large scale high resolution parameter data for those components, and reconstruct a specific mind correspondingly? That is not clear.

Working on SIM Today

The most active route to SIM in terms of ongoing projects and persons involved is Whole Brain Emulation (WBE). I coined the term in early 2000, to end confusion about the popular term "mind uploading". Mind uploading refers to a process of transfer of a mind from a biological brain to another substrate. WBE caught on. The less specific term "brain emulation" is now sometimes used in neuroscience projects that do not address the scope of a whole brain. Emulation implies running an exact copy of the functions of a mind on another processing platform. It is analogous to the execution of a computer program that was written for a specific hardware platform (e.g., a Commodore 64 computer) in a software emulation of that platform, utilizing different computing hardware (e.g., a Macintosh computer).

Whole brain emulation differs from typical model simulations in computational neuroscience, because the functions and parameters used for the emulation come from one original brain. The emulated circuitry is identical, a specific outcome of development and learning. Connections, connection strengths and response functions are meaningful, implementing the characteristics of the specific original mind. But, there are also many similarities between WBE and ambitious large-scale simulations such as the Blue Brain project (Markram 2006), at least in terms of the computational implementation of components of the neural architecture (e.g., template models of neurons). Blue Brain is composed of neural circuitry that was generated by stochastic sampling from distributions of reasonable parameter values that were identified by studying many different animals. The result has gross aspects recognizable in the brains of rats, and exhibits typical large-scale oscillatory and propagating activity.

Large-scale simulations can be trained so that their output is behaviorally interesting. Unfortunately, any highly complex and over-parametrized system can implement functions to carry out a specific task in many different ways. Successfully performing the task does not prove that the system faithfully represents the implementation found in a specific brain.

Imagine a task to regulate the temperature of a dish washing machine. There are 50 different programs written for the task. There are differences between each of the implementations, and while all of the programs can carry out the task there is some variance in the result (e.g., different delay times in the hysteresis loop turning on and off the heating element, different algorithms interpreting sensor data). A simulation is like writing a 51st program, using knowledge of the typical observable behavior and implementation hints obtained by sampling random lines of code from each of the 50 existing programs. An emulation, as we use the term, is a line-by-line re-write of one of the programs.

We consider whole brain emulation the most conservative approach to SIM. If we understood a lot more about the way the mind works and how brain produces mind then we might have far more creative or effective ways to achieve a transfer (an "upload") from a biological brain to another substrate. The resulting implementation would be more efficient, taking the greatest advantage of a new processing substrate. We might call this "compilation" rather than emulation, as when a smart compiler is used to generate efficient executable code.

Today, we do not know enough to achieve SIM at the level of a compilation. We do understand enough about neurons, synaptic receptors, dendritic computation, diffuse messengers and other modulating factors that we can concurrently undertake projects to catalog details about the range of those fundamental components and to identify and re-implement neuro-anatomy and neuro-physiology in another computational substrate. WBE is like copying each speck on the canvas of a masterpiece instead of attempting to paint a copy using broad strokes carried out by a another artist. Figure 12.2 shows one moment in such a process, as carried out using the Automatic Tape-Collecting Lathe Ultramicrotome (ATLUM) built for this purpose by Ken Hayworth and Jeff Lichtman.

WBE involves building tools such as the ATLUM. Using those tools will teach us more about the brain, even though a full understanding of brain and mind is neither prerequisite nor goal of WBE. SIM through WBE deliver backup and fault tolerance, plus complete access to every operation in the emulation. That access enables exploration. We can then incrementally and reversibly augment our capabilities. Ultimately, we can do what our creations can do and intimately benefit from the advances.

A Decade of Developments

Since 2000, several important developments have turned SIM into an objective that can be feasibly achieved in the foreseeable future. The transistor density and storage available in computing hardware have increased between 50 and 100 fold, at an exponential rate. More recently, increases in the number of processing cores in CPUs and GPUs indicate a rapid drive toward parallel computation, which better resembles neural computation. An example of those efforts is the neuromorphic integrated circuit developed by IBM within the DARPA SyNAPSE project (*Systems of Neuromorphic Adaptive Plastic Scalable Electronics*), led by Dharmendra Modha.

In the same decade, the field of large-scale neuroinformatics brought systematic study to computational neuroscience with a focus on more detail and greater scale. This was driven by several highly ambitious projects, such as the Blue Brain project and by new organizations such as the International Neuroinformatics Coordinating Facility (INCF). Methods of representation and implementation at scale and resolution are essential to WBE.

In recent years, studies have begun to test our hypotheses of scope and resolution as they apply to data acquisition and re-implementation in WBE. Briggman, Helmstaedter and Denk demonstrated both electrical recording and reconstruction from morphology obtained by electron microscope imaging of the same retinal tissue (Briggman et al. 2011). They were able to determine the correct functional

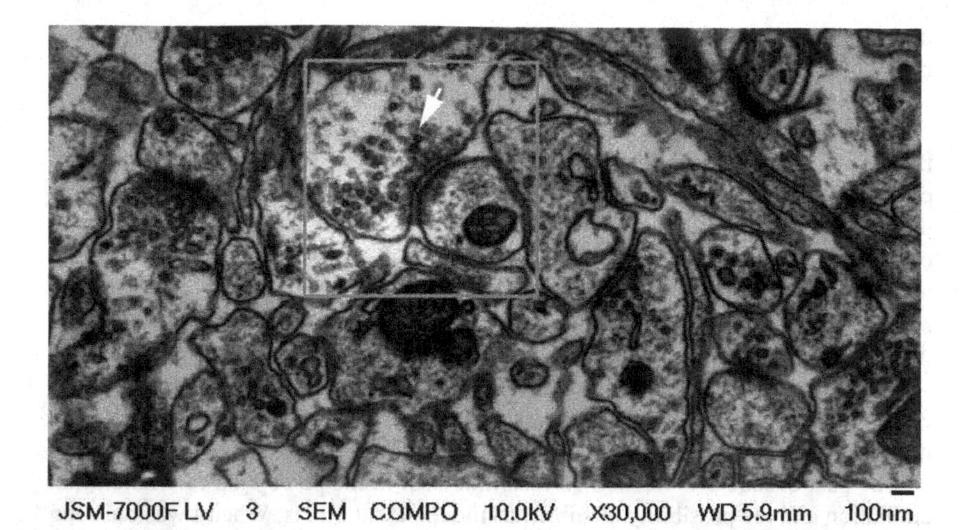

Fig. 12.2 Electron microscope image taken at 5 nm resolution from a slice of brain tissue. The red rectangle contains the outlines of a synaptic terminal, with neurotransmitter containing vesicles indicated by the arrow. Stacks of images such as this are used to reconstruct the detailed morphology of the neuronal network

layout from the morphology. Bock et al. carried out similar studies of neurons of the visual cortex (Bock et al. 2011).

The so-called optogenetic technique was developed by Karl Disseroth and Ed Boyden (Boyden et al. 2005), which enables very specific excitation or inhibition in vivo by adding light sensitivity to specific neuronal types. These, and similar techniques enable testing of hypotheses about the significance of specific groups of neurons in the context of a mental function.

The Blue Brain Project led by Henry Markram is a prime example of work in recent years that carries out very specific hypothesis testing about brain function in a manner that is useful for WBE and SIM. This year, David Dalrymple has commenced work to test the hypothesis: "Recording of membrane potential is sufficient to enable whole brain emulation of C. Elegans." The results may demonstrate when molecular level information is or is not needed and will elicit follow-up studies in vertebrate brains with predominantly spiking neurons and chemical synapses. These are the beginnings of systematic hypothesis testing for the development of SIM.

An increasing number of projects are explicitly building the sort of tools that are needed to acquire data from a brain at the large scope and high resolution required for WBE. There are at least three different versions of the ATLUM (Hayworth et al. 2007). Ken Hayworth is presently working on its successor, using focused ion beam scanning electron microscopy (FIBSEM) to improve accuracy, reliability and speed of structural data acquisition from whole brains at a resolution of 5 nm (Hayworth 2011). The Knife-Edge Scanning Microscope (KESM) developed by Bruce McCormick is able to acquire neuronal fiber and vasculature morphology from entire mouse brains at 300 nm resolution (McCormick and Mayerich 2004).

A number of labs, including the MIT Media Lab of Ed Boyden, are aiming at the development of arrays of recording electrodes with tens of thousands of channels. To go beyond this in vivo, recent collaborations have emerged to develop ways of recording the connectivity and the activity of millions and billions of neurons concurrently from within the brain. There are a range of different approaches to the design of such internal recording agents. One design takes advantage of biological resources that already operate at the requisite scale and density, such as the application of viral vectors for the delivery of specific DNA sequences as markers for the synaptic connection between two neurons (Zador 2011). Another takes advantage of existing expertise in integrated circuit technology to build devices with the dimensions of a red blood cell.

The past decade also marked an essential shift in the perception of whole brain emulation and the possibility of substrate-independent minds. When I was building a roadmap and a network of researchers aimed at SIM in 2000, it was difficult to present and discuss the ideas within established scientific institutions. Whole brain emulation was science fiction, beyond the horizon of feasible science and engineering. That is not true anymore. Leading investigators, including Ed Boyden, Sebastian Seung, Ted Berger and George Church now regard high resolution connectomics and efforts towards whole brain emulation as serious and relevant goals for research and technology in their laboratories.

Structural Connectomics and Functional Connectomics

In the brain, processing and memory are both distributed and involve very large numbers of components. The connectivity between those components is as important to the mental processing being carried out as the characteristic response functions of each individual component. This is the structure–function entanglement in neural processing.

From a tool development perspective, it is tempting to focus primarily on the acquisition of one of those dimensions, either the detailed structure or the collection of component functions. We should be able to look at the detailed morphology of neuronal cell bodies, their axonal and dendritic fibers, and the morphology of synapses where connections are made. Perhaps we could identify the correct component response functions from that information. To classify components based on their morphology and derive specific parameter values we need extensive catalogs and mapping models that are injective (one-to-one) so that there is no ambiguity about possible matches. Despite promising results by Bock et al. (2011) and Briggman et al. (2011) it is not yet certain that this can be done.

Alternatively, it may be possible to carry out solely functional data acquisition and to deduce a functional connectivity map. Pick a resolution at which you regard the elements (e.g., individual neurons with axon and dendrites) as a black box that processes I/O according to some transfer function (Friedland and Bernard 2005). For the relevant signals (e.g., membrane potential), measure all discernible input and output. A transfer function may be derived that generates the whole range of input–output relationships observed.

Observe how the elements operate in concert. The manner in which one element is affected by sets of other elements suggests a functional connectivity map. Unfortunately, this approach is limited by the completeness of observations that can be achieved. If the time during which measurements are taken is relatively small or does not include a sufficiently thorough set of events then latent function may be missed. In some cases, sensitivity analysis can be used to address this problem by applying patterns of stimulation to put a system through its paces. But the brain is plastic! A significant amount of stimulation will change connection strengths and responses.

Even if solely structural or solely functional data acquisition could provide all the necessary information for WBE those approaches by themselves carry an engineering risk. It is unwise to reconstruct an entire complex system without incremental validation. Better to obtain data about both function and structure for cross-validation (e.g., similar to the validation carried out in the study by Briggman et al. 2011).

We designate research and tool development in the two domains Structural Connectomics and Functional Connectomics. Structural connectomics includes leading efforts by Ken Hayworth and Jeff Lichtman (Harvard) with the ATLUM (Hayworth et al. 2007). The ATLUM is a solution to the problem of collecting all of the ultra-thin slices from the volume of a whole brain for imaging by electron microscopy. Winfried Denk (Max-Planck) and Sebastian Seung (MIT) popularized the search for the human connectome and the Denk group has contributed to milestones such as the reconstructions by Briggman et al. (2011). The laboratory of Bruce McCormick (now led by Yoonsuck Choe, Texas A and M) also addressed automated collection of structure data from whole brains, but at the resolution obtainable with light microscopy. The resulting Knife-Edge Scanning Microscope (KESM) can image the volume of a brain in a reasonable amount of time, but cannot directly see individual synapses.

Groups led by Anthony Zador and Ed Callaway have chosen an entirely different route to obtain high resolution full connectome data. As mentioned earlier, Zador proposes using viral vectors to deliver unique DNA payloads to the pre- and post-synaptic neurons of each synaptic connection (Zador 2011). Neuronal cell bodies are extracted and DNA is recovered from each. By identifying the specific DNA sequences within, it should be possible to find matching pairs that act as pointers between connected neurons.

Functional connectomics includes new ground-breaking work by Yael Maguire and the lab of George Church (Harvard). The aim is to create devices with the dimensions of a red blood cell (8 micrometers in diameter), based on existing integrated circuit fabrication capabilities and on infrared signaling and power technology. A collaboration between Ed Boyden (MIT), Konrad Kording (Northwestern U.), George Church (Harvard U.), Rebecca Weisinger (Halcyon

Molecular) and myself (Halcyon Molecular & Carboncopies.org) is exploring an alternative approach that seeks to record functional events in biological media at all neurons, resembling a kind of "molecular ticker-tape".

There are ongoing efforts in the Ed Boyden lab to move to micro-electrode arrays with thousands of recording channels that incorporate light-guides for optogenetic stimulation. A stimulation-recording array of that kind can explore hypotheses of great relevance to WBE. Peter Passaro (U. Sussex) is working on an automation scheme for research and data acquisition aimed at WBE. Suitable modeling conventions are inspired by neuro-engineering work by Chris Eliasmith (U. Waterloo). Meanwhile, Ted Berger (USC) is continuing his work on cognitive neuroprosthetics, which forces investigators to confront challenges in functional interfacing that are also highly relevant to WBE.

SIM as a Singularity Event

If uploading to substrate-independent minds is possible in the context of events that we consider a technological singularity then this can affect how such a singularity will appear to us. Consider the position taken by Bela Nagy and collaborators, that the singularity is a phase transition. A transition of some kind is necessary before the time at which a singularity (described as a hyper-exponential function) is indicated. A very clear example of a situation that demands a phase transition is one in which the main challenges are so changed that our evolved abilities are insufficient to cope with them. For homo sapiens, such a situation might indeed be as singular as a great meteor impact was to the dinosaurs.

So what may the singularity look like with SIM? Are events as difficult to predict or prepare for as in the case of a technological singularity brought about by artificial intelligence emerging from a comparatively unchanging human species (the Vingean prediction horizon)?

As Anders Sandberg points out, an ability to cheaply copy mental capital (as in the case of AI) may indeed lead to extremely rapid growth. Another source of rapid growth lies in the straightforward ways in which a sufficiently advanced AI can self-improve: faster computation, larger memory capacity. Notice that both of these avenues of growth also apply to SIM. Conceptual improvements are more difficult to identify at this point, and it is not immediately clear if SIM and AI would benefit equally from those, or if one or the other would have distinct advantages.

Developments are not taking place in isolation: Without making bold statements about exactly which mathematical function best approximates the course of growth or change, we can regard the singularity as a horizon in our planning. It may even be gradual, step-wise, as encountered in prior history, driven by advances in all fields and their application not only to machines but also to humans. We may not be able to see details at some distance. But we can still make educated deductions about the universe to be, which is where the end-perspective approach that we took at the beginning of this paper applies.

Existential Risk

Arguments about the precise nature of a singularity can quickly distract from the most important issues involved. Those issues are about navigating the balance of **opportunities** and **risks**. The opportunities are about the advancement of technology in pursuit of our wishes and objectives for the future. The risks can be existential.

Dennis Bray, for example, points out that machines presently lack several key components of human intelligence in areas that require a strong grasp of context. He believes that computers will be empty of function as long as they do not have an equivalent of development and learning. But of course, learning algorithms are already a standard feature of machine learning and AI. But even if a substantial argument against human level or greater AI could be made, and if a refutation of a Vingean Singularity could be justified on those grounds, it would not be a cause to quench consideration of the balance between opportunities and risk.

How does a singularity with SIM differ from one driven primarily by AI? Does the development of SIM itself bear existential risk? Consider that early SIM have minds that are essentially human, with behavior that is familiar to us. Also, human minds are not engineered from the ground-up to be iteratively self-improving by designing their own successors.

Even though SIM will have consequences for human society, at least it gives us the option to participate in an advancement that is aimed explicitly at humans. This is quite different than a circumstance in which humans have to deal with the fact that another, somewhat alien intelligence that they cannot join takes over the reins. There are different types of existential risk to the status quo if we continue to advance technology. Some risks are less desirable. In the case of SIM, risk is reduced further by participation. In the same way that SIM embraces the requirements for success in a competitive universe, a human species that embraces SIM can sustain its successful development.

Some paths towards SIM carry more risk than others. In the end though, the most important consideration is simply that there is no probable scenario whereby a lesser intelligence is guaranteed safety as well as the ability to grow and flourish in an environment where a significantly superior and more adaptable intelligence is present.

As mentioned near the beginning of this paper, notions of constructing a single friendly AGI that would remain friendly are so far entirely unconvincing. If the AGI has plasticity, if it can be modified by learning, by accident, incident or error, then any so-called utility function will drift. And plasticity is necessary, since the reason to create general AI in the first place is so that it may gain knowledge from new information and adapt to new challenges. If AGI is brought about by bootstrapping then drift is inherently guaranteed and *by-design*. Even the notion of a "singleton" in AGI, a possible sole guardian without need for competitive behavior is a slippery concept. At the very least, the AGI will entertain multiple competing problem-solving algorithms. In due course, these create distinctions, even if those do not quite amount to multiple personality disorder. The question is a matter of perspective, even within our own brains.

Realistic Routes

We have empirical evidence from the present situation involving the 7 billion most intelligent minds on Earth for the degree of effectiveness of a balance of power. That balance is imperfect on the micro-scale and over small time-intervals, but has repeatedly been restored throughout history without the catastrophic potential of runaway feedback scenarios. Of course, it is clear that such a balance serves those who are the dominant species involved. A prerequisite is therefore that we remain within that set, not left behind. Ultimately, the strongest way to reduce existential risk and to avoid irrelevance is to merge with our own tools and embrace their capabilities. If you do not become a substrate-independent mind (e.g., through whole brain emulation) then you are effectively choosing not to be as competitive. The consequences follow.

We presented our Red Queen's Hypothesis of Balance of Intelligences earlier, which is of course the description of an arms race with the inclusion of collaboration between agents as a means to place checks on rapidly advancing leaders until others can catch up. Concerns about arms race scenarios in AGI generally focus on the problem of a possible "hard take-off"—an incredibly rapid increase in the capabilities of a system after it reaches some threshold level, without adequate controls by resource constraints. It is not clear if such a take-off could indeed break a balance that was maintained by a sufficiently large pool of agents, before bumping into actual resource constraints.

The hard take-off scenario for AI is not a brand new concept, of course. It is closely related to the evolutionary theory of *Punctuated Equilibria*, championed by Niles Eldredge and Jay Gould (Eldredge and Gould 1972), as well as the theory of Quantum Evolution (Simpson 1944) that is similar but applies itself at higher level and scope. We can posit without great controversy that in societies of intelligence, as in all systems subject to developmental processes, competition and selection, gradual advances that take place in Balance must take turns with punctuations of such equilibrium. That is necessary when either the challenges change so that they no longer fit the prevailing course of progress or when constraints to the previously "inexpensive" and relatively predictable growth are reached. In general, this necessity is recognized for specific technologies or approaches, and represented within models such as the Technology S-Curve. It is up to us to be so flexible that we participate with the new direction of development. For that, it is useful to note that even a rapid turn of events does not have infinite speed and that acceleration is not effortless, in fact, it has "energy" requirements.

What are the physical requirements of a super-intelligent AI undergoing iterations of self-improvement? How much would the need to gather resources slow down the advance? That these practical questions exist at least points to the possibility that such factors—where we can actually exert control—may be incorporated in a plan of advancement. In practical terms, *we should focus on plans that emphasize levers where our actions exert control and affect the course*, rather than fantastic optimal solutions that may be impossible in principle (e.g., the "friendly" AI guardian scenario) or organizationally beyond our grasp (e.g., managing the cooperation of all groups working on AI). We aim to address practical plans in greater detail in future publications.

AGI researcher Ben Goertzel often compares soft versus hard take-off scenarios for human level artificial intelligence and beyond. Often left out of the discussion about preconditions for each scenario is the question, what exactly does it mean to improve one's intelligence?

First we can ask: How does one measure intelligence? A common approach is to regard it as a comparison between the performance of different systems on a set of tasks. Those tasks may be of a more or less general variety, in which case you are also measuring the generality of the intelligence. So, what does it mean to improve one's intelligence in such a case? It means that you need to come up with a more effective way of carrying out the tasks, e.g., of solving certain types of problems.

How does that happen? Are there any limiting factors to the rate at which one might then improve? Is there a difference between AI and SIM in the way they can improve? Would all sides benefit from across-the-board step-wise granular improvements, or does one side or one system take-off out of control?

Earlier, we mentioned efforts at life-extension. It is important not to lose sight of existential risk while in pursuit of longer life. At the same time, we should be practical when selecting approaches, and not waste time on optima that do not lie on feasible paths. A satisfactory "friendly" AI solution may be such an unfeasible theoretical optimum. On a path forward, we should know which things we do control and which we cannot practically control. That distinction helps us focus efforts by constraining the solution space. On the one hand, for example, we may not be able to control a coordination between AI researchers if all it takes to break ranks is to tempt a few individuals with promises of exceptional rewards. On the other hand, there may be little to lose by purposefully accelerating work on WBE (instead of AI), which is a variable that we can indeed influence.

If SIM is not achieved when other technological advances drive the rate of change to the level that we now think of as the Singularity then it is likely that we will no longer play an active influential role in significant global or cosmic developments. A scenario in which we cannot participate as members of the dominant intelligent set of actors is one in which we do not determine the course of events. We can debate about whether this results in an actual downfall of the human species or if it merely implies its domestication under the auspices of a more advanced keeper. But it is simply not reasonable to assume that we could use a truly more advanced general intelligence (of our own creation or not) constrained for our own purposes. It is as if the mice in a laboratory considered the human experimenter a useful tool applied to their goals.

References

Bock, D. D., et al. (2011). Network anatomy and in vivo physiology of visual cortical neurons. *Nature, 471*, 177–182.

Boyden, E. S., Zhang, F., Bamberg, E., Nagel, G., & Deisseroth, K. (2005). Millisecond-timescale, genetically targeted optical control of neural activity. *Nature Neuroscince, 8*(9), 1263–1268.

Briggman, K. L., Helmstaedter, M., & Denk, W. (2011). Wiring specificity in the direction-selectivity circuit of the retina. *Nature, 471*, 183–188.

Chalmers, D. (2010). A philosophical analysis of the possibility of a technological singularity or "intelligence explosion" resulting from recursively self-improving AI. John Locke Lecture, 10 May, Exam Schools, Oxford.

de Grey, A., & Rae, M. (2007). *Ending aging: The rejuvenation breakthroughs that could reverse human aging in our lifetime*. New York: St. Martin's Press.

Dennett, D. (2005). *Darwin's dangerous idea* (pp. 352–360). New York: Touchstone Press.

Eldredge, N., & Gould, S. J. (1972). Punctuated equilibria: An alternative to phyletic gradualism. In T. J. M. Schopf (Ed.), *Models in Paleobiology* (pp. 82–115). San Francisco: Freeman Cooper.

Friedland and Bernard. (2005). Control system design: An introduction to state space methods. Dover. (ISBN 0-486-44278-0).

Gildert, S. (2010). Pavlov's AI: What do super intelligences really want? Humanity + @ Caltech, Pasadena.

Goertzel, B. (2010). AI for increased human healthspan. Next big future, August 14.

Good, I. J. (1965). Speculations concerning the first ultraintelligent machine. In Franz L. Alt, Morris Rubinoff, (Ed.), Advances in Computers, 6 (pp. 31–88). Academic Press.

Hayworth, K. J., Kasthuri, N., Hartwieg, E. Lichtman, J. W. et al. (2007). Automating the collection of ultrathin brain sections for electron microscopic volume imaging. Program No. 534.6, Neuroscience Meeting, San Diego.

Hayworth, K. J. (2011). Lossless thick sectioning of plastic-embedded brain tissue to enable parallelizing of SBFSEM and FIBSEM imaging. High resolution circuit reconstruction conference 2011. Janelia Farms, Ashburn.

Koene, R. A. (2011). Pattern survival versus gene survival, KurzweilAI.net, February 11, 2011. http://www.kurzweilai.net/pattern-survival-versus-gene-survival.

Kurzweil, R. (2005). The singularity is near. (pp. 135–136). Penguin Group.

Markram, H. (2006). The blue brain project. *Nature Reviews Neuroscience, 7*, 153–160.

McCormick, B. Mayerich, D. M. (2004). Three-dimensional imaging using knife-edge scanning microscopy. In proceedings of the microscopy and micro-analysis conference 2004, Savannah.

Moravec, H. (1998). Robot: Mere Machine to Transcendent Mind. Oxford University Press.

Paredis, J. (1997). Coevolving cellular automata: Be aware of the red queen! In proceedings of the seventh international conference on genetic algorithms ICGA97.

Ridley, M. (1995). The red queen: Sex and the evolution of human nature. Penguin books, ISBN 0-14-024548-0.

Sandberg, A. Bostrom, N. (2008). Global catastrophic risks survey. Technical Report 2008/1, Future of humanity institute, Oxford University.

Simpson, G. G. (1944). *Tempo and mode in evolution*. New York.: Columbia Univ. Press.
Solomonoff, R. J. (1985). The time scale of artificial intelligence: reflections on social effects. *Human Systems Management, 5*, 149–153.
Tegmark, M. (2011). The future of life: A cosmic perspective, presented at the singularity summit 2011, Oct. 15, New York.
Vinge, V. (1981). True names and other dangers, Baen books, ISBN-13: 978-0671653637.
Vinge, V. (1993). The coming technological singularity, Vision-21: Interdisciplinary science and engineering in the era of cyberspace, proceedings of a symposium held at NASA lewis research center (NASA Conference Publication CP-10129).
Zador, A. (2011). Sequencing the connectome: A fundamentally new way of determining the brain's wiring diagram, Project proposal, Paul G. Allen foundation awards grants.

Chapter 12A
Philip Rubin on Koene's "Embracing Competitive Balance: The Case For Substrate-Independent Minds and Whole Brain Emulation"

Building Brains

When I read about the singularity, brain emulation, and similar concepts that push us to consider the extrapolation of recent developments in science and technology and possibilities of a future in which science fiction could become reality, I often come away with a mix of fascination and considerable frustration. I am drawn to the enthusiasm that stems, in part, from rapid developments and the enormous potential in areas like genomics and proteomics, quantum physics, materials science, nanontechnology, microelectronics, neuroscience, and many other domains. At the same time, I am frustrated by the hubris and by the lack of adequate consideration for the complexities that make the work in many scientific disciplinary areas so difficult, challenging, and often rewarding.

What is a brain? How could we emulate it? Well, before we take on that challenge, how about considering a "simpler" one. What is chair? How do we emulate it? For us humans, a chair is something that we might sit on. For a mouse, it could provide shelter from the rain. To an elephant it is, perhaps, something to step on and crush. Thus, the way in which a physical object is used, considered, and possibly characterized in an emulative process, can depend on what it affords to a living entity in a dynamic, interactive process.

If our goal is to "build" or model a brain by reverse engineering it, we need to know a bit about what its function is, just as it would help to know how a radio or an Arduino is intended to be used before starting to reverse engineer them. But functionality in the brain spans many levels. Things like meaning, perception, and emotion are often secondary considerations when thinking about building a brain. Our focus is often on the extremes—either at the lowest levels, driven perhaps by the mechanistic and reductive tendencies that our scientific tradition and its successes force us in, or on the highest levels, such as consciousness, perhaps because it is so elusive and alluring. But the problems frequently can be both harder and more mundane than this. When considering the brain we need to ponder multiple dimensions and scales, from neuron to neighborhood, with consideration of the temporal, spatial, cultural, and conceptual extents that these entail.

I remain an optimist and an enthusiast regarding understanding brain, mind, and behavior, but I also believe that problems in domains like neuroscience and the behavioral, cognitive and social sciences, are deliciously hard ones. Making progress in these areas can require more than just an understanding of how primitives and fundamental low-level entities, such as neurons, or genes, or words, function at their most basic levels, interact, combine, and form aggregates and networks. We also must consider the context within which these entities arise and exist. To my mind this requires including in the scientific/technological enterprise concepts like: meaning, abstraction, culture, embodiment, temporality, multimodality, animal-environment synergy, ecological validity, complexity, recursion, affordance, and more. It is disappointing to me that so many of the forwarding-looking ideas underlying the potential technological rapture avoid the richness and nuance of these areas and concepts. It does not bode well for the future that many want to see and the progress that may be attainable. March 2012

Chapter 13
Brain Versus Machine

Dennis Bray

Abstract Many biologists, especially those who study the biochemistry or cell biology of neural tissue are sceptical about claims to build a human brain on a computer. They know from first hand how complicated living tissue is and how much there is that we still do not know. Most importantly a biologist recognizes that a real brain acquires its functions and capabilities through a long period of development. During this time molecules, connections, and large scale features of anatomy are modified and refined according to the person's environment. No present-day simulation approaches anything like the complexity of a real brain, or provides the opportunity for this to be reshaped over a long period of development. This is not to deny that machines can achieve wonders: they can perform almost any physical or mental task that we set them—faster and with greater accuracy than we can ourselves. However, in practice present day intelligent machines still fall behind biological brains in a variety of tasks, such as those requiring flexible interactions with the surrounding world and the performance of multiple tasks concurrently. No one yet has any idea how to introduce sentience or self-awareness into a machine. Overcoming these deficits may require novel forms of hardware that mimic more closely the cellular machinery found in the brain as well as developmental procedures that resemble the process of natural selection.

> *It is not impossible to build a human brain and we can do it in 10 years.*
>
> Henry Markram. TED Conference Oxford, 2009.

D. Bray (✉)
Department of Physiology, Development, Neuroscience,
University of Cambridge, Downing Street, Cambridge, CB2 3DY, UK
e-mail: db10009@cam.ac.uk

A. H. Eden et al. (eds.), *Singularity Hypotheses*, The Frontiers Collection,
DOI: 10.1007/978-3-642-32560-1_13,

At the 2010 Singularity Summit, Salk professor Terry Sejnowski gave a talk entitled 'Reverse Engineering of the Brain is within Reach'. He presented evidence from a variety of studies, many from his own group, in which circuits in particular regions of the human brain were reproduced on a computer using software equivalents of neurons. Toward the end of his presentation, he showed a video clip of a network of simulated glutaminergic neurons firing in complex patterns, apparently replicating the activity of a region of brain cortex.

In the ensuing open discussion a young man, lying on a gurney and helped to the microphone by his wife, asked the following question. The progressive degenerative motor neuron disease ALS (Lou Gehrig's) arises through disorders of glutamate release, he said, which is precisely the kind of neuronal activity modelled by Sejnowski. So, the questioner continued, if the action of these nerve cells is so well understood, how is it that ALS remains incurable? A fair question, you might think, and one that could be applied to more than just ALS. A dolorous coterie of conditions, including schizophrenia, Huntingdon's disease, and Alzheimer's, presently defy all attempts at a cure. And this surely is odd. If we were to take the term 'reverse engineering' literally—shouldn't we be able diagnose what is wrong in these and other clinical conditions, and fix them?

The reverse engineering claim will also receive a thumbs–down from another quarter—research neurobiologists. Anyone who devotes their professional life to understanding long-term potentiation, say, or the molecular structure of the synaptic endplate will view attempts to build brains on a computer as irrelevant at best. How can you represent a neuronal synapse—a complex, structure containing many hundred different kinds of protein, each a chemical prodigy in its own right and arranged in a mare's nest of interactions—with a few lines of code? Where in such a model would one find the synthesis and turnover of crucial molecules and the dynamic growth and shrinkage of synaptic structure that accompany learning? Where would be the local synthesis of synaptic proteins and the modulating effect of microRNAs; the menagerie of different neuronal types not to mention glial cells; diffusing hormones, oxygen, and blood flow? How can such simplistic computer representations be taken seriously when they ignore the baroque intricacies of neuroanatomy that require years of development and learning to attain their mature form?

The brain-builders know all this, of course, but choose to ignore it. Theirs is the big picture, the intoxicating vision of a thinking machine made in our own image. The dream of creating human life is deep in our psyche and has fascinated thinkers for centuries. Every age had its view how this should be done, from golems made of clay of early Judaism to the mechanical automata of eighteenth century artisans. The twentieth century opened with electric talking dolls made for the Edison Company, and closed with humanoid robots such as Ichiro Kato's WABOT, Cynthia Breazel's KISMET, and Honda's ASIMO. Each machine was made using the most advanced technology available at the time, and in this respect today's models are no different. Contemporary models are swept along by spectacular progress in the neurosciences and the stupendous capacity now available to perform computations. That is why, proponents argue, the time has finally come. This time, they say, it is for real.

Well, what do you think? Is the age-old quest close to being achieved so that humans will soon (in ten years or so) create a genuine image of themselves? Or will we find ourselves at the end of the decade consigning yet another generation of computer "brains" to the museum of charming failures?

No one can see into the future. But it seems to me that we can gain insight into this question using a tactic familiar to molecular biologists of separating the *functional* aspects of the problem from the *structural*. Thus, in the first case we can ask: How much and to what degree can artificial machines perform the same (mental or intellectual) functions as human beings? And in the second: How closely does the machinery (that is, the internal structure and workings) of automata resemble that of the human brain? The two aspects are linked of course but if we draw them apart we will see that they produce very different answers.

How Much and to What Degree Can Artificial Machines Perform the Same Functions as Human Beings?

The success of a computer called Watson in the television contest called Jeopardy! aired 16 February 2011 was yet another nail in the coffin of human superiority. We have taken on board the fact that computers outperform us in complex mathematical calculations and know them to be much, much better at storing and retrieving data. We accept that they can beat us at chess—once regarded as the apogee of human intellect. But this one is more insidious. It is not so much that an inert lump of metal knows that the Vedic religion was written in Sanskrit, or that Kathmandu is further north than Manila or Jakarta, or that Michael Caine will feature in a forthcoming movie *Dark Knight Rises*—or by extension that it is aware of a whole world of trivia. It is not even that it managed to beat two humans previously shown to excel in this contest. What really hurts is that the machine produced its answers in response to questions posed in colloquial English, making sense of cultural allusions, metaphors, puns, and jokes and even replying in kind. If Alan Turing had been given printed transcripts of the three contestants, would he have spotted the odd man out, I wonder?

Watson's success prompts us to ask: Just what are these silicon protégés capable of? Computers drive cars and pilot spy planes; recognize people by their fingerprints and faces, their voices, or gaits; simulate (for the purposes of movie entertainment) fire, water, explosions, clothing, hair, muscle, or skin. Robots help with the care of children and the elderly, move things around, build cars, play the violin.

The list seems endless and it is probably easier to think what these intelligent machines *cannot* do. Most limitations come under the category of 'everyday activities' or 'common sense'. Thus, understanding speech and maintaining a conversation is still a problem, pace Watson. So is human–level vision. It will be some time before a robot plays a good game of tennis or cooks a gourmet dinner from scratch. Robots that can multitask—walk, talk, recognize faces, and find prime numbers all at the same time—are even further away. But given the phenomenal progress to date and the way intelligent machines just get faster and smarter every year, one would be rash to declare that any particular function is permanently beyond reach.

In other words, intelligent machines *could* be capable of at least reproducing human performance in any intellectual task we care to specify within a decade.

Note, however, that the way they do it will have little to do with the brain. Robots that assemble automobiles or fly spy planes, computers that find prime numbers or play chess, are custom built for a particular purpose. They represent the best solution to a specific problem based on current engineering practice and this is rarely if ever the one discovered by biology. Some broad similarities in organization might be imposed by the nature of the task, but in general software engineers neither know nor care about anatomy or neurophysiology.

The honourable exception to this statement is the field of *bio-inspired computing* in which programs are explicitly designed to mimic natural processes such as evolution or immunity. Their declared aim is usually to discover new approaches—new algorithms—that would not otherwise come to mind. Thus, cellular automata, genetic algorithms, wireless sensor networks, image-rendering techniques, neural networks, and others, were devised to address particular technological challenges. However, even a cursory examination of any one of these approaches reveals that its link to biology is notional at best. Yes, the original idea came from the living world, but the present instantiation bears only a superficial resemblance to anything biological.

Consider neural networks—arguably the most developed and widespread example of bio–inspired computing. In the 1940s, McCulloch and Pitts made a mathematical formulation of nerve cells and showed that they could be connected up in different ways to perform basic logical operations. A few years later, in 1946, Donald Hebb suggested that synapses could be strengthened by simultaneous pre and postsynaptic activity. These simple notions became the underlying fabric of theoretical models and, as computers became more available and powerful, resulted in sets of idealized "neurons" being woven into networks.

This led naturally to the concept of neural networks and their use in pattern recognition. But as practical applications of neural networks become ever more widespread and successful their relationship to the human brain became increasingly tenuous. The *idea* of a web of modifiable elements selected to perform a final task is undoubtedly correct: something like this almost certainly occurs throughout the brain (and inside individual cells, as well). But one would be hard put to identify a specific example of a circuit of neurons in an actual brain that acts like a canonical neural network. First of all, individual neurons are not the simple all-or-none ciphers with unchanging properties envisaged by McCulloch and Pitts. They are sophisticated elements that display an almost infinite range of possible parameters and change continually with activity. Every nerve cell in every brain is in some respect unique. Secondly, we rarely if ever have any notion of all of the inputs and outputs to the network are, what the cost function is, nor the rules by which connection strengths are modified. So while neural networks, genetic algorithms, cognitive power grids, and the rest, represent powerful abstractions of biological processes, they are much too generic and vague to provide a blueprint for a real biological system.

How Closely Does the Machinery of Automata Resemble that of the Human Brain?

This brings us to the second question asked at the outset. What can we say of those brave souls who seek to reproduce not just the function of the brain but also its structure? It is understood, first of all, that they plan to do this using dry solid–state circuitry based on silicon rather than real wet tissue. Biologists regularly take slices of brain and immerse them in nutrient media; the explants remain healthy for days and display apparently normal electrical properties. Similarly, cells taken from particular parts of the brain such as the hippocampus can grow long processes in culture, establish synaptic connections and fire action potentials. But the principal use of such techniques is analytical rather than synthetic. Cultures are simply a convenient place to ask specific questions about the development and function of the real brain. Brain-builders, by contrast, believe that they already know enough. Their vision is to put together networks of nerve cells on a computer in sufficient numbers to create a thinking entity.

The general features of this approach are illustrated in a model developed by Eugene Izhikevich and Gerald Edelman in 2008—one of the first of its kind (Izhikevich and Edelman 2008). The authors based their simulation on a detailed survey of the anatomy of a particular region of the brain (the thalamocortical projection) and set out to reproduce its main features on a computer. Multiple copies of a microcircuit were arrayed over contours representing the cortical anatomy, each microcircuit comprising eight types of excitatory neuron and nine types of inhibitory neuron linked in a pattern that resembled that in the brain. The neurons had a somatic compartment and a number of dendritic compartments and responded to incoming signals in a way that reproduced generic features of the real brain in terms of spiking frequency, electrotonic spread, integration of signals, and the activity dependent reinforcement of synapses. The full simulation displayed an extreme sensitivity to initial conditions and a propensity to generate oscillations and develop travelling waves of excitation, somewhat similar to the waves recorded from a functioning mammalian brain on an electroencephalograph.

The Izhevich/Edelman simulations employed up to one million neurons and one billion synapses, which seemed impressive at the time. However, this achievement has now been dwarfed in numerical terms by a team from IBM Research led by Dharmendra Mohda. This group announced at the IEEE/SC2009 supercomputer conference in 2009 that they had built a simulation containing 10^9 neurons and 10^{13} synapses, comparable to or even exceeding the numbers in a cat brain. The prodigious scale of this simulation (awarded the prestigious Gordon Bell Prize) came at the expense of a severely reduced internal complexity. Individual ‘neurons’ in the IBM simulation were simple devices not unlike McCulloch Pitts neurons lacking axons and dendrites. Their ‘synapses’ were represented by a single function with fixed input/output relationship. This rudimentary composition is so different to the real situation that their claim to have ‘achieved the complexity of a cat brain’ provoked widespread sceptical comment, even among computer scientists.

The limitations of the cat-brain simulation are revealed when we ask what it can do. The model is highly sensitive to starting conditions and shows a tendency to oscillate, developing spontaneous waves at a range of frequencies. But from a functional standpoint this electronic citadel containing billions of elements is actually less effective than an electric kettle. Far from reproducing a cat brain it cannot perform the simplest tasks—smell food, for example, tell day from night, or distinguish up from down. And how could it be otherwise, since the program has essentially no communication with the outside world? Connections in the program are exclusively internal, established at the time of manufacture according to some probabilistic recipe and remain unchanged thereafter. By contrast the synapses of a real mammalian brain are acquired step-by-step during neuronal development and continually refined during the lifetime of the animal according to stimuli received from the environment. Nature and nurture are both important. Growth cones of growing nerve processes are guided to future synaptic targets by a plethora of substrate–bound and diffusing signals. They then undergo extensive pruning and selection based on functional criteria during which their activity becomes shaped to the requirements of the animal. Synaptic plasticity endures into adulthood and accompanies the acquisition of new motor skills and memories. The ‘wiring diagram’ of any mammalian brain is consequently unique. It is distinct not only from the brain of any other animal, no matter how closely related, but also from the very same brain at a different time in the past or future.

At the other end of the scale to the wholesale agglomeration of switch like elements, other investigators build increasingly detailed and accurate representations of neurons. Well–established computer packages such as NEURON and GENESIS allow the electrical activities of individual neurons to be rendered in mathematically precise terms. Further refinements can attain an impressive level of realism, as in recent representation of a pyramidal cell from layer 5b of the mammalian cortex (Hay 2011). This simulated cell has realistic morphology and faithfully replicates back propagating action potentials, activated Ca^{2+} spike firing, the perisomatic firing response to current steps, as well as the experimental variability of these properties. In principal one could link multiple simulated neurons of this kind into circuits of increasing size and sophistication, although the computational cost of such an endeavour would be astronomical.

There would also be the small problem of not actually knowing which connections to make. Neurobiologists actively investigating specific regions of the brain such as the visual system or the hypothalamus are the best qualified to answer this question. They can build detailed simulations that embody features of neuroanatomy, physiology and even molecular processes. They have an unrivalled ability to test and refine these simulations against experimental data—confronting their computational idealizations with the pragmatic reality of real flesh. For example, take an ongoing investigation of the olfactory bulb—part of the mammalian brain concerned with the sense of smell—by a team of researchers at the National Centre for Biological Research in Bangalore led by Upinder Bhalla. Each of the ten million or so neurons in the rodent nose carries a molecularly distinct receptor type that responds to a particular odour. Axons from olfactory neurons extend into the olfactory bulb and converge onto a smaller number of glomeruli—globular tangles of axons and dendrites where the initial stages of processing occur. Within each glomerulus, axons from about 40,000 olfactory neurons, each expressing the same receptor, form synaptic connections with two principal types of neuron, termed mitral and tufted cells. The mitral and tufted cells are extensively connected to each other through their dendrites and to neighbouring granule cells. Axons carrying outputs of this processing leave via the lateral olfactory tract for other regions of the brain. Upi Bhalla and his team analyze this system using a combination of experimental and theoretical approaches. Recordings from extracellular electrodes and optogenetic probes (in which electrogenic proteins are expressed in particular types of cells and then activated by light with great spatial and temporal precision) allow them to monitor multiple neurons simultaneously. They obtain data from well-defined olfactory stimuli applied to awakened animals and incorporate them into multiscale cellular and network computer models, which include essential elements of biochemical signalling.

There is every hope that enterprises such as this will lead to enhanced understanding of how odours are encoded in the mammalian olfactory cortex and how this arises from the biophysics and connectivity of the olfactory bulb. It may even help us to a broader appreciation of the mechanism by which networks in general settle towards their steady–state computational properties. But it is important to appreciate that even in this relatively well-ordered region of the brain we still do not have a detailed, synapse–by–synapse description of the circuitry. The cells are too numerous and their processes are too intertwined (in addition to the neurons already mentioned the olfactory bulb also receives "top-down" information from such brain areas as the amygdala, neocortex, and hippocampus). The identity of synapses in light or electron micrographs is too uncertain and their physiology inaccessible to systematic recording. The best anyone can do at this stage is to represent the anatomical and physiological features of the olfactory bulb and associated regions in a generic sense—that is, to assign probable neuronal shapes and frequencies of connections. Synaptic strengths, dendritic conductances and other properties are then assigned on a statistical basis according to *current theories* of how the overall system operates. For example, the initial encoding of a particular odour is of a spatial nature, due to the array of olfactory neurons. However, later stages have temporal features imposed by intermittent sampling coupled to respiration and it is believed that computations here may have a more temporal aspect (Khan et al. 2010). The difference between spatial and temporal encoding in terms of synaptic connections is enormous. Clearly, a model of the olfactory bulb or any other region of the brain can never be more accurate than our present understanding of the physiology.

Just how perplexing real synaptic coding can be is illustrated in a recent study of the mouse auditory cortex (Chen 2011). Two-photon imaging techniques allowed calcium levels in individual dendritic spines of a living brain to be visualized and their activities monitored. Applied to this region of the brain responsible for the analysis of sound, the method revealed, as expected, synapses responding to both the level of sound (decibels) as well as individual frequencies. But the *patterns* of firing—which synapses became active under particular conditions—were highly heterogeneous and failed to conform to any obvious rule or pattern. An individual neuron might have spines responding to different tone frequencies but these were not distributed in any obvious order. Moreover, not every active spine led to the firing of the neuron as a whole and many computational processes appeared to take place locally, within a dendritic tree—a conclusion previously reached in (Branco and Hausser 2010).

So we are still a long way from reproducing the detailed structure of a human brain in any artificial device. We still do not know the detailed circuitry of any region of the brain sufficiently well to reproduce its structure at the level of individual dendritic branches and synapses. Even if we did have enough information at the cellular level it would still be impossible to include all of the relevant molecular species and processes. Moreover, the plasticity of an actual brain means that its connectivity continually changes in response to features of the world, past and present, experienced by the animal. Machines do not do this.

Why Bother with the Brain?

If computers, robots and other intelligent machines are so fast, so versatile, and improving at such a phenomenal rate, then why should we even *want* to build models of the brain? Given that biology is fiendishly complicated, isn't our best course to disregard the idiosyncratic, jury-rigged contrivances created by evolution and engineer our way into the future using established strategies and methods? The answer, I believe, is no. Quite apart from our natural interest in this mass of wet tissue and our biological curiosity about how it functions, we also have much to learn from the human brain from a computational standpoint. It has unique abilities that set it apart from any intelligent machine yet constructed.

Thanks to our brains, we not only play chess, find prime numbers and store lots of data. We also navigate through the world, know what to say and do, and perform myriad vital functions, many of them concurrently. Human intelligence has broad, deep, and subtle features; its powers of pattern recognition and command of language cannot yet be matched by a machine (pace Watson). As already mentioned, brains change continually even in adulthood, so that each individual is distinct not only from any other individual but also different from the same brain in the past.

There is also the tricky matter of self-awareness and motivation. The only hint of awkwardness in IBM computer Watson's television performance came when he made a slightly odd bet of "6,435 dollars". The unwarranted precision of this figure—no doubt perfectly rational from the standpoint of probabilities—was sufficiently distinct from the casual, rounded-off figure usually provided by contestants as to produce a ripple of laughter in the audience. Watson did not care, of course… and that's the point. It would take a competent programmer five minutes to modify the computer's bidding strategy so that it tended to offer round numbers. But what would not be so easy would be to change Watson so that he *cared* that he was being laughed at. The reason we hate to be laughed at, especially in public, has do with our sense of personal dignity and self-respect. Evolutionary biologists can no doubt offer explanations in terms of social standing and, ultimately, reproductive success. But Watson has no personal dignity, no sense of personal worth. He… it…. is just a hunk of metal capable of performing clever tasks at someone else's behest.

The enduring failure to install plasticity, multitasking, and anything resembling sentience in electronic devices leads me to question our basic assumptions. Could our fundamental approach to building these machines be simply wrong? For example, all of the brain simulations so far described have been based on circuits of nerve cells and synapses operating in a feed–forward mode. But we know that real neurons are widely connected to different regions of the brain and both send and receive signals, often in oscillatory fashion, over a range of time scales. As already mentioned, they continually modify their electrical properties and anatomy according to features of the environment. No brain simulations so far include these features.

It could even be that our preoccupation with electrical events (which arises perhaps from ease of measurement) is misplaced and we need to incorporate other processes occurring within cells at the molecular level into our simulations. Living cells have the capacity to probe their environment, process incoming sensory information, and generate appropriate and often highly subtle biochemical responses independently of electrical signals (Bray 2009). At the very least, we know that biochemical events modulate the size and frequency of action potentials and the sensitivity and dynamic response of synapses. Changes in microanatomy of brain structures such as dendritic spines, can occur within minutes and thereby refine physiological performance (Berning 2012). Perhaps cell biology needs to be part of our simulations?

Possibly the most important missing element is evolution. When you come down to it, the force that drives our actions, and those of every motile organism no matter how humble, is the need to survive and reproduce. Somehow, in its blind tinkering, evolution stumbled on a way to channel behaviour along paths that are optimal from the standpoint of survival, leading to the emergence of ever richer and more complex life forms. But this primeval drive is totally lacking in intelligent machines. Watson is a stand–alone, one–off upstart. He cannot reproduce. He has never had to compete with similar machines for survival in the real world. Unless and until an element of natural selection is added to the design of intelligent machines it is unlikely they will ever compete directly with living brains.

Summary

- It must be obvious that there is no realistic prospect of ever 'reverse engineering' the human brain in the strict sense of the term. No one could—literally—take a brain apart into all of its essential components and then put them together again, in the same way that one might reverse engineer a radio or other electronic device. Even the notion, sometimes explored in theory, of replacing each and every synapse by an equivalent silicon device is out of reach—not because it requires anything outside physics and chemistry but because the biological systems are so idiosyncratic and historical. So far as we know, building a brain requires years of development inside an actively functioning body with continual exposure to and interaction with the outside world. No one to my knowledge has ever suggested how this might be achieved with a silicon device.
- By contrast, the harnessing of intelligent machines to intellectual and physical tasks useful to humans has already been incredibly successful and is without discernable limits. In theory, they could be able to do almost any physical or mental task that we set them—faster and with greater accuracy than we can ourselves.
- However, in practice present day intelligent machines still fall behind biological brains in a variety of tasks, such as those requiring flexible interactions with the surrounding world and the performance of multiple tasks concurrently. No one yet has any idea how to introduce sentience or self-awareness into a machine. Overcoming these deficits may require novel forms of hardware that mimic more closely the cellular machinery found in the brain as well as developmental procedures that resemble the process of natural selection.

References

Izhikevich, E., & Edelman, G. M. (2008). Large-scale model of mammalian thalamocortical systems. *Proceedings of the National academy of Sciences of the United States of America, 105*(9), 3593–3598.

Hay, E. (2011). Models of neocortical layer 5b pyramidal cells capturing a wide range of dendritic and perisomatic active properties. *PLoS Computational Biology, 7*(7), e1002107.

Khan, G., Parthasarathy, K., & Bhalla, U. S. (2010). Odor representations in the mammalian olfactory bulb. *WIREs Systems Biology and Medicine, 2*, 603–611.

Chen, X. (2011). Functional mapping of single spines in cortical neurons in vivo. *Nature, 475*, 501–505.

Branco, T., & Hausser, M. (2010). The single dendritic branch as a fundamental functional unit in the nervous system. *Current Opinion in Neurobiology, 20*, 494–502.

Bray, D. (2009). *Wetware: a computer in every living cell* (p. 267). New Haven: Yale University Press.

Berning, S., et al. (2012). Nanoscopy in a living mouse brain. *Science, 335*, 551. 3 Feb.

Chapter 13A
Randal Koene on Bray's "Brain Versus Machine"

Emulation Versus Understanding

In his article, "Brain versus Machine", Dennis Bray discusses and critiques claims that "reverse engineering of the brain is within reach", and he compares this with developments in artificial intelligence. It is clear from his exposition that Bray does not agree with the optimistic outlook espoused by Terry Sejnowski at the 2010 Singularity Summit. As in all disagreements though, it is important to be careful about the subject matter that is the apparent cause of disagreement, the terminology used and how that terminology is understood. We will see that a more precise use of terms constrains the sweeping conclusions drawn by Bray.

What is meant by "reverse engineering" the brain? And does the validity of the reverse engineering claim affect the topics subsequently addressed in Bray's article? The Merriam-Webster dictionary says: "Reverse engineer: to disassemble and examine or analyze in detail (as a product or device) to discover the concepts involved in manufacture usually in order to produce something similar." According to Bray, we should be able to diagnose and cure ALS, schizophrenia, Huntington's disease and Alzheimer's if we are indeed shortly able to reverse engineer the brain. But the definition makes no claims about repairing existing systems. Even in the case of simple consumer electronics, repairing is much harder than replacing (which is what reverse engineering to "produce something similar" enables). For example, we do not tend to fix malfunctioning integrated circuits, even though we can design them. A fix would require additional sophisticated tools.

There is data acquisition about a system; there is the replication of a functioning system based on such data; and there is intervention in an existing system, with the aim to modify or correct its operation. As written, these three different accomplishments are likely listed in order of increasing difficulty. It is for this reason that I have in prior writings and presentations made an effort to compare side by side the probable degrees of difficulty in efforts to repair biology and thereby greatly extend life or to acquire relevant brain data in order to replicate function in a whole brain emulation. Given the number of possible points of break-down and the tendency of interventions to lead to unexpected downstream side-effects in vivo, the former may be much more difficult than the latter.

It is true that neuroscience has not given us a strong understanding of the various strategies at different levels that together make up the mind. Sejnowski's optimism comes from the bottom up. He, like many others in the rapidly expanding fields of computational neuroscience and neuroinformatics, deals predominantly with functions carried out at the neuronal mechanistic level. That is where neuroscience has spent most of the last 100 years learning to identify elements and measure compounds and signals.

Bray's assertion that experts in neurophysiology would view attempts to build brains on a computer as irrelevant is a bit of a strawman argument. Computational modeling is used widely in neurophysiology (see for example the Computational Neurophysiology lab at Boston University, at which I used to work). The scale of this work in neuroinformatics is increasing rapidly. Any representation, any model is an effort at system identification where you pick your level of detail to match the input and output you are interested in. For many representations of brain mechanisms we are interested in the timing of spikes of activity at individual neurons. Sensory input arrives as trains of spikes, spikes drive muscles (e.g., for speech), and inter-spike timing is crucial for memory formation at synapses.

Obviously, no one insists on representing a synapse with a line of code, as Bray posits. A much more likely approach is to represent each neuron by what is known as a compartmental model, where each compartment can be thought of as an electric circuit and implements the Hodgkin-Huxley equations for membrane channel dynamics. Through system identification, you translate underlying physics into functions. For example, the modulating effects of microRNA that Bray points to implement "mode switches" at many neurons in a diffuse manner. You can identify these modes by observing behaving neurons and a functional version of such a switch could be a 1-bit flag.

There are really two different questions:

1. What does it take to acquire data from a brain and replicate its unique function?
2. What does it take to understand the system strategy from the top down to the cells so that you could build similar systems based on that strategy?

Bray's arguments focus on question 2. But we have already achieved excellent results aimed at question 1, such as those published by Briggman et al. and by Bock et al. (both in Nature, 2011). Each team demonstrated proofs of principle for system identification by reconstructing detailed individual neural circuitry, one in retina and one in visual cortex. Kozloski and Wagner (2011) showed how to take this to large-scale neural tissue simulation. Some researchers are indeed attempting to learn about emergent properties of neural networks by using large numbers of generic cell simulations. That is an endeavor separate from the building of tools with which to acquire sufficient structural and functional data from a specific piece of neural tissue to solve the system identification task within that constrained context. Examples of such tools are coming out of the labs of Winfried Denk (Max Planck Institute) and Jeff Lichtman (Harvard University). Although Bray refers to work by Izhikevich and Edelman in 2008, he omits references to some of Eugene Izhikevich's more famous work, developing neuron representations now commonly known as Izhikevich Neurons. Those are a good example of system identification. They can produce the output of a wide range of different types of neurons without having to model any of the deeper neurophysiology.

Just as modern astronomy became possible by developing telescopes, what our goal requires is also the development of the right measurement instruments. If you capture the input–output perspective of each neuron with correct system identification then you capture everything that the system of neurons is responsible for, including a mind's sense of dignity, sense of self-worth, respect, humor, and so forth.

Chapter 14
The Disconnection Thesis

David Roden

In this essay I claim that Vinge's idea of a technologically led intelligence explosion is philosophically important because it requires us to consider the prospect of a posthuman condition succeeding the human one. What is the "humanity" to which the posthuman is "post"? Does the possibility of a posthumanity presuppose that there is a 'human essence', or is there some other way of conceiving the human-posthuman difference? I argue that the difference should be conceived as an emergent disconnection between individuals, not in terms of the presence or lack of essential properties. I also suggest that these individuals should not be conceived in narrow biological terms but in "wide" terms permitting biological, cultural and technological relations of descent between human and posthuman. Finally, I consider the ethical implications of this metaphysics If, as I claim, the posthuman difference is not one between kinds but emerges diachronically between individuals, we cannot specify its nature a priori but only a posteriori. The only way to evaluate the posthuman condition would be to witness the emergence of posthumans. The implications of this are somewhat paradoxical. We are not currently in a position to evaluate the posthuman condition. Since posthumans could result from some iteration of our current technical activity, we have an interest in understanding what they might be like. It follows that we have an interest in making or becoming posthumans.

The Posthuman Impasse

In a 1993 article "The Coming Technological Singularity: How to survive in the posthuman era" the computer scientist Vernor Vinge ar-gued that the invention of

D. Roden (✉)
Faculty of Arts, Department of Philosophy, The Open University, Walton Hall,
Milton Keynes, MK7 6AA, UK
e-mail: david.roden@open.ac.uk

A. H. Eden et al. (eds.), *Singularity Hypotheses*, The Frontiers Collection,
DOI: 10.1007/978-3-642-32560-1_14,

a technology for creating entities with greater than human intelligence would lead to the end of human dominion over the planet and the beginning of a posthuman era dominated by intelligences vastly greater than ours (Vinge 1993).

According to Vinge, this point could be reached via recursive improvements in the technology. If humans or human-equivalent intelligences could use the technology to create superhuman intelligences the resultant entities could make even more intelligent entities, and so on. Thus, a technology for intelligence creation or intelligence amplification would constitute a singular point or "singularity" beyond which the level of mentation on this planet might increase exponentially and without limit.

The form of this technology is unimportant for Vinge's argument. It could be a powerful cognitive enhancement technique, a revolution in machine intelligence or synthetic life, or some as yet unenvisaged process. However, the technology needs to be "extendible" in as much that improving it yields corresponding increases in the intelligence produced. Our only current means of producing human-equivalent intelligence is non-extendible: "If we have better sex… it does not follow that our babies will be geniuses" (Chalmers 2010, p. 18).

The "posthuman" minds that would result from this "intelligence explosion" could be so vast, according to Vinge, that we have no models for their transformative potential. The best we can do to grasp the significance of this "transcendental event", he claims, is to draw analogies with an earlier revolution in intelligence: the emergence of posthuman minds would be as much a step-change in the development of life on earth as the "The rise of humankind".

Vinge's singularity hypothesis—the claim that intelligence-making technology would generate posthuman intelligence by recursive improvement—is practically and philosophically important. If it is true and its preconditions feasible, its importance may outweigh other political and environmental concerns for these are predicated on human invariants such as biological embodiment, which may not obtain following a singularity.

However, even if a singularity is not technically possible—or not imminent—the Singularity Hypothesis (SH) *still* raises a troubling issue concerning our capacity to evaluate the long-run consequences of our technical activity in areas such as the NBIC technologies (Nanotechnology, Biotechnology, Information Technology, and Cognitive Science). This is because Vinge's prognosis presupposes a weaker, more general claim to the effect that our technical activity in NBIC areas or similar might generate forms of life which might be significantly alien or "other" to ours. I refer to this more general thesis as "Speculative Posthumanism".

If we assume Speculative Posthumanism it seems we can adopt either of two policies towards the posthuman prospect. Firstly, we can *account* for it: that is, assess the ethical implications of contributing to the creation of posthumans through our current technological activities.

However, Vinge's scenario gives us reasons for thinking that the differences between humans and posthumans could be so great as to render accounting impossible or problematic in the cases that matter. The differences stressed in Vinge's essay are cognitive: posthumans might be so much smarter than humans that we could not understand their thoughts or anticipate the transformative effects of posthuman technology. There might be other very radical differences. Posthumans might have experiences so different from ours that we cannot envisage what living a posthuman life would be like, let alone whether it would be worthwhile or worthless one.

For this reason, we may just opt to *discount* the possibility of posthumanity when considering the implications of our technological activity: considering only its implications for humans or for their modestly enhanced transhuman cousins. We can refer to the latter using Ray Kurzweil coinage "MOSH": Mostly Original Substrate Human (Agar 2010, pp. 41–20).

However, humans and MOSH's have a prima facie duty to evaluate the outcomes of their technical activities of these differences with a view to maximizing the chances of achieving the good posthuman outcomes or, at least, avoiding the bad ones. It is, after all, their actions and their technologies that will antecede a posthuman difference-maker such as a singularity while the stakes for humans and MOSH's will be very great indeed.

From the human/MOSH point of view some posthuman dispensations might be transcendently good. Others could lead to a very rapid extinction of all humans and MOSH's, or something even worse. Charles Stross' novel *Accelerando* envisages human and MOSH social systems being superseded by Economics 2.0: a resource allocation system in which supply and demand relationships are computed too rapidly for those burdened by a "narrative chain" of personal consciousness to keep up. Under Economics 2.0 first person subjectivity is replaced "with a journal file of bid/request transactions" between autonomous software agents, while inhabited planets are pulverized and converted to more "productive" ends (Stross 2006, p. 177).

This post-singularity scenario is depicted as comically dreadful in Stross' novel. It is bad for humans and for their souped-up transhuman offspring who prove equally redundant amid such virulent efficiency. However, as the world-builder of *Accelerando*'s fictional posthuman future, Stross is able to stipulate the moral character of Economics 2.0. If we were confronted with posthumans, things might not be so easy. We cannot assume, for example, that a posthuman world lacking humans would be worse than one with humans but no posthumans. If posthumans were as unlike humans as humans are unlike non-human primates, a fair evaluation of their kinds of life might be beyond us.

Thus *accounting* for our contribution to making posthumans seems obligatory but may also be impossible with radically alien posthumans, while discounting our contribution is irresponsible. We can call this double bind: "the posthuman impasse".

If the impasse is real rather than apparent, then there may be no principles by which to assess the most significant and disruptive long-term outcomes of current developments in NBIC (and related) technologies.

One might try to circumvent the impasse by casting doubt on Speculative Posthumanism. It is conceivable that further developments in technology, on this planet at least, will never contribute to the emergence of significantly nonhuman forms of life.

However, Speculative Posthumanism is a weaker claim than SH and thus more plausible. Vinge's essay specifies one recipe for generating posthumans. But there might be posthuman difference-makers that do not require recursive self-improvement (we will consider some of these in due course). Moreover, we know that Darwinian natural selection has generated novel forms of life in the evolutionary past since humans are one such. Since there seems to be nothing special about the period of terrestrial history in which we live it seems hard to credit that comparable novelty resulting from some combination of biological or technological factors might not occur in the future.

Is there any way round the impasse that is compatible with Speculative Posthumanism? I will argue that there is, though some ethicists may prefer the *discounting* option to my proposal. However, to understand how the impasse can be avoided we must consider what Speculative Posthumanism entails in more detail.

As a first step towards this clarification, I will gloss the speculative posthumanist claim as the schematic possibility claim SP:

> (SP) Descendants of current humans could cease to be human by virtue of a history of technical alteration.

SP has notable features which, when fully explicated, can contribute to a coherent philosophical account of posthumanity.

Firstly, the SP schema defines posthumanity as the result of a process of technical *alteration*. Value-laden terms such as "enhancement" or "augmentation" which are more commonly used in debates about transhumanism and posthumanism are avoided. I shall explain and justify this formulation in section Value Neutrality.

Secondly, it represents the relationship between humans and posthumans as a historical successor relation: *descent*. "Descent" is used in a "wide" sense insofar as qualifying entities might include our biological descendants or beings resulting from purely technical mediators (e.g., artificial intelligences, synthetic life-forms, or uploaded minds). The concept of Wide Descent will be further explained in section Wide Descent.

Wide Descent also bears on one of the harder problems confronting a general account of the posthuman: what renders posthumans *nonhuman*? Is Speculative Posthumanism committed to a "human" or MOSH *essence* which all posthumans lack, or are there other ways of conceiving the difference?

I will argue that the account of Wide Descent, together with more general metaphysical considerations, militates against essentialism. I will propose, instead, that human-posthuman difference be understood as a concrete *disconnection* between individuals rather than as an abstract relation between essences or kinds. This anti-essentialist model will allow us to specify the circumstances under which accounting would be possible.

Value Neutrality

SP states that a future history of a general type is metaphysically and technically possible. It does not imply that the posthuman would *improve on* the human or MOSH state or that there would be a commonly accessible perspective from which to evaluate human and posthuman lives. Posthumans may, as Vinge writes, be "simply too different to fit into the classical frame of good and evil" (Vinge 1993).

It could be objected that the value-neutralization of the historical successor relation in the SP schema is excessively cautious and loses traction on what distinguishes humans from their hypothetical posthuman descendants: namely, that posthumans would be in some sense "better" by virtue of having greater capacities.

One of the most widely used formulations of the idea of the posthuman—that of transhumanist philosopher Nick Bostrom—is non-neutral. He defines a posthuman as a "being that has at least one posthuman capacity" by which is meant "a central capacity greatly exceeding the maximum attainable by any current human being without recourse to new technological means". Candidates for posthuman capacities include augmented "healthspan", "cognition" or "emotional dispositions" (Bostrom 2009).

While this is not a purely metaphysical conception of the posthuman it is, it might be argued, not so loaded as to beg ethical questions against critics of radical enhancement. As Allen Buchanan points out, "enhancement" is *a restrictedly value-laden notion* insofar as enhancing a capacity implies making it function more effectively but does not imply improving the welfare of its bearer (Buchanan 2009, p. 350).

Moreover, it could be objected that "alteration" is so neutral that a technical process could count as posthuman engendering if it resulted in wide descendants of humans with capacities far below that of normal humans (I address this point in section Modes of Disconnection below).

However, it is easy to see that the value-ladenness of "enhancement" is not restricted enough to capture some conceivable paths to posthumanity. To be sure, posthumans might result from a progressive enhancement of cognitive powers—much as in Vinge's recursive improvement scenario. Alternatively, our posthuman descendants might have capacities we have no concepts for while lacking some capacities that we can conceive of.

In a forthcoming article I consider the possibility that shared "non-symbolic workspaces"—which support a very rich but non-linguistic form of thinking—might render human natural language unnecessary and thus eliminate the cultural preconditions for our capacity to frame mental states with contents expressible as declarative sentences (Philosophers call such states "propositional attitudes"—e.g. the *belief* that Snow is White or the *desire* to vote for Obama in next presidential election). If propositional attitude psychology collectively distinguishes humans from non-humans, users of non-symbolic workspaces might acquire a non-propositional psychology and thus cease to be human (As I show in section

Disconnection and Anti-Essentialism being "human distinguishing" in this manner does not have to entail being part of a human essence).

It is not clear that process leading to this relatively radical cognitive alteration would constitute an augmentation history in the usual sense—since according to my scenario it could involve the loss of one central capacity (the capacity to have and express propositional attitudes) and the acquisition of an entirely new one. Yet it is arguable that it could engender beings so different from us in cognitive structure that they would qualify as posthuman according to SP (See section Modes of Disconnection).

The Borg from the TV series *Star Trek* are a more popular variation on the theme of the "value-equivocal" posthuman. While the Borg seem like a conceivable kind of posthuman life, they result from the inhibition of the kind of cognitive and affective capacities whose flowering Bostrom treats as constitutive of the posthuman. The Borg-Collective, it is implied, possesses great cognitive powers and considerable technical prowess. However, the Collective's powers emerge from the interactions of highly networked "drones", each of whom has had its personal capacities for reflection and agency suppressed.

Wide Descent

As advertised earlier, SP uses a notion of *wide descent* to understand our relationship to prospective posthumans.

I will elaborate the distinction between wide descent and narrow descent below in term of a distinction between a *narrow* conception of the human qua species and a *wide* conception of the human. Whereas Narrow Humanity can be identified, if we wish, with the biological species *Homo sapiens*, Wide Humanity is a technogenetic construction or "assemblage" with both narrowly human and narrowly non-human parts.

There are two principle justifications for introducing wide descent and the correlative notion of *Wide Humanity*:

The Appropriate Concept of Descent for SP is Not Biological

Exclusive consideration of biological descendants of humanity as candidates for posthumanity would be excessively restrictive. Future extensions of NBIC technologies may involve discrete bio-technical modifications of the reproductive process such as human cloning, the introduction of transgenic or artificial genetic material or very exotic processes such as personality uploading or "mind-cloning". Thus, entities warranting our concern with the posthuman could emerge via modified biological descent, recursive extension of AI technologies (involving human and/or non-human designers), quasi-biological descent from synthetic organisms, a convergence of the above, or via some technogenetic process yet to be envisaged.

It follows that when considering the lives of hypothetical posthuman descendants we must understood "descent" as relationship that is *technically mediated to an arbitrary degree*.

"Humanity" is Already the Product of a Technogenetic Process

A plausible analogy for the emergence of posthumans, as Vinge observes, is the evolutionary process that differentiated humans from non-human primates. But there are grounds for holding that the process of becoming human (hominization) has been mediated by human cultural and technological activity. One widely employed way of conceiving hominization is in terms of cultural niche construction. Niche-construction occurs where members of a biological population actively alter their environment in a –way that alters the selection pressures upon it. For example, it has been argued that that the invention of dairy farming technology (around 10,000 BC) created an environment selective for genes that confer adult lactose tolerance. Thus, the inventors of animal husbandry unwittingly reconfigured the bodies of their descendants to survive in colder climes (Kevin et al. 2000; Buchanan 2011, p. 204). The anthropologist Terrence Deacon proposes that the emergence of early symbolic practices produced a symbolically structured social environment in which the capacity to acquire competence in complex symbol systems was a clear selective advantage. Thus, it is possible that the selection pressures that made humans brains adept at language learning were a consequence of our ancestors' own social activity even as these brains imposed a learnability bottleneck on the cultural evolution of human languages (Deacon 1997, pp. 322–6, 338).

If this model is broadly correct, hominization has involved a confluence of biological, cultural and technological processes. It has produced socio-technical "assemblages" in which humans are coupled with other active components: for example, languages, legal codes, cities, and computer mediated information networks.[1]

[1] The term "assemblage" is used by the philosopher Manuel Delanda to refer to any *emergent* but *decomposable* whole and belongs to the conceptual armory of the particularist "flat" ontology I will propose for Speculative Posthumanism in section Disconnection and Anti-Essentialism below. Assemblages are *emergent* wholes in that they exhibit powers and properties not attributable to their parts but which causally depend upon their parts. Assemblages are also *decomposable* insofar as all the relations between their components are "external": each part can be detached from the whole to exist independently (Assemblages are thus opposed to "totalities"

Biological humans are currently "obligatory" components of modern technical assemblages. Technical systems like air-carrier groups, cities or financial markets depend on us for their operation and maintenance much as an animal depends on the continued existence of its vital organs. Technological systems are thus intimately coupled with biology and have been over successive technological revolutions.

However, this dependency runs in the other direction: the distinctive social and cognitive accomplishments of biological humans require a technical and cultural infrastructure. Our capacity to perform mathematical operations on arbitrarily large numbers is not due to an innate number sense but depends on our acquisition of routines like addition or long division and our acculturation into culturally stored numeral systems. Our species-specific language ability puts us in a unique position to apply critical thinking skills to thoughts expressed in public language, to co-ordinate social behavior via state institutions, or record information about complex economic transactions (Clark 2004, 2006). Philosophers such as Donald Davidson and Robert Brandom have gone further, arguing that our capacity to think in and express propositional attitudes depends on our mastery of public language. Davidson argues that the ability to have beliefs (and hence other propositional attitudes such as desires or wishes) requires a grasp of what belief is since to believe is also to understand "the possibility of being mistaken". This in turn requires us to grasp that others might have true or false beliefs about the same topic. Thus, no belief can be adopted by someone not already involved in evaluating her own and others' attitudes in common linguistic coin (Davidson 1984).

These considerations lend support to the claim that the emergence of biological humans has been one aspect of the technogenesis of a planet-wide assemblage composed of biological humans locked into networks of increasingly "lively" and "autonomous" technical artifacts (Haraway 1989). It is this wider, interlocking system, and not bare-brained biological humans, that would furnish the conditions for the emergence of posthumans. Were the emergence of posthumans to occur, it would thus be a historical rupture in the development of this extended socio-technical network.

However, while the emergence of posthumans *must* involve the network, the degree to which it would involve modifications of biological humans is conceptually open (as argued above). Posthumans may derive from us by some technical process that mediates biological descent (such as a germ-line cognitive enhancement) or they may be a consequence of largely technological factors.

I shall refer to this wider network as the "Wide Human" (WH). An entity is a *wide human* just so long as it depends for its continued functioning on the Wide Human while contributing to its operations to some degree. Members of the

(Footnote 1 continued)
in an idealist or holist sense). This is the case even where the part is functionally necessary for the continuation of the whole (DeLanda 2006, p. 184).

biological species Homo sapiens, on the other hand, are *narrowly* human. Thus, domesticated animals, mobile phones and toothbrushes are wide humans while we obligatory biologicals are both *narrowly* and *widely* human.

Having outlined the patient and the generic process of becoming posthuman, we now state a recursive definition of Wide Human descent:

An entity is a wide human descendant if it is the result of a technically mediated process:

A) Caused by a part of WH—where the ancestral part may be wholly biological, wholly technological or some combination of the two.
B) Caused by a wide human descendant.

A is the "basis clause". It states what belongs to the initial generation of wide human descendants without using the concept of wide descent. B is the recursive part of the definition. Given *any* generation of wide human descendants it specifies a successor generation of wide human descendants.

It is important that this definition does not imply that a wide human descendant *need be human* in either wide or narrow senses. Any part of WH ceases to be widely human if its wide descendants go "feral": acquiring the capacity for independent functioning and replication outside the human network. SP entails that with becoming posthuman this would occur as a result of some history of technical change.

Becoming posthuman would thus be an unprecedented discontinuity in the hominization process. WH has undergone revolutions in the past (like the shift from hunter-gatherer to sedentary modes of life) but no part of it has been *technically altered* so as to function outside of it.

It follows that a wide human descendent is a posthuman if and only if:

i It has ceased to belong to WH (The Wide Human) as a result of technical alteration.
ii Or is a wide descendant of such a being.

I refer to this claim as the *disconnection thesis*.

Disconnection and Anti-Essentialism

My formulation of what it means to cease to be human will seem strange and counter-intuitive to some. We are used to thinking of *being human* not as a part-whole relation (being a part of WH in this case) but as instantiating a human nature or "essence".

An essential property of a kind is a property that no member of that kind can be without. If humans are necessary rational, for example, then it is a necessary truth that if x is human, then x is rational.[2]

To say that a human essence exists is just to say that there is a set of individually necessary conditions for humanity.

Anthropological essentialism (the claim that there is a human essence) implies that the technically mediated loss of even one of these would export the loser from humanity to posthumanity. As metaphysical formula go, this has the immediate appeal of simplicity.

It also provides *a nice clear method for resolving the posthuman impasse*. We can call this the "apophatic method": after the method of apophatic or "negative" theology. Apophatic theologians think that God is so mysterious that we can only describe Him by saying what He is not (Dale 2010). By extension, anthropological essentialism, if true, would allow us to identify each path to posthumanity with the deletion of some component of the human essence. This, in turn, would allow us to adjudicate the value of these paths by considering the ethical implications of each loss of an anthropologically necessary property.

For example, an essentialist may claim on either a posteriori or a priori grounds that humans are necessarily *moral persons* with capacities for deliberation and autonomous agency. If so, one sure route to posthumanity would be to lose those moral capacities. Put somewhat crudely, we could then know that *some* conceivable posthumans are non-persons. If persons are, as Rawls claims, sources of moral value and non-persons are not then *this* posthuman state involves the loss of unconditional moral status (Rawls 1980). This particular path to posthumanity would, it seems, involve unequivocal loss.

The Disconnection Thesis does not entail the rejection of anthropological essentialism but it renders any reference to essential human characteristics unnecessary. The fact that some wide human descendant no longer belongs to the Wide Human implies nothing about its intrinsic properties or the process that brought about its disconnection. However, we can motivate the disconnection thesis and its mereological (part-whole) conception of wide humanity by arguing against essentialism on general grounds.

The most plausible argument for abandoning anthropological essentialism is naturalistic: essential properties seem to play no role in our best scientific explanations of how the world acquired biological, technical and social structures and entities. At this level, form is not imposed on matter from "above" but emerges via generative mechanisms that depend on the amplification or inhibition of differences between particular entities (For example, natural selection among biological species or competitive learning algorithms in cortical maps). If this picture holds generally, then essentialism provides a misleading picture of reality.

The philosopher Manuel Delanda refers to ontologies that reject a hierarchy between organizing form and a passive nature or "matter" as "flat ontologies".

[2] Another way of putting this is to say that in any possible world that humans exist they are rational. Other properties of humans may be purely "accidental"—e.g. their colour or language. It is not part of the essence of humans that they speak English, for example. Insofar as speaking English is an accidental property of humans, there are possible worlds in which there are humans but no English speakers.

Whereas a hierarchical ontology has categorical entities like essences to organise it, a flat universe is "made exclusively of unique, singular individuals, differing in spatio-temporal scale but not in ontological status" (DeLanda 2002, p. 58).

The properties and the capacities of these entities are never imposed by transcendent entities but develop out of causal interactions between particulars at various scales. Importantly for the present discussion, *a flat ontology recognizes no primacy of natural over artificial kinds* (Harman 2008).

It is significant that one of Delanda's characterizations of flat ontology occurs during a discussion of the ontological status of biological species in which he sides with philosophers who hold that species are individuals rather than types or universals (DeLanda 2002, pp. 59–60). For example, Ernst Mayr's "biological species concept" (BSC) accounts for species differences among sexually reproducing populations in terms of the reproductive isolation of their members. This restricts gene recombination and thus limits the scope for phenotypic variation resulting from gene flows, further reinforcing discontinuities between conspecifics (Okasha 2002, p. 200).

Motivated by such anti-essentialist scruples, the bioethicist Nicolas Agar has argued that differences between humans and prospective posthumans can be conceived in terms of membership or non-membership of a reproductively isolated population as conceived by the BSC (Agar 2010, p. 19). Posthumans would arise where (and only where) radical enhancement created reproductive barriers between the enhanced and the mainstream human population.

Agar's proposal illustrates one variant of the flat ontological approach. However, importing the BSC neat from the science of the evolutionary past is problematic when considering the ontology of technogenetic life forms. Biotechnologies such as the artificial transfer of genetic material across species boundaries could make the role of natural reproductive boundaries less significant in a posthuman or transhuman dispensation (Buchanan 2009, p. 352). If these alternative modes of genetic transmission became routinely used alongside regular sex, the homeostatic role of reproductive barriers would be significantly reduced.

While BSC has a clear application to understanding speciation in sexually reproducing life forms, the BSC has no applicability to non-sexually reproducing life forms. Likewise, the distinction between the genetics lab and nature cannot be assumed relevant in a posthuman world where biotechnology or post-biological forms of descent dominate the production of intelligence and the production of order more generally. The flat ontological injunction not to prioritise natural over artificial sources of order provides a more reliable methodological principle than Agar's misguided ethical naturalism.

The distinction between Wide and Narrow Humanity broached earlier in this paper accommodates this possibility by distinguishing between the Narrow Human (which can be understood in terms of the BSC) and the socio-technical assemblage WH which fully expresses human societies, cultures and minds.

WH has the same *ontological status* as species like *Homo sapiens*—both are complex individuals rather than kinds or essences. However, WH is constituted by causal relationships between biological and non-biological parts, such as languages, technologies and institutions. A disconnection event would be liable to involve technological mechanisms without equivalents in the biological world and this should be allowed for in any ontology that supports Speculative Posthumanism.

Modes of Disconnection

As mentioned above, Vinge considers the possibility that disconnection between posthumans and humans may occur as a result of differences in the cognitive powers of budding posthumans rendering them incomprehensible and uninterpretable for baseline humans.

For example, he speculates in passing that rich informational connections between posthuman brains (or whatever passes for such) may be incompatible with a phenomenology associated with a biographically persistent subject or self (Vinge 1993).

If non-subjective phenomenology among posthumans is possible then Vinge's concern that such form of existence might not be evaluable according our conceptions of good or evil seem warranted. Human ethical frameworks arguably require that candidates for our moral regard have the capacity to experience pain. Most public ethical frameworks have maximal conditions. For example, liberals valorise the capacity for personal autonomy that allows most humans "to form, to revise, and rationally to pursue a conception of the good" (Rawls 1980, p. 525).

Autonomy presumably has threshold cognitive and affective preconditions such as the capacity to evaluate actions, beliefs and desires (practical rationality) and a capacity for the emotions, and affiliations informing these evaluations. However, the capacity for practical reason at issue in our conception of autonomy might not be accessible to a being with non-subjective phenomenology. Such an entity could be incapable of experiencing itself as having a life that might go better or worse for it.

We might not be able to coherently imagine what these impersonal phenomenologies are like (e.g. to say of them that they are "impersonal" is not to commit ourselves regarding the kinds of experiences might furnish). This failure may simply reflect the centrality of *human* phenomenological invariants to the ways humans understand the relationship between mind and world rather than any insight into the necessary structure of experience (Metzinger 2004), p. 213. Thomas Metzinger has argued that our kind of subjectivity comes in a spatiotemporal pocket of an embodied self and a dynamic present whose structures depends on the fact that our sensory receptors and motor effectors are "physically integrated within the body of a single organism". Other kinds of life—e.g. "conscious interstellar gas clouds" or (more saliently) post-human swarm intelligences composed of many mobile processing units—might have experiences of a radically impersonal nature (Metzinger 2004 p. 161).

Disconnection may take other forms, however. All that is required for a disconnection from the Wide Human recall is that some part of this assemblage becomes capable of going wild and following an independent career. This is not true of current types of artificial intelligence, for example, which need to be built, maintained by narrow humans and powered by other human artefacts. This is why beings that are artificially "downgraded" so that their capacities are less than human are unlikely to generate a disconnection event (See section Value Neutrality)—though this possibility cannot be entirely precluded.

A disconnection could ensue, then, wherever prospective posthumans have properties that make their feasible forms of association disjoint from humans/MOSH forms of association.

I suggested in sectionValue Neutrality that propositional attitude psychology might distinguish humans from non-humans. However, as our excursus into flat ontology shows, the capacity to form propositional attitudes such as the *belief* that Lima is in Peru need not be thought of a component of a human essence but as a filter or "sorting mechanism" which excludes non-humans from human society much as incompatibilities in sexual organs or preferences create reproductive barriers between Mayr-style biological species (Agar 2010, p. 19–28). Wide successors to humans who acquired a non-propositional format in which to think and communicate might not be able to participate in our society just as our unmodified descendants might not be able to participate in theirs. They would, in this case, "bud off" from the Wide Human, just as a newly isolated species buds off from its predecessors. Such disconnections could happen by degrees and (unlike in a Vingean singularity) relatively slowly relative to the individual lifetimes. There might also be cases where the disconnection remains partial (for example, some non-propositional thinkers might retain a vestigial capacity to converse with humans and MOSH's).

Are Disconnections Predictable?

I do not claim that speculations in the previous section reliably predict the nature and dynamics of a disconnection event. For example, we do not know whether greater than human intelligence is possible or whether it can be produced by an "extendible" technological method (Chalmers 2010).

Nor, at this point, can we claim to have knowledge about the feasibility of the other disconnection events that we have speculated about (e.g. the replacement of propositional attitude psychology with some non-linguistic cognitive format).

These scenarios are merely intended to illustrate the *ontological thesis* that posthuman-human difference would be a discontinuity resulting from parts of the Wide Human becoming so technically altered that they could split off from it. The intrinsic properties exhibited by these entities are left open by the disconnection thesis.

This epistemological caution seems advisable given that the advent of posthumanity is a (currently) hypothetical future event whose nature and precipitating causes are unprecedented *ex hypothesis*. There are many conceivable ways in which such an event might be caused. Even if a Vinge-style singularity is conceivable but not possible some unrelated technology might be a possible precursor to a disconnection. Disconnections are not defined by a particular technical cause (such as mind-uploading) but purely by an abstract relation of wide descent and the property of functional and replicative independence. Disconnection *can be multiply realized by technologies which have little in common other than (a) feasibility and (b) that disconnection is one of their possible historical effects*. Thus, speculating about how currently notional technologies might bring about autonomy for parts of WH affords no substantive information about posthuman lives (even if it may enable a metaphysically and ethically salutary exploration of the scope for posthuman difference).

Assuming that a conceivable technology (For example, controlled nuclear fusion—other than by gravitational confinement in a star) does not violate physical principles the only sure demonstration of feasibility is the production of a working model or prototype. Thus, we can have no reliable grounds for holding that conceivable precursors to a disconnection are feasible precursors so long as the relevant technologies are underdeveloped. However, once a feasible precursor has been produced the Wide Human could be poised at the beginning of a disconnection process since the capacity to generate disconnection would be a realized technological power.[3] We may be in a position to know which, if any, of the "usual suspects" (Nanotechnology, Biotechnology, Information Technology, Cognitive Science) might bring about a disconnection only when the potential for disconnection is in prospect.

Thus it is plausible to suppose that any disconnection (however, technically realized) will be an instance of what Paul Humphrey terms *diachronic emergence* (Humphrey 2008). A diachronically emergent behaviour or property occurs as a result of a temporally extended process, but cannot be inferred from the initial state of that process.[4] It can only be derived by allowing the process to run its course (Bedau 1997).

If disconnections are diachronically emergent phenomena their morally salient characteristics and effects will not be predictable prior to their occurrence. While this constrains our ability to prognosticate about disconnections, it leaves other aspects of their epistemology quite open. As Humphrey reminds us, diachronic emergence is a one-time event. Once we observe a formerly diachronically emergent event we are in a position to predict tokens of the same type of emergent property from causal antecedents that have been observed to generate it in the past.

---

[3] Absent defeaters (See Chalmers 2010).

[4] Where the emergent property occurs at the same time as the microstates on which it depends, we have an instance of "synchronic emergence" (Humphrey 2008, pp. 586–7).

Most importantly, that disconnections would be diachronically emergent has no implications for the uninterpretability or "alienness" of posthumans since their nature is left open by the disconnection thesis.

The anti-essentialist flat ontology I have recommended as a basis for the disconnection thesis, gives us grounds to be wary of terms like "uninterpretability" or "alienness". To be sure, posthumans might be strange in ways that we cannot currently imagine. Their human or MOSH contemporaries might struggle unsuccessfully to understand their thoughts or motives. However, the fact that interpretative success is not guaranteed does not entail that relatively alien posthumans would be humanly uninterpretable. There are, after all, many things that we do not understand that we might understand under ideal conditions.

An utterly incomprehensible being ("a radical alien") would not belong to this set. Such a being would be humanly uninterpretable. The inability to understand it would be a necessary or essential part of the human/MOSH cognitive essence. But if, as proposed, we reject taxonomic essences, we must hold that there are no such modal facts of this nature.

It follows that there are no grounds for some holding posthumans to be humanly uninterpretable in principle (i.e. to be radical aliens) since the set of humanly uninterpretable things is not defined. Posthuman thinking may still be so powerful or so strangely formatted that it could defy the interpretative capacities of wide human descendants *not altered to an equivalent degree*. But this would depend on the contingencies of disconnection—which are, as yet, unknown. As pointed out in section Modes of Disconnection, disconnection—like speciation—may come by degrees. If the technology exists to create posthumans, then the same technology might support "interfaces" between human and posthuman beings such as the bi-formatted propositional/non-propositional thinkers mentioned above. Thus, where conditions favour it "Posthuman Studies" may graduate from speculative metaphysics to a viable cultural research program.[5]

Resolving the Impasse

What are the implications of the disconnection thesis for attempts to negotiate the ethical bind of the posthuman impasse? The impasse is a way of formulating the ethical concern that the posthuman consequences of our own technical activity may be beyond our moral compass. I have conceded that posthumans might be very different from us in diverse ways, but have argued that there is no basis for concluding that posthumans would be beyond evaluation.

[5] Vinge alludes to this possibility in his far-future space epic *A Fire Upon the Deep* (Vinge 1992). *In Fire* posthumans so powerful as to be god-like in comparison with the most enhanced transhuman exist on a computationally extreme fringe of space known as "the Transcend" where they are studied by "applied theologians" from observatories on the margins of the Milky Way.

As argued in section Are Disconnections Predictable?, we may be in a better position to undertake a value-assessment once a disconnection has occurred. Thus, if we have a moral (or any other) interest in *accounting* for posthumans we have an interest in bringing about the circumstances in which accounting can occur. Thus, we have an interest in contributing to the emergence of posthumans or becoming posthuman ourselves where this is liable to mitigate the interpretative problems of disconnection.

It could be objected, at this point, that we may also have countervailing reasons for *preventing* the emergence of posthumans and not becoming posthuman ourselves.

We have acknowledged that some disconnections could be very bad for humans. Since disconnection could go very wrong, it can be objected that the precautionary principle (PP) trumps the accounting principle. Although there is no canonical formulation, all versions of the PP place a greater burden of proof on arguments for an activity alleged to have to potential for causing extensive public or environmental harm than on arguments against it (Cranor 2004; Buchanan: pp. 199–200). In the present context the PP implies that even where the grounds for holding that the effects of disconnection will be harmful are comparatively weak, the onus is on those who seek disconnection to show that it will not go very wrong. However, the diachronically emergent nature of disconnection implies that such a demonstration is not possible prior to a disconnection event. Thus, one can use the PP to argue that accounting for disconnection (assessing its ethical implications) is not morally obligatory but morally wrong.

One might conclude at this point that we have substituted one impasse (the conflict between accounting and discounting) for a second: the conflict between the principle of accounting and the PP. However, this will depend on the different attitudes to uncertainty expressed in different versions of the PP. If the principle is so stringent as to forbid technical options whose long-range effects remain uncertain to any degree, then it forbids the development of disconnection-potent technology. However, this would forbid almost any kind of technological decision (including the decision to relinquish a technology).[6] Thus a maximally stringent PP is self-vitiating (Buchanan 2011), pp. 200–1.

It follows that the PP should require reasonable evidence of possible harm before precautionary action is considered. A selective precautionary approach to the possibility of disconnection would require that suspect activities be "flagged" for the potential to produce bad disconnections (even where this evidence is not authoritative). But if disconnections are diachronically emergent phenomena, the evidence to underwrite flagging will not be available until the process of technical change is poised for disconnection.

To take a historical analogy: the syntax of modern computer programming languages is built on the work on formal languages developed in the Nineteenth

[6] Given our acknowledged dependence on technical systems, the long-run outcomes of relinquishment may be as disastrous as any technological alternative.

Century by carried out by mathematicians and philosophers like Frege and Boole. Lacking any comparable industrial models, it would have been impossible for contemporary technological forecasters to predict the immense global impact of what appeared an utterly rarefied intellectual enquiry. We have no reason to suppose that we are better placed to predict the long-run effects of current scientific work than our Nineteenth Century forebears (if anything the future seems more rather than less uncertain). Thus, even if we enjoin selective caution to prevent worst-case outcomes from disconnection-potent technologies, we must still place ourselves in a situation in which such potential can be identified. Thus seeking to contribute to the emergence of posthumans, or to become posthuman ourselves, is compatible with a reasonably constrained PP.

References

Agar, N. (2010). *Humanity's end: Why we should reject radical enhancement*. Cambridge: MIT Press.

Bedau, M. A. (1997). Weak emergence. *Philosophical Perspectives, 11*, 375–399.

Bostrom, N. (2009). Why I want to be a posthuman when I grow up. In G. Bert & C. Ruth (Eds.), *Medical enhancement and post humanity* (pp. 107–137). Dordrecht: Springer.

Buchanan, A. (2009). Moral status and human enhancement. *Philosophy and Public Affairs, 37*(4), 346–381.

Buchanan, A. E. (2011). *Beyond humanity? : The ethics of biomedical enhancement*. Oxford: Oxford University Press.

Chalmers, D. J. (2010). The singularity: A philosophical analysis. *Journal of Consciousness Studies, 17*, 7–65.

Clark, A. (2004). *Natural born-cyborgs: Minds, technologies and the future of human intelligence*. Oxford: OUP.

Clark, A. (2006). Language, embodiment and the cognitive niche. *Trends in Cognitive Science, 10*(8), 370–374.

Cranor, C. F. (2004). Toward understanding aspects of the precautionary principle. *Journal of Medicine and Philosophy, 29*(3), 259–279.

Dale, T. (2010). Trinity. Stanford Encyclopedia of Philosophy. Edited by Edward N. Zalta.http://plato.stanford.edu/archives/fall2009/entries/trinityAccessed 28 Sep 2011.

Davidson, D. (1984). Thought and Talk. In: *Inquiries into Truth and Interpretation*, (pp. 155–170). Oxford: Clarendon Press.

Deacon, T. (1997). *The symbolic species: The co-evolution of language and the human brain*. London: Penguin.

DeLanda, M. (2002). *Intensive science and virtual philosophy*. London: Continuum.

Delanda, D. (2006). *A new philosophy of society: Assemblage theory and social complexity*. London: Continuum.

Haraway, D. (1989). A manifesto for cyborgs: Science, technology, and socialist feminism in the 1980s. In W. Elizabeth (Ed.), *Coming to terms* (pp. 173–204). London: Routledge.

Harman, G. (2008). DeLanda's ontology: Assemblage and realism. *Continental Philosophy Review, 41*(3), 367–383.

Humphreys, P. (2008). Computational and conceptual emergence. *Philosophy of Science, 75*(5), 584–594.

Kevin, N. L., John, O. S., & Marcus, W. F. (2000). Niche construction, biological evolution, and cultural change. *Behavioural and Brain Sciences, 23*(1), 131–146.

Metzinger, T. (2004). *Being no one: The self-model theory of subjectivity*. Cambridge: MIT Press.
Okasha, S. (2002). Darwinian metaphysics: Species and the question of essentialism. *Synthese, 131*(2), 191–213.
Rawls, J. (1980). Kantian constructivism in moral theory. *Journal of Philosophy, 77*(9), 515–572.
Stross, C. (2006). *Accelerando*. London: Orbit.
Vinge, V. (1992). *A fire upon the deep*. New York: Tor.
Vinge, V. (1993). The coming technological singularity: how to survive in the post-human era, *vision-21:interdisciplinary science and engineering in the era of cyberspace*. http://www.rohan.sdsu.edu/faculty/vinge/misc/singularity.html Accessed 8 Dec 2007.

Part IV
Skepticism

Chapter 15
Interim Report from the Panel Chairs: AAAI Presidential Panel on Long-Term AI Futures

Eric Horvitz and Bart Selman

The AAAI 2008-09 *Presidential Panel on Long-Term AI Futures* was organized by the president of the Association for the Advancement of Artificial Intelligence (AAAI) to bring together a group of thoughtful computer scientists to explore and reflect about societal aspects of advances in machine intelligence (computational procedures for automated sensing, learning, reasoning, and decision making). The panelists are leading AI researchers, well known for their significant contributions to AI theory and practice. Although the final report of the panel has not yet been issued, we provide background and high-level summarization of several findings in this interim report.

AI research is at the front edge of a larger computational revolution in our midst—a technical revolution that has been introducing new kinds of tools, automation, services, and new access to information and communication. Efficiencies already achieved via computational innovations are beyond the scope of what people could have imagined just two decades ago. It is clear that AI researchers will spearhead numerous innovations over the next several decades. Panelists overall shared a deep enthusiasm and optimism about the future influence of AI research and development on the world. Panelists expect AI research to have great positive influences in many realms, including healthcare, transportation, education, commerce, information retrieval, and scientific research and discovery.

The panel explored a constellation of topics about societal influences of AI research and development, reviewing potential challenges and associated opportunities for additional focus of attention and research. Several topics were

E. Horvitz (✉)
Distinguished Scientist & Co-Director, Microsoft Research Redmond,
One Microsoft Way, Redmond, WA 98052, USA
e-mail: horvitz@microsoft.com

B. Selman
Professor, Department of Computer Science, Cornell University,
4148 Upson Hall, Ithaca, NY 14853, USA

A. H. Eden et al. (eds.), *Singularity Hypotheses*, The Frontiers Collection,
DOI: 10.1007/978-3-642-32560-1_15,

highlighted as important areas for future work; there was a sense that, for these issues, increased sensitivity, attention, and research would help to ensure better outcomes. The panel believed that identifying and highlighting potential "rough edges" that might arise at the intersection of AI science and society would be beneficial for directing ongoing reflection, as well as for guiding new research investments. The study had three focus areas and associated subgroups.

Subgroup on Pace, Concerns, Control

The first focus group explored concerns expressed by lay people—and as popularized in science-fiction for decades—about the long-term outcomes of AI research. Panelists reviewed and assessed popular expectations and concerns. The focus group noted a tendency for the general public, science-fiction writers, and futurists to dwell on *radical* long-term outcomes of AI research, while overlooking the broad spectrum of opportunities and challenges with developing and fielding applications that leverage different aspects of machine intelligence.

Popular perspectives on the outcomes of AI research include expectation that there will be one or more disruptive outcomes. These include that notion that the research will somehow lead to the advent of utopia or catastrophe. The utopian perspective is perhaps best captured in the writings of Ray Kurzweil and others, who speak of a forthcoming "technological singularity". At the other end of the spectrum, some people are concerned about the "rise of intelligent machines", fueled by popular novels and movies, that tell stories of the loss of control of robots. Whether forecasting utopian or catastrophic outcomes, the radical perspectives are frightening to people in that they highlight some form of radical change on the horizon—often founded on a notion of the loss of control of the computational intelligences that we create.

The panel of experts was overall skeptical of the radical views expressed by futurists and science-fiction authors. Participants reviewed prior writings and thinking about the possibility of an "intelligence explosion" where computers one day begin designing computers that are more intelligent than themselves. They also reviewed efforts to develop principles for guiding the behavior of autonomous and semi-autonomous systems. Some of the prior and ongoing research on the latter can be viewed by people familiar with Isaac Asimov's Robot Series as formalization and study of behavioral controls akin to Asimov's Laws of Robotics. There was overall skepticism about the prospect of an intelligence explosion as well as of a "coming singularity", and also about the large-scale loss of control of intelligent systems. Nevertheless, there was a shared sense that additional research would be valuable on methods for understanding and verifying the range of behaviors of complex computational systems to minimize unexpected outcomes. Some panelists recommended that more research needs to be done to better define "intelligence explosion", and also to better formulate different classes of such accelerating intelligences. Technical work would likely lead to enhanced understanding of the likelihood of such phenomena, and the nature, risks, and overall outcomes associated with different conceived variants.

The group suggested outreach and communication to people and organizations about the low likelihood of the radical outcomes, sharing the rationale for the overall comfort of scientists in this realm, and for the need to educate people outside the AI research community about the promise of AI for enhancing the quality of human life in numerous ways, coupled with a re-focusing of attention on actionable, shorter-term challenges.

Subgroup on Shorter-Term Challenges

A second subgroup focused on nearer-term challenges, examining potential "rough edges", where AI research touches society, that may be addressed via new vigilance, sensitivity, and, more generally, with investment in additional focused research. Several areas for future research were identified as having valuable payoff in the shorter-term. These include the promise of redoubling research on using AI methods to enhance peoples' privacy. There already has been interesting and valuable work in the AI research community on methods for enhancing privacy while enabling people and organizations to personalize services. Other shorter-term opportunities include the value of making deeper investments in methods that enhance interactions and collaborations between people and machine intelligence. The panel's deliberation included discussion of the importance of endowing computing systems with deeper competencies at working in a complementary manner with people on the joint solution of tasks, and in supporting fluid transitions between automated reasoning and human control. The latter includes developing methods that make machine learning and reasoning more transparent to people, including, for example, giving machines abilities to better explain their reasoning, goals, and uncertainties. Another focus of discussion centered on the prospect that people, organizations, and hostile governments might harness a variety of AI advances for malevolent purposes. To our knowledge, such efforts have not yet occurred, yet it is not difficult to imagine how future computer malware, viruses, and worms might leverage richer learning and reasoning, accessing an increasing number of channels of information about people. AI methods might one day be used to perform relatively deep and long-term learning and reasoning about individuals and organizations—and then perform costly actions in a sophisticated and potentially secretive manner. There was a shared sense that it would be wise to be vigilant and to invest in proactive research on these possibilities. Proactive work includes new efforts in security, cryptography, and AI research in such areas as user modeling and intrusion detection directed at this potential threat, in advance of evidence of such criminal efforts.

Subgroup on Ethical and Legal Issues

A third subgroup focused on ethical and legal questions. This subgroup reflected about ethical and legal issues that could become more salient with the increasing commonality of autonomous or semi-autonomous systems that might one day be charged with making (or advising people on) high-stakes decisions, such as medical therapy or the targeting of weapons. The subgroup's deliberation included reflection about the applicability of current legal frameworks. As an example, the group reviewed potential issues with assignment of liability associated with costly, unforeseen behaviors of autonomous or semi-autonomous decision making systems. Other reflection and discussion centered on potential ethical and psychological issues with human responses to virtual or robotic systems that have an increasingly human appearance and behavior. For example, the group reflected about potential challenges associated with systems that synthesize believable affect, feelings, and personality. What are the implications of systems that emote, that express mood and emotion (e.g., that appear to care and nurture), when such feelings do not exist in reality? Discussion centered on the value of investing more deeply in research in these areas, and of engaging ethicists, psychologists, and legal scholars.

Meeting at Asilomar

After several months of discussion by email and phone, a face-to-face meeting was held at Asilomar, at the end of February 2009. Asilomar was selected as a site for the meeting primarily because it is simply a fabulous place for a reflective meeting. We also selected the site given the broad symbolism of the location. The AAAI Panel on Long-Term AI Futures resonated broadly with the 1975 Asilomar meeting by molecular biologists on recombinant DNA—in terms of the *high-level goal of social responsibility for scientists*. The AAAI panel co-chairs also alluded to the goal of generating a report on an assessment and recommendations that would be similar to the 1975 recombinant DNA report in terms of the crispness, digestibility, and design for consumption by scientists and the public alike. However, the symbolism stops there: The context and need for the AAAI study differs significantly and in multiple ways from the context of the 1975 meeting on recombinant DNA. In 1975, molecular biologists needed urgently to address a fast-paced set of developments that had recently led to the ability to modify genetic material. The 1975 meeting took place amidst a recent moratorium on recombinant DNA research. In stark contrast to that situation, the context for the AAAI panel is a field that has shown relatively graceful, ongoing progress. Indeed, AI scientists openly refer to progress as being somewhat disappointing in its pace, given hopes and expectations over the years. However, we are seeing ongoing advances in the prowess of AI methods and an acceleration in the fielding of real-world applications (some quite large in scale), a natural increase of reliance on automation, the coming availability of sophisticated methods to a wider set of developers, extending well outside the research community (e.g., in the form of a variety of toolkits), and a growing interest and focus among non-experts on radical outcomes of AI research. On the latter, some panelists believe that the AAAI study was held amidst a *perception* of urgency by non-experts (e.g., a book and a forthcoming movie titled "*The Singularity is Near*"), and focus of attention, expectation, and concern growing among the general population.

The panel has identified multiple opportunities for proactive reflection, focused research, and ongoing sensitivity and attention. We believe that focusing effort as a community of AI scientists on potential societal issues and consequences will ensure the best outcomes for AI research, enabling society to reap the maximal benefits of AI advances.

Chapter 15A
Itamar Arel on Horwitz's "AAAI Presidential Panel on Long-Term AI Futures"

The term Artificial Intelligence (AI), which has been coined over 50 years ago, was followed by era of optimism and, in retrospect, great naiveness regarding the field's prospect. Despite the many impressive achievement in building complex computerized systems, such as robots that can drive a car and programs that play chess at grandmaster level, the holy grail of building a machine with human-level intelligence remains an unfulfilled dream. In many ways, the report of the 2009 AAAI Presidential Panel on Long-Term AI is a solid reflection of both the disappointment and frustration from the lack of a much-anticipated AI breakthrough. In the absence of the latter, one is inevitably confined to the current, somewhat narrow, interpretation of AI systems. However, it is important to understand that while a conceptual AI breakthrough has yet to materialize, it may very well occur in the not so distant future. When such a breakthrough does take place it will undoubtedly have profound impact on our lives in ways which are difficult to imagine at this time.

The scientific fronts that many feel offer the most promise when it comes to revolutionizing the field of AI are computational neuroscience and neuropsychology. Great advances have been made in these areas over the past couple of decades with many anticipating continuing emergence of novel insight, which will contribute to our understanding of the mammalian brain. Recent progress in developing experimentally-grounded cognitive models, such as accurate descriptions of cortical circuitry, not only deepens our knowledge of how the brain works but also inspires researchers to propose new ideas pertaining to intelligent systems design. Hence, biological inspiration, rather than explicitly reverse engineering biological circuits, seems a promising approach for moving forward in mimicking cognitive functionality using machines. An example of a machine learning niche that emerged as a result of recent neuroscientific findings is deep machine learning, which employs hierarchical architectures for multi-modal perception in a manner resembling that of the neocortex.

As one would assert from reading this book, there are different interpretations of the singularity, many of which pertain to machines reaching a critical intelligence level beyond which predicting the future of humanity becomes impractical. Regardless of whether human-level intelligent machines will be the result of a singular AI breakthrough or not, now is the time to consider its various implications, particularly in guaranteeing that such a technology is not exploited maliciously. Rather than discarding the possibility of human-level machine intelligence, we may begin by asking key questions that assume its plausibility. Questions such as: how much time will humanity have before it will be impossible to control the evolutionary trajectory of this new life form on earth? Would it be possible to prevent the technology from reaching adversarial entities, and if so how can we effectively enforce policies that achieve this important goal? Human-level intelligence does not necessarily imply human-like intelligence; to that end, how would these new creatures behave, particularly as they interact with humans? Would it be possible at all to prevent a catastrophe for humanity as a result, for example, of a grand existential conflict of interests? The questions above cannot be comprehensively answered at this time, suggesting that open-mindedness to multiple futuristic scenarios is the logical position to take. Simply dismissing such scenarios may prove one day to be a historical mistake of epic proportions.

Chapter 15B
Vernor Vinge on Horvitz's "AAAI Presidential Panel on Long-Term AI Futures"

Points in the Panel report I especially agree with:

- Explicit AI research is just part of a very large human undertaking. (The demand for progress in computation, communication, and automation is coming from almost all directions.)
- Improving software and hardware for collaborations is important. (In my opinion, it is the most important measure—along with the ongoing conversation about these issues—for assuring a good outcome.)

The Singularity is very different from concerns such as climate change. For one thing, talk about the Singularity is only in part about avoiding disaster. No one need cry in the wilderness for help on this: as time passes, more and more people will be involved in the endeavor, mostly for happy reasons.

Chapter 16
Why the Singularity Cannot Happen

Theodore Modis

Abstract The concept of a Singularity as described in Ray Kurzweil's book cannot happen for a number of reasons. One reason is that all natural growth processes that follow exponential patterns eventually reveal themselves to be following S-curves thus excluding runaway situations. The remaining growth potential from Kurzweil's "knee", which could be approximated as the moment when an S-curve pattern begins deviating from the corresponding exponential, is a factor of only one order of magnitude greater than the growth already achieved. A second reason is that there is already evidence of a slowdown in some important trends. The growth pattern of the U.S. GDP is no longer exponential. Had Kurzweil been more rigorous in his fitting procedures, he would have recognized it. Moore's law and the Microsoft Windows operating systems are both approaching end-of-life limits. The Internet rush has also ended—for the time being—as the number of users stopped growing; in the western world because of saturation and in the underdeveloped countries because infrastructures, education, and the standard of living there are not yet up to speed. A third reason is that society is capable of auto-regulating runaway trends as was the case with deadly car accidents, the AIDS threat, and rampant overpopulation. This control goes beyond government decisions and conscious intervention. Environmentalists who fought nuclear energy in the 1980s, may have been reacting only to nuclear energy's excessive *rate* of growth, not nuclear energy per se, which is making a comeback now. What may happen instead of a Singularity is that the rate of change soon begins slowing

Theodore Modis is a physicist, futurist, and international consultant. He is also the founder of Growth Dynamics, an organization specializing in strategic forecasting and management consulting.http://www.growth-dynamics.com/

T. Modis (✉)
Growth Dynamics, Via Selva 8, Massagno, 6900 Lugano, Switzerland
e-mail: tmodis@gmail.com

A. H. Eden et al. (eds.), *Singularity Hypotheses*, The Frontiers Collection,
DOI: 10.1007/978-3-642-32560-1_16, © Springer-Verlag Berlin Heidelberg 2012

down. The exponential pattern of change witnessed up to now dictates more milestone events during year 2025 than witnessed throughout the entire 20th century! But such events are already overdue today. If, on the other hand, the change growth pattern has indeed been following an S-curve, then the rate of change is about to enter a declining trajectory; the baby boom generation will have witnessed more change during their lives than anyone else before or after them.

Background

In 2005 together with four other members of the editorial board of *Technological Forecasting and Social Change* I was asked to review Ray Kurzweil's book *The Singularity Is Near.* The task dragged me back into a subject that I was familiar with. In fact, ten years earlier I had thought I was the first to have discovered it only to find out later that a whole cult with increasing number of followers was growing around it. I took my distance from them because at the time they sounded nonscientific. I published on my own adhering to a strictly scientific approach (Modis 2002a, 2003). But to my surprise the respected BBC television show HORIZON that became interested in making a program around this subject found even my publications "too speculative". In any case, for the BBC scientists the word singularity is reserved for mathematical functions and phenomena such as the big bang.

Kurzweil's book constitutes a most exhaustive compilation of "singularitarian" arguments and one of the most serious publications on the subject. And yet to me it still sounds nonscientific. Granted, the names of many renowned scientists appear prominently throughout the book, but they are generally quoted on some fundamental truth other than the direct endorsement of the so-called singularity. For example, Douglas Hofstadter is quoted to have mused that "it could be simply an accident of fate that our brains are too weak to understand themselves." Not exactly what Kurzweil says. Even what seems to give direct support to Kurzweil's thesis, the following quote by the celebrated information theorist John von Neumann "the ever accelerating process of technology…gives the appearance of approaching some essential singularity" is significantly different from saying "the singularity is near". Neumann's comment strongly hints at an illusion whereas Kurzweil's presents a far-fetched forecast as a fact.

What I want to say is that Kurzweil and the singularitarians are indulging in some sort of pseudo-science, which differs from real science in matters of methodology and rigor. They tend to overlook rigorous scientific practices such as focusing on natural laws, giving precise definitions, verifying the data meticulously, and estimating the uncertainties.

The work I present here uses a number of science-based approaches to argue against the possibility that the kind of singularity described by Kurzweil in his book will take place during the 21st century. I will concentrate on the near future because horizons of several hundred years permit and are more appropriate for fantasy scenarios that tend to satisfy the writer's urge for sci-fi prose.

There are No Exponentials. There are Only S-Curves

Exponential Versus S-Curve

The law describing natural growth has been put into mathematical equations called growth functions. The simplest mathematical growth function is the so-called logistic. It is derived from the law, which states that the rate of growth is proportional to both the amount of growth already accomplished and the amount of growth remaining to be accomplished. If either one of these quantities is small, the rate of growth will be small. This is the case at the beginning and at the end of the process. The rate is greatest in the middle, where both the growth accomplished and the growth remaining are sizable. This natural law described with words here has been cast in a differential equation (the logistic equation) the solution of which gives rise to an S-shaped pattern (S-curve). In Eq. (16.1) below M is the value of the final ceiling, t_0 is the time of the midpoint and α reflects the steepness of the rising slope.

$$X(t) = \frac{M}{1 + e^{-\alpha(t-t_0)}} \tag{16.1}$$

It is easy to see that for t large and positive the population $X(t)$ tends to M. Similarly for t large and negative the expression reduces to a simple exponential.

It is a remarkably simple and fundamental law. It has been used by biologists to describe the growth in competition of a species population, for example, the number of rabbits in a fenced off grass field. It has also been used in medicine to describe the diffusion of epidemic diseases. J. C. Fisher and R. H. Pry refer to the logistic function as a diffusion model and use it to quantify the spreading of new technologies into society (Fisher and Pry 1971). One can immediately see how ideas or rumors may spread according to this law. Whether it is ideas, rumors, technologies, or diseases, the rate of new occurrences will always be proportional to how many people have it and to how many don't yet have it.

The analogy has also been pushed to include the competitive growth of inanimate populations such as the sales of a new successful product. In the early phases, sales go up in proportion to the number of units already sold. As the word spreads, each unit sold brings in more new customers. Sales grow exponentially. It is this early exponential growth which gives rise to the first bend of the S-curve. Business looks good. Growth is the same percentage every year—exponential growth—and hasty planners prepare their budgets that way. Growth, however, cannot be exponential. Explosions are exponential. Natural growth follows

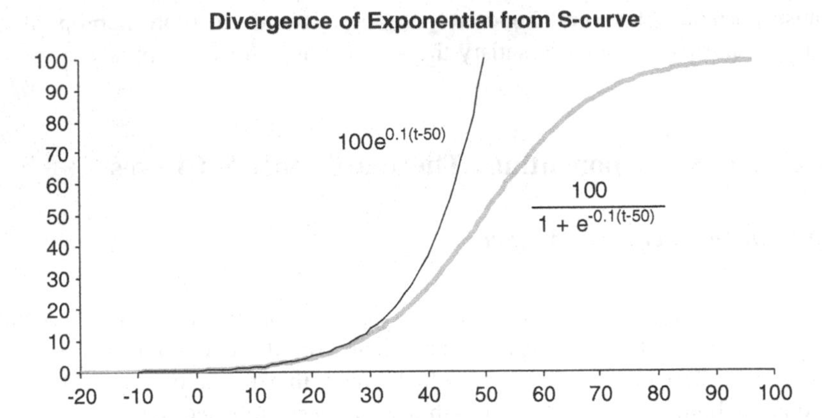

Fig. 16.1 The construction of a theoretical S-curve (*gray line*) and the exponential (*thin black line*) it reduces to as time goes backward. The formulae used are shown in the graph

S-curves, but all S-curves behave like exponentials early on. Kurzweil himself addresses this issue, which he calls "The Criticism from Malthus" toward the end of his book. He admits that the exponential patterns he publicizes will eventually turn into S-curves, but dismisses the fact because, as he claims, this will happen so far into the future—after the Singularity—that for all practical purposes it becomes irrelevant. (Elsewhere, Kurzweil acknowledges that there are smaller S-curves that saturate early, but argues that they are replaced by other small S-curves thus cascading indefinitely. He does not seem to be aware of the fact that there is a fractal aspect to such cascades and constituent S-curves are bound by envelope S-curves, which themselves saturate with time, see discussion in section Fractal Aspects of Natural Growth below.)

Let us see whether it is possible to estimate when exponential trends can be expected to turn into S-curves. First let us see at what time the S-curve deviates from the exponential pattern in a significant way, see Fig. 16.1. Table 16.1 below quantifies the deviation between a logistic and the corresponding exponential pattern as a fraction of the S-curve's final ceiling. By "corresponding" exponential I mean the limit of Eq. (16.1) as $t \to -\infty$.

In Table 16.1 we appreciate the size of deviation between exponential and S-curve patterns as a function of how much the S-curve has proceeded to completion. Obviously beyond a certain point the difference becomes flagrant. When exactly this happens maybe subject to judgment so Table 16.1 is there to quantitatively help readers make up their mind. Most readers will agree that a 15 % deviation between exponential and S-curve patterns is significant because it makes it clear that the two processes can no longer be confused. This happens when the S-curve that corresponds to the exponential has reached about 13 % of its ceiling level. In other words, the future ceiling that caps this growth process is about 7 times today's level.

This moment when an exponential pattern begins deviating significantly from an S-curve also defines an upper limit for Kurzweil's so called "exponential

Table 16.1 The deviation between exponential and S-curve patterns as a function of how much the S-curve has proceeded to completion

Deviation (%)	Penetration (%)
11.1	10.0
12.2	10.9
13.5	11.9
15.0	13.0
16.5	14.2
18.3	15.4
20.2	16.8
22.3	18.2
24.7	19.8
27.3	21.4
30.1	23.1
33.3	25.0
36.8	26.9
40.7	28.9
44.9	31.0
49.7	33.2
54.9	35.4
60.7	37.8
67.0	40.1
74.1	42.6
81.9	45.0
90.5	47.5
100.0	50.0

knee". Of course, exponential patterns do not have knees (this can be trivially demonstrated with a logarithmic plot where the exponential pattern becomes a straight line from $t = -\infty$ to $t = +\infty$). What Kurzweil sees as the moment an exponential "abruptly" rises will move toward the future (or the past) as we increase (or decrease) the vertical scale of a linear plot. But if we interpret Kurzweil's knee as the moment when a growth process still following an exponential pattern begins having *very serious impact* on society—a subjective definition carrying large uncertainties—then there will be at least a 7-fold increase remaining before the process stops growing.

It has been theoretically demonstrated that fluctuations of a chaotic nature begin making their appearance as we approach the ceiling of an S-curve. This is evidence of the intimate relationship that exists between growth and chaos (the logistic equation in discrete form becomes the chaos equation) (Modis 2007). In an intuitive way these fluctuations can be seen as the stochastic search of the system for equilibrium around a homeostatic level.

But it has been argued that fluctuations of a chaotic nature may also precede the steep rising phase of the logistic (Modis and Debecker 1992). The intuitive understanding of these fluctuations is "infant" mortality. In all species survival is uncertain during the early phases of the life cycle.

Infant mortality and common sense can help us establish a lower limit for Kurzweil's knee. Any natural growth process that has achieved less than 10 % of its final growth potential cannot possibly have a serious impact on society. In fact 10 % growth is usually taken as the limit of infant mortality. A tree seedling of height less than 10 % of the tree's final size is vulnerable to rabbits and other herbivores or simply to be stepped on by a bigger animal.

Below we look at some real cases. They all corroborate a lower limit of the order of 10 % below which the impact on society cannot be considered *very serious*.

US Oil Production

A real case is the production of oil in the United States, shown in Fig. 16.2. Serious oil production in the United States began in the second half of the 19th century. During the first one hundred years or so cumulative oil production followed an exponential pattern. But soon it became clear that the process followed a logistic growth pattern—S-curve—and rather closely, see Fig. 16.2. If we try to fit an exponential function to the data, we obtain a good fit—comparable in quality with the logistic fit—only on the range 1859–1951. As we stretch this period beyond 1951 the quality of the exponential fit progressively deteriorates. S-curve and exponential begin diverging from the 1951 onward, which corresponds to around 20 % penetration level of the S-curve.[1]

If we were to position a "knee" *a la* Kurzweil on this exponential pattern, it could by some thinking be in the early 1930s time by which almost all horses had been replaced by cars in personal transportation maximizing the demand for oil (Modis 1992). At that time the penetration level of the S-curve was around 7 %. Consequently on this growth process, which seemed exponential for the better part of one hundred years, there is a ceiling waiting at about 14 times the knee's production level.

Moore's Law

The celebrated Moore's Law is a growth process that has been evolving along an exponential growth pattern for four decades. The number of transistors in Intel microprocessors has doubled every two years since the early 1970s. But it is now unanimously expected that this growth pattern will eventually turn into an S-curve and reach a ceiling. On page 63 of his book Kurzweil claims that Moore's law is one of the many technological exponential trends whose knee we are approaching.

---

[1] The fitted exponential here is not quite the same as the exponential that the S-curve tends to as $t \to \infty$.

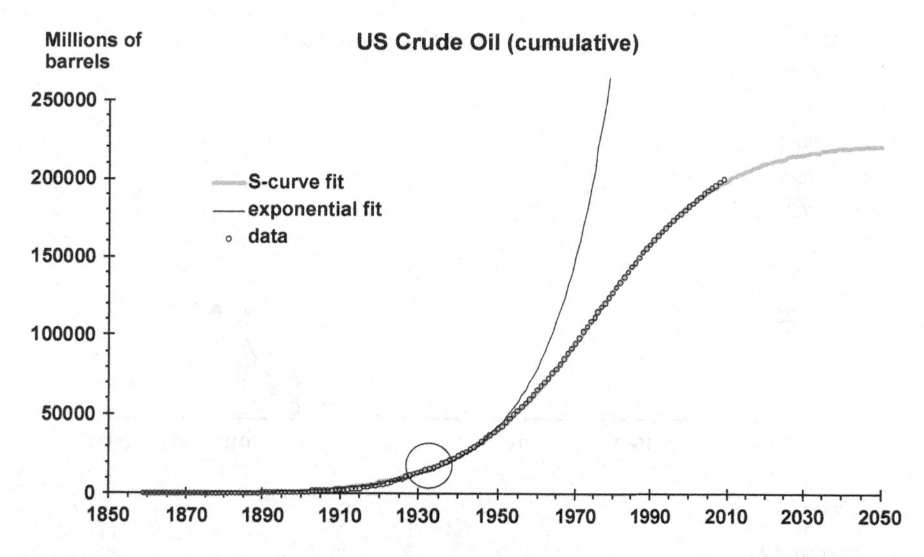

Fig. 16.2 S-curve and exponential fit on yearly data points. The circle indicates a possible position for the "knee" of the exponential; it lies at the 7 % penetration level of the S-curve. *Data Source* U.S. Energy Information Administration (*EIA*)

But he also agrees that Moore's law will reach the end of its S-curve before 2020. Moore himself agrees, "no physical thing can continue to change exponentially forever," he says and positions an end for this phenomenon around 2015. But he still expects three more generations (we should mention that in 1995 Moore had consistently expected 5 more generations.)

It must be pointed out—particularly for those who claim that every time people thought a limit was reached in the past new ways were found to cram more transistors together—that in the very first formulation of Moore's law in 1965 the doubling was every year. David House, an Intel executive, raised it to 18 months, and later in 1975 Moore himself raised it to two years. These successive adjustments may not constitute proof but the fact that we are dealing with an eventual S-curve cannot be disputed. Given that we are dealing with an S-curve, the slowing down must be gradual so that three generations may bring an overall increase with respect to today's numbers by a factor smaller than $2^3 = 8$. But even if the factor is 8, today's level (which Kurzweil argues is the exponentials "knee") corresponds to around 12.5 % of the S-curve's ceiling.

Figure 16.3 shows an S-curve fit on the data, which has been constrained to reach a ceiling by the late 2020s (a conservative constraint). The corresponding exponential is also shown as was done in Fig. 16.1. The expected announcement by Intel of the Poulson processor in 2012 argues in favor of the S-curve and against the exponential trend.

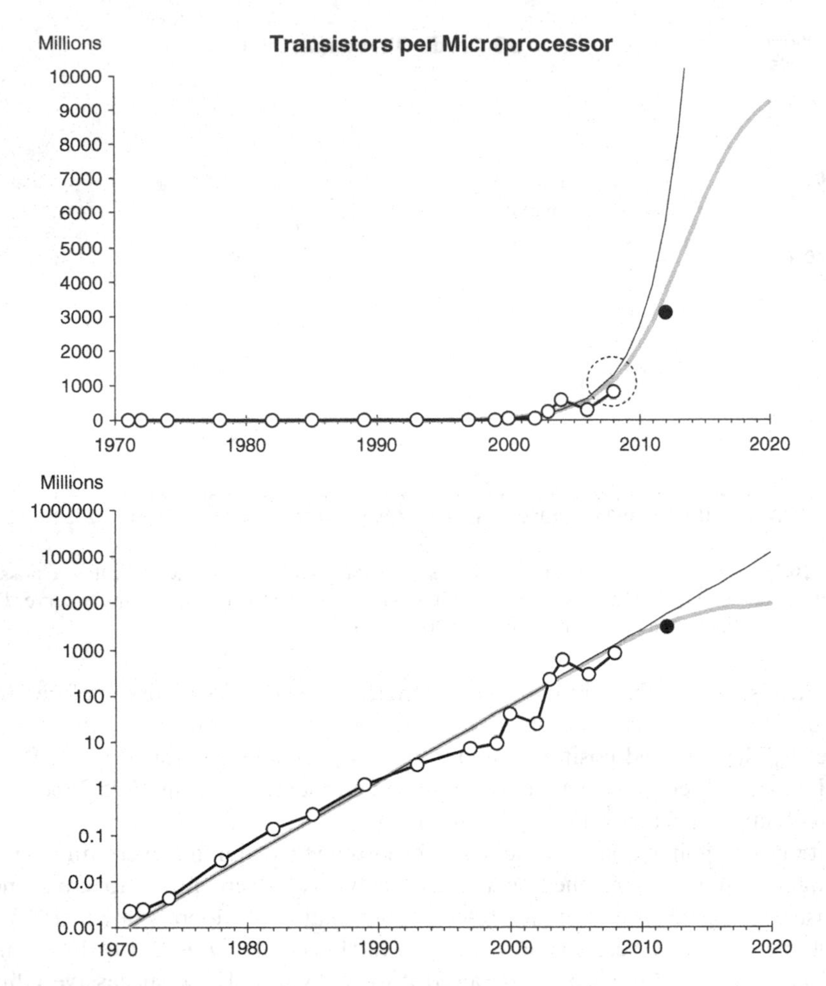

Fig. 16.3 Moore's law with linear scale (*above*) and logarithmic scale (*below*). Exponential (*black line*) and S-curve (*gray line*) begin to diverge in the *top graph* around the *big dotted circle* (penetration of 11 %) and a moment that could serve as a candidate for the "knee". The *black dot* indicates the expected announcement by Intel of the Poulson processor. *Data sources* Intel and Wikipedia

World Population

The evolution of the world population during the 20th century followed an S-curve in an exemplary way. Figure 16.4 shows an agreement between data and curve that is astonishing if we consider the variety of birth rates and death rates around the world, the multitude of stochastic processes that impact the evolution of the world population such as epidemics, catastrophes, wars (twice at world level), important climate changes, etc., and on the other hand the simplicity of the curve's description, namely only three parameters (plus a parameter for a pedestal in this case).

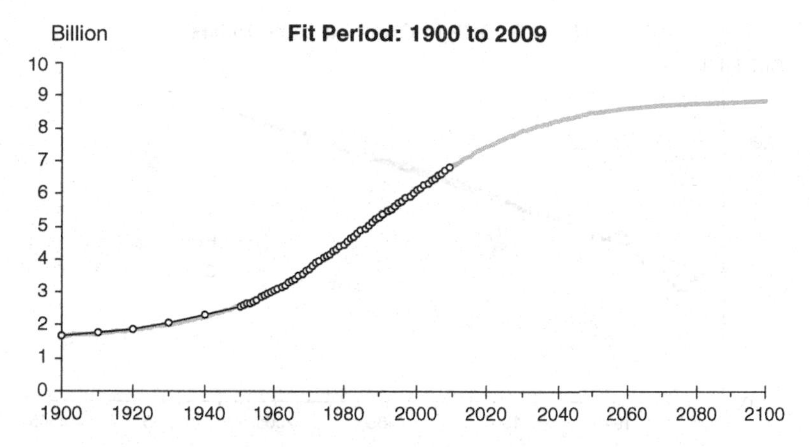

Fig. 16.4 The evolution of the world population during the last 109 years has followed an exemplary S-curve. *Data sources* United Nations Department of Economic and Social Affairs (UN DESA) and U.S. Census Bureau

And yet again, the evolution of the world population has often been likened to an explosion following an exponential trend. Where could a "knee" for this exponential be positioned? Looking at Fig. 16.4, any "knee" would have to be positioned before the 1980s by which time the trend significantly deviated from an exponential pattern.

By some historians the population explosion began in the West, around the middle of the 19th century. The number of people in the world had grown from about 150 million at the time of Christ to somewhere around 700 million in the middle of the 17th century. But then the rate of growth increased dramatically to reach 1.2 billion by 1850. In this case the exponential "knee" would have occurred when world population reached 8 % of its final ceiling.

In the above three examples—US Oil Production—, Moore's law, and World Population—we have seen that the "knee" of the exponential curve tends to occur at a threshold situated between 7 and 13 % of the ceiling of the corresponding S-curve, which translates to a factor of at most 14 between the level of the "knee" and the level of the final ceiling. This factor has been estimated rather conservatively and corroborates the previously mentioned corollary of natural-growth studies: infant mortality.

Such growth potential on any of the variables purported to contribute to a singularity by mid 21st century would fall short of becoming alarming.

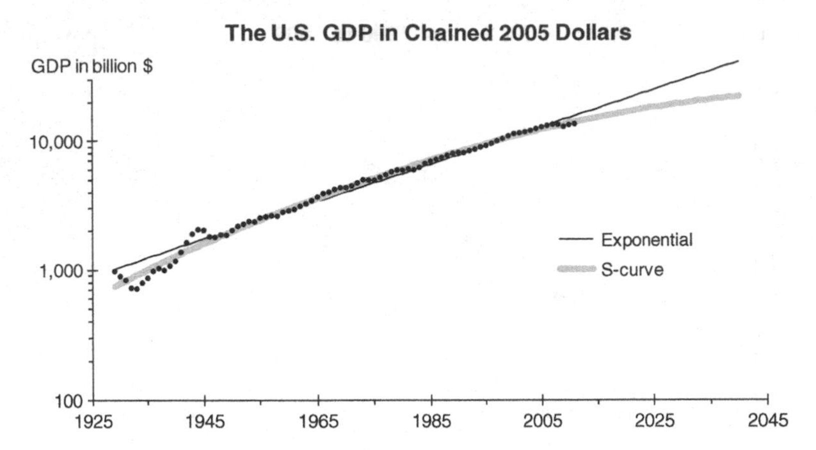

Fig. 16.5 Exponential (*thin black line*) and logistic (*thick gray line*) fits on the evolution of the real U.S. GDP. This graph can be directly compared with the one on Page 98 from Kurzveil's *The Singularity Is Near. Data source* U.S. Department of Commerce, Bureau of Economic Analysis

There is Already Some Evidence for Saturation

The U.S. GDP

The evolution of the gross domestic product (GDP) in America in constant dollars had followed an exponential growth pattern for a while but has deviated from it some time ago and now has almost reached the midpoint of an S-curve. Kurzveil was too quick to pronounce its evolution as exponential. On logarithmic scale Kurzveil's straight wide band accommodated the gentle curving of the time-series data and his criterion of correlation—coefficient r^2—was close enough to 1.0. But high correlation between two curves does not mean one is a good representation of the other. Just think of two straight lines, one almost vertical the other almost horizontal; they will be 100 % correlated (correlation coefficient 1.0) and yet one will be a very poor representation of the other. A closer examination reveals that a logistic function fits better the evolution of the U.S. GDP than an exponential one, see Fig. 16.5. And if we judge the fits by their Chi Square (χ^2), a more appropriate criterion than correlation coefficients, the logistic fit comes out three times better than the exponential one (χ^2 of 1112 instead of 3250). Conclusion: the U.S. GDP will certainly not contribute to the building up toward a singularity event around 2045 because from 2013 onward its rate of growth will progressively slow down.

The End of the Internet Rush

One of the "explosive" variables in Kurzweil book is that of the diffusion of the Internet, but the graph on Page 79 of his book shows clear evidence that we are dealing with an S-curve developed about half-way to its ceiling. What is being witnessed instead is the end of the Internet rush (Modis 2005). In an article published in 2005 I demonstrate that the number of Internet users will not grow much in the near future. In the US the ceiling of the S-curve has already been reached at 72 % of the population, and in the E.U. it should not rise much above today's 67 %. In the rest of the world today's 18 % will grow to a ceiling of 33 % in ten years.

It would be unreasonable to expect the percentage of the rest of the world to remain at this low level forever. The rest of the world includes such countries as Japan, Korea, Honk Kong, and Australia where the number of Internet users is already practically at maximum. But the rest of the world also includes Africa, China, and India, where one can be certain that the number of Internet users will eventually grow by a large factor. However, it will be some time before the necessary infrastructures are put in place there to permit large-scale Internet diffusion.

For the time being one may infer that the boom we have been witnessing in Internet expansion is over. The parts of the world that were ready for it have practically filled their niches whereas the parts of the world that were not ready for it need much preparatory work (infrastructures, nourishment, education, etc.) and will therefore grow slowly.

The final percentage of Internet users may also reflect cultural differences. A percentage of 72 % in the US compared to 67 % in the E.U. might partially reflect missing infrastructures in some of the lesser-developed E.U. countries but most likely also reflects the different life styles. European society admits less change than American society. For example, there are fewer cars per inhabitant in Europe, and the Europeans never went to the moon. They will probably end up using the Internet somewhat less than the Americans.

We can make a rough estimate of when a follow-up Internet growth phase should be expected. To a first approximation logistics that cascade harmoniously show periods of low and high growth of comparable duration (Modis 2007). Accordingly, and given that Internet has had a decade of rapid growth, a decade of low growth can reasonably be expected before a new S-curve begins. Contrary to the image of an explosive uncontrollable growth process we are witnessing piecemeal growth with stagnating periods in between that offer fertile ground for control and adaptability.

In the next section we quantify the cascade of S-curves in a *natural* succession and the relationship between their life cycles.

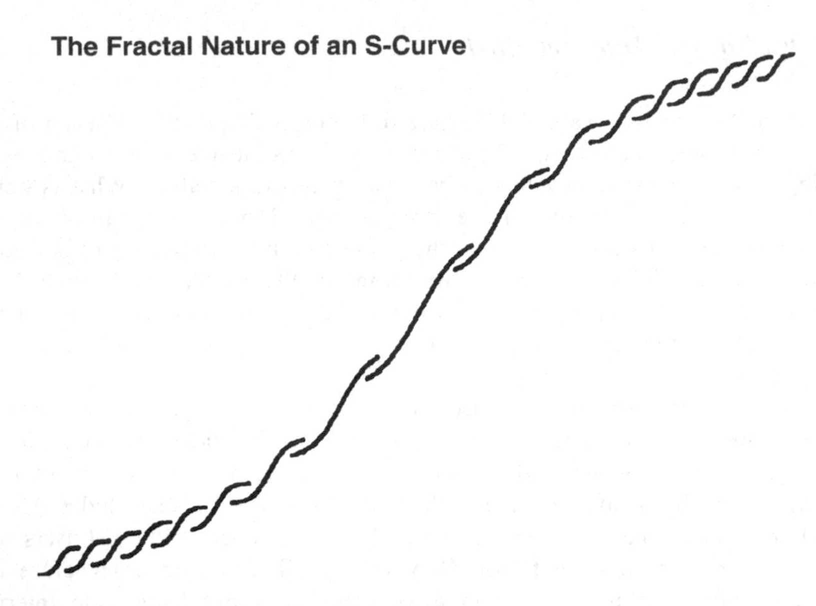

Fig. 16.6 A large S-curve decomposed into smaller ones

Fractal Aspects of Natural Growth

Sustained growth is not a steady and uniform process. It consists of successive S-shaped steps, each of which represents a well-defined amount of growth. A new S-curve begins where the previous one left off. Every step is associated with a niche that opened following some fundamental change (a mutation, a major innovation, a technological break-through, etc.).

Successive growth stages depicted by cascading logistic curves may outline an overall growth process that is itself amenable to a logistic description. Two such examples from published work are the discovery of stable chemical elements, which spanned three centuries and can be broken down into four rather distinct regularly-spaced growth phases (Solla and Derek 1936), and the diffusion of Italian railways which came in three waves (Marchetti 1986).

The graph of Fig. 16.6 has been constructed in a rigorous quantitative way (Modis 1994). Notice that the step size and the associated time span (life cycle) of the constituent S-curves first progressively increases but then progressively decreases going over a maximum step and longest life cycle around the middle of the envelope S-curve. Life cycles become longer during the high-growth period and shorter during the low-growth periods.

The phenomenon of shrinking product life cycles, an important concern of marketers, can be quantitatively linked to the saturation of the enveloping process. Table 16.2 relates the shortening of product life cycles to the level of saturation reached.

Table 16.2 The relation between the shortening of life cycles and saturation

Life-cycle length (*relative to longest*)	Level of saturation (*percent of ceiling*)
0.17	3.1
0.19	4.0
0.20	5.2
0.22	6.9
0.24	9.1
0.30	12.8
0.41	20.0
0.70	31.4
1.00	50.0
0.70	68.6
0.41	80.0
0.30	87.2
0.24	90.9
0.22	93.1
0.20	94.8
0.19	96.0
0.17	96.9

The idea that a growth process can, on closer examination, reveal similar but smaller cascading growth processes, suggests a fractal nature for the logistic curve. The implication is that further "zooming-in" may reveal an even finer structure of logistic cascades.

The fractal nature of logistics permits one to estimate the level of overall saturation from observing life cycle trends. This is a powerful approach. Table 16.2, first published in Modis 1994, can be used by anyone to determine the position on the envelope curve from observations on constituent curves. For example, a non-specialist, such as a laborer loading boxes into trucks off the dock of a computer manufacturer, may notice that the names on the boxes change three times as frequently as they used to and deduce that the technology behind these products is about 87 % exhausted. This image may be naive, but the approach offers valuable insights for those situations in which the tracking of the overall envelope is rather imponderable, for example, the evolution of the Windows operating systems.

Windows Operating Systems

Microsoft has been regularly releasing new operating systems. New software developments triggered by hardware improvements make it necessary to introduce major changes to the operating systems. This is a typical characteristic of all new industries like microchips; they are mutational. But as the industry matures, "mutations" become rare and life cycles become longer. Figure 16.7 shows the

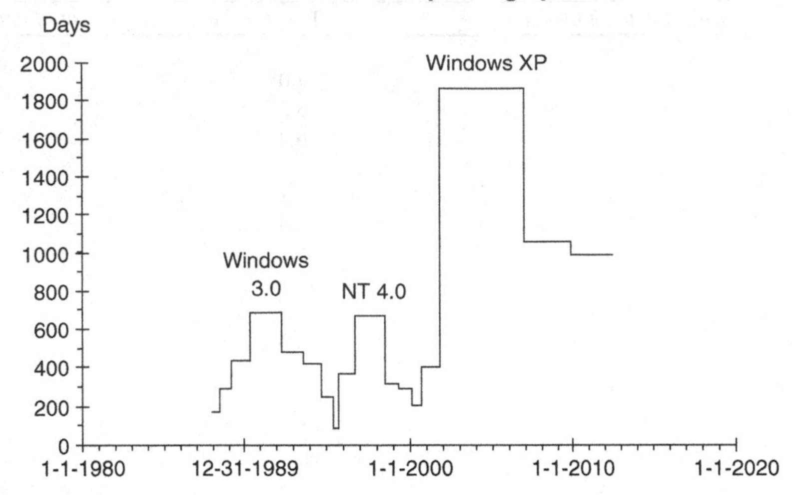

Fig. 16.7 Lifetimes of Windows Operating Systems as defined by the time period to the next release. Labels highlight the three longest-lived ones. *Data sources* Microsoft and Wikipedia

evolution of the life cycle of Microsoft operating system as defined by the time to the next launching announcement.

Windows XP was the operating system with the longest life cycle. Vista's life cycle was 57 % of that of Windows XP and the life cycle of Windows 7 will be 53 % if Windows 8 comes in 2012. From Table 16.2 we can estimate that the Windows technology will be between 68.6 and 80 % exhausted by that time.

With Windows XP at the center of the envelope S-curve for Windows operating systems and the life cycle being symmetric to a first approximation, we can reasonably expect an end for this growth process by the late 2020s. This coincides with our previous estimate for the end of Moore's law.

Just as with Internet users, here again we are facing an upcoming lull in the rate of growth of such "explosive" processes as microchip and PC operating system improvements. This lull should also last about 20 years, duration comparable to the duration of the rapid-growth phase of the envelope S-curve. From then (ca 2030) onward a new sequence of cascading S-curves may slowly enter the scene as per Fig. 16.6. But once again, we have a hard time accommodating singular events due to "explosive" trends like these by mid 21st century.

Society Auto-Regulates Itself

There have been many documented cases where society has demonstrated a wisdom and control unsuspected by its members. In this section we will see four such examples portraying society as a super species capable of auto-regulating and safeguarding itself from runaway trends.

Car Accidents

Car accidents have received much attention and at times provoked emotional reactions. In the 1960s cars had been compared to murder weapons. To better understand the mechanisms at play we must look at the history of car accidents. We need accurate data and the appropriate indicator. Deaths are better defined and recorded than lesser accidents. Moreover, the car as a public menace is a threat to society, which may "feel" the pain and try to keep them under control. Consequently, the number of deaths per one hundred thousand inhabitants per year becomes a better indicator than accidents per mile, or per car, or per hour of driving (Marchetti 1983).

The data shown in Fig. 16.8 are for the United States starting at the beginning of the 20th century. What we observe is that deaths caused by car accidents grew along an exponential trend that led into an S-curve that reached a ceiling in the mid 1920s, when deaths reached twenty-four per one hundred thousand per year. From then onward car accidents have stabilized even though the number of cars continued to grow. A homeostatic mechanism seems to enter into action when this limit is reached, resulting in an oscillating pattern around the equilibrium position. The peaks may have produced public outcries for safety, while the valleys could have contributed to the relaxation of speed limits and safety regulations. What is remarkable is that for over sixty years there has been a persistent auto-regulation on car safety in spite of tremendous variations in car numbers and performance, speed limits, safety technology, driving legislation, and education.

Why the number of deaths is maintained constant and how society can detect excursions away from this level? Is it conceivable that someday car safety will improve to the point of reducing the level of automobile accidents to practically null the way it was before cars were invented? American society has tolerated this level of accidents for the better part of a century. A Rand analyst has described it as follows: "I am sure that there is, in effect, a desirable level of automobile accidents—desirable, that is, from a broad point of view, in the sense that it is a necessary concomitant of things of greater value to society." (Williams 1958). Abolishing cars from the roads would certainly eliminate car accidents, but at the same time it would introduce more serious hardship to citizens.

A homeostatic equilibrium represents a state of well-being. It has its roots in nature, which develops ways of maintaining it. Individuals may come forward from time to time as advocates of an apparently well-justified cause. What they do not suspect is that they may be acting as unwitting agents to deeply rooted necessities for maintaining the existing balance, which would have been maintained in any case. An example is Ralph Nader's crusade for car safety, *Unsafe at Any Speed*, published in the 1960s, by which time the number of fatal car accidents

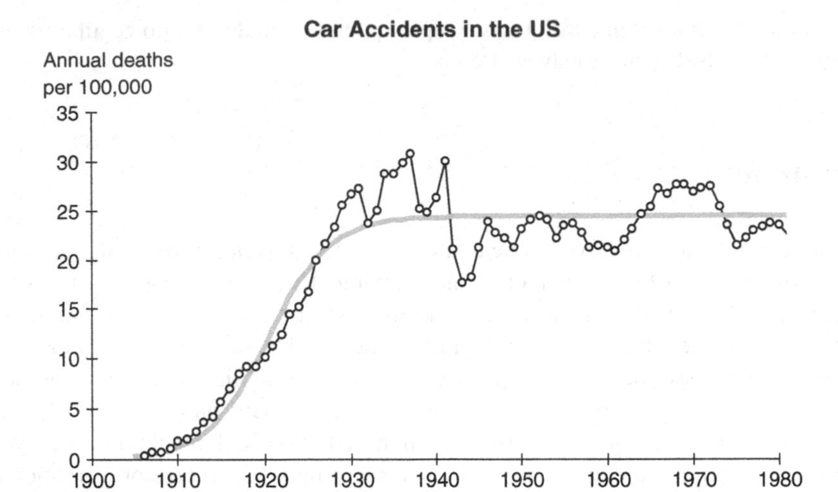

Fig. 16.8 The annual number of deaths from motor vehicle accidents per 100,000 population has been fluctuating around 24 since the mid 1920s. The peak in the late 1960s provoked a public outcry that resulted in legislation making seat belts mandatory. Data after 1980 show a decline in the number of deadly car accidents, but this is due the fact that travelers have been replacing the automobile by other means of transportation and in particular the airplane, see discussion in Modis 1992, 2002b. *Data source* Statistical Abstract of the United States

had already demonstrated a forty-year-long period of relative stability. But examining Fig. 16.8 more closely, we see that the late 1960s show a relative peak in accidents, which must have been what prompted Nader to blow the whistle. Had he not done it, someone else would have. Alternatively, a timely social mechanism might have produced the same result; for example, an "accidental" discovery of an effective new car-safety feature.

Another such example of society's ability to auto-regulate and safeguard itself is the spreading of AIDS in the United States.

The AIDS Niche

At the time of the writing of my first book *Predictions,* AIDS was diffusing exponentially claiming a progressively bigger share of the total number of deaths every year, and forecasts ranged from pessimistic to catastrophic. Alarmists worried about the survival of the human species. But finally the AIDS "niche" in the U.S. turned out to be far smaller than that feared by most people. In this case the variable studied was death victims from AIDS as a percentage of all deaths.

The S-curve I fitted on the data up to and including 1988 had indicated a growth process that would be almost complete by 1994. The ceiling for the disease's growth, projected as 2.3 % of all deaths was projected to be reached in the late

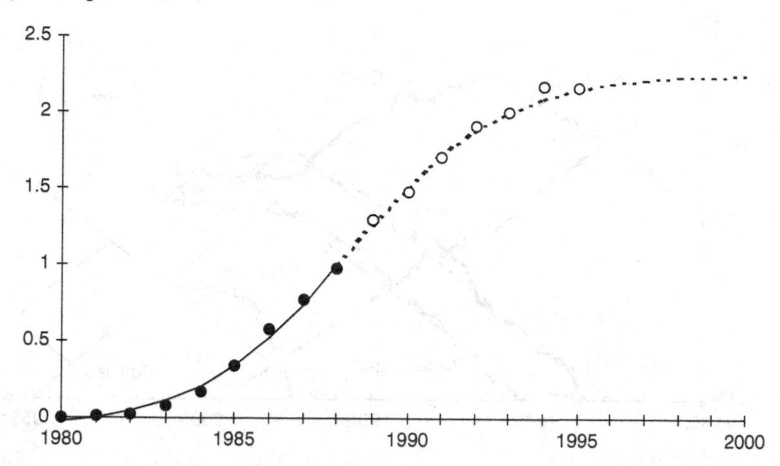

Fig. 16.9 Deaths from AIDS in the United States. The ceiling of the S-curve fitted on data up to and including 1988 (*black dots*) is 2.3. Data from years 1989–1998 (*open circles*) confirm that the AIDS niche in the United States was essentially completed by 1995 (Modis 1992 and Modis 2002b). *Data source* HIV/AIDS Surveillance, Centers for Disease Control, U.S. Department of Health and Human Services, Atlanta, GA

1990s, see Fig. 16.9. In other words my conclusion at that time was that a place had been reserved for AIDS in American society just above the 2 % level of all deaths.

The little circles in the figure confirm the S-curve trend and the completion of this microniche by 1995. By the late 1990s questions were being raised why forecasts had overestimated the AIDS threat by so much.

There seems to have been a mechanism that limited AIDS in a natural way even in the absence of adequate medication. As if there were other, more important causes of death. This mechanism may have reflected the control exercised by American society through subconscious collective concern. The natural-growth pattern that the disease followed from its beginning correctly anticipated that AIDS would not spread uncontrollably. Eventually of course there would be effective medication for the disease and the number of victims would decline. Those who had predicted imminent doom in the absence of a miracle drug in the 1980s had failed to take into account the natural competitive mechanisms which regulate the split of the overall death rate among the different causes of death, safeguarding all along an optimum survival for society.

After 1995 the number of deaths from AIDS progressively declined in what could be described as another natural process—a downward S-curve—reflecting the development of progressively effective medication. What Fig. 16.9 spells out as shown is society's ability to safeguard its wellbeing in the absence of effective medication and miracle drugs.

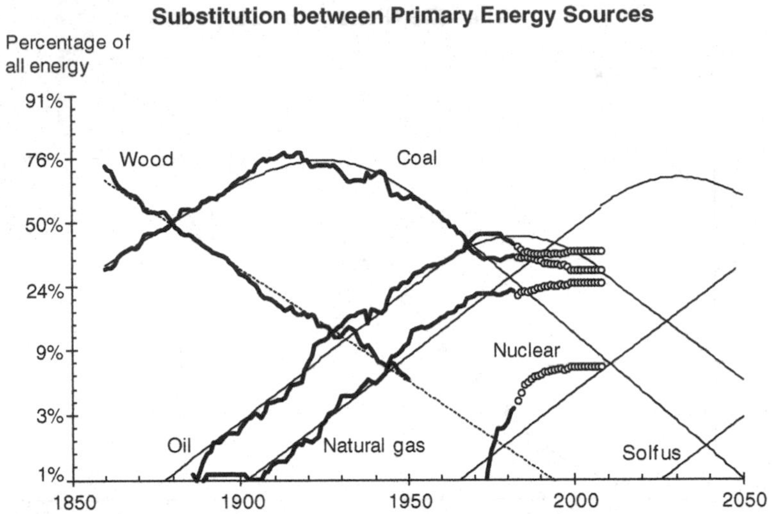

Substitution between Primary Energy Sources

Percentage of all energy

91% 76% 50% 24% 9% 3% 1%

Wood Coal Nuclear Oil Natural gas Solfus

1850 1900 1950 2000 2050

Fig. 16.10 Data, fits, and projections for the shares of different primary energy sources consumed worldwide. For nuclear, the smooth straight line is not a fit but a trajectory suggested by analogy. The futuristic source labeled "Solfus" may involve solar energy and thermonuclear fusion. The *small circles* show how things evolved since 1982 when this graph was first put together

A more subtle example of society's ability to auto-regulate and safeguard itself is primary-energy substitution and the advent of nuclear energy.

Nuclear Energy

During the last one hundred years, wood, coal, natural gas, and nuclear energy have been the main protagonists in supplying the world with energy. More than one energy source is present at any time, but the leading role passes from one to the other. Other sources of energy (such as hydro, wind, etc.) have been left out because they command too small a market share.

In the early 19th century and before, most of the world's energy needs were satisfied through wood burning and to a lesser extent animal power not considered here, see Fig. 16.10. The substitution process shows that the major energy source between 1870 and 1950 was coal. Oil became the dominant player from 1940 onward, as the automobile matured, together with petrochemical and other oil-based industries. The vertical scale of Fig. 16.10 is *logistic* transforming S-curves into straight lines. All straight sections in this figure would show up as S-shaped on a graph with a linear vertical scale.

It becomes evident from this picture that a century-long history of an energy source can be described quite well—thin smooth lines—with only two constants, those required to define a straight line. (The curved sections are calculated by subtracting the straight lines from 100 %.) The destiny of an energy source, then, seems to be cast during its early stages of phasing-in, as soon as the two constants describing the straight line can be determined.

A detailed description and the many ramifications of Fig. 16.10 are discussed in detail in the literature (Marchetti 1987; Modis 1992, 2009). Here I want to draw the reader's attention to the subtle ways with which society imposes its will. This graph was originally put together in 1988 and indicated that natural gas would progressively replace oil. Its update twenty years later (little circles) shows that the persistent consumption of coal has been at the expense of natural gas. This may not only be due to aggressively developing countries such as China who use coal ravenously. Developed countries such as the UK and the US have also proven reluctant to give up coal. Whoever the culprit, the widening gap between the persistent level of coal use and coal's naturally declining trajectory becomes a source of pressure to the system, which is likely to manifest itself through the voice of environmentalists.

Environmentalists in any way have been very vocal in their support of natural gas. I wonder, however, what has really been their role in the greening of natural gas. The importance of gas in the world market has been growing steadily for the last ninety years, independent of latter-day environmental concerns. The voice of environmentalists resembles Ralph Nader's crusade in the 1960s for car safety, while the number of deaths from car accidents had already been pinned around twenty-four annually per one hundred thousand population for more than forty years.

Environmentalists have also taken a vehement stand on the issue of nuclear energy. This primary energy source entered the world market in the mid 1970s when it reached more than 1 % share. The rate of growth during the first decade seems disproportionately rapid, however, compared to the entry and exit slopes of wood, coal, oil and natural gas, all of which conform closely to a more gradual rate. At the same time, the opposition to nuclear energy also seems out of proportion when compared to other environmental issues. As a consequence of the intense criticism, nuclear energy growth has slowed considerably, and it is not surprising to see the little circles in Fig. 16.10 approach the straight line proposed by the model. One may ask what the prime mover here was—the environmental concerns that succeeded in slowing the rate of growth or the nuclear-energy craze that forced environmentalists to react?

The coming to life of such a craze is understandable. Nuclear energy made a world-shaking appearance in the closing act of World War II by demonstrating the human ability to access superhuman powers. I use the word superhuman because the bases of nuclear reactions are the mechanisms through which stars generate their energy. Humans for the first time possessed the sources of power that feed our sun, which was often considered as a god in the past. At the same time mankind acquired independence; nuclear is the only energy source that would remain indefinitely at our disposal if the sun went out.

The worldwide diffusion of nuclear energy during the 1970s followed a rate that could be considered excessive, compared to previous energy sources. The market share of nuclear energy grew along a trajectory much steeper than the earlier *natural* ones of oil, natural gas, and coal. This abnormally rapid growth—possibly responsible for a number of major nuclear accidents—may have been what triggered the vehement reaction of environmentalists, who succeeded in slowing it down. Appropriately, as the technology matured, the number of major nuclear accidents was drastically reduced. However, environmentalists are far from having stopped nuclear energy. Ironically, under pressing concerns of CO_2 pollution, their opposition to nuclear energy had considerably weakened until the accident at Fukushima nuclear plant. I believe they will again reduce their opposition as public opinion cools off and better safety measures are put in place.

The changeable behavior of environmentalists suggests that there are other more fundamental forces at play while environmentalists behave more like puppets. These forces do not involve governments and their policies that usually become shaped after the fact in response to public outcries.

World Population: The Big Picture

Another example of society's wise and subtle ways of controlling human behavior is the slowing down in the rate of growth of the world population during the 20th century. The phenomenon has sometimes been erroneously attributed to people having become aware of the perils of Earth's overpopulation and reacted accordingly with adequate birth control. But it is only China that has imposed nationwide birth controls via legislation and that accounts for only 20 % of the world's population. The main reason the world population has slowed down is that rising standards of living offer people more highly preferred things to do than having children. The flattening of the S-curve shown earlier in Fig. 16.4 is very little a consequence of top-down conscious decision-making. It is mostly a consequence of subconscious bottom-up forces shaping the patterns of our lives.

But let us zoom back and consider a much greater historical window. Figure 16.11 shows world population data since the time of Christ (before that time estimates are rather unreliable). A dramatic exponential pattern belongs to an S-curve penetrated only to 23 % with an eventual ceiling of 1,750,000,000,000 by year 2700. Obviously the uncertainties involved on the level of the ceiling estimated from data that cover only the beginning of the S-curve are very large. From a detailed Monte-Carlo study on error estimation we obtain up to $\pm$ 75 % uncertainty on this number with confidence level of 90 % (Debecker and Modis 1994).

The year 2700 is a far-fetched horizon date for making forecasts and such statements are more appropriate to fiction than to scientific writing, but can there be a grain of truth? At first glance such a conclusion may seem absurd by today's standards and in view of the section World Population above. But is it really

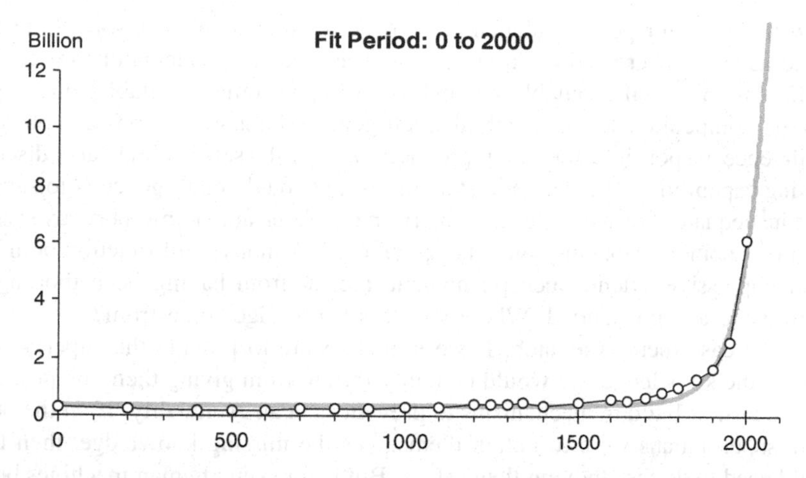

Fig. 16.11 The world population since the time of Christ has followed the early part of an S-curve, compatible with an exponential. The slowdown during the 20th century is not visible with 50 year time bins. *Data source* United Nations Population Division (U.N. 1999)

absurd? Could the S-curve of Fig. 16.4 be followed up by other S-curves in the paradigm of the section Fractal Aspects of Natural Growth?

Altogether possible, claims Cesare Marchetti, who calculated that Earth's carrying capacity is around 10^{12} people. His is not a forecast but scientific calculations taking into account availability of resources, energy, housing, and the environment (Marchetti 1979). If he is right that there is no fundamental law violated by reaching such a number, you can be sure that this will *eventually* happen (no niche in nature that could be filled to completion was ever left occupied only partially). But as we saw earlier, people's subconscious behavior during the 20th century assures us that such a thing would take place slowly, controllably, and avoiding catastrophe.

More than Just Cerebral Intelligence

This section addresses the claim that intelligent machines will eventually take over as a new species—posthumans—reducing humans to the equivalent of monkeys for us today.

Intelligence according to the singularitarians is measured by the speed of calculation. I believe that the astronomical numbers of FLOPS (floating-point operations per second) forecasted by Kurzweil as the ultimate computing power, namely10^{50} and beyond, will fall short by a large factor, at least 25 orders of magnitude, mostly because such computing power will no longer be desired. You can get too much of a good thing, for example there is no longer demand for cars to go faster, or for pocket calculators to become thinner/smaller, something that would have never crossed the mind of early-car and early-calculator users.

But assuming unfathomable computing power becomes available, this would only be competition to our cerebral intelligence. Humans also possess physical intelligence responsible for our reproduction, growth, self-healing, and disease-fighting capabilities. Human understanding of the body's intelligence is to say the least inadequate. For example, the enteric part of our autonomic nervous system has more memory capability than the spinal cord. A mouse will function brainless to an impressive extent. Such phenomena are far from having been thoroughly investigated and understood. Where will this knowledge come from?

In any case there is a catch. If we humans were to provide the superhumans with *all* the knowledge, we would certainly refrain from giving them the power to overtake us, or build in mechanisms to prevent such an eventuality. If on the other hand, superhumans were to obtain themselves the missing knowledge, then they would need to do the studying themselves. But before superhuman machines begin dissecting us and putting us under microscopes—as we do with monkeys—they will first need physical bodies themselves, which they should be able to fabricate and maintain. One cannot argue that advance robotics will produce machines that can do that because there is a catch. In order for these robots to be able to move around, gather resources, and carry out research to acquire the missing knowledge they would first need to have in their system the vey knowledge they set out to obtain through studies.

On the other hand, intellectual power all by itself would not achieve much. Besides some evidence for occasional correlation, it is well known that in general among the most intelligent people you are not likely to find: the richest, the happiest, the most normal (by definition!), the best-adjusted, the most good-natured, the most trustworthy, the most creative, the best artists, the most powerful, the most popular, or the most famous. In short, fast thinking is not the ultimate desirable quality, and thinking faster is not necessarily better in an absolute sense.

All in all, superhumans enabled to develop thanks to supercomputing power would certainly not pose a threat to humans by mid 21st century. In fact, I wouldn't hesitate to extend this reassuring message to the far-fetched horizon date of 2,700 that we mentioned earlier.

What May Happen Instead

A central graph early in Kurzweil's book, which he uses as a platform to launch the whole Singularity development, displays a set of data I had painstakingly collected earlier for my own publications (Modis 2002a, 2003). The set of data basically consists of thirteen independent timelines for the most significant turning points—milestones—in the evolution of the Universe. The emerging overall trend displays an unambiguous crowding of milestones in recent times. The thinking behind my article was that the spacing of the most important milestones could

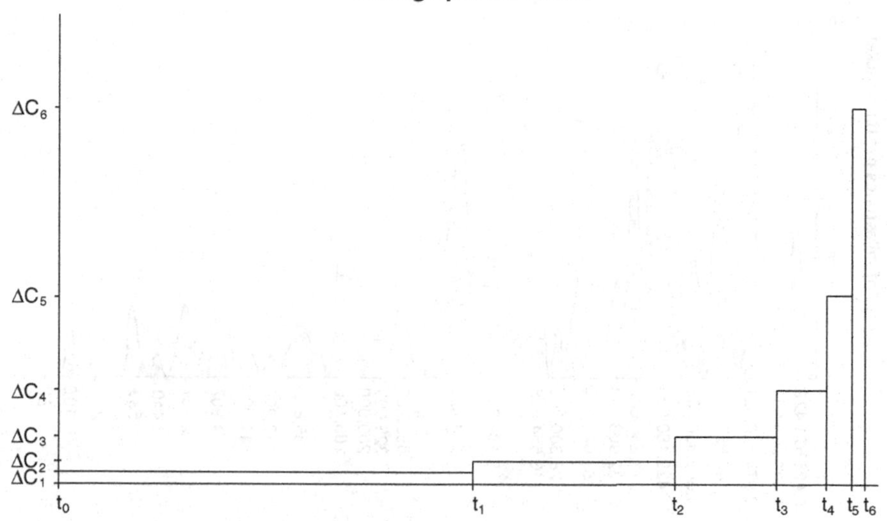

Fig. 16.12 To the extent that milestones of equal importance appear more frequently, the respective change they introduce increases. The area of each rectangle represents importance and remains constant. The scales of both axes are linear

serve to quantify the evolution of change and complexity and therefore enabling us to forecast it.

It is reasonable to assume that the greater the change associated with a given milestone, or the longer the ensuing stasis, the greater the milestone's importance will be.

$$\text{Importance} = (\text{change introduced}) \times (\text{duration to the next milestone}) \qquad (16.1)$$

Following each milestone there is change introduced in the system. At the next milestone there is a different amount of change introduced. Assuming that milestones are approximately of equal importance, and according to the above definition of importance we can conclude that the change ΔC_i introduced by milestone i of importance I is

$$\Delta C_i = \frac{I}{\Delta t_i} \qquad (16.2)$$

where Δt_i the time period between milestone i and milestone i + 1 (Fig. 16.12).

My dataset has a number of weaknesses. Only twelve out of the fifteen timelines used were independent. One timeline had been given to me without dates and I introduce them myself; another consisted of my own guesses. Both were heavily biased by the other twelve in my disposal. Moreover, some data were simply weak by their origin (e.g., an assignment post on the Internet by a biology professor for his class).

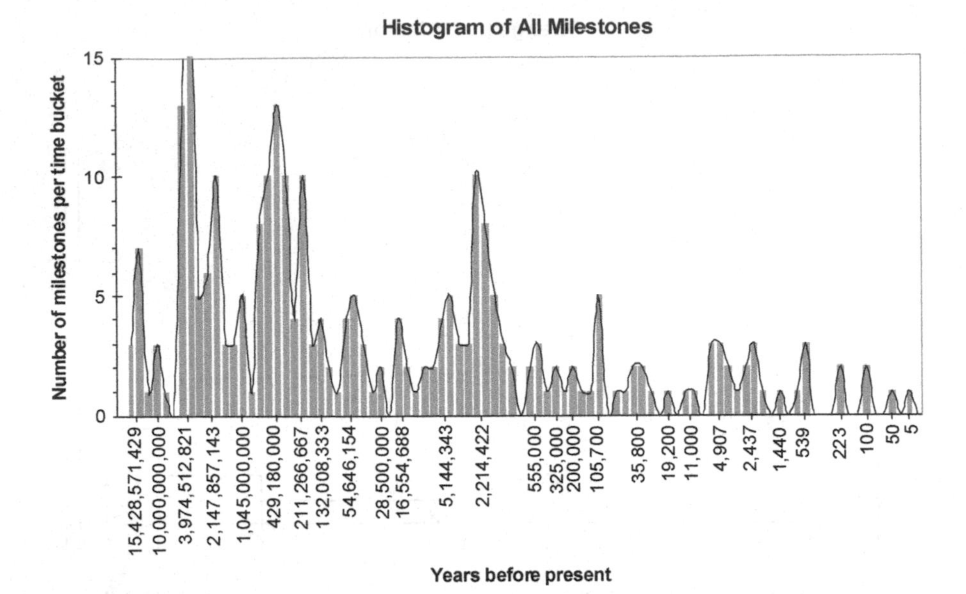

Fig. 16.13 A histogram of all milestones with logarithmic time buckets suggested by clusters of milestone dates. The thin black line is superimposed to outline the clusters of milestones that define the dates of a canonical milestones set. On the horizontal axis we read the dates of these canonical milestones. Present is defined as year 2000. The width of each peak becomes a measure of the uncertainty on the date of the canonical milestone (and consequently also the uncertainty on the change it brings)

As a matter of fact only one timeline (Sagan's Cosmic calendar) covers the entire range (big bang to 20th century) with dates. A second complete set (by Nobel Laureate Boyer) was provided to me without dates. All the other timelines coming from various disciplines covered only restricted time windows of the overall timeframe, which results in uneven weights for the importance of the milestones as each specialist focused on his or her discipline.

In any case, the grand total of all milestones came to 302 and in a histogram—Fig. 16.13—revealed clusters of milestones with peaks that defined a canonical set of milestones. Present time was taken as year 2000. Within all sources of uncertainties mentioned above, I tried to quantitatively study the evolution of complexity and change in the Universe.

Figure 16.14 shows that the evolution of change with milestone number has so far followed an exponential-growth pattern, which could also be an S-curve (logistic) or its life cycle (first derivative) as all three behave exponentially early on. Given that the data depict change *per milestone* I fit to the latter shown by the thick gray line. The implication of doing so is that the total amount of change in the Universe will be finite by the time the Universe ends.

The quality of the fit being a little better and the position of the last point both argue in favor of the logistic life cycle rather than the corresponding exponential.

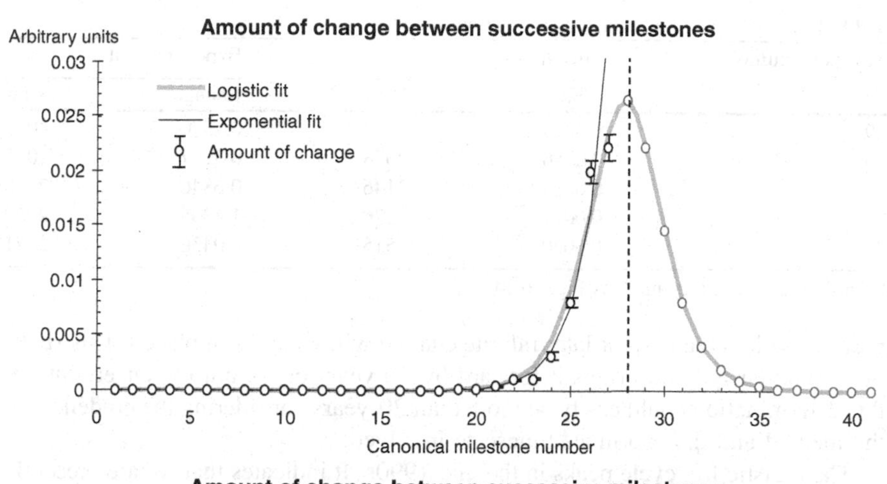

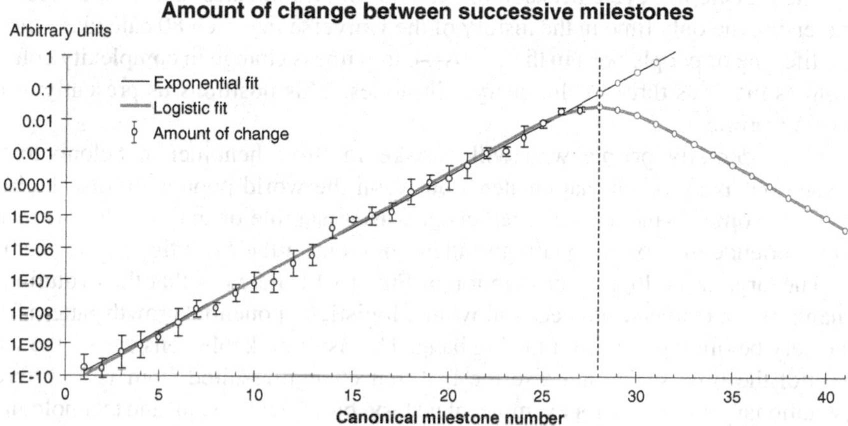

Fig. 16.14 Exponential and logistic life-cycle fits to the data of the canonical milestone set. The *vertical axis* depicting the amount of change per milestone is linear (*graph at the top*) and logarithm (*graph at the bottom*). The intermittent *vertical line* denotes the present. The *gray circles* on the forecasted trends indicate change from future milestones. The change associated with the most recent milestone, No 28, will not be known before the appearance of milestone No 29. The error bars have been calculated from the spread of entries clustered around each canonical milestone date

But these are weak arguments. More serious impact have the forecasts for the change expected by future milestones. Table 16.3 lists these forecasts for the next five milestones. The logistic fit expects milestones to begin appearing less frequently in the future whereas the exponential fit expects them at an accelerating pace. In particular, the logistic fit has next milestone appearing in 2033 and the one after that in 2078. In contrast, the exponential fit has next milestone in 2009 (remember zero was defined as year 2000 in this study), and the one after that in year 2015. By year 2022 the exponential fit forecasts a new milestones every

Table 16.3 Forecasts for change as a function of time

Milestone number	Logistic fit		Exponential fit	
	Change[a]	Year	Change[a]	Year
29	0.0223	2033	0.1540	2009
30	0.0146	2078	0.3247	2015
31	0.0081	2146	0.6846	2018
32	0.0041	2270	1.4435	2020
33	0.0020	2515	3.0436	2021

[a] In the same arbitrary units as Fig. 16.14

6 days and less than a year later infinite change will have taken place![2] This spells out "Singularity" and brings it forward by 20 years or so, but the uncertainty of this determination could easily be more than 20 years considering the crudeness of the method and the enormous timescale involved.

The logistic life cycle peaks in the mid 1990s. It indicates that we are presently traversing the only time in the history of the Universe in which 80 calendar years—the lifetime of people born in the 1940s—can witness change in complexity coming from as many as three evolutionary milestones. This positions us presently at the world's prime!

Coincidentally people who will partake in this phenomenon belong to the mysterious baby boom that creates a bulge on the world population distribution.[3] As if by some divine artifact a larger-than-usual sample of individuals was meant to experience this exceptionally turbulent moment in the evolution of the cosmos.

The large-scale logistic description of Fig. 16.14 indicates that the evolution of change in the Universe has been following a logistic/exponential growth pattern from the very beginning, i.e. from the big bang. This is remarkable considering the vastness of the time scale, and also the fact that change resulted from very different evolutionary processes, for example, planetary, biological, social, and technological. The fitted logistic curve has its inflection point—the time of the highest rate of change—in mid 1990. Considering the symmetry of the logistic-growth pattern, we can thus conclude that the end of the Universe is roughly another 15 billion years away. Such a conclusion is not really at odds with some scientific thinking that places the end of our solar system some 5 billion years from now.

We have obviously been concerned with an anthropic Universe here because we have to a large extend overlooked how change has been recently evolving in other parts of the Universe. Still, I believe that such an analysis carries more weight than just the elegance and simplicity of its formulation. The celebrated physicist John Wheeler has argued that the very validity of the laws of physics

[2] The pattern of a decaying exponential is asymptotic, i.e. it needs infinite time to reach zero, but its definite integral between x and ∞ is finite.

[3] It has been often argued that the baby boom was due to soldiers coming back from the fronts of WWII. This is wrong because the phenomenon began well before the war and lasted until the early 1970 s. The effect of WWII was only a small and narrow wiggly dent in the population's evolution.

depends on the existence of consciousness.[4] In a way, the human point of view is all that counts! It reminds me of a whimsical writing I once saw on a tee shirt: "One thing is certain, Man invented God; the rest is debatable".

Epilogue

The exponential pattern of milestones in Fig. 16.14, which provides a central argument for the Singularity, resembles—and to some extent is affected by—the patterns of Moore's law (Fig. 16.3) and world population (Fig. 16.11). All three show many orders of magnitude growth along exponential trends. But Figs. 16.3 and 16.11 have both avoided the ominously rising trend and have done so in a *natural* way.

The last two milestones with present defined as year 2000 are:

- 5 years ago: Internet/human genome sequenced
- 50 years ago: DNA/transistor/nuclear energy.

The next such world-shaking milestone should be expected—even by common sense—around 2033 rather than of around 2009 because despite a steady stream of significant recent discoveries, there is still no obviously candidate in 2012. That was not the case with the last two milestones: the significance of the Internet became clear simultaneously with its diffusion, and the significance of nuclear energy had become clear long before it was materialized.

An Afterthought

Playing the Devil's Advocate

Could it be that on a large scale there may be no acceleration at all? Could it be that the crowding of milestones in Fig. 16.13 is simply a matter of perception? The other day I was told that I should have included Facebook as a milestone. "It is just as important as the Internet," she told me. Would Thomas Edison have thought so? Will people one thousand years from now, assuming we will survive, think so? Will they know what Facebook was? Will they know what the Internet was?

It is natural that we are more aware of recent events than events far in the past. It is also natural that the farther in the past we search for important events the fewer of them will stick out in society's collective memory. This by itself would suffice to explain the exponential pattern of our milestones. It could be that as

[4] John Wheeler was a renowned American theoretical physicist. One of the later collaborators of Albert Einstein, he tried to achieve Einstein's vision of a unified field theory.

importance fades with the mere distancing from the present it "gives the appearance", in John von Neumann's words, that we are "approaching some essential singularity". But this has nothing to do with year 2045, 2025, today, von Neumann's time—the 1950s—or any other time in the past or the future for that matter.

Appendix: The Canonical Milestones

The dates generally represent an average of clustered events not all of which are mentioned in this table. That is why some events e.g. asteroid collision appears dated too recent. Highlighted in bold is the most outstanding event in the cluster. Present time is taken as year 2000.

No.	Milestone	Years ago
1	**Big bang** and associated processes	1.55×10^{10}
2	**Origin of milky way**/first stars	1.0×10^{10}
3	**Origin of life on Earth**/formation of the solar system and the Earth/oldest rocks	4.0×10^{9}
4	**First eukaryots**/invention of sex (by microorganisms)/atmospheric oxygen/oldest photosynthetic plants/plate tetonics established	2.1×10^{9}
5	**First multicellular life** (sponges, seaweeds, protozoans)	1.0×10^{9}
6	**Cambrian explosion**/invertebrates/vertebrates/plants colonize land/first trees, reptiles, insects, amphibians	4.3×10^{8}
7	**First mammals**/first birds/first dinosaurs/first use of tools	2.1×10^{8}
8	**First flowering plants**/oldest angiosperm fossil	1.3×10^{8}
9	**Asteroid collision**/first primates/mass extinction (including dinosaurs)	5.5×10^{7}
10	**First humanoids**/first hominids	2.85×10^{7}
11	**First orangutan**/origin of proconsul	1.66×10^{7}
12	**Chimpanzees and humans diverge**/earliest hominid bipedalism	5.1×10^{6}
13	First stone tools/first humans/ice age/*homo erectus*/origin of spoken language	2.2×10^{6}
14	**Emergence of Homo sapiens**	5.55×10^{5}
15	**Domestication of fire**/*Homo heidelbergensis*	3.25×10^{5}
16	**Differentiation of human DNA types**	2.0×10^{5}
17	**Emergence of "modern humans"** earliest burial of the dead	1.06×10^{5}
18	**Rock art**/protowriting	3.58×10^{4}
19	**Invention of agriculture**	1.92×10^{4}
20	**Techniques for starting fire**/first cities	1.1×10^{4}
21	**Development of the wheel/writing**/archaic empires	4907
22	**Democracy**/city states/the Greeks/Buddha	2437
23	**Zero and decimals invented**/Rome falls/Moslem conquest	1440
24	**Renaissance (printing press)**/discovery of new world/the scientific method	539
25	**Industrial revolution (steam engine)**/political revolutions (French, USA)	223
26	**Modern physics**/radio/electricity/automobile/airplane	100
27	**DNA/transistor/nuclear energy**/W.W.II/cold war/sputnik	50
28	**Internet/human genome sequenced**	5

References

Debecker, A., & Modis, T. (1994). Determination of the Uncertainties in S-curve Logistic Fits. *Technological Forecasting and Social Change, 46*, 153–173.

de Solla Price, & Derek, J. (1936). *Little science, big science*. New York: Columbia University Press.

Fisher, J. C., & Pry, R. H. (1971). A simple substitution model of technological change. *Technological Forecasting and Social Change, 3*(1), 75–88.

Marchetti, C. (1979). On 1012: A check on earth carrying capacity for man. *Energy, 4*, 1107–1117.

Marchetti, C. (1983). The automobile in a systems context: The past 80 years and the next 20 years. *Technological Forecasting and Social Change, 23*, 3–23.

Marchetti, C. (1986). Fifty-year pulsation in human affairs: Analysis of some physical indicators. *Futures, 17*(3), 376–388.

Marchetti, C. (1987). Infrastructures for movement. *Technological forecasting and social change, 32*(4), 146–174.

Modis, T., & Debecker, A. (1992). Chaos like states can be expected before and after logistic growth. *Technological Forecasting and Social Change, 41*, 111–120.

Modis, T. (1992). *Predictions: Society's telltale signature reveals the past and forecasts the future*. New York: Simon & Schuster.

Modis, T. (1994). Fractal aspects of natural growth. *Technological Forecasting and Social Change, 47*, 63–73.

Modis, T. (2002a). Forecasting the growth of complexity and change. *Technological Forecasting and Social Change, 69*(4), 337–404.

Modis, T. (2002b). *Predictions: 10 years later*. Geneva: Growth Dynamics.

Modis, T. (2003). The limits of complexity and change. *The Futurist, 37*(3), 26–32. (May-June).

Modis, T. (2005). The end of the internet rush. *Technological Forecasting and Social Change, 72*(8), 938–943.

Modis, T. (2007). The normal, the natural, and the harmonic. *Technological Forecasting and Social Change, 74*(3), 391–399.

Modis, T. (2009). Where has the energy picture gone wrong? *World Future Review, 1*(3), 12–21. (June-July).

Williams, J. D. (1958). The nonsense about safe driving. *Fortune, LVIII*(3), 118–119. (September).

U.N. (1999). Population division department of economic and social affairs. The world at six billion. United Nations Secretariat.

Chapter 16A
Vernor Vinge on Modis' "Why the Singularity Cannot Happen"

The cosmic reach of Modis' essay is especially interesting. It's the first time I've seen any but an extreme techno-optimist undertake trend analysis on such a scale.

I count myself as one of the extreme techo-optimists, but I have only intuition and hope for what things are like at the largest scales. Closer to home: the Technological Singularity is exactly the rise, through technology, of superhuman intelligence. So the critical question is what level of computation and software is sufficient to achieve this transition and will our progress reach that level? I see the Singularity as something oncoming along a variety of research paths, with good progress on almost all fronts. Leaving aside catastrophic failures (e.g, nuclear war), a singularity-free future might be the most bizarre outcome of all… finally enough time to rationalize all our legacy software?

Chapter 16B
ARay Kurzweil on Modis' "Why the Singularity Cannot Happen"

My 1999 book *The Age of Spiritual Machines* generated several lines of criticism, such as *Moore's law will come to an end, hardware capability may be expanding exponentially but software is stuck in the mud, the brain is too complicated, there are capabilities in the brain that inherently cannot be replicated in software,* and several others. I specifically wrote *The Singularity is Near* to respond to those critiques.

Many of the critics of *The Singularity is Near* fail to respond to the actual arguments I make in the book, but instead choose to mischaracterize my thesis and then attack the mischaracterization. In his essay "Why the Singularity Cannot Happen," Theodore Modis takes this one step further by borrowing ideas from my book to criticize the straw man thesis that he incorrectly attributes to it.

Modis' argument is summarized in his first few sentences. He writes "One reason [that the concept of a Singularity as described in Ray Kurzweil's book cannot happen] is that all natural growth processes that follow exponential patterns eventually reveal themselves to be following S-curves thus excluding runaway situations. The remaining growth potential from Kurzweil's "knee" which could be approximated as the moment when an S-curve begins deviating from the corresponding exponential is a factor of only one order of magnitude greater than the growth already achieved. A second reason is that there is already evidence of a slowdown in some important trends".

Modis essentially ignores that I make the exact same point about S-curves in almost the same language in my book.

He then goes on to cite the U.S. GDP, Moore's law, the Microsoft Windows operating system, the number of users of the Internet, car accidents, AIDS cases, population growth, and nuclear power as examples of exponential growth patterns that did not go on forever (as if I claim that every exponential inherently goes on indefinitely).

Let me summarize what it is that I am saying in *The Singularity is Near* because none of the examples that Modis gives have any relevance to my thesis.

A primary definition of the Singularity is a future time when we will substantially enhance our own intellectual capabilities by merging with the intelligent technology we have created. The Singularity is the result of the Law of Accelerating Returns. The LOAR is the primary thesis of the book, and it states that fundamental measures of price-performance and capacity of information technologies grow exponentially and do so by spanning multiple paradigms. A particular paradigm, such as Moore's law, will follow an S-curve, but the basic measures of an information technology transcend each specific S-curve by spanning multiple paradigms.

The LOAR certainly does not state that every exponential trend goes on indefinitely. Almost all of the cases that Modis goes on to discuss in great detail have nothing to do with the LOAR or the Singularity. GDP, car accidents, AIDS cases, population and nuclear power are not information technologies and have nothing to do with my thesis.

Modis is implying that my thesis is the following: computers have shown exponential growth; every example of exponential growth must go on forever; and therefore computer capability will continue growing indefinitely. That represents a basic misrepresentation.

Even those cases which have some relevance to information technology are misstated by Modis. The number of users of the Internet is not a basic measure of communications power. The units of a basic measure would be in bits or bits per constant unit of currency, not in numbers of people. Obviously the number of people doing anything is going to saturate.

Nor is the number of transistors on a chip a basic measure. That represents part of one paradigm, but the basic measure of price-performance of computing is calculations per second per constant dollar. This measure has not shown any slow down. This trend has been going on unabated for well over a century and in fact is speeding up, not slowing down. Here is the logarithmic scale graph updated through 2008. Note that a straight line on a logarithmic scale represents exponential growth and the trend has been and continues to be better than exponential.

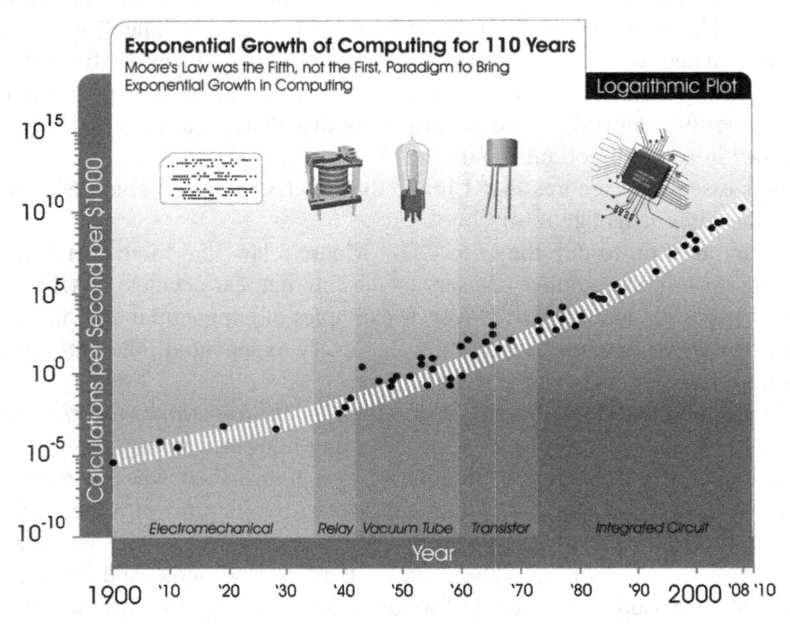

The paradigm of Moore's law is only the vertical region at the right. Engineers were shrinking vacuum tubes in the 1950s to continue the law of accelerating returns (as it pertains to the price-performance of computing). Indeed that paradigm ran out of steam which led to transistors and that led to integrated circuits (and the paradigm of Moore's law). In the book, which came out in 2005, I describe that the sixth paradigm will be three-dimensional computing and that is indeed now underway. Today, many chips for applications in MEMs, image sensing, and memory utilize three-dimensional stacking technologies, and other approaches are being perfected.

I discuss the relationship of the LOAR and the S-curves of individual paradigms in five different sections of *The Singularity is Near*. Here are brief excerpts from these sections:

> A specific paradigm (a method or approach to solving a problem; for example, shrinking transistors on an integrated circuit as a way to make more powerful computers) generates exponential growth until its potential is exhausted. When this happens, a paradigm shift occurs, which enables exponential growth to continue.
>
> The life cycle of a paradigm. Each paradigm develops in three stages:
>
> 1. Slow growth (the early phase of exponential growth)
> 2. Rapid growth (the late, explosive phase of exponential growth), as seen in the S-Curve figure below.
> 3. A leveling off as the particular paradigm matures. The progression of these three stages looks like the letter "S," stretched to the right. The S-curve illustration shows how an ongoing exponential trend can be composed of a cascade of s-curves…

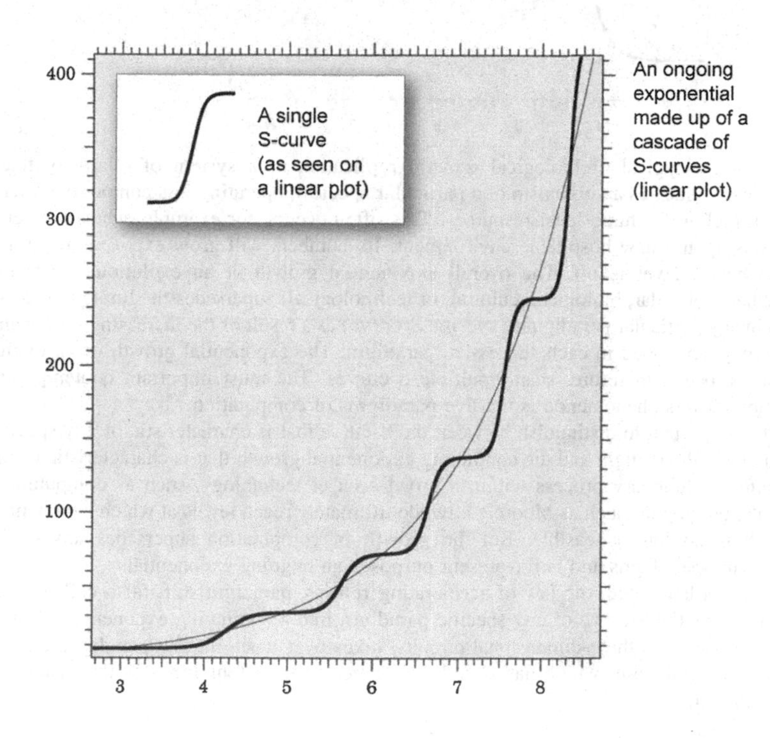

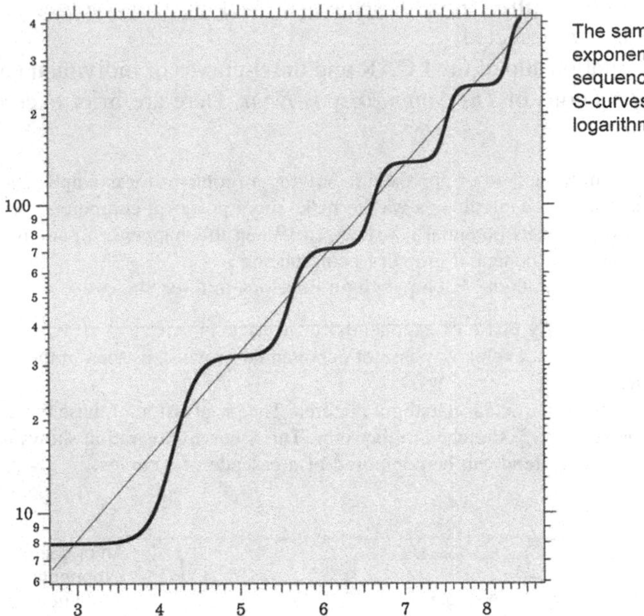

The same exponential sequence of S-curves on a logarithmic plot

> S-curves are typical of biological growth: replication of a system of relatively fixed complexity (such as an organism of a particular species), operating in a competitive niche and struggling for finite local resources. This often occurs, for example, when a species happens upon a new hospitable environment. Its numbers will grow exponentially for a while before leveling off. The overall exponential growth of an evolutionary process (whether molecular, biological, cultural, or technological) supersedes the limits to growth seen in any particular paradigm (a specific S-curve) as a result of the increasing power and efficiency developed in each successive paradigm. The exponential growth of an evolutionary process, therefore, spans multiple S-curves. The most important contemporary example of this phenomenon is the five paradigms of computation…
>
> It is important to distinguish between the S-curve that is characteristic of any specific technological paradigm and the continuing exponential growth that is characteristic of the ongoing evolutionary process within a broad area of technology, such as computation. Specific paradigms, such as Moore's Law, do ultimately reach levels at which exponential growth is no longer feasible. But the growth of computation supersedes any of its underlying paradigms and is for present purposes an ongoing exponential.
>
> In accordance with the law of accelerating returns, paradigm shift (also called innovation) turns the S-curve of any specific paradigm into a continuing exponential. A new paradigm, such as three-dimensional circuits, takes over when the old paradigm approaches its natural limit, which has already happened at least four times in the history of computation.

Modis acknowledges that I mention S-curves but he states this in a confusing and misleading way. Modis writes, "Kurzweil acknowledges that there are smaller S-curves that saturate early, but argues that they are replaced by other small S-curves thus cascading indefinitely. He does not seem to be aware of the fact that there is a fractal aspect to such cascades and constituent S-curves are bound by envelope S-curves, which themselves saturate with time".

Modis' discussion of the "fractal aspect" of paradigms is not relevant to the discussion and does not contradict my conclusions. The reality is that there has already been a cascade of paradigms in computation and in other examples of basic measures of information technology. As a result these measures have followed smooth exponential trajectories for lengthy periods of time. Modis goes on to completely confuse individual paradigms with the ongoing exponential that spans paradigms. He uses Moore's law as synonymous with the LOAR and in any event misstates the measures of Moore's law. His other examples of exponential growth slowing down such as car accidents and AIDS cases are completely irrelevant to the discussion.

I discuss in the book ultimate limits to the ongoing exponential growth of the price-performance of computation based on the physics of computation (that is, the amount of matter and energy required to compute, remember, or transmit a bit of information). Based on this analysis we have trillions fold improvement yet to go. This is not unprecedented as we have indeed already made trillions fold improvement since the advent of computing, a several billion fold improvement just since I was an undergraduate.

There is one other misstatement in Modis' essay that I will comment on. He writes "Intelligence according to the singularitarians is measured by the speed of calculation." This ignores about a hundred pages of the book where I talk about improvements in software independently of improvements in hardware. I clearly discuss hardware speed and memory capacity as a necessary but not sufficient conditions for achieving human-level intelligence (and beyond) in a machine.

In *The Singularity is Near*, I address this issue at length, citing different methods of measuring complexity and capability in software that demonstrate a similar exponential growth. One recent study ("Report to the President and Congress, Designing a Digital Future: Federally Funded Research and Development in Networking and Information Technology" by the President's Council of Advisors on Science and Technology) states the following: "Even more remarkable—and even less widely understood—is that in many areas, *performance gains due to improvements in algorithms have vastly exceeded even the dramatic performance gains due to increased processor speed.* The algorithms that we use today for speech recognition, for natural language translation, for chess playing, for logistics planning, have evolved remarkably in the past decade… Here is just one example, provided by Professor Martin Grötschel of Konrad-Zuse-Zentrum für Informationstechnik Berlin. Grötschel, an expert in optimization, observes that a benchmark production planning model solved using linear programming would have taken 82 years to solve in 1988, using the computers and the linear programming algorithms of the day. Fifteen years later—in 2003—this same model could be solved in roughly 1 min, an improvement by a factor of roughly 43 million. Of this, a factor of roughly 1,000 was due to increased processor speed, whereas a factor of roughly 43,000 was due to improvements in algorithms! Grötschel also cites an algorithmic improvement of roughly 30,000 for mixed integer programming between 1991 and 2008". I cite many other examples like this in the book.

My primary thesis which I discuss in one of the major chapters of the book is that we are making exponential gains in reverse-engineering the methods of the human brain and using these as biologically inspired paradigms to create intelligent machines. Modis makes no mention of these arguments. Instead he misrepresents my position as stating that computational speed alone is sufficient to achieve human-level intelligence. I update discussion of our progress in understanding human intelligence in a book that will be published by Viking in October 2012 titled *How to Create a Mind, The Secret of Human Thought Revealed.*

Chapter 17
The Slowdown Hypothesis

Alessio Plebe and Pietro Perconti

Abstract The so-called *singularity hypothesis* embraces the most ambitious goal of Artificial Intelligence: the possibility of constructing human-like intelligent systems. The intriguing addition is that once this goal is achieved, it would not be too difficult to surpass human intelligence. While we believe that none of the philosophical objections against strong AI are really compelling, we are skeptical about a singularity scenario associated with the achievement of human-like systems. Several reflections on the recent history of neuroscience and AI, in fact, seem to suggest that the trend is going in the opposite direction.

Introduction

The so-called *singularity hypothesis* embraces the most ambitious goal of Artificial Intelligence: the possibility of constructing human-like intelligent systems. The intriguing addition is that once this goal is achieved, it would not be too difficult to surpass human intelligence. A system more clever than humans should also be better at designing new systems as well, leading to a recursive loop towards ultraintelligent systems (Good 1965), with an acceleration reminiscent of mathematical *singularities* (Vinge 1993). According to David Chalmers, the singularity hypothesis is to be taken very seriously. If "there is a singularity, it will be one of the most important events in the history of the planet. An intelligence explosion has enormous potential benefits: a cure for all known diseases, an end to poverty,

A. Plebe (✉) · P. Perconti
Department of Cognitive Science, v. Concezione 8, 98121, Messina, Italy
e-mail: alessio.plebe@unime.it

P. Perconti
e-mail: perconti@unime.it

A. H. Eden et al. (eds.), *Singularity Hypotheses*, The Frontiers Collection,
DOI: 10.1007/978-3-642-32560-1_17, © Springer-Verlag Berlin Heidelberg 2012

extraordinary scientific advances, and much more. It also has enormous potential dangers: an end to the human race, an arms race of warring machines, the power to destroy the planet" (Chalmers 2010, p. 9).

Back when AI suffered from a significant lack of results with respect to the claims put forth by some of its most fervid enthusiasts, and faced strong philosophical criticism (Searle 1980; Dreyfus and Dreyfus 1986), skepticism about the possibility of it achieving its main goal spread, leading to a loss of interest in the singularity hypothesis as well. Our opinion is that, despite the limited success of AI, progress in the understanding of the human mind, coming especially from modern neuroscience, leaves open the possibility of designing intelligent machines. We also believe that none of the philosophical objections against strong AI are really all that compelling.

This, however, is not our main point. What we will address instead, is the issue of a singularity scenario associated with the achievement of human-like systems. With this respect, our view is skeptical. Reflection on the recent history of neuroscience and AI suggests to us instead, that trends are going in the opposite direction. We will analyze a number of cases, with a common rate pattern of discovery: important achievements in simulating aspects of human behavior become on one hand, examples of progress, and on the other, a point of slowdown, by revealing how complex the overall functions are of which, they are just a component. There is no knockdown argument for posing that the slowdown effect is intrinsic to the development of intelligent artificial systems, but so far, there is good empirical evidence for it. Furthermore, the same pattern seems to characterize the recent inquiry concerning the core notion of intelligence.

We will present two lines of reasoning in this paper. First, we will provide a simple formalization of the slowdown hypothesis in mathematical terms, showing in an abstract way what the causes of the slowdown are, and their effects on the evolution of AI research. The aim of the formalization is not to propose a mathematical model of some kind of automatism inscribed in the logic of scientific discovery, but simply to show in formal terms how a given field of research (in this case, AI) could end up in a slow down progress (under the circumstances we will discuss in what follows). We will then move inside various domains of AI, observing how the history of their scientific development provides support for our hypothesis. One of these will be the field of artificial vision, where the long history of research and the rich body of evidence obtained make it a significant case in point.

Furthermore, we will discuss how the recent inquiry concerning the core notion of intelligence seems to show a similar pattern, with a series of new and far reaching fields of research that have grown around the initial one, such as the role played by consciousness in the social nature of intelligence. On the whole, we will argue that the slowdown effect is due both to reasons that are internal to the logic of scientific discovery, and to the changes in the expectations held in regard to a much idealized subject of inquiry: "intelligence".

Formalization of the Slowing Down

In this section we will try to give a mathematical formalization of the reasons why the research enterprise for an artificial intelligence is characterized by the slowdown effect. Let us call Δ the normalized distance between the performance of an artificial system and that of one held as point of reference, assuming that $\Delta = 0$ means equally valid performances. In very general terms Δ can be the sum of distances over a set $\mathcal{P}$ of simple elementary processes p, producing some measurable performance b_p, assumed to be 1 when fully intelligent, and 0 when absolutely dull.

$$\Delta = \sum_{p \in \mathcal{P}} 1 - b_p \tag{17.1}$$

With a leap of faith in progress, we can imagine that the performance b_p will become better and better as long as research efforts accumulate in time, for all possible single processes p, therefore $b_p(t)$ is a function of time t, continuously increasing towards 1

The core of our reasoning is a very common phenomena found in the research of any process involved in human intelligence. It is the fact that a higher and a more detailed knowledge of a cerebral process often spawns a new field of investigation, that is discovered to be a necessary component of the overall intelligent behavior. Process spawning can arise for different reasons. For example, a deeper investigation might reveal that a process, previously thought to be atomic, is in fact the result of two almost independent subprocesses, each deserving its own research specificity, or, while searching for a known brain process, a new and different one, that was previously unknown, is discovered. Still yet, a known process that had little empirical evidence and that could not be reproduced artificially, might begin to become more clear thanks to some new scientific discovery, leading to a new research direction.

A very crude simplification is to take the same trend of performance in time for all processes and assume that some of them, after a certain research time T, will spawn a new process.

We can rewrite Eq. (17.1) in this way:

$$\Delta(t) = \sum_{i=0}^{\left\lceil (\phi 2)^{\left\lfloor \frac{t}{T} \right\rfloor} \right\rceil} e^{\frac{t - \left\lfloor \log_{\phi 2} i \right\rfloor}{\alpha}} \tag{17.2}$$

where α is the rate of improvement in performance, and T is the amount of research time after which a current process may spawn a new one, and ϕ is the fraction of processes that actually do spawn after T elapsed. Operators $\lfloor \cdot \rfloor$ and $\lceil \cdot \rceil$ are respectively the floor and ceiling functions. In Fig. 17.1 two examples are shown, with different parameter values. For the dark plot it is assumed that a new

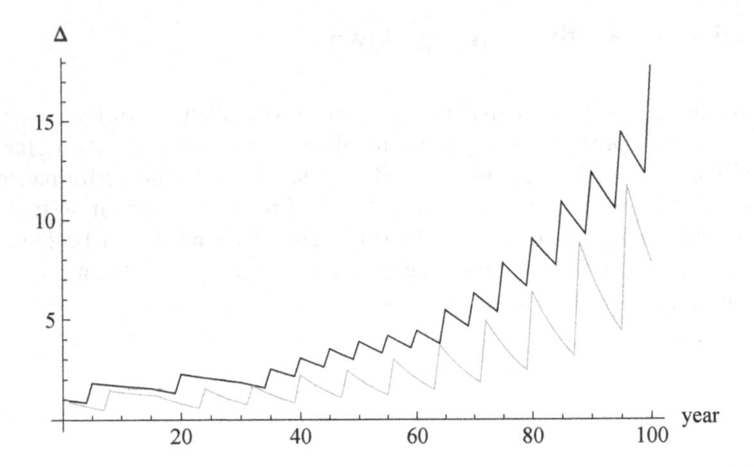

Fig. 17.1 Examples of Δ trends according to Eq. (17.2). Parameters for the *dark gray plot* are $\alpha = 25, T = 5, \phi = .6$, for the *light gray curve* $\alpha = 10, T = 8, \phi = .7$

process will come about every five years, while for the light plot this event happens every eight years. It is immediately apparent how the spawning phenomena prevents Δ from converging to 0, the level of perfect intelligent behavior, on the contrary it diverges in time towards increasingly worsening values.

It may seem paradoxical in Fig. 17.1 that at the very beginning of research Δ would be smaller than after a century, during which research has expanded in many directions. In fact, it is to be expected, if you take into account that in Eq. (17.1) Δ is an estimate of the intelligence level reached by an artificial system with respect to the set of processes $\mathcal{P}$ only. It is not an absolute measure. In principle Eq. (17.1) could give an estimate of the absolute general intelligence, using a theoretical $\widetilde{\mathcal{P}}$, the set of all possible processes necessary for a general intelligent system, including many processes for which no research has yet begun. For all those unexplored $\widetilde{p}$, it holds $b_{\widetilde{p}}(t) = 0$ all the time. Equation (17.3) can be rewritten as:

$$\Delta(t) = \left|\widetilde{\mathcal{P}}\right| - \left\lceil (\phi 2)^{\left\lfloor \frac{t}{T} \right\rfloor} \right\rceil + \sum_{i=0}^{\left\lceil (\phi 2)^{\left\lfloor \frac{t}{T} \right\rfloor} \right\rceil} e^{\frac{t - \left\lfloor \log_{\phi 2} i \right\rfloor}{\alpha}} \tag{17.3}$$

In practice, however, there is no way of knowing $\widetilde{\mathcal{P}}$ in advance, due to the fact that precisely knowing all the processes contributing to a general intelligent agent would mean knowing almost everything about intelligence. What happens instead is that every time a new component is discovered to play an important role in intelligence, almost immediately or shortly afterward, a new research effort for simulating this process artificially, begins. Let us take the example of consciousness: the focus on this problem in philosophy and the awareness of its crucial influence in how the mind works, has triggered a growing amount of research on machine consciousness.

There are clearly many factors neglected in the simple formulation of Eq. (17.2) but it reflects real research trends in artificial intelligence. Some will not influence overall trends in a significant way . For example, the birth of new processes would not be synchronous, but each parent process may spawn a new one at a different time. The effect, however, would be just that of having a less regular curve derived by Eq. (17.2), but if T is the average time of spawning of all the processes, the long term trends will be similar. Some of the neglected factors would indeed make the forecast of (17.2) worse, and some better, for the sake of fairness we will discuss the inclusion of a few terms only from among those that warrant more optimism.

A reasonable argument would be that it is unrealistic to believe that all processes p simulating intelligence equally contribute in the summation of Eq. (17.2). One may argue that the first studied processes are more important, and as long as research continues, and new fields are spawned, the new fields are components that are gradually less and less crucial to the overall goal of reaching intelligent behavior. Along the same line, one may challenge as unrealistic the expectation that the spawning process will continue forever, and at the same rate as when the AI enterprise began.

We can take into account these two factors, with a formulation that is slightly more complex than that in (17.2), as follows:

$$\phi(t) = \phi_\infty + (1 - \phi_\infty) e^{\frac{-t}{\gamma}} \tag{17.4}$$

$$\Delta(t) = \sum_{i=0}^{\left\lceil (2\phi(t))^{\left\lfloor \frac{t}{T} \right\rfloor} \right\rceil} i^{-\beta} e^{\frac{t - \left\lfloor \log_{2\phi(t)} i \right\rfloor}{\alpha}} \tag{17.5}$$

where β is the decay in importance of the processes added late on the overall performance of the system, and γ is the decay of the number of processes that during their advanced research stage may spawn a new field of research.

In Fig. 17.2 two examples of the evolution in time of Δ with the new formulation are shown. As previously for the dark plot, it is assumed that a new process will come about every five years, while for the light plot this event happens every eight years.

Contrary to the plots in Fig. 17.1, in this cases Δ does not move towards worse values, it remains around the value of 1 during the 100 years of the simulation, but again the spawning of new processes hampers the continuous decrease towards the optimal value of 0.

A possible objection to the formalization here presented, might be that in principle Δ can only reach asymptotically its best value 0, even in absence of process multiplication, while the singularity hypothesis postulates that human intelligence cannot only be approximated, but even surpassed. For this reason we defined the value 1 of the measured performance b_p of a process p as the reference best value, without a strict reference to human intelligence, therefore, it can be

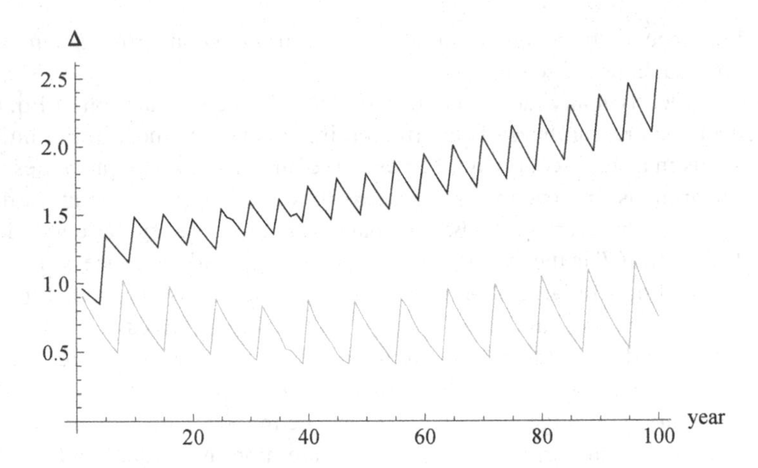

Fig. 17.2 Examples of Δ trends according to Eq. (17.5). Parameters for the *dark gray plot* are $\alpha = 25, T = 5, \beta = .9, \phi_\infty = .7, \gamma = 10$, for the *light gray curve* $\alpha = 10, T = 8, \beta = .8, \phi_\infty = .7, \gamma = 100$

held to be more than the average human performance. The argument that the fundamental speed up in the singularity hypothesis is based on the ability of the intelligent system to design and implement a newly created system automatically, will be discussed in section Machines Designing Machines.

In concluding our sketch of a formal justification of the slowdown effect, we would like to mention certain aspects, missing in Eqs. (17.2) and (17.5), that will make the development of a fully intelligent system even slower. All processes p are treated as independent, each with its own continuous progress in time, by proceeding in this way we neglected the problem of the interactions between the many processes involved in general intelligent behavior. More realistically, a new process often requires its own research and development, but it also requires the effort to understand and simulate the interfaces between this new process, and at the very least, the one that exists with its parent, not mentioning those between the many other related processes involved. Often, understanding the interactions between processes is more demanding than simulating the processes itself. Moreover, it seems that the singularity hypothesis requires an exponential growth of computational and design capacities. We argue, on the contrary, that even the case of an indefinite linear growth is questionable, and that this sounds as an *a fortiori* argument supporting the slowdown hypothesis.

There are also cases in which a new field of investigation may reveal novel solutions for many other ongoing research investigations, a paramount example is the first connectionist approach to modeling neural networks. On the other hand, not all newly initiated research directions turn out to be fruitful. The history of AI, as any other scientific domain, is full of new attempts that initially seemed promising, and later revealed themselves to be wrong or useless. An example is research of the so called $2\frac{1}{2}$-*dimensional sketch* in vision. As a side effect of a successful line of research that affects many others, such as that of connectionism, mentioned above, several older processes may die, substituted by others based on the new paradigm. Summing up, the interactions between processes would certainly make the evolution of Δ more complex than the abstract formulations here suggested, in ways likely to enhance the slowdown effect even more.

Scientific Idealization: The "Zooming in" and "Zooming out" Effect

The suspicion that the maturation of brain function simulations is characterized by the slowdown effect emerged before attempting to formalize its mechanism, from the observation of a typical pattern in the social history of science, as shown by both the typical pathway of scientific idealization and the recent history of several scientific domains.

Idealization plays an important role in scientific inquiry. In a sense, every scientific enterprise is based on a sort of scientific idealization, that is, "the intentional introduction of distortion into scientific theories" (Weisberg 2007, p. 639). Let's take into consideration the well-known case of Galileo's use of the inclined plane to study the force of gravity. According to Aristotle, a continually acting force would be necessary to keep a body moving horizontally at a uniform velocity. Galileo believed that if there was be no air resistance and no friction, and if a perfectly round and smooth ball was rolled along a perfectly flat inclined endless plane, the speed of the body would accelerate in a predictable way. Of course, such a scenario does not exist in the real world. In fact, it is an idealized state of affairs. In order to arrive at knowledge about gravitational acceleration from the observation of a falling body, Galileo supposed that the inclined plane was an idealized frictionless object. The aim of this theoretical move is to make the problem computationally tractable. Galilean idealization is a computational advantage in elaborating theories with strong predictive power. It is even possible to compute the gravitational acceleration taking into account the influence of the friction, but it is a known fact that Galileo's theoretical move to imagine a frictionless plane allowed significant achievements in the study of classical mechanics. Galilean idealization chooses only certain traits among the plethora of features a given phenomenon is endowed with. It is a deliberative act by the scientist, which is eventually justified by the research program he adopts. In other words, the price the scientist has to pay in order to explain a given phenomenon, is to leave aside or deliberately ignore some of its (not crucial, one has to hope) features.

The gap between the real and the ideal object is a matter of how much the scientist's attitude is flexible in regards to the scientific idealization. In the history of science the process of idealization has changed continuously in its scope. It is like a "zooming in" and "zooming out" effect which depends on how much of the features of the object are neglected or not. The rationale of this "zooming in" and "zooming out" effect consists in allowing the scientific enterprise to respond to social influences in a flexible way. If there is a significant social pressure to include a given trait into the scientific explanation, the community of researchers can modify the "zooming in" of the idealization and incorporate that feature.

Besides this mechanism of regulation of the relationship between science and society, the "zooming in" and "zooming out" effect also depends on the internal dynamics of the logic of scientific discovery. This is exactly what happened to the concept of intelligence from the time of Alan Turing's pioneering studies. The quest for Artificial Intelligence is an attempt to produce a human-like creature from a rather restricted idea of what intelligence actually is. In this perspective, in fact, intelligence is conceived as a computational feature of a disembodied mind. All that really matters according to this perspective is the sensibility to a set of formal characteristics and a good information processing device. The computational nature of this conception of intelligence leads AI scholars to believe that they are dealing with cumulative progress. The singularity hypothesis is based on the assumption that AI findings are cumulative. This expectation, however, depends on the stability of the idealization "zooming". If the amount of features we are interested in grows in a remarkable way, the cumulative effect vanishes. In fact, the growth of knowledge involves increases in a horizontal direction rather than in a vertical cumulative one. As long as scientists devoted their efforts to abilities such as arithmetic computations, the cumulative effect was remarkable. Calculating machines have long excelled their masters, and this has happened without the help of the machines' capacity to design other machines. While these results are increasingly promising, scientific idealization of intelligence has deeply changed. In the last decades, the studies on intelligence have become increasingly more focused on many of the aspects of the phenomenon ignored in the past. Intelligence is now considered as a multifaceted cognitive process, with a proliferation of proposals of new kinds of intelligence, from emotional to musical, from spatial to social.

Gardner (2006) famously argued there are many kinds of intelligence, including numerical intelligence, language mastering, body-kinesthetic, memory, and of spatial perception. On the whole, intelligence now appears to be a more ecological capacity than it did in the past: it is deeply influenced by motor schemata and is constituted by subsymbolic (and perhaps encapsulated) processes. In this way, however, the number of aspects one has to take into account become increasingly more significant, and the pathway of scientific discovery heads more towards a slowdown rather than towards a singularity effect.

The Case of Artificial Vision

Let consider now the case of artificial vision, the research aimed at designing artificial systems capable of a visual perception comparable with humans or other animals. This is an interesting benchmark, because its history dates back almost as long as AI, and because it has been and currently is, the most understood cognitive function in the brain.

In Fig. 17.3 is a sketch of process spawning, the phenomenon at the basis of our Eq. (17.5), in the case of vision, showing how the overall domain tends to branch into many autonomous fields of research. There are clearly many different criteria for splitting the domain of artificial vision into singular processes, we have used the principle of only including simulations that target visual behavior clearly identified in visual neuroscience, simulated in a way that adheres to the knowledge of the equivalent brain process. It is far from being exhaustive or objective, even inside the mentioned principle, in that the choice of spawning a new research field and citing a specific work as the beginning of that field, is largely subjective.

The point is not in the details of which processes are included or not, but rather in the fact that as long as artificial vision progresses, more and more new areas are discovered that are important components to be addressed in order to achieve an *intelligent* enough vision.

The scenario of natural language understanding also appears to be well described by our Eq. (17.5), with many new independent research fields that have started as this discipline has progressed, with one main difference from vision. Our current understanding of how our brain processes language is far from being as complete as that of the knowledge we have on the processing stages in vision. The precise brain areas involved in language and the characterization of the computational functions of those areas remain obscure. It is therefore, impossible to sketch a tree of stable processes spawning new ones where language is concerned. Lacking relationships with analogous brain processes, it is not possible today to know which of the many ongoing research fields in understanding natural language will continue to progress and to spawn new components, or that instead will lead to a dead end.

The history of efforts in simulating aspects of intelligence offers another recurrent clue related to the slowdown effect, not directly captured by the formalism given in section Formalization of the Slowing Down. In a number of cases a common pattern of discovery can be detected: an important achievement in simulating aspects of human behavior become on one hand, an example of progress, on the other, gives the illusion of easy progress, while its follow-up reveals how complex the overall functions are of which, it is just a component.

While the discovery of receptive cells in cortical area V1 (Hubel and Wiesel 1959) was a major breakthrough in the understanding of the visual system, that gave confidence in believing that this achievement would be a first step towards artificial vision comparable to that of humans, today it is clear that the computation done by V1 is but a small fraction, and the simplest, of that involved in the whole vision process. Not only, in V1, there is an overlap of processes that are much more complex than just the selectivity to orientation and ocularity.

A puzzle in the early era of neural computation was the simulation of language, requiring syntactic processing. Jerry (Elman 1990) made another breakthrough with his recurrent network, that exhibited syntactic and semantic abilities. It was a toy-model, with a vocabulary of just a few words, however, it was then presumed

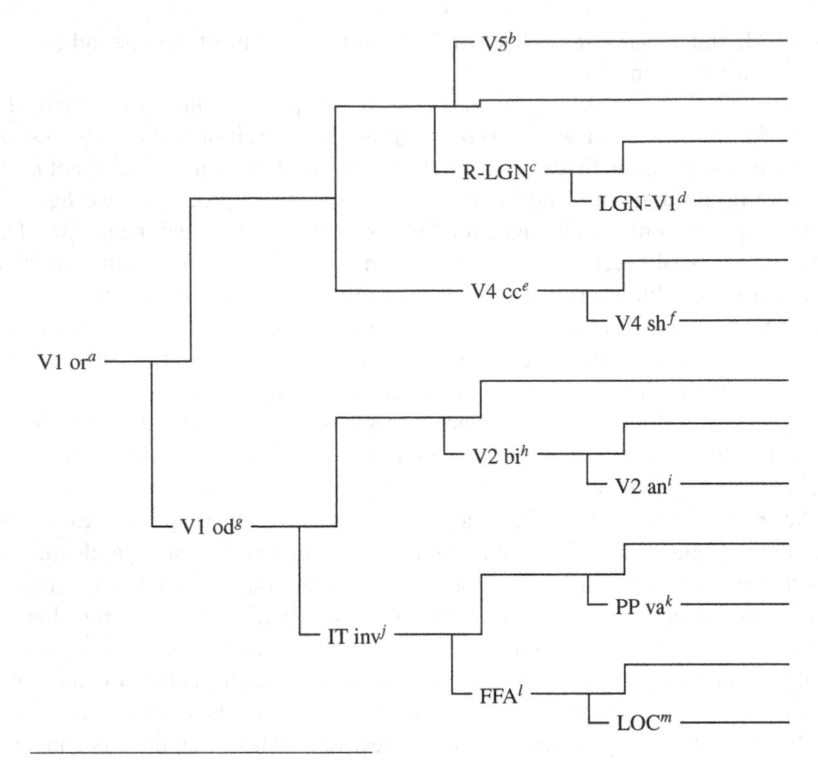

[a]orientation selectivity in V1, see (von der Malsburg 1973)
[b]models of motion perception in V5, see (Simoncelli and Heeger 1992)
[c]Lateral Geniculate Nuclues model, see (McCormick and Huguenard 1992)
[d]geniculo-cortical loop, see (Adorjan, Piepenbrock, and Obermayer 1999)
[e]color constancy in V4, see (Dufort and Lumsden 1991)
[f]shape selectivity in V4, see (Cadieu et al. 2007)
[g]ocular dominance in V1, see (von der Malsburg and Willshaw 1976)
[h]binocular disparity in V2, see (Zhaoping 2005)
[i]angle selectivity in V2, see (Plebe 2007)
[j]invariance in InferoTemporal Cortex, see (Rolls 1992)
[k]visual attention in the Posterior Parietal Complex, see (Deco 2001)
[l]Face Fusiform Area simulation, see (Jiang et al. 2006)
[m]Lateral Occipital Complex simulation, see (Plebe and Domenella 2006)

Fig. 17.3 Sketch of research field branching in the case of vision

that it would open the road to fast progress in simulating language. In the twenty years that have followed, no other model has achieved results that are comparable to Elman's. Minor improvements were gained at the price of much more complex systems (Miikkulainen 1993).

The biggest success in mathematical modeling of brain functions has been the H-H model of neural polarization (Hodgkin and Huxley 1952). Decades later a powerful simulator became available, based on the core equations of the HH model (Wilson and Bower 1989). Oddly enough, no mathematical model of similar importance for the brain has been developed since. Today, mathematical models are lacking or are oversimplified and limited for the most important phenomena at a cellular level, such as synaptic transmission (Rudolph and Desrexhe 2005), synaptic facilitation and depression (Dittman et al. 2000) or dendritic growth and axon arborization (van Pelt et al. 2010), and the reliability of numerical simulations of biological neurons, in itself, has been questioned (Rudolph and Desrexhe 2007).

From "Intelligence" to the (Rest of the) Mind

The variety of things we call "intelligence" and that we would like to give an account for within a scientific framework, is much greater now than in the past. We now know much more about intelligence, but this knowledge does not accumulate as before. On the contrary, it is lost in many different pathways, offering a more detailed understanding of "intelligence", but a less cumulative one than in the past. The point, however, is that since the very beginning of the adventure of Artificial Intelligence, the attempt to simulate intelligence has actually involved the whole mind. The privilege granted to the property of intelligence was based on the belief that it is the crucial property of the mind. Understanding intelligence is understanding the mystery of the mind. In reality, the true aim of Artificial Intelligence has never been intelligence, but the mind as a whole. The challenge was to produce a mind like that of humans, and do it artificially. To be like that of human beings, however, intelligence must be endowed with many skills that traditionally were not even associated with it, including the ability to experience emotions appropriate to circumstances and to reason metaphorically. This latter feature, for example, opened the way for attempts to develop computational models of metaphor. The road has proven difficult and at the same time fascinating because it could also shed some light on the metaphorical nature of thought and language. This, in turn, has drawn attention to the role that metaphors and frames have in many aspects of social life, as in the case of political discourse.

Consider the case of moral reasoning. It is a crucial kind of intelligence, if we want to create something that resembles human performance in a significant way. For this reason, researchers, including Wendell Wallach and Colin Allen, have devoted their efforts to the investigation of machine ethics. Is it possible to design machines endowed with ethical principles? That is, that have the capacity to reflect upon different alternatives or to compute a procedure for discovering a way to resolve the ethical dilemmas machines might encounter (Wallach and Allen 2008; Anderson and Anderson 2011)? In this perspective, developing an ethic for machines appears as an interdisciplinary endeavor. This circumstance, however, calls into question many other issues, including the normativity of moral judgments and the sensitivity to the social context of such judgments. The general point is that intelligence has a social nature and when observed in its human form, requires some kind of consciousness. It seems that the kind of intelligence typical of the singularity hypothesis is inspired by methodological solipsism, a widespread tenet in classical cognitive science. According to this way of thinking, the mind is considered as something that pertains to a given individual and consists of a set of skills that can be gradually amplified so as to exceed those of humans. But, if we accept the major claims of current cognitive science, we are also driven to consider the mind as a social and ecological thing. Externalists argue that to specify the content of many mental states one must take into consideration its reference rather than the manner in which, it is given to the mind (Menary 2010). According to theorists of embodied cognition, mental contents are determined by the way the body acts in the environment (Shapiro 2011). Moreover, the success of the account of social cognition, with its idea of the core ability to interpret behavior as a consequence of the mental states of its performer, has finally shown the limits of solipsism (Tomasello 2009). In sum, intelligence is no longer conceived as a mere individual property. In this way, the ghost of normativity and its intractable character appear on the horizon of experimental science.

Overall, if we observe the manner in which the science of mind has concretely evolved in recent decades, it seems that we are faced with a systematic tendency to investigate new areas of research originating from previous ones. In this way, as soon as results are achieved in a certain field, new questions arise and new areas of investigation open up.

Machines Designing Machines

There is a peculiar aspect of intelligence, which is crucial in the singularity hypothesis, and that is worth underlining. The key idea is that a machine that is more intelligent than humans will be better than humans at designing machines. Designing machines or developing algorithms, is a very special ability only a select few of humans have, and are not included in the common meaning of general intelligence. One may argue that it is possible to escape the slowing down implicit in Eq. (17.5), and move towards singularity, simply because it is not interested in general intelligence, but rather in the very specialized aspect of designing algorithms. How many examples of this aspect of intelligence can be found in artificial systems, even at very early stages? Not one. As far as we know, there is no example of any artificial system able to design new algorithms. The computer science domain dubbed "automatic programming" of "generative programming" or "automatic code generation" have nothing to do with designing algorithms. They are just tools that help human programmers write their code, at a higher abstraction level, for example , or by using templates and prototypes. So far, no sign of the aspect of intelligence that is most crucial to the singularity hypothesis, has been seen yet.

It is interesting to analyze under the perspective of aspects of intelligence, some of the most exciting achievements of in AI. Playing chess was one of the first challenges undertaken by AI (Shannon 1950), and was the one that (Dreyfus 1972) bet computers would never even come close to being able to do as well as human beings. The historic victory of IBM's Deep Blue over world chess champion Gary

Kasparov, therefore looked like a momentous one. Despite the positive reaction and renewed enthusiasm for the perspective of a strong AI, Deep Blue's victory involves a very marginal aspect of general intelligence. Unlike previous chess computer programs, Deep Blue's architecture was entirely focused on highly efficient database mining (Campbell et al. 2002). It worked on a database of about 700,000 grandmaster games, and at each move the current position was searched over the entire database for the closest one, at the speed of 200 million positions per second. A similar approach allowed a more recent success to Watson, another IBM supercomputer that won the American quiz show *Jeopardy!*. This system includes a structured and complete version of Wikipedia, that can be searched at a speed of about 500 GB per second. In this case the search is based on a sophisticated analysis carefully specialized for the type of clues used in the *Jeopardy!* challenge (Ferrucci et al. 2010). The 30-clue session is organized into six categories, that range from broad subject headings like "The European Union" to less informative puns like "One buck or less", to specific like "Cambridge". Wide samples of *Jeopardy!* questions were analyzed to classify the so-called LAT (*Lexical Answer Type*), which is a word in the clue that indicates the type of the answer, independent of assigning semantics to that word. For example in the chess category clue "Invented in the 1500s to speed up the game, this maneuver involves two pieces of the same color" the LAT is in the string "this maneuver", and the answer is "castling". Related to LAT is the focus of the clue, that is the part of the question that, if replaced by the answer, makes the question a meaningful standalone statement. For example in the clue category "Cambridge": "In 1546 this king founded Trinity College, the largest of Cambridge's colleges" the focus is "this king" and the answer is "Henry VIII". In the Watson system the result of the question analysis is used to interrogate its huge knowledge base adopting a variety of search techniques, generating many candidate answers that are further filtered and ranked. The system is impressively complex and efficient, however it is clearly highly tailored to the *Jeopardy!* quiz interaction, and would be almost unable to maintain a simple, ordinary conversation.

We do not claim that knowledge base mining is not intelligent, on the contrary an aspect of human intelligence is certainly the ability to retrieve from long term memory what is relevant to the current stream of thought. However, this aspect becomes rather shallow and limited when the whole system is highly specialized and with a single focus, like browsing chess grandmaster games only, or answering clues in the *Jeopardy!* format only.

There is another reason why AI champions, such as the two cases here analyzed, are unlikely to be steps on the way to a general intelligence machine. These systems are not only highly customized to fulfill their goals, they also lack any reference to how the brain works when pursuing the same goals. The extreme search of performance obliged the designers to abandon any attempt to implement processes with biological plausibility. As a result, the solutions cannot be counted as processes to sum with others in the direction of a general intelligence, they cannot be affected by the continuous progress of neuroscience, and cannot spawn new processes for new components of intelligence. IBM itself, seems well aware of this, and its current most promising line of research in AI is in a completely opposite direction to that of Deep Blue and Watson: that of mimicking in detail the functions of biological neurons in computer chips, for building the future *Cognitive computing* (Modha etal. 2011). The $21 million DARPA funded project SyNAPSE (*Systems of Neuromorphic Adaptive Plastic Scalable Electronics*) is an ambitious program to engender a revolutionary neuromorphic chip comprising one million neurons and 10 billion synapses per square centimeter. Several tech commentators have suggested this may be the beginning of singularity.[1]

We are convinced that this is one of the main roads to intelligent machines. However, the fastest hardware for simulating neural circuits would be useless, if we do not yet have clear ideas of what those circuits should be computing. Given how limited our knowledge still is on the computations done in our brain, to process language, for example, or sustain consciousness, we suspect that the path will be slow and problematic. Chalmers is right when he claims that if there is a singularity, it will be a turning-point in human history. However, if the slowdown hypothesis has some basis, then perhaps we should also worry about its possible chilling effect on the course of research on the mind. In fact, if it were to confirm the trend of research on the mind to slow its progress due to the persistent tendency of making the explanation of a given aspect of the mind depend on understanding other mental phenomena, then we risk appreciating the magnificent complexity of the human mind, but without knowing how to cope with it.

[1] http://www.techjournalsouth.com/2011/08/ibms-brain-like-cognitive-chips-can-learn-video/

Chapter 17A Eliezer Yudkowsky on Plebe and Perconti's "The Slowdown Hypothesis"

The hypothesis presented for a curve of diminishing returns of optimization power in versus intelligence out is incompatible with the historical case of natural selection, in which it did not take a hundred times as long to go from Australopithecus to humans as it did to go from the first brainstto Australopithecus, but rather the reverse. Many people have postulated logarithmic returns or other such diminishing returns to intelligence. They are easy to postulate.

It is much harder to make them fit the observed facts of either the evolution of intelligence (for talk about diminishing returns to brain size, genome size, or optimization pressure on the brain) or the history of technology (for talk about diminishing returns to knowledge or intelligence). Specifically exponential theories of progress are probably wrong, of course; Moore's Law has already broken down. But the historical cases we've observed are for roughly constant input processes producing increasing (though not always exponential!) outputs. Constant evolutionary pressure has produced, not exponential, but increasing outputs from hominid intelligence. A fourfold increase in hominid brains has not produced exponential returns, but to characterize the resulting returns as sublinear seems rather odd. In a nuclear pile, neutron multiplication factors are strictly linear—each neutron giving rise to 1.0006 output neutrons on average, for example—and the resulting pile of neutrons sparking other fissions would produce an exponential meltdown if not for external braking processes such as cadmium rods. For the novel phenomenon of recursively self-improving intelligence, where AI intelligence in is a direct function of AI intelligence out, rather than the AI intelligence being produced by a constant external optimization pressure such as human programmers, to fail to go FOOM once a threshold level of intelligence is reached, we need all these observed curves to exhibit a sudden sharp turnaround the moment they are past the level of human intelligence, and produce extremely sharply diminishing curves of intelligence-out versus optimization power in. Simply put, nobody has ever devised a realistic model of optimization power in versus optimization power out which both accounts for the observed curves of hominid history and human technology, which fails to exhibit an intelligence explosion once intelligences are designing new intelligences and a feedback loop is added from design intelligence to output intelligence. In fact, nobody has ever tried to develop such a model, and all attempts to postulate the lack of an intelligence explosion have done so by making up models which either completely ignore the new feedback loop and simply project normal economic growth out into the indefinite future without considering that AIs creating AIs might be in any way qualitatively different from a world of humans making external gadgets without tinkering with brain designs; or which simply ignore the observed parameters of evolutionary history and technological history in favor of making up plausible-sounding mathematical models in isolation which would have vastly mispredicted the observed course of history over the last ten million or ten thousand years, predicting observed diminishing returns rather than increasing ones. This paper falls into the second class.

References

Adorjan, P., Piepenbrock, C., & Obermayer, K. (1999). Contrast adaptation and infomax in visual cortical neurons. *Reviews in the Neurosciences, 10*, 181–200.

Anderson, M., & Anderson, S. L. (Eds.), (2011). *Machine Ethics*. Cambridge: Cambridge University Press.

Cadieu, C., Kouh, M., Pasupathy, A., Connor, C. E., Riesenhuber, M., & Poggio, T. (2007). A model of V4 shape selectivity and invariance. *Journal of Neurophysiology, 98*, 1733–1750.

Campbell, M., Hoane, A. J., & Hsuc, F. (2002). Deep blue. *Artificial Intelligence, 134*, 57–83.

Chalmers, D. (2010). The singularity: A philosophical analysis. *Journal of Consciousness Studies, 17*, 7–65.

Deco, G. (2001). Biased competition mechanisms for visual attention in a multimodular neurodynamical system. In S. Wermter, J. Austin, & D. Willshaw (Eds.), *Emergent neural computational architectures based on neuroscience: towards neuroscience-inspired computing* (pp. 114–126). Berlin: Springer-Verlag.

Dittman, J. S., Kreitzer, A. C., & Regehr, W. G. (2000). Interplay between facilitation, depression, and residual calcium at three presynaptic terminals. *Journal of Neuroscience, 20*, 1374–1385.

Dreyfus, H. (1972). *What Computers Can't Do: A Critique of Artificial Reason*. New York: Harper and Row Pub. Inc.

Dreyfus, H. L., & Dreyfus, S. E. (1986). *Mind Over Machine: The Power of Human Intuition and the Expertise in the Era of the Computer*. New York: The Free Press.

Dufort, P. A., & Lumsden, C. J. (1991). Color categorization and color constancy in a neural network model of V4. *Biological Cybernetics, 65*, 293–303.

Elman, J. L. (1990). Finding structure in time. *Cognitive Science, 14*, 179–221.

Ferrucci, D., Brown, E., Chu-Carroll, J., Fan, J., Gondek, D., Kalyanpur, A. A. et al. (2010). Building Watson: An overview of the DeepQA project. *The AI magazine, 31*, 59–79.

Gardner, H. (2006). *Multiple Intelligences: New Horizons*. New York: Basic Books.

Good, I. J. (1965). Speculations concerning the first ultraintelligent machine. In F. L. Alt & M. Rubinoff (Eds.), *Advances in Computers* (Vol. 6, pp. 31–88). New York: Academic Press.

Hodgkin, A. L., & Huxley, A. F. (1952). A quantitative description of ion currents and its applications to conduction and excitation in nerve membranes. *Journal of Physiology, 117*, 500–544.

Hubel, D., & Wiesel, T. (1959). Receptive fields of single neurones in the cat's striate cortex. *Journal of Physiology, 148*, 574–591.

Jiang, X., Rosen, E., Zeffiro, T., VanMeter, J., Blanz, V., & Riesenhuber, M. (2006). Evaluation of a shape-based model of human face discrimination using fMRI and behavioral techniques. *Neuron, 50*, 159–172.

McCormick, D. A., & Huguenard, J. R. (1992). A model of the electrophysiological properties of thalamocortical relay neurons. *Journal of Neurophysiology, 68*, 1384–1400.

Menary, R. (Ed.), (2010). *The Extended Mind*. Cambridge: MIT Press.

Miikkulainen, R. (1993). *Subsymbolic Natural Language Processing: and Integrated Model of Scripts, Lexicon and Memory*. Cambridge: MIT Press.

Modha, D. S., Ananthanarayanan, R., Esser, S. K., Ndirango, A., Sherbondy, A. J., & Singh, R. (2011). Cognitive computing. *Communications of the Association for Computing Machinery, 54*, 62–71.

Plebe, A. (2007). A model of angle selectivity development in visual area V2. *Neurocomputing, 70*, 2060–2066.

Plebe, A., & Domenella, R. G. (2006). Early development of visual recognition. *BioSystems, 86*, 63–74.

Rolls, E. (1992). Neurophysiological mechanisms underlying face processing within and beyond the temporal cortical visual areas. *Philosophical transactions of the Royal Society B, 335*, 11–21.

Rudolph, M., & Desrexhe, A. (2005). An extended analytic expression for the membrane potential distribution of conductance-based synaptic noise. *Neural Computation, 17*, 2301–2315.

Rudolph, M., & Desrexhe, A. (2007). An extended analytic expression for the membrane potential distribution of conductance-based synaptic noise. *Neural Computation, 17*, 2301–2315.

Searle, J. R. (1980). Mind, brain and programs. *Behavioral and Brain Science, 3*, 417–424.

Shannon, C. (1950). Programming a computer for playing chess. *Philosophical Magazine, 41*, 256–275.

Shapiro, L. (2011). *Embodied Cognition*. London: Routledge.

Simoncelli, E. P., & Heeger, D. J. (1992). A computational model for perception of two-dimensional pattern velocities. *Investigative Opthalmology and Visual Science Supplement, 33*, 1142.

Tomasello, M. (2009). *Why We Cooperate*. Cambridge: MIT Press.

van Pelt, J., Carnell, A., de Ridder, S., Mansvelder, H. D., & van Ooyen, A. (2010). An algorithm for finding candidate synaptic sites in computer generated networks of neurons with realistic morphologies. *Frontiers in Computational Neuroscience, 4*, 1–17.

Vinge, V. (1993). *The Coming Technological Singularity: How to Survive in the Post-human Era". Proc. Vision 21: Interdisciplinary Science and Engineering in the Era of Cyberspace* (pp.11–22). NASA: Lewis Research Center.

von der Malsburg, C. (1973). Self-organization of orientation sensitive cells in the striate cortex. *Kybernetic, 14*, 85–100.

von der Malsburg, C., & Willshaw, D. J. (1976). A mechanism for producing continuous neural mappings: ocularity dominance stripes and ordered retino-tectal projections. *Experimental Brain Research, 1*, 463–469.

Wallach, W., & Allen, C. (2008). *Moral Machines: Teaching Robots Right from Wrong*. Oxford: Oxford University Press.

Weisberg, M. (2007). Three kinds of idealization. *The Journal of Philosophy, 12*, 639–661.

Wilson, M. A., & Bower, J. M. (1989). The simulation of large-scale neural networks. In C. Koch& I. Segev (Eds.), *Methods in Neuronal Modeling* (pp. 291–333). Cambridge: MIT Press.

Zhaoping, L. (2005). Border ownership from intracortical interactions in visual area V2. *Neuron, 47*, 143–153.

Chapter 18
Software Immortals: Science or Faith?

Diane Proudfoot

Techno-Supernaturalism

According to the early futurist Julian Huxley, human life as we know it is 'a wretched makeshift, rooted in ignorance'. With modern science, however, 'the present limitations and miserable frustrations of our existence could be in large measure surmounted' and human life could be 'transcended by a state of existence based on the illumination of knowledge' (1957a, p. 16). What we need, Huxley claimed, is 'a new religion' (1957b, p. 309)—an 'evolution-centred religion' that utilizes 'the findings of science' (1964, p. 223, 1957b, p. 305).

Supernaturalism has been rendered 'untenable' by scientific progress, in Huxley's view:

> The supernatural hypothesis, taken as involving both the god hypothesis and the spirit hypothesis ..., appears to have reached the limits of its usefulness as an interpretation of the universe and of human destiny, and as a satisfactory basis for religion. It is no longer adequate to deal with the phenomena, as revealed by the advance of knowledge and discovery (1957c, pp. 284, 285).

We are in a 'new religious situation', he argued: 'There should no longer be any talk of conflict between science and religion. ... On the contrary, religion must now ally itself wholeheartedly with science' (ibid., pp. 287, 288).

Ray Kurzweil's account of the 'Singularity' and post-Singularity life is a prototype of the 'new religion' that Huxley advocated. Modern technological futurists replace supernaturalism with what I shall call *techno-supernaturalism*—a fusion of science and religion. 'What awaits is not oblivion but rather a future which, from our present vantage point, is best described by the words

D. Proudfoot (✉)
Department of Philosophy and Political Science,
University of Canterbury, Christchurch, New Zealand
e-mail: diane.proudfoot@canterbury.ac.nz

A. H. Eden et al. (eds.), *Singularity Hypotheses*, The Frontiers Collection,
DOI: 10.1007/978-3-642-32560-1_18,

"postbiological" or even "supernatural"', Hans Moravec says (1988, p. 1). The Singularity will 'infuse the universe with spirit',[1] and evolution will move toward the conception of God in 'every monotheistic tradition'—'infinite knowledge, infinite intelligence, infinite beauty, infinite creativity, infinite love' (Kurzweil 2006d, p. 389). Kurzweil asks '[H]ow do we contemplate the Singularity?', and answers 'As with the sun, it's hard to look at directly; it's better to squint at it out of the corner of our eyes' (ibid., p. 371). St Anselm wrote of God in similar terms: 'I cannot look directly into [the light in which God dwells], it is too great for me … it is too bright … the eye of my soul cannot bear to turn towards it for too long'.[2] Techno-supernaturalists also predict a post-Singularity future that is remarkably similar to the post-salvation (or post-spiritual liberation) life promised by major world religions: a software-based existence that is 'immortal', 'truly meaningful', and 'blissful'. Modern physics and computer science, they claim, can give us 'transcendence', 'resurrection', 'souls', 'spirit', and 'heaven'. Becoming non-biological is 'an essentially spiritual undertaking', according to Kurzweil (2006d, p. 389).

Proponents of the Singularity hypothesis claim that their approach is very different from that of (what Kurzweil calls) 'traditional' religions (Kurzweil 2006d, p. 370). Although techno-supernaturalists describe a 'supernatural' future, they have no time for 'ornate dualis[m]', claiming that 'transcendence' can be found in the material world and 'spiritual machines' will appear later this century (ibid., p. 388; Kurzweil 1999). Their forecasts are based, they say, on scientific evidence rather than faith. And in their view traditional religions attempt to rationalize death as giving meaning to life, whereas death is really a 'tragedy' that technology will soon postpone indefinitely (Kurzweil 2006d, pp. 326, 372–374).[3] My aim in this essay is to show that techno-supernaturalism is much like supernaturalist religion, philosophically speaking. I analyse similarities between techno-supernaturalism and 'old' religions (in the section 'The New Good News'), and argue that the former's account of spiritual resurrection faces severe philosophical problems (in the section 'The Perils of Being a Pattern'). Techno-supernaturalism, moreover, seems to be based as much on faith as on science (see the section 'Doctrine and Faith'), and also may fail as a terror management strategy (see the section 'Terror Management'). If death is a tragedy, then—despite the confidence of Kurzweil and others—our only options seem still to be tragedy or mysticism.

[1] The Singularity will infuse the universe with 'spirit' in the sense that, Kurzweil predicts, we will be able to convert much of the matter of the universe into 'computronium'—the 'ultimate computing substrate' (2007b). Moravec too hypothesizes that the entire universe might be converted into 'an extended thinking entity, a prelude to even greater things' (1988, p. 116).

[2] Anselm 1078/1973, Chap. 16 (p. 257).

[3] Kurzweil calls the belief that death gives meaning to life the 'deathist meme' (Olson and Kurzweil 2006).

The New Good News

Techno-supernaturalism adds the advantages of being software-based to Huxley's vision of the future. Kurzweil—along with Nick Bostrom, Moravec, Frank Tipler, and other technological futurists—makes dramatic predictions, including the following. Human beings will be 'enhanced', using advances in biotechnology and nanotechnology; by the 2030s, with nanobot implants, we will be 'more non-biological than biological' (Kurzweil 2007a, p. 19). Humans will become 'trans-humans' (e.g. Bostrom 2005b). Human-level AI will be achieved spectacularly soon—by 2029, Kurzweil claims[4]—and superhuman-level AI will follow shortly thereafter. This will lead, by '2045, give or take' (Kurzweil in Else 2009), to the Singularity—'technological change so rapid and profound it represents a rupture in the fabric of human history' (Kurzweil 2001).[5] This change includes the emergence of 'immortal software-based humans' (ibid.)—'posthumans', 'ex-humans', or 'postbiologicals'.

Specifically, high-resolution neuro-imaging, coupled with the reverse-engineering of the human brain and the discovery of 'the software of intelligence', will enable us to simulate an individual human brain: a human being's 'personal mind file' can then be 'reinstantiated' in a non-biological substrate (Kurzweil 2001):[6]

> Ultimately software-based humans will be vastly extended beyond the severe limitations of humans as we know them today. They will live out on the Web, projecting bodies whenever they need or want them, including virtual bodies in diverse realms of virtual reality, holographically-projected bodies … and physical bodies comprising nanobot swarms and other forms of nanotechnology. (Kurzweil 2006d, p. 325)

The rewards of being software-based include, techno-supernaturalists claim, being able to 'think a thousand times faster' and 'transmit oneself as information at the speed of light' (Moravec 1988, p. 112; Bostrom 2005a, p. 7). A software-based person will be able to send her mind to a robot body on a faraway star, 'explore, acquire new experiences and memories, and then beam [her] mind back home' (Moravec 1988, p. 114). Software-based persons will have capabilities that humans now 'can only dream about', and with proper data storage they will be immortal (Tipler 2007, p. 76; Kurzweil 2006b, p. 44).

---

[4] In 2001 Kurzweil predicted that we would be able to build hardware matching the computational capacity of the human brain by 2010, and in 2006, he predicted software enabling a machine to match a human's cognitive capacities by 2029—i.e., by 2029 machines will be able to pass the Turing test (Kurzweil 2001, 2006a). Tipler predicts human-level AI by 2030 (2007, p. 251).

[5] According to Goertzel (2007b), with a coordinated effort we could reach the Singularity even earlier—by approximately 2016.

[6] Other futurists, who are not techno-supernaturalists, make similar claims about the possibility or indeed feasibility of uploading: see e.g. Goertzel 2007a, b.

Simulation Resurrection

Huxley said that we need ‘a new religious terminology and a reformulation of religious ideas and concepts in a new idiom’ (1964, p. 224). Techno-supernaturalism and supernaturalism have a shared metaphysic of personhood and of the persistence of individual persons; typically they identify the person with the mind or soul and hold that an individual survives if (and only if) his or her mind survives (see the section ‘The Perils of Being a Pattern’). Both downplay the human body, on the ground that biology imposes limitations on the mind. For both techno-supernaturalism and theology influenced by dualism (e.g. Platonic or Aristotelian), cognition is independent of any particular bodily process. The ideas are analogous, but the terminology different. According to the 12th century rabbi Moses Maimonides, for example, the body is ‘only the carrier of the soul’[7] and ‘the world-to-come is made up of souls without bodies, like the angels’ (1191/1985, pp. 214, 220).[8] A human being is ‘empty and deficient’ in comparison with angels (c. 1178/1981, p. 39b). In place of ‘incorporeal’ souls, techno-supernaturalists have the *human-soul program*—‘an immaterial entity generated by the activity of neurons in a human brain’ (Tipler 2007, p. 70)—and in place of spiritual resurrection, they have *simulation resurrection* (e.g. Tipler 1994).

Tipler argues that conceiving of the soul as ‘a sort of white, ghostly substance that permeates the human body, to be released at death’ is a mistake, since (on this conception) if the soul existed it ‘would indeed *be* a substance, and hence material’ (2007, p. 70). In contrast, he says, if the soul is ‘nothing but a program being run on a computer called the brain’ (1994, p. xi), it *is* immaterial—and in this sense a ‘spiritual’ entity (2007, pp. 70, 80). This reasoning, however, merely begs the question against those dualists who claim that there are two sorts of substance, material and immaterial, and that the soul consists of the latter. Further, in the sense of ‘immaterial’ that can be predicated of substance, programs are surely not immaterial—they are abstract, like concepts and propositions, rather than immaterial.

God will ‘destroy death forever’, certain traditional scriptures say—for techno-supernaturalists, on the other hand, *godlike artificial intelligences* will ensure immortality for human beings.[9] Techno-supernaturalists predict ‘[w]holesale resurrection’ by artificial ‘[s]uperintelligent archaeologists armed with wonder-instruments’ (Moravec 1988, pp. 122, 123). It is in principle possible, they say,

---

[7] According to Maimonides, a human being is composed of a ‘substance’ and a ‘form’ (c. 1178/1981, p. 38a). The afterlife will be made up of ‘separated souls’, which are ‘divested of anything corporeal’ (1191/1985, pp. 215, 216). Angels are ‘forms without substance’ (c. 1178/1981, p. 39a).

[8] Maimonides also endorsed the doctrine of physical resurrection. Prior to the world-to-come, God can return the soul to the body, enabling the individual to live another long life. ‘Life in the world-to-come follows the Resurrection’, Maimonides said (1191/1985, p. 217).

[9] Isaiah 25:8. *Tanakh*: A New Translation of THE HOLY SCRIPTURES According to the Traditional Hebrew Text. Philadelphia: The Jewish Publication Society, 1985.

that 'all human beings that have ever lived and will live from now to the end of time can be resurrected' (Tipler 1994, p. 248). Artificial 'Minds' that are 'vast and enduring' and 'unimaginably powerful' will be able to reconstruct all information about past humans and simulate their minds, bodies, histories, and environments (Moravec 1999, pp. 167, 168; Tipler 1994, pp. 219, 227). Posthumans will be 'living memories' in these machines—'more secure in their existence, and with more future than ever before, because they have become valued houseguests of transcendent patrons' (Moravec 1999, p. 167). In short, it is in post-Singularity artificial intelligences—rather than in God—that we will 'live, and move, and have our being'.[10]

Some techno-supernaturalists make explicit connections to established religions. According to St Paul, after death a human being will have a 'spiritual body' to replace his or her physical body.[11] Tipler says that this is 'completely accurate'—the 'spiritual body' is just the posthuman's simulated body, which is a 'vastly improved' and 'undying' body (1994, p. 242).[12] In Tipler's view, simulation resurrection also provides *physical* resurrection (the Hebrew Bible says, 'Let corpses arise! Awake and shout for joy, You who dwell in the dust'[13]). 'There would be nothing "ghostly" about the simulated body, and nothing insubstantial about the simulated world in which the simulated body found itself', he claims (1994, p. 242).

The notion of simulation as physical resurrection raises vividly the fundamental question about 'simulation resurrection'—is it a type of resurrection or a mere simulacrum of resurrection? Techno-supernaturalists claim that simulated people will be 'as real as you or me, though imprisoned in the simulator' (Moravec 1988, p. 123).[14] Simulation is recreation, they argue, just because '[t]o a simulated entity, the simulation *is* reality' (Moravec 1999, p. 168). Likewise, simulation is physical resurrection, just because to the simulated person his or her body is 'real' and 'solid' (Tipler 1994, p. 242). However, it is only if a simulation of a cognizing subject *is* itself a cognizing subject that there is any entity to whom 'the simulation is reality', or to whom his or her body is 'real'. The techno-supernaturalists' reasoning plainly begs the very question at issue. Moreover, their claim that programs are immaterial—i.e. not substances—contradicts their other claim that there is nothing 'insubstantial' about a simulated world.

---

[10] Acts 17:28. *The Holy Bible*, King James Version.

[11] 1 Corinthians 15:44. *The Holy Bible*, New Revised Standard Version. New York: Oxford University Press, 1989.

[12] See too Steinhart 2008.

[13] Isaiah 26:19. *Tanakh*: A New Translation of THE HOLY SCRIPTURES According to the Traditional Hebrew Text.

[14] If true, how can we know that *this* life isn't a simulation (Tipler 1994)? The notion of simulation resurrection leads to the 'simulation argument' (see Bostrom 2003b). On sceptical arguments based on simulation-resurrection (or 'matrix') thought-experiments, see further Weatherson 2003; Chalmers 2005; Brueckner 2008; Bostrom 2009b; Bostrom and Kulczycki 2011. On the simulation argument with a theological twist, see Steinhart 2010.

God, Heaven and Transcendence in the New Religion

What will software-based life be like? Techno-supernaturalists claim that science now offers '*precisely* the consolations in facing death that religion once offered' (Tipler 1994, p. 339). According to Maimonides, life in the world to come is 'everlasting' and 'entirely blissful'—it is the 'ultimate and perfect reward, the final bliss which will suffer neither interruption nor diminution' (c. 1178/1981, pp. 92a, 91b, 92a). According to techno-supernaturalists, posthumans will feel 'surpassing bliss' (Bostrom 2008a).[15] Posthuman life is 'a higher state of being ... Beyond dreams. Beyond imagination':

> There is a beauty and joy here that you cannot fathom. It feels so good that if the sensation were translated into tears of gratitude, rivers would overflow. ... It's like a rain of the most wonderful feeling, where every raindrop has its own unique and indescribable meaning (Bostrom 2008a)

This certainly sounds like *heaven*, and indeed Tipler says that life after death will take place in 'an abode that closely resembles the Heaven of the great world religions' (1994, p. xi).

Many established religions claim that transcendence, typically through salvation or liberation, makes possible a *meaningful* life. Likewise, according to techno-supernaturalists, the 'freeing of the human mind from its severe physical limitations of scope and duration [is] the necessary next step in evolution'—it 'will make life more than bearable; it will make life truly meaningful' (Kurzweil 2001; 2006d, p. 372). Software-based persons will be able to find 'virtually inexhaustible sources [of] meaning' in creative and intellectual pursuits (Bostrom 2008b, p. 135).

Some futurists also hold on to a concept of *God*. For Huxley, the word 'God' is 'one name for the Universe'—identifying God with the universe, he said, frees God from 'the anthropomorphic disguise of personality' (1927, pp. 17, 18). Kurzweil also recommends that we not 'restrict' our view of God, and encourages the use of 'new metaphors' to capture the idea of God—'to attempt to express what is inherently not fully expressible in our finite language' (2002, p. 218). Techno-supernaturalism thus will even have room for *ineffability*. On God, Kurzweil remarks:

> Once we saturate the matter and energy in the universe with intelligence, it will "wake up," be conscious, and sublimely intelligent. That's about as close to God as I can imagine. (2006d, pp. 375)

For techno-supernaturalists, humans create God, rather than the traditional other way around.[16]

---

[15] Moravac does not share the view of the posthuman future as *heaven*—as he points out, '[s]uperintelligence is not perfection' (1988, p. 125). See further the section 'Doctrine and Faith'.

[16] According to Kurzweil, super-intelligent humans may engineer new universes (2007b)—another behaviour typically attributed to God.

Supernaturalism, Huxley said, is 'repugnant and indeed intellectually immoral to a growing body of those trained in the scientific tradition' (1957b, p. 285). Techno-supernaturalism asks many of the same 'big' questions as established religions, and provides similar answers. Can it escape the acute conceptual difficulties for supernaturalism?

The Perils of Being a Pattern

Predictions of human-level and superhuman-level AI in the near future are frequently based solely on expected increases in computing power (see e.g. Joy 2000; Moravec 1998). According to Bostrom, if 'Moore's Law' continues to hold, 'the speed of artificial intelligences will double at least every two years. Within fourteen years after human-level artificial intelligence is reached, there could be machines that think more than a hundred times more rapidly than humans do' (Bostrom 2003a, p. 763; see also 2006). However, Gordon Moore's detailed (1965, 1975) projections were confined to computational resources at the chip level. Moore's projections provide no reason to think that relative increases in computer speed will be matched (let alone exactly matched) by increases in the speed of computer thought.

Kurzweil bases his predictions on the 'Law of Accelerating Returns', which (he claims) governs the rate of progress of *any* evolutionary process—and thus software as well as hardware. His predictions about the evolution of technology—including brain scanning and simulation—have been vigorously criticized (e.g. Allen and Greaves 2011; Ayres 2006; Hawkins 2008; Modis 2006; Nordmann 2008; Devezas 2006). Likewise his claims about the emergence of human-level and superhuman-level artificial intelligence (e.g. McDermott 2006; Horgan 2008; Zorpette 2008; Hofstadter 2008).[17] And as to the Singularity, Moore himself says that it will never occur (2008).

Kurzweil's confidence that human-level AI will be achieved by the end of the 2020s (e.g. Kurzweil 2011) runs counter to the many in AI who regard the goals of 'human-level' AI and 'artificial general intelligence' as ill-defined (e.g. McDermott 2007; Sloman 2008) or unproductive (e.g. Whitby 1996; Ford and Hayes 1998)—and to those working instead in 'narrow' AI on task-specific systems or 'mindless intelligence' (e.g. Pollack 2006). Some researchers even fear that Singularitarianism may harm AI, by exaggerating its successes—for this reason, Drew McDermott wishes that Kurzweil 'would stop writing these books!' (2006, p. 1233). But even if these various criticisms are unfounded, and even if Kurzweil's predictions concerning the evolution of technology and of cognitive and computational neuroscience are broadly accurate, these predictions cannot by themselves justify his forecast about the evolution of immortal software-based persons. For this, techno-

[17] See too criticisms of Kuzweil's claims about the history of computing (Proudfoot 1999a, b).

supernaturalism requires a suitable theory of what it is to be a *person*, and for a person to *survive*.

According to Tipler, 'the word "person" refers to the total individual human mind', and 'the human mind—and the human soul—is a very complex computer program' (1994, pp. 125, 156). Kurzweil claims similarly that a person is 'a profound pattern (a form of knowledge) which is lost when he or she dies' (2006d, p. 372)—this is the pattern of 'information' comprising an individual's 'mind file'. For techno-supernaturalists, death is a tragedy because it involves a loss of information—its worst feature is 'the wanton loss of knowledge and function' (Moravec 1988, p. 121).

With respect to survival, Kurzweil claims, 'I am principally a pattern that persists in time' (2006d, p. 386). Moravec advocates much the same theory of persistence conditions for persons, which he calls 'pattern-identity':

> Pattern-identity ... defines the essence of a person ... as the *pattern* and the *process* going on in my head and body, not the machinery supporting that process. If the process is preserved, I am preserved. The rest is mere jelly. (1988, p. 117; see also 1992, p. 18)

Tipler too claims that '[t]he pattern is what is important, not the substrate': the 'identity of two entities which exist at different times lies in the (sufficiently close) identity of their patterns' (1994, pp. 127, 227). Tipler's position is in fact internally inconsistent. He claims that the identity of a person over time 'is ultimately a question of physics'; two persons 'in the same quantum state are the same person' (1994, pp. 229, 234). Yet he also says that 'if the essential personality is simulated, this is good enough to be identified with the original person' (ibid., p. 226). As necessary and sufficient conditions, these different analyses can come into conflict. Two persons could both be the one person, by virtue of having the same 'essential personality'—but not be the one person, because they are in different quantum states.

The theory of persistence conditions for persons provided by techno-supernaturalists can be expressed as follows: a human being *A* and a (future) uploaded computer file *B* are the one person if (and only if) *A's brain and B instantiate the same pattern*—or if (and only if) *A*'s brain and *B* are, as Kurzweil puts it, 'functionally equivalent'.[18]

At the heart of techno-supernaturalism is simulation resurrection, and at the heart of simulation resurrection is the just-described account of personhood and survival.[19] Can this account—'patternism', to use Kurzweil's term—successfully ground techno-supernaturalism's move from technology to immortality?

---

[18] Patternism typically addresses the brain, despite Moravec's reference to brain and body.

[19] Goertzel (2007b) also uses the term 'pattern' (and 'patternist philosophy of mind'), claiming that the mind is a 'set of patterns'. According to Goertzel, 'the mind can live on via transferring the patterns that constitute it into a digitally embodied software vehicle'. What lives on is a 'digital twin'.

Patternism's Pitfalls

Patternism is hard to pin down. A 'patternist', Kurzweil says, 'views patterns of information as the fundamental reality' (2006d, p. 5). So, in order to 'upload' a 'mind file', do we copy brain organisation at the level of atoms and sub-atomic particles, or of neurons and sub-neuronal connections, or of the computations (that futurists typically assume are) realized in the brain? Techno-supernaturalists give no clear answer.

Moravec appears to opt for physics, since he illustrates the 'pattern-identity position' by means of a hypothetical 'matter transmitter' that duplicates atoms (1988, p. 117). Kurzweil opts for biology when he says that the mind file consists of 'the pattern that we call our brain (together with the rest of our nervous system, endocrine system, and other structures ...)' (2006d, p. 325). Yet he also emphasizes information-processing and says that uploading captures a person's 'entire personality, memory, skills, and history' (2006d, p. 199)—this suggests that what matters in uploading is simulating the person's psychology, in whatever way that correlates (or does not correlate) to biology. Bostrom proposes 'whole brain emulation'—simulation 'at a relatively fine-grained level of detail' (Sandberg and Bostrom 2008, p. 7). But what level is this? Techno-supernaturalism is typically vague on this point. For example, Kurzweil says 'If we emulate in as detailed a manner as necessary everything going on in the human brain and body ... why wouldn't it be conscious?' (2006d, p. 375). Is this to simulate functional blocks and hyperblocks of neurons, or individual neurons, or individual chemical transactions in the synapses? If all techno-supernaturalists can say about the required level of simulation is that it is the level *sufficient for intelligence* (or consciousness), their theorizing about simulation is no more than a promissory note.[20] In addition, Bostrom assumes that 'brain activity is Turing-computable, or if it is uncomputable, the uncomputable aspects have no functionally relevant effects on actual behaviour' (Sandberg and Bostrom 2008, p. 15). This assumption is mere speculation: it may be that 'functionally relevant' brain activity is not simulable by Turing machine in any interesting sense (see Copeland 2000; Proudfoot and Copeland 2011).

Why be a patternist? What matters for personhood, according to Tipler, is 'not the shape and form of a being but rather whether or not he-she-it can talk to you on a human-level'—and so, 'if they can talk to us, human downloads are people' (2007, p. 75). Specifically, a person just is 'a computer program which can pass the Turing test' (Tipler 1994, p. 124). What matters for persistence, techno-supernaturalists claim, must be a person's 'pattern', since survival of the body is 'irrelevant'—'Every atom present within us at birth is likely to have been replaced half way through our life. Only our pattern, and only some of it at that, stays with us until our death' (Tipler 1994, p. 227; Moravec 1988, p. 117; see also Kurzweil 2006d, p. 383).

[20] Similarly, Bostrom says that a brain scan must be detailed enough to capture the 'features that are functionally relevant to the original brain's operation' (Bostrom and Yudkowsky 2011). But *which* features are these?

However, these arguments are unconvincing. Turing proposed satisfactory performance in the imitation game as a criterion of human-level intelligence in machines, not of personhood. Indeed, some critics of the Turing test argue that it is logically possible that a program could pass the test yet not be conscious, or have even the in-principle capacity for consciousness—if so, intuitively such a program would not be a person. And with respect to survival, the fact that the microstructure of *A*'s body changes over time (by the progressive shedding and regeneration of cells, say) does not entail that *A*'s body as a macro-object does not persist. Viewed in this way, it is true of *A*'s one and only one body that previously it had microstructure 1 and now it has microstructure 2. This leaves it open whether *A*'s body is crucial to *A*'s survival.

In addition to the above, patternism, as a theory of what it is to be a person, faces two broad challenges. The first is the claim that 'the shape and form of a being' *does* matter to personhood. For example, philosophers have argued that the concept of a person is logically connected to the concept of the body—we predicate 'personal' (including psychological) properties only of an entity that also has bodily properties.[21] Also, numerous researchers in philosophy, cognitive science, and AI have argued that only an embodied and situated entity is capable of cognition (e.g. Brooks 1991, 1995, 1999; Dreyfus 1992, 2007; Dreyfus and Dreyfus 1986; Clark 1997; Shapiro 2011). Using this reasoning, a 'mind file' that 'lives out on the Web' either is not a person or is a person denied the rich cognitive experiences of posthuman life.

The second challenge is the familiar 'hard problem' of *consciousness* (Chalmers 1995). Techno-supernaturalists typically endorse some version of the computational theory of mind; Bostrom, for example, claims that it suffices for 'the generation of subjective experiences that the computational processes of a human brain are structurally replicated in suitably fine-grained detail' (2003b, p. 244; see also Sandberg and Bostrom 2008). However, a variety of 'qualia'-related thought experiments have been designed to counter functionalist theories of mind, and these also apply here. For example, it might be argued, we can conceive of a machine that replicates 'the computational processes of a human brain'—and is even embodied and situated—but is a zombie. Although behaviourally and computationally indistinguishable from a human being, the machine lacks phenomenal consciousness.[22] Bostrom allows that an 'artificial intellect' that has no phenomenal consciousness—a 'sapient zombie'—may nevertheless be a person (Bostrom and Yudkowsky 2011). However, being a person is of little value to an uploaded mind file, if that upload has no access to the 'surpassing bliss' and 'rain of the most wonderful feeling' offered by posthuman life.

---

[21] The locus classicus is Strawson 1959. Philosophers have also argued that, since human beings are animals, the appropriate persistence conditions for human persons are those for biological organisms (e.g. Olson 1997).

[22] On zombie thought-experiments, see Chalmers 1995; Block 1995; McCarthy 1995; Dennett 1995; Flanagan and Polger 1995; Sloman 2010.

Me, Myself, and I

As a theory of the persistence of persons, patternism faces the notorious *duplication* (or re-duplication) problem.[23] Let us suppose that we simulate the pattern of a human being *A*'s brain, and after *A*'s death simultaneously upload *two* duplicate files, *B* and *C*. According to patternism, *A* and *B* are one person, by virtue of the 'functional equivalence' of *A*'s brain and *B*. Likewise *A* and *C* are one person. This is problematic. If *A* is identical to *B* and *A* is identical to *C*, then (by transitivity of identity) *B* is identical to *C*.[24] So *B* and *C* are one person—even if *B* lives 'out on the Web' while *C* is in a robot body on a faraway star, and each is subject to very different inputs. The *one* person is in *two* places at once. This looks like a contradiction—and if the contradiction is real rather than merely apparent, patternism is in dire straits. The duplication problem is hard; if *A* and *C* are *not* one person, then by parity neither are *A* and *B* one person, and so *A* does not survive death.

The duplication problem arises, a techno-supernaturalist might say, only if there is *in fact* more than one instantiation, biological or non-biological, of *A*'s mind file at any one time. A 'vast and enduring' and 'unimaginably powerful' superintelligent 'Mind' would (it might be argued) know this, and so would upload only one duplicate at a time. But techno-supernaturalists themselves claim that several 'back-up' copies are required, to guard against hardware and software problems; Moravec says, for example, 'With enough widely dispersed copies, your permanent death would be highly unlikely' (1988, p. 112). And it would be risky to upload a copy only when needed, since bugs, copying errors, or other problems might cause it to fail. Moreover, attempting to solve the duplication problem by limiting the number of actual duplicates is, inappropriately, a merely practical reply to a logical problem.[25] It is like trying to solve the 'unexpected examination' paradox (a 'surprise' exam scheduled to take place next week could not be on Friday as by Thursday the students would expect it, or by the same reasoning on Thursday ... or *any* day next week![26]) merely by forbidding all unexpected exams.

According to Moravec, there is no problem in supposing that both *B* and *C* are *A*—the assumption that 'one person corresponds to one body' is simply 'confusing and misleading' (1988, p. 118). Uploading two or more simulations of your brain merely results in 'two or more thinking, feeling versions of you', he says (ibid., p. 112).[27]

---

[23] The classic statement of the duplication problem is found in Williams 1973a, p. 77; 1973b, p. 19. Making bodily continuity a necessary condition of persistence of persons still allows an analogous problem arising from 'fission' (see Parfit 1987, pp. 254–261).

[24] Here, as elsewhere in this essay, I suppress the symmetry step $A = B \vdash B = A$.

[25] Likewise, if a back-up of *A* is a mere copy, then it is a mere copy even if *in fact* it is the only backup: a mere copy that is actually created has no more claim to be *A* than any other back-up that might have been created. (Using the standard distinction, *A*'s duplicate is qualitatively, but not numerically, identical to *A*.)

[26] See Sainsbury 2009, pp. 107–109.

[27] Cf. Steinhart's notion of a 'variant' (2002, pp. 311, 312).

This is to treat a person as a *type* (and so A, B, and C are type-identical, even though they are different instantiations of the type). Knowing that a duplicate exists might be a 'comfort' if you were in danger, Moravec says; if instead you still attempt to protect yourself, this response is merely 'an evolutionary hangover from your one-copy past, no more in tune with reality than fear of flying is an appropriate response to present airline accident rates' (ibid., p. 119).

Nevertheless, Moravec allows that over time two 'active' copies of A will become distinct persons, as they acquire different experiences and memories (1988, p. 119). Let us call these new persons B^* and C^*; B becomes B^* and C becomes C^*. What differentiates B and B^*? Moravec says only that this question is 'about as problematical as the questio[n] "When does a fetus become a person?"' (ibid., p. 119). Differentiating B and B^* is not the only unsolved problem; Moravec's strategy leads again to what looks like contradiction. If (as he claims) B and C are the one person, B and B^* are the one person, and C and C^* are the one person, then (by transitivity of identity) B^* and C^* are the one person. But, Moravec claims, B^* and C^* are *distinct* persons. Introducing the notion of 'versions' of a person can avoid these contradictions if A merely *survives as* a future version B or C (and B merely survives as B^*).[28] This is to deny the assumption, not (as Moravec claims) that 'one person corresponds to one body', but that the relation between the pre-mortem person and the post-mortem person is *identity*. However, the cost is that, whatever it is that survives my death, it is not *me*.[29]

The enormous philosophical literature on identity of persons over time is replete with technical distinctions. It may be that, by using the notion of (for example) relative identity or persistence in four dimensions, the patternist can find a way around the duplication problem—without abandoning the claim that the pre-mortem person and the post-mortem person are one and the same. But the patternist has a difficult task ahead. Bostrom, on the other hand, sidesteps the issue. In his view, it may be that the posthuman's 'mode of being' is so radically different from that of the human from whom the posthuman originated that these two are not the one person (2005a, p. 8).[30] This avoids the duplication problem, but only because it jettisons entirely the notion of simulation resurrection.

Even if humans will not survive death, Bostrom claims, we should still try to create posthumans. He says, 'Preservation of personal identity ... is not everything. We can value other things than ourselves' (2005a, p. 9). We need the 'posthuman realm' if we are to make progress, since '[t]here are limits to how much can be achieved by low-tech means such as education, philosophical

---

[28] On the notion of A's 'surviving as' (rather than being identical to) both B and C, see Parfit 2008. On 'survival as' a digital 'ghost', see Steinhart (2007, 2010). Chalmers (2010) also suggests this move.

[29] The proponent of replacing identity with survival-as regards the cost as minimal—'this way of dying is about as good as ordinary survival', Parfit claims (1987, p. 264).

[30] Bostrom gives mixed signals on the question of survival. He also claims that, as an uploaded mind file, one will have 'the ability to make back-up copies of oneself (favorably impacting on one's life-expectancy)' (2005a, p. 7).

contemplation, moral self-scrutiny' (ibid., p. 9). There is another advantage to simulation, in Bostrom's view: it would take only a very short time to copy an upload, leading to 'rapidly exponential growth in the supply of highly skilled labour' (2009a, p. 207; see also Bostrom 2004). This may seem a poor substitute for personal survival in the techno-supernaturalists' version of heaven.

Careful, That Upgrade Might Kill You

Can a posthuman—whether or not identical to a prior human being—be immortal? According to techno-supernaturalists, posthuman life is truly meaningful because it enables continuous growth; a software-based person might 'undergo a cyclical rejuvenation, acquiring new hardware and software in periodic phases' or 'update the contents of its mind and body continuously, adding and deleting' (Moravec 1988, p. 5). However, can a software-based person survive 'rejuvenation'? For the patternist, the substrate holding the pattern is irrelevant; so, *if the pattern changes, what's left*?

In Moravec's view, the software-based person will not survive (extensive) rejuvenation. He says, 'In time, each of us will be a completely changed being ... Personal death as we know it differs from this inevitability only in its relative abruptness' (1988, p. 121). Although with software and hardware changes we 'must die bit by bit', this is not so terrible—'who among us would wish to remain static, possessing for a lifetime the same knowledge, memories, thoughts, and skills we had as children?' (ibid., pp. 121, 122). The situation, however, looks much worse than Moravec suggests. A person who is nothing more than a pattern does not die 'bit by bit' when additions and deletions are made. Instead there is a series of *different* patterns, each with a very short lifespan. The choice for post-humans, it seems, is exactly to 'remain static' or die.

According to Tipler, a software-based person can survive change—if (and only if) the pattern before change and the pattern after change are 'sufficiently close'. Bostrom appears to use a similar notion, since he says 'If most of who someone currently is, including her most important memories, activities, and feelings, is preserved, then adding extra capacities on top of that would not easily cause the person to cease to exist' (2005a, p. 9). However, this approach simply raises another question: under what conditions are two patterns *sufficiently* close? Techno-supernaturalists do not tell us. Moreover, unlike the identity relation, the 'sufficiently close' relation is not transitive (i.e., x may be sufficiently close to y, and y sufficiently close to z, without x being sufficiently close to z). Let us suppose that, by virtue of the sufficient closeness of their patterns, A and B are the one person despite change, and B and C are the one person despite change. It follows (again by the transitivity of identity) that A and C are the one person. Nevertheless, let us suppose, A's pattern and C's pattern are not 'sufficiently close'; so, using Tipler's version of patternism, A and C are *not* the one person. Contradiction again.

Perhaps fuzzy logic, widely applied in AI (e.g. in knowledge representation) can help techno-supernaturalists here.[31] Applying this approach to the case of the pre-change person A and post-change person B, the statement that A is identical to B can be *true enough*, even if not completely (i.e. determinately) true. It is not 100% true that the person before change and the person after change are identical, but not 100% false either. In fuzzy logic, numerical degrees of truth or falsity (between 0 and 1) are assigned to statements.[32] If there is no difference between A's pattern and B's pattern, it is completely true (i.e. the truth value is 1) that A and B are the one person; and if the difference between A's pattern and B's pattern is minimal, it is almost completely true that A and B are the one person—0.99 true, say. With more thoroughgoing change to A, it is nevertheless more true than not that A and B are the one person (0.7, say). With major change—or incremental change over a very long period—it will become almost completely false that A survives (and in the limit completely false).

Fuzzy patternism (as I shall call it) certainly fits what techno-supernaturalists say about identity. Moravec claims that for posthumans the '[b]oundaries of personal identity will be very fluid—and ultimately arbitrary and subjective—as strong and weak interconnections between different regions rapidly form and dissolve' (1999, p. 165). Kurzweil tells us, '[J]ust wait until we're predominantly nonbiological … finding boundaries [between persons] will be even more difficult' (2006d, p. 387). Fuzzy patternism is likewise in accord with Tipler's claim that the identity of two persons at different times lies in the sufficient closeness of their patterns; and it removes the contradiction facing this claim, simply because indeterminate identity is *not* transitive (because the truth-value of the consequent of $A = B, B = C \vdash A = C$ is not constrained by the truth-values of the antecedents).[33] In fuzzy patternism Tipler's truth-condition for identity statements is reformulated as follows: it is true enough that A and B are the one person if (and only if) A's pattern is sufficiently close to B's pattern (where 'true enough' is itself a fuzzy notion). Since, as we have already seen, the 'sufficiently close' relation is not transitive, it follows that Tipler-style personal identity is not transitive. But in fuzzy logic there is nothing paradoxical about that.

Fuzzy patternism has another advantage for the techno-supernaturalist—it supplies the (formal) basis for a resolution of the duplication problem. In the duplication case, where A has duplicates B and C existing at the same time but in different places, $A = B$ and $A = C$ are equally true (e.g. both true to degree 0.5) but $B = C$ is completely false. So no contradiction arises. The challenge for fuzzy patternism, though, is to make sense of the idea that an identity statement is 'half true'. So far this way of defending patternism has only reached first base, sketching a merely formal semantics. The task is to convert this into a philosophically

[31] Jack Copeland suggested this strategy to me, and I am indebted to him for helpful discussion of this point.

[32] See e.g. Zadeh (1975); Goguen (1969).

[33] See Copeland (1997).

meaningful semantics by providing an adequate interpretation of the formalism. Techno-supernaturalists who wish to follow the fuzzy route out of contradiction must meet this challenge. They must also provide an account of minimal versus major pattern change—*which* upgrades should A allow, if she wishes to survive?

Taking the route of fuzzy patternism introduces another problem. In the philosophical literature there is a famous and powerful objection to the idea of indeterminately true identity statements (Evans 1978), which runs as follows. Let us assume (for the purposes of deriving a contradiction) that *A* is indeterminately identical to *B*, i.e. that *A* is identical to *B* but not determinately so (in Evans's notation, $\nabla(A = B)$. From this it follows, using the principle of property abstraction, that *B* has the property *being indeterminately identical to A* (in Church-style lambda notation, $\lambda x[\nabla(x = A)]\ B$). *A*, however, does not have this property, because *A* is *determinately* identical to *A*. *B*, therefore, has a property that *A* lacks, and so by Leibniz's Law *A* is not identical to *B*. In sum, if *A* is indeterminately identical to *B*, then *A* is *not* identical to *B*—this consequent, Evans says, 'contradict[s] the assumption, with which we began'. So, although fuzzy patternism avoids the contradictions inherent in (what we might call) naïve patternism, it itself leads to another prima facie paradox. And in any case, is the notion of the partial truth of identity statements sufficient to underwrite *resurrection* and *immortality*? If it is only partly true that the pre-mortem *A* is identical to the post-mortem *B*, then that might suffice for a 'version' of *A* to survive *A*'s death, or for *A* to 'survive as' *B* (see above). But intuitively *A* himself or herself survives death only if it is *determinately* true that *A* and *B* are the one person.

The various paradoxes for patternism—naïve or fuzzy—are at the foundations of techno-supernaturalism. They are akin to the problem of evil for supernaturalist religions—not necessarily unsolvable, but the believer needs a very good answer.

Doctrine and Faith

Techno-supernaturalists claim that their predictions make 'no appeal, anywhere, to faith' (Tipler 1994, p. 16), but is this true?

Kurzweil's Principles of Faith

Kurzweil recognizes conceptual difficulties for patternism. For example, he concedes that, if his own mind file is reinstantiated in 'a more durable substrate' while he is still alive, that software-based person is not him. The 'copy may look and act just like me, but it's nonetheless *not* me', he says—even if he is destroyed and the reinstantiated file remains (2006d, p. 384; see also 2001). But if this is so, then patternism is false. Yet Kurzweil does not abandon the theory, saying merely 'Despite these dilemmas my personal philosophy remains based on patternism'

(2006d, p. 386). This 'personal philosophy' is belief despite perceived falsehood—one of the characterizations of faith. According to Voltaire, for example, 'Faith consists in believing, not what appears to be true, but what appears to our understanding to be false' (1764/1971, p. 208).

How to tell if Kurzweil's mind file has been successfully reinstantiated? According to Kurzweil, the answer is a 'Ray Kurzweil' form of the Turing test; a human judge must be convinced that 'the uploaded re-creation is indistinguishable from the original specific person' (2006d, p. 200).[34] However, he also claims that the Turing test fails as a 'consciousness detector'—in his view, it is an 'objective' test and nothing objective (or 'scientific') can guarantee that an entity is conscious (Brooks et al. 2006; see also Kurzweil 2006d, pp. 377–380). Kurzweil distinguishes 'apparent consciousness' from 'really having subjective experience'; the former suffices to pass Turing's test, but the latter, he says, is crucial to successful uploading (Brooks et al. 2006). Yet he does not abandon the Turing test as a means of verifying uploads, saying merely 'My own philosophical take is if an entity seems to be conscious, I would accept its consciousness. But that's a philosophical and not a scientific position' (ibid.). Kurzweil's 'philosophical take' is belief despite perceived absence of justification. This is another characterization of faith—according to Martin Luther, for example, the believer 'must not consult reason and mind how a doctrine sounds and whether it is consistent with reason. He must say forthwith: "I do not care whether it agrees with reason or not"'.[35]

Resurrection: Why Bother?

Even if godlike artificial intelligences were to emerge, and even if simulation resurrection were unproblematic, what would give future 'Minds' the information necessary for wholesale simulation resurrection? There will be 'clues in solar-system quantities to deduce and recreate the most microscopic details of preceding eras', Moravec says (1999, p. 167). But these 'clues' will be consistent with an indefinite number of different possible human pasts—how will the Minds know which is the actual past? Techno-supernaturalists do not tell us, seeming merely to rely on the quasi-omniscience of post-Singularity machines.

Also, why would these intelligences choose wholesale human resurrection? According to Moravec, future AIs will be 'so vast and enduring that rare infinitesimal flickers of interest by them in the human past will ensure that our entire history is replayed in full living detail' (1999, p. 168). Yet why would they be interested in us, even if only briefly? According to Tipler, they will seek 'total knowledge' for the sake of their survival, and this will make resurrection 'inevi-

---

[34] On person-specific Turing tests, see further Steinhart 2007.

[35] Luther c. 1530–2/1959, p. 78. For Luther, belief *can* be justified—by faith itself.

table' (1994, pp. 219, 220; see also 2007, p. 79).[36] But why would their survival require knowledge of how long-dead humans behaved—any more than our survival now requires knowledge of how our primeval ancestors acted? And if 'total' knowledge were required, these machines would surely also simulate the very *worst* conditions for humans—rather than solely the 'ideal fantasy worlds' that Tipler envisages (1994, p. 241).

Tipler claims that superhuman-level artificial intelligences are likely to act out of 'a sense of obligation'; our lives are much poorer than theirs, yet their capacities derive from our efforts (2007, pp. 79, 80). However, it may be that, for post-Singularity AIs, humans are a primitive life-form to whom they feel no more obligation than we now feel to millipedes. We are their 'ultimate parents' and "Honor thy father and mother" is a universal moral principle', Tipler says (ibid., pp. 79, 80)—but this is a human attitude, and the Minds are superhuman. Even posthumans (i.e. post-Singularity artificial intelligences with direct human ancestors) may be impossible to second-guess.[37] According to Moravec, as software-based persons 'our thinking procedures might be totally liberated from any traces of our original body. … [T]he bodiless mind that results, wonderful though it may be in its clarity of thought and breadth of understanding, would be hardly human' (1999, p. 172). Ex-humans will become 'very unhuman disembodied superminds' (Moravec 1992, p. 20). And even if posthumans do retain human characteristics, they still may not honour their 'ultimate parents'—any more than we now honour our closest relations among the primates.

Just as many supernatural religions anthropomorphize spirits and gods—and just as current researchers in AI frequently anthropomorphize their machines (see Proudfoot 2011)—so techno-supernaturalists anthropomorphize post-Singularity AIs. Speculations concerning their interests and choices are as difficult to ground as conjectures concerning God's unexpressed desires, and as much a matter of faith.

It Will be Bliss, Trust Us

Even if wholesale simulation resurrection were to occur, *would* it be heaven?

Techno-supernaturalists claim that, as software-based entities, we will 'expand our cognitive and emotional capabilities, as well as the depth and richness of our … experiences many fold, ultimately by factors of trillions' (Kurzweil 2004). This

[36] According to Tipler (1994), God is the 'Omega Point'—the 'completion' of all finite existence; the Omega point 'loves us' and for this reason will give us immortality (pp. 12, 14). Again this is unjustified anthropomorphism.

[37] Steinhart (2008) argues that posthumans (since they have been perfected) will be sensitive to their 'ethical and epistemic obligations', and so will simulate 'all lesser civilizations'. However, this is still to anthropomorphize beings that are more like angels than humans. In response to the argument from evil, for example, many theologians and philosophers have insisted that we *cannot* deduce the moral attitudes of the divine—following this reasoning, there may be a 'noseeum' reason why posthumans will not recognize (or observe) Tipler's 'universal' moral principle.

will lead to more 'numerous and meaningful' projects (Bostrom 2008b, p. 132), 'deepen our intellectual lives' (Bostrom 2009a, p. 201), and may even enable us to answer 'the traditional big philosophical questions' that stump dumb humans (Bostrom 2005a, p. 6). However, expanded cognitive and affective capabilities offer at most the possibility of richer experiences (and so on); they do not guarantee them. Even if computer viruses and hackers could be eliminated—Moravec himself points out the possibility of 'software parasites' and 'shockingly original gremlins' (1988, p. 133; see also Kurzeweil (2006c) on the 'grey goo' threat)—superhuman intelligence is no surety of a meaningful life. Perhaps, on the contrary, the posthuman's greater capacities enable a greater despair, and a more acute insight into the meaninglessness of life.

Even in a very long life, boredom 'will not be an issue', Kurzweil also assures us (2004). But if unenhanced humans with limited cognitive resources can become bored within a short time, why not posthumans with greater cognitive resources but a very long time? According to Moravec, there will always be new and interesting problems for posthumans to solve (1988, p. 149), but even problem-solving may become tedious eventually. Moravec claims that the software-based person can simply alter his or her program to delay the onset of boredom (ibid., p. 114). (Likewise, according to Bostrom, controlling brain activity may guarantee happiness (2009a, p. 201).) However, there is a strong argument that software change annihilates the software-based person—just as a Turing machine is no longer the same Turing machine if a few more lines are added to its instruction table (see the section 'The Perils of Being a Pattern'). Fuzzy patternism may be able to resist this argument, but fuzzy patternism has its own problems.

According to techno-supernaturalists, posthumans will experience 'surpassing bliss' and will be 'posthumanly happy'—they will have experiences 'more blissful than those that humans are capable of' (Bostrom 2008b, p. 120). But what does this mean? Techno-supernaturalists cannot tell us, they say. The words 'surpassing bliss' are 'invented to describe human experience', whereas what the posthuman will feel is 'far beyond human feelings' (Bostrom 2008a). Posthuman life is a state of being that is beyond 'dreams' and 'imagination', and offers a beauty and joy that humans 'cannot fathom' (see the section 'The New Good News'). In this respect techno-supernaturalism is exactly like supernaturalism. Maimonides too promised 'bliss' but said that a 'clear comprehension of the bliss in the life hereafter is unattainable to any man':

> As to the blissful state of the soul in the world to come, there is no way on earth in which we can comprehend or know it. For in this earthly existence we only have knowledge of physical pleasure … [T]here is no comparison between the bliss of the soul in the life hereafter and the gratification offered to the body on earth (c. 1178/1981, p. 91a)

Yet if posthuman life is intelligible only to posthumans (or the righteous in the world to come), techno-supernaturalism's account of the future is mere handwaving, with little substantive content.

According to the Catechism of the Catholic Church, the doctrine of the resurrection of the dead 'exceeds our imagination and understanding: it is accessible only to faith'.[38] The techno-supernaturalists' prediction of software-based immortality, despite their claim that it is based on science, also rests on faith.

Terror Management?

For techno-supernaturalists, death is a tragedy—'the darkness that enshrouds all life' (Bostrom 2008a). 'Any death prior to the heat death of the universe is premature if your life is good' (ibid). Never fear, they say—post-Singularity technology will eliminate all human deaths.

Hume said that people are 'anxious concerning their future fortune' and for this reason 'acknowledge a dependence on invisible powers'[39]; and Freud said that we believe in an afterlife, and in supernatural beings who provide it, in order to 'exorcise the terrors of nature'.[40] According to modern Terror Management Theory, the combination of a biological tendency to self-preservation and the awareness of death produces 'potentially debilitating terror'; the remedy is to believe in immortality and in 'guarantees of safety' from powerful invisible beings.[41] The Singularity hypothesis can be seen as a new-and-improved therapy for death anxiety, based on AI and neuroscience rather than on revelation.[42]

As a terror management strategy, however, techno-supernaturalism is no more successful than traditional religion. The promises—of we know not exactly what—are still based on faith. And however good (or bad) the posthuman future might be, techno-supernaturalism requires a philosophically worked-out account of survival to underwrite its lavish claims of resurrection and immortality.[43] Don't stop paying your life insurance premiums yet.

References

Allen, P., & Greaves, M. (2011). The singularity isn't near. *Technology Review*. October 12, 2011, http://www.technologyreview.com/blog/guest/27206.

Anselm. (1078/1973). *The prayers and meditations of St Anselm*. (Translated and with an introduction by Sister B. Ward). London: Penguin Books.

Ayres, R. U. (2006). Review [of *The singularity is near*]. *Technological Forecasting & Social Change, 73*, 95–100.

---

[38] *Catechism of the Catholic Church, with modifications from the Editio Typica* (New York: Doubleday, 1994), Part One, Chapter Three, Article 11, 1000 (p. 282).

[39] Hume (1757/1956), p. 30.

[40] Freud (1949), p. 30.

[41] Solomon et al. (2004), pp. 16, 17.

[42] Of course, this does not falsify the Singularity hypothesis—any more than it does the claims of supernaturalist religion.

[43] I am grateful to Jack Copeland and to Eric Steinhart for their valuable comments on an earlier draft of this paper.

Block, N. (1995). On a confusion about the function of consciousness. *Behavioral and Brain Sciences, 18*(2), 227–287.

Bostrom, N. (2003a). When machines outsmart humans. *Futures, 35*, 759–764.

Bostrom, N. (2003b). Are we living in a computer simulation? *Philosophical Quarterly, 53*(211), 243–255.

Bostrom, N. (2004). The future of human evolution. In C. Tandy (Ed.), *Death and anti-death: Two hundred years after Kant, fifty years after Turing* (pp. 339–371). Palo Alto, CA: Ria University Press http://www.nickbostrom.com/fut/evolution.html.

Bostrom, N. (2005a). Transhumanist values. *Journal of Philosophical Research, Special Supplement*: *Ethical Issues for the Twenty-First Century*, F. Adams (Ed.), (pp. 3–14). Charlottesville, VA: Philosophy Documentation Center.

Bostrom, N. (2005b). A history of transhumanist thought. *Journal of Evolution and Technology 14*(1). April, 2005, http://jetpress.org/volume14/freitas.html.

Bostrom, N. (2006). How long before superintelligence? *Linguistic and Philosophical Investigations, 5*(1), 11–30.

Bostrom, N. (2008a). Letter from Utopia. *Studies in Ethics, Law, and Technology, 2*(1). http://www.bepress.com/selt/vol2/iss1/art6.

Bostrom, N. (2008b). Why I want to be a Posthuman when I grow up. In B. Gordijn & R. Chadwick (Eds.), *Medical enhancement and posthumanity* (pp. 107–137). Dordrecht: Springer.

Bostrom, N. (2009a). The future of humanity. In J. K. Berg Olsen, E. Selinger & S. Riis (Eds.), *New waves in philosophy of technology* (pp. 186–215). New York: Palgrave Macmillan.

Bostrom, N. (2009b). The simulation argument: Some explanations. *Analysis, 69*(3), 458–461.

Bostrom, N., & Kulczycki, M. (2011). A patch for the simulation argument. *Analysis, 71*(1), 54–61.

Bostrom, N., & Yudkowsky, E. (2011). The ethics of artificial intelligence. In W. Ramsey & K. Frankish (Eds.), Draft for *Cambridge handbook of artificial intelligence*. Accessed November 8, 2011, http://www.fhi.ox.ac.uk/selected_outputs_journal_articles.

Brooks, R. A. (1991). Intelligence without representation. *Artificial Intelligence, 47*, 139–159.

Brooks, R. A. (1995). Intelligence without reason. In L. Steels & R. A. Brooks (Eds.), *The artificial life route to artificial intelligence*. Hillsdale: Lawrence Erlbaum.

Brooks, R. A. (1999). *Cambrian intelligence: The early history of the new AI*. Cambridge, MA: MIT Press.

Brooks, R.A., Kurzweil, R., & Gelernter, D. (2006). Gelernter, Kurzweil debate machine consciousness. KurzweilAI.net. December 6, 2006, http://www.kurzweilai.net/articles/art0688.html?printable=1.

Brueckner, A. (2008). The simulation argument again. *Analysis, 68*(3), 224–226.

Chalmers, D. (1995). Facing up to the problem of consciousness. *Journal of Consciousness Studies, 2*(3), 200–219.

Chalmers, D. (2005). The matrix as metaphysics. In C. Grau (Ed.), *Philosophers explore the matrix* (pp. 132–176). Oxford: Oxford University Press.

Chalmers, D. (2010). The singularity: A philosophical analysis. *Journal of Consciousness Studies, 17*, 7–65.

Clark, A. (1997). *Being there: Putting mind, body, and world together again*. Cambridge, MA: MIT Press.

Copeland, B. J. (1997). Vague identity and fuzzy logic. *Journal of Philosophy, 94*(10), 514–534.

Copeland, B. J. (2000). Narrow versus wide mechanism. *Journal of Philosophy, 97*(1), 5–32.

Dennett, D.C. (1995). The unimagined preposterousness of zombies. *Journal of Consciousness Studies, 2*(4), 322–326.

Devezas, T. C. (2006). Discussion [of *The singularity is near*]. *Technological Forecasting & Social Change, 73*, 112–121.

Dreyfus, H. L. (1992). *What computers* still *can't do*. Cambridge, MA: MIT Press.

Dreyfus, H. L. (2007). Why Heideggerian AI failed and how fixing it would require making it more Heideggerian. *Artificial Intelligence, 171*, 1137–1160.

Dreyfus, H. L., & Dreyfus, S. E. (1986). *Mind over machine*. Oxford: Blackwell.

Else, L. (2009). Ray Kurzweil: A singular view of the future. *New Scientist Opinion,* 2707. May 6, 2009. http://new.scientist.com.

Evans, G. (1978). Can there be vague objects? *Analysis, 38*(4), 208.

Flanagan, O., & Polger, T. (1995). Zombies and the function of consciousness. *Journal of Consciousness Studies, 2*(4), 313–321.

Ford, K., & Hayes, P. (1998). On conceptual wings: Rethinking the goals of artificial intelligence. *Scientific American Presents, 9*(4), 78–83.

Freud, S. (1949). *The future of an illusion*. London: The Hogarth Press.

Gershenson, C., & Heylighen, F. (2005). How can we think the complex? In K. Richardson (Ed.), *Managing organizational complexity: Philosophy, theory and application* (pp. 47–61). Information Age Publishing.

Goertzel, B. (2007a). Human-level artificial general intelligence and the possibility of a technological singularity. A reaction to Ray Kurzweil's *The Singularity is Near,* and McDermott's critique of Kurzweil. *Artificial Intelligence, 171*, 1161–1173.

Goertzel, B. (2007b). Artificial general intelligence: Now is the time. KurzweilAI.net. April 9, 2007, http://www.kurzweilai.net/artificial-general-intelligence-now-is-the-time.

Goguen, J. A. (1969). The logic of inexact concepts. *Synthese, 29*, 325–373.

Hawkins, J. (2008). [interviewed in] Tech luminaries address singularity, [and in] Expert view. *IEEE Spectrum*. June, 2008. http://spectrum.ieee.org/computing/hardware/tech-luminaries-address-singularity.

Heylighen, F. (2012). A brain in a vat cannot break out: Why the singularity must be extended, embedded and embodied. *Journal of Consciousness Studies, 19*(1–2), 126–142.

Hofstadter, D. (2008). [interviewed in] Tech luminaries address singularity. *IEEE Spectrum*. June, 2008. http://spectrum.ieee.org/computing/hardware/tech-luminaries-address-singularity.

Horgan, J. (2008). The consciousness conundrum. *IEEE Spectrum, 45*(6), 36–41.

Hume, D. (1757/1956). *The natural history of religion*, H. E. Root (Ed.), London: Adam & Charles Black.

Huxley, J. (1927). *Religion without revelation*. London: Ernest Benn Ltd.

Huxley, J. (1957a). Transhumanism. In J. Huxley, *New bottles for new wine* (pp. 13–17). London: Chatto and Windus.

Huxley, J. (1957b). Evolutionary humanism. In J. Huxley, *New bottles for new wine* (pp. 279–312).

Huxley, J. (1964). The new divinity. In J. Huxley, *Essays of a humanist* (pp. 218–226). London: Chatto and Windus.

Joy, W. (2000). Why the future doesn't need us. *Wired Magazine, 8*(4). http://www.wired.com/wired/archive/8.04/joy.pr.html.

Kurzweil, R. (1999). *The age of spiritual machines: When computers exceed human intelligence*. New York: Viking Press.

Kurzweil, R. (2001). The law of accelerating returns. KurzweilAI.net. March 7, 2001, http://www.kurzweilai.net/articles/art0134.html?printable=1.

Kurzweil, R. (2002). The material world: "Is that all there is?" Response to George Gilder and Jay Richards. In J. W. Richards (Ed.), *Are we spiritual machines? Ray Kurzweil vs. the critics of strong A.I.* Seattle: Discovery Institute.

Kurzweil, R. (2004). A dialogue on reincarnation. KurzweilAI.net. January 6, 2004, http://www.kurzweilai.net/articles/art0609.html?printable=1.

Kurzweil, R. (2006a). Why we can be confident of turing test capability within a quarter century. KurzweilAI.net. July 13, 2006, http://kurzweilai.net/meme/frame.html?main=/articles/art0683.html.

Kurzweil, R. (2006b). Reinventing humanity: The future of machine-human intelligence. *The Futurist, 40*(2), 39–46.

Kurzweil, R. (2006c). Nanotechnology dangers and defenses. *Nanotechnology Perceptions, 2*, 7–13.

Kurzweil, R. (2006d). *The singularity is near: When humans transcend biology*. New York: Penguin Books.
Kurzweil, R. (2007a). Let's not go back to nature. *New Scientist, 2593*: 19.
Kurzweil, R. (2007b). Foreword to *The intelligent universe*. KurzweilAI.net. February 2, 2007, http://www.kurzweilai.net/articles/art0691.html?printable=1.
Kurzweil, R. (2011). Don't underestimate the singularity. *Technology Review*. October 19, 2011, http://www.technologyreview.com/blog/guest/27263.
Luther, M. (c. 1530-2/1959). *Luther's Works, Vol. 23: Sermons on the Gospel of St. John Chapters 6–8*. J. Pelikan & D. E. Poellot (Eds.). St. Louis, Missouri: Concordia Publishing House.
Maimonides, M. (c. 1178/1981). *Mishneh torah: The book of knowledge*. In M. Hyamson (Ed.), New, corrected edition. Jerusalem and New York: Feldheim Publishers.
Maimonides, M. (1191/1985). Essay on resurrection. In *Crisis and leadership: epistles of Maimonides* (A. Halkin, translation and notes; D. Hartman, discussion). Philadelphia: Jewish Publication Society of America.
McCarthy, J. (1995). Todd Moody's zombies. *Journal of Consciousness Studies, 2*(4), 345–347.
McDermott, D. (2006). Kurzweil's argument for the success of AI. *Artificial Intelligence, 170*, 1183–1186.
McDermott, D. (2007). Level-headed. *Artificial Intelligence, 171*, 1183–1186.
Modis, T. (2006). Discussion [of *The singularity is near*]. *Technological Forecasting & Social Change, 73*, 104–112.
Moore, G. E. (1965). Cramming more components onto integrated circuits. *Electronics, 38*(8), 114–117. (The article is reprinted in *Proceedings of the IEEE*, 86(1), 82–85, 1998).
Moore, G. E. (1975). Progress in digital integrated electronics. *Technical Digest, IEEE International Electron Devices Meeting, 21*, 11–13.
Moore, G. E. (2008). [interviewed in] Tech luminaries address singularity. *IEEE Spectrum*. June 2008, http://spectrum.ieee.org/computing/hardware/tech-luminaries-address-singularity.
Moravec, H. (1988). *Mind children: The future of robot and human intelligence*. Cambridge, MA: Harvard University Press.
Moravec, H. (1992). Pigs in cyberspace. In B. R. Miller & M. T. Wolf (Eds.), *Thinking robots, an aware internet, and cyberpunk librarians: The 1992 LITA President's Program, presentation by Hans Moravec, Bruce Sterling, and David Brin*. Chicago, Illinois: Library and Information Technology Association.
Moravec, H. (1998). When will computer hardware match the human brain? *Journal of Evolution and Technology, 1*. http://www.jetpress.org/volume1/moravec.htm.
Moravec, H. (1999). *Robot: mere machine to transcendent mind*. Oxford: Oxford University Press.
Nordmann, A. (2008). Singular simplicity. *IEEE Spectrum, 45*(6), 60–63.
Olson, E. (1997). *The human animal: Personal identity without psychology*. Oxford: Oxford University Press.
Olson, S., & Kurzweil, R. (2006). Sander Olson interviews Ray Kurzweil. KurzweilAI.net. February 3, 2006, http://www.kurzweilai.net/articles/art0643.html?printable=1.
Parfit, D. (1987). *Reasons and persons*. Oxford: Oxford University Press.
Parfit, D. (2008). Personal identity. In J. Perry (Ed.), *Personal identity* (2nd ed.). Berkeley: University of California Press.
Pollack, J. B. (2006). Mindless intelligence. *IEEE Intelligent Systems, 21*(3), 50–56.
Proudfoot, D. (1999a). How human can they get? *Science, 284*(5415), 745.
Proudfoot, D. (1999b). Facts about artificial intelligence. *Science, 285*(5429), 835.
Proudfoot, D. (2011). Anthropomorphism and AI: Turing's much misunderstood imitation game. *Artificial Intelligence, 175*, 950–957.
Proudfoot, D. (2012). Software immortals: Science or faith. In A. Eden, J. Søraker, J. Moor, & E. Steinhart (Eds.), *The singularity hypothesis: A scientific and philosophical analysis*, The Frontiers Collection. Springer.

Proudfoot, D., & Copeland, B. J. (2011). Artificial intelligence. In E. Margolis, R. Samuels, & S. P. Stich (Eds.), *The Oxford handbook to philosophy and cognitive science* (pp. 147–182). New York: Oxford University Press.
Pyszczynski, T., Greenberg, J., & Solomon, S. (1999). A dual-process model of defense against conscious and unconscious death-related thoughts: An extension of terror management theory. *Psychological Review, 106*(4), 835.
Richards, J. W. (Ed.). (2002). *Are we spiritual machines? Ray Kurzweil vs. the critics of strong A.I.* Seattle: Discovery Institute.
Sainsbury, R. M. (2009). *Paradoxes* (3rd ed.). Cambridge: Cambridge University Press.
Sandberg, A., & Bostrom, N. (2008). Whole brain emulation: A roadmap. Technical report #2008-3, Future of Humanity Institute, Oxford University. http://www.fhi.ox.ac.uk/reports/2008-3.pdf.
Shapiro, L. (2011). *Embodied cognition*. Milton Park: Routledge.
Sloman, A. (2008). The well-designed young mathematician. *Artificial Intelligence, 172*, 2015–2034.
Sloman, A. (2010). Phenomenal and access consciousness and the 'Hard' problem: A view from the designer stance. *International Journal of Machine Consciousness, 2*(1), 117–169.
Solomon, S., Greenberg, J., & Pyszczynski, T. (2004). The cultural animal: Twenty years of terror management theory and research. In J. Greenberg, S. L. Koole, & T. Pyszczynski (Eds.), *Handbook of experimental existential psychology*. New York: The Guilford Press.
Steinhart, E. (2002). Indiscernible persons. *Metaphilosophy, 33*(3), 300–320.
Steinhart, E. (2007). Survival as a digital ghost. *Minds and Machines, 17*, 261–271.
Steinhart, E. (2008). Teilhard de Chardin and Transhumanism. *Journal of Evolution and Technology, 20*(1), 1–22. http://jetpress.org/v20/steinhart.htm.
Steinhart, E. (2010). Theological implications of the simulation argument. *Ars Disputandi, 10*, 23–37.
Strawson, P. F. (1959). *Individuals: An essay in descriptive metaphysics*. London: Methuen.
Tipler, F. J. (1994). *The physics of immortality: Modern cosmology, God and the resurrection of the dead*. New York: Doubleday.
Tipler, F. J. (2007). *The physics of christianity*. New York: Doubleday.
Voltaire (1764/1971). *Philosophical dictionary*. (T. Besterman, Edited and Translated). Harmondsworth, Middlesex: Penguin Books.
Weatherson, B. (2003). Are you a sim? *Philosophical Quarterly, 53*(212), 425–431.
Whitby, B. (1996). The turing test: AI's biggest blind alley? In P. Millican & A. Clark (Eds.), *The legacy of Alan Turing*, (Vol. I Machines and thought). Oxford: Oxford University Press.
Williams, B. (1973a). Are persons bodies? In B. Williams, *Problems of the self: Philosophical papers* (1956–1972). Cambridge: Cambridge University Press.
Williams, B. (1973b). Bodily continuity and personal identity. In B. Williams, *Problems of the self: Philosophical papers* (1956–1972). Cambridge: Cambridge University Press.
Zadeh, L. A. (1975). Fuzzy logic and approximate reasoning. *Synthese, 30*, 407–428.
Zorpette, G. (2008). Waiting for the rapture. *IEEE Spectrum, 45*(6), 32–35.

Chapter 18A
Francis Heylighen on Proudfoot's "Software Immortals: Science or Faith?"

The Continuity of Embodied Identity

I enjoyed reading Diane Proudfoot's essay on "technological supernaturalism", i.e. the belief that human individuals will be resurrected as immortal software entities by some future, God-like artificial intelligence(s) (Proudfoot 2012). Proudfoot thoroughly deconstructs the many dubious assumptions underlying this philosophy, as propounded by authors such as Kurzweil, Bostrom, Moravec and Tipler.

I particularly liked her arguments showing that this purportedly scientific vision is almost wholly parallel to the traditional religious vision in which our souls are promised an eternal life in heavenly bliss after our mortal bodies have passed away. The "terror management" theory (Pyszczynski et al. 1999) that she refers to indeed provides a plausible explanation for why people, whether religiously or scientifically inspired, seem to be drawn so strongly to the idea that their personhood would somehow survive physical death. But we may not even need such a psychological explanation for this glaring similarity between technological and religious supernaturalism: to me it seems obvious that the former is directly inspired by the latter. For example, while Tipler initially presented his ideas as purely scientific inferences, in further writing (Tipler 2007) he made it clear that he is a devout Catholic who takes doctrine rather literally. The motivation to rationalize a pre-existing faith may be less obvious in the case of more humanistic thinkers, like Bostrom or Moravec. But even a staunch atheist cannot avoid being influenced by such a pervasive meme as the belief in an afterlife, and may be tempted to defuse its power to convert people to religion by reinterpreting it scientifically.

After pointing out where I agree with Proudfoot, let me now indicate where we part ways. In my view, her paper falls in the common trap of what may be called "analytic nitpicking". Philosophers from the analytic tradition investigate issues by making fine-grained distinctions between the different possible meanings of a concept, and then applying logic to draw out the implications of each of these possible interpretations, in particular in order to show how a particular interpretation may lead to some inconsistency or counter-intuitive result. But these "technical distinctions"—to use Proudfoot's phrase—are in general considered meaningful only by philosophers: scientists and practitioners typically do not care, because these distinctions tend to lack operational significance. A classic example is the zombie thought experiment about consciousness (Chalmers 1995): if a zombie by definition behaves indistinguishably from a normal human, then according to Leibniz's principle of the identity of the indistinguishables, a zombie must be a human. The zombie argument therefore fails to clarify anything about consciousness.

Proudfoot applies the analytic method to the problem of personal identity: in how far can an "uploaded", software personality be identical to the original flesh-and-blood person that it is supposed to resurrect? She argues that various interpretations of the identity concept all lead to problems—such as lack of transitivity or the apparently nonsensical conclusion that two independent software instantiations, A and B, are actually one person. I consider this nitpicking because the identity concept, like practically any concept used in real life, is essentially vague and fluid. The recurrent error made by analytic philosophers is to assume that distinctions are absolute and invariant, while in the complex reality that surrounds us distinctions tend to vary across times, observers and contexts (Gershenson and Heylighen 2005).

Apparently universal rules about the logical notion of identity (such as $A = B$, $B = C$, therefore $A = C$), hence, are unlikely to be applicable to the much more fluid notion of personal identity. Proudfoot is to some degree aware of these difficulties, and therefore considers the alternative model of fuzzy logic. But fuzzy logic is still a kind of logic, and therefore built on invariant (albeit fuzzy) distinctions. The nature of personal identity is precisely that it is not invariant. It is not only the case—as the authors cited by Proudfoot point out—that since I was born about every atom in my body has changed, but also that about every bit of knowledge, experience or emotion in my mind has changed. My personality is substantially different from the personality I had when I was born, or even when I was 5, 10, 15, or 20 years old....

The only thing that allows me to state that the Francis Heylighen of today is somehow still the same as the Francis Heylighen of 40 years ago is *continuity*: during that time, there was a continuing distinction between Francis Heylighen and the rest of the world, even while the nature of that distinction was changing. This continuity was not one of consciousness (which waxed and waned along with my sleep-wake cycle), but of the rough outline of my body and personality. This continuity is precisely what lacks in the resurrection scenarios of the technological supernaturalists. In such scenario, my body and personality break down at my biological death, while my personality (or at least a software equivalent of it) is recreated by a super-intelligent AI many decades later, in a completely different (non-physical) environment.

Proudfoot is right to question the claim that the resurrected personality would be identical to my original personality (together with the more outlandish claims that the AI would feel compelled to resurrect every person that ever lived, or that the information about all these personalities would have survived the inevitable thermodynamic dissipation). However, rather than wandering through "technical distinctions" about identity, she should better have focused on the most glaring difference: the resurrected personality would lack both my body and my environment. While she mentions the *situated and embodied* perspective on cognition merely in passing, for me it is crucial: the ability to interact with the environment via bodily sensors and effectors is a defining feature of the notions of person, mind, consciousness or intelligence. As I have developed this point in more depth in my criticism of the common view of the Singularity as the emergence of a disembodied super-intelligence (Heylighen 2012), I won't go into further details here.

However, note that this philosophy does not deny the possibility of attaining some sort of technological immortality: continuity of identity can in principle be maintained by gradually replacing my different body parts by various electronic circuits—as long as these maintain (or augment) my ability to interact with the world via high-bandwidth sensors and effectors. But now we are entering the domain of practical implementation, leaving behind both the metaphysical speculations of the techno-supernaturalists and the Platonic nitpicking of the analytic philosophers...

References

Chalmers, D. (1995). Facing up to the problem of consciousness. *Journal of Consciousness Studies, 2*(3), 200–219.

Gershenson, C., & Heylighen, F. (2005). How can we think the complex? In K. Richardson (Ed.), *Managing organizational complexity: Philosophy, theory and application* (pp. 47–61). Information Age Publishing.

Heylighen, F. (2012). A brain in a vat cannot break out: Why the singularity must be extended, embedded and embodied. *Journal of Consciousness Studies, 19*(1–2), 126–142.

Proudfoot, D. (2012). Software immortals: Science or faith. In A. Eden, J. Søraker, J. Moor, & E. Steinhart (Eds.), *The singularity hypothesis: A scientific and philosophical analysis*, The Frontiers Collection. Springer.

Pyszczynski, T., Greenberg, J., & Solomon, S. (1999). A dual-process model of defense against conscious and unconscious death-related thoughts: An extension of terror management theory. *Psychological Review, 106*(4), 835.

Tipler, F. J. (2007). *The physics of christianity*. New York: Doubleday.

Chapter 19
Belief in The Singularity is Fideistic

Selmer Bringsjord, Alexander Bringsjord and Paul Bello

Abstract We deploy a framework for classifying the bases for belief in a category of events marked by being at once *weighty, unseen, and temporally removed* (*wutr*, for short). While the primary source of wutr events in Occidental philosophy is the list of miracle claims of credal Christianity, we apply the framework to belief in The Singularity, surely—whether or not religious in nature—a wutr event. We conclude from this application, and the failure of fit with both rationalist and empiricist argument schemas in support of this belief, not that The Singularity won't come to pass, but rather that regardless of what the future holds, believers in the "machine intelligence explosion" are simply fideists. While it's true that fideists have been taken seriously in the realm of religion (e.g. Kierkegaard in the case of some quarters of Christendom), even in that domain the likes of orthodox believers like Descartes, Pascal, Leibniz, and Paley find fideism to be little more than wishful, irrational thinking—and at any rate it's rather doubtful that fideists should be taken seriously in the realm of science and engineering.

S. Bringsjord (✉)
Department of Computer Science, Department of Cognitive Science,
Lally School of Management and Technology, Rensselaer Polytechnic Institute (RPI),
Troy, NY 12180, USA
e-mail: Selmer.Bringsjord@gmail.com

A. Bringsjord
Motalen Inc, Troy, NY 12180, USA
e-mail: brings623@gmail.com

P. Bello
Human and Bioengineered Systems, Code 341 Office of Naval Research
875 N Randolph Street, Arlington, VA 22203, USA
e-mail: paul.bello@navy.mil

A. H. Eden et al. (eds.), *Singularity Hypotheses*, The Frontiers Collection,
DOI: 10.1007/978-3-642-32560-1_19, © Springer-Verlag Berlin Heidelberg 2012

Introduction; Plan

We deploy a framework for classifying the bases for belief in a category of events marked by being at once *weighty, unseen, and temporally removed* (= *wutr*). While the primary source in Occidental philosophy of such events is credal (= orthodox) Christianity,[1] we follow Dennett (2007) in viewing philosophizing as equally applicable to religion and science, and apply this framework to the dominant basis ($\mathcal{A}$) for belief in The Singularity, surely—whether or not itself religious in nature—a wutr event. We conclude from this application not that The Singularity won't come to pass, but rather that regardless of what the future holds, the failure of a fit between $\mathcal{A}$ and either rationalist or empiricist argument schemas in support of this belief implies that believers in the "machine intelligence explosion" are simply fideists. While it's true that fideists have been taken seriously in the realm of religion (e.g. Kierkegaard 1986 in the case of some quarters of Christendom), even in that domain the likes of believers like Descartes, Pascal, Leibniz, and Paley,[2] in line as they are with Christian orthodoxy and hence rationalism, find fideism to be little more than wishful, irrational thinking—and at any rate it's rather doubtful that fideists should be taken seriously in the realm of science and engineering.

Preliminaries

To make the situation a bit more tidy before we begin in earnest, we take a series of preliminary steps.

First, we acknowledge the initial oddness of speaking of belief *in an event*. Traditionally, of course, the targets of belief (and knowledge) are propositions—though we certainly do say such things as that "Knox believes in General Washington," in which case we are pointing to belief in a *person*. The situation before us is easily and quickly made cleaner: When we say that some believe in The Singularity, where this is an event, we simply mean that some believe that The Singularity will *occur*. In a parallel that will form a persistent theme in the inquiry herein, when someone says that they believe in The Resurrection, they are reporting their belief that the event in question (Jesus rising from the dead)

---

[1] Which shouldn't be confused with the denomination known as 'Greek Orthodox'—a denomination that does though happen to itself be orthodox/credal in our sense. An elegant characterization of orthodox Christianity is provided by Chesterton (2009). Along the same lines, and no doubt paying homage to his intellectual and spiritual hero, is Lewis's (1960) *Mere Christianity*. A more mechanical and modern characterization is obtained by simply following Swinburne (1981) in identifying orthodox Christianity with the union of the propositional claims in its ancient creeds (e.g. Apostle's, Nicene, Athanasian), which then declaratively speaking within this limited scope harmonizes Catholicism and Protestantism.

[2] And—see footnote 1—Chesterton, Lewis, and Swinburne.

happened. Note that belief in an event is thus paired with belief that a certain proposition is true. The same kind of association is in play in the case of belief in a person, since Knox clearly believes such things as that Washington is competent.[3]

In a second preliminary step, note that the properties *being weighty*, *being unseen*, and *being temporally removed* are here applied to *events*; and we invoke the already-seen abbreviation of this three-part adjective: *wutr*. We assume that the property of being temporally removed is clear enough to obviate any sustained analysis. This property applies to an event if it's purported occurrence is either beyond the recent past or immediate future. Hence, the aforementioned Resurrection is temporally removed. So is WWI, the American Revolutionary War, the death of Adolf Hitler, the falling of a vast part of California into the Pacific Ocean due to a major earthquake, the Second Coming, and the arrival in 2020 of aliens superior in intelligence to most currently alive Norwegians. Clearly, The Singularity is temporally removed. We devote section The World of the Weighty, Unseen, and Temporally Removed to characterizing the first two properties in 'wutr.'

In a third preliminary move, we denote by $\mathcal{S}$ the event associated with The Singularity (the arrival on Earth of computing machines more intelligent, indeed *vastly* more intelligent, than human persons,[4] and denote by **S** the corresponding *proposition* that this event will in the near future come to pass. By 'near future' we mean to encompass any length of time short of a century; hence we charitably adopt a temporally latitudinarian stance with respect to those confident that $\mathcal{S}$ will occur. On this stance, we are of course allowing much more time than any reasonable interpretation of 'foreseeable future,' and this is a phrase often used to frame predictions that advanced computing-machine intelligence will or will not arrive. For instance, Turing (1950) famously declared that he could foresee a time when humans not only routinely ascribed intelligence and other mental attributes to computers, but also when his test (the so-called 'Turing Test') would be passed; indeed he specifically predicted that by 2000 a level of such intelligence on par, linguistically speaking, with that possessed by humans would arrive.[5] This prediction turned out to fall completely flat, as we all know by now.[6] In a second

---

[3] We are happy to agree that believing in a person includes more than mere propositional belief, but this topic isn't germane to our objectives herein.

[4] We recognize that The Singularity has now come to be associated with a *group* of events (e.g. the group often is taken to include the ability of human persons to exist in forms that are not bio-embodied), but to maintain a reasonable scope in the present paper we identify $\mathcal{S}$ with only the "smart-machine" prediction, which is quite in line with e.g. the sub-title of the highly influential (Kurzweil 2000): "When Computers Exceed Human Intelligence." There is also in alignment with the *locus classicus*: (Good 1965).

[5] We recognize that Turing's optimism was constrained by certain conditions regarding how long a computing machine's prowess on his test would last, but such niceties can be safely left aside.

[6] As a matter of fact, Turing, like—as we shall see—those predicting the coming $\mathcal{S}$, would seem to be guilty of the same fatal sin: failing to give a rationalist (or even an empiricist) argument for the prediction in question. One of us rather long ago happily conceded that the Turing Test

example, this one falling on the side of pessimism, Floridi (2005) has argued that a certain ingenious test of self-consciousness cannot in the *foreseeable future* be passed by a computing machine.[7] Finally, note that Chalmers (2010) recounts a number of the time-indexed predictions about when $\mathcal{S}$ will supposedly occur; our allowing a full century is in this context hyper-charitable. As confirmation of this, consider: Good (1965): $\mathcal{S}$ by 2000; Vinge (1993): the explosion between 2005–2030; Yudkowsky (1996): 2021; and Kurzweil (2000): 2030. On the other hand, Chalmers himself appears to believe that $\mathcal{S}$ will occur within *centuries* (note the plural). Since we wish to retain the concept of a foreseeable future, this is too large a range for us to use herein.

Plan for the Remainder

With these preliminary matters settled, we announce our planned sequence for the remainder: In the next (section The World of the Weighty, Unseen, and Temporally Removed), our exposition aided by consideration of the claims of credal Christianity,[8] we briefly characterize the category of the weighty and unseen, into which, as will be seen, surely The Singularity falls. Then in (section The Tripartite Framework) we briefly summarize the three main epistemic positions of *empiricism*, *rationalism*, and *fideism*. These positions are sketched with help from the basic but eminently sensible and non-partisan epistemological framework erected by Chisholm (1977). Next (section Belief in The Singularity is Fideistic), we present our proof-by-cases argument for the claim that belief in **S** is fideistic—an argument that will in turn require at least some study of the dominant basis for believing that **S** holds; that is, some study of the aforementioned $\mathcal{A}$, a basis due originally to (1965), and ably modernized recently by Chalmers (2010). A brief conclusion wraps up the paper.

(Footnote 6 continued)
will be passed (Bringsjord 1992), but this concession was not accompanied by any timeline whatsoever—and if there *had* been a timeline, it would have been an exceedingly conservative one.

[7] A counter-argument can be found in (Bringsjord 2010).

[8] This is as good a spot as any to say that we could mine the supernatural event-claims of Islam and Judaism instead of those in credal Christianity, but we aren't that familiar with these other two monotheistic religions, and Western philosophy, for better or for worse, has certainly focused on the event-claims of Christianity of the other two historical monotheistic religions.

The World of the Weighty, Unseen, and Temporally Removed

We've already commented on the property *temporally removed*. What is meant by 'w' and by 'u' in the composite adjective 'wutr'? We haven't the space here to give a rigorous definition, and such a thing isn't needed anyway, because illuminating examples abound in philosophy, especially in the philosophy of (again, Occidental) religion, which typically relates to such things as whether God exists, and whether he really has intervened, and will intervene, directly in our world. Philosophy of religion typically targets those things which are in turn the targets of faith, and as such, things which are at once weighty and unseen. Here is the writer of Hebrews (11:1) in the New Testament: "Now faith is being sure of what we hope for and certain of what we do not see." The context of this passage indicates that what is believed in faith targets things both weighty enough to be earnestly hoped for, and invisible—things that, in short, are miraculous. As philosophical treatments of miracles indicate, miracles are by definition weighty and (to nearly all, anyway) unseen. For example, as is noted by Mcgrew (2010), we would hardly count as a miracle, or even a purported miracle, some stray, minor deviation from physical laws in a remote corner of the inanimate universe.

Likewise, the context of sustained historic treatments of faith and reason, such as Leibniz's (1998) *Theodicy*, point to events both weighty and unseen; namely, what Leibniz calls the "oracles" of God; that is, the "major" miracles claimed by the creeds of Christianity. These events are paradigmatic examples of profundity and (at least from the perspective of generations living long after the times at which they are to have occurred) invisibility. And the same source of an ostensive definition of w-&-u is found in contemporary treatments, for instance in the work of Oxford philosopher Richard Swinburne (1981, 2010)—work explicitly devoted to substantiating the credal claims of orthodox Christianity. So, these are the examples we provide to clarify the weighty-and-unseen: the Resurrection, the virgin birth, and so on. We are in no way saying that The Singularity is supernatural in nature; we are saying that in *structure*, and specifically with respect to wutr, The Singularity ($=\mathcal{S}$) parallels the—to again use the Leibnizian term—oracles of Christianity.

Put in terms of propositions (i.e., the underlying content of declarative sentences traditionally signaled in English by "that" phrases), and generalizing to some degree, we can say that *propositions* are weighty-and-unseen when they directly and immediately entail the existence of some being(s), and/or the occurrence of some event, which is at once by nearly any metric such that were it to obtain, or were it known to be arriving in the future, (1) would cause rational agents to significantly alter their beliefs and their behavior, and (2) involves beings as of yet invisible. The proposition **S** that The Singularity will come to pass within a century certainly seems to qualify as w-&-u with flying colors, for this proposition makes reference to a profound event, and to a being or beings (immeasurably smarter-than-human computing machines) that are invisible as of now, and perhaps invisible even after they arrive on the scene. So condition (2) is satisfied. What about condition (1)? Anyone who knows even a smidgeon of the literature on The Singularity knows those who expect it often adjust their "cognitive maps." They consider for instance how best to prepare for and perhaps to a degree manage $\mathcal{S}$. Chalmers (2010) is an example of such level-headed cerebration.

The Tripartite Framework

Now, what is the framework we have available? By our lights, the basis for believing some wutr proposition *P* conforms to only one of three normative views: namely, *rationalism*, *empiricism*, or *fideism*. In order to flesh out these bases, we turn to a discrete continuum of epistemic "strength" provided by Chisholm (1977). There are of course any number of ways to unpack the trio, but it's safe to say that Chisholm's scheme is eminently reasonable, and that the result that we obtain (belief in **S** is fideistic) would be generated by *any* epistemologically sensible unpacking of the three concepts in question.[9]

Chisholm's spectrum of the strength of a proposition for a rational human mind is a nine-point one, and ranges from 'certainly false' to 'certain.' At the halfway point are propositions said to be *counterbalanced*. There are then four positive strength factors working up from there: first *probable*, then *beyond reasonable doubt*, then *evident*, and finally *certain*. Certain propositions include the indubitable truths of formal logic (e.g. *modus ponens*, 1 = 1, etc.), and presumably "Cartesian" truths such as "I exist," and "It seems to me that I'm sad." What kind of thing is evident? For the most part, the evident would be populated by those propositions we affirm on the strength of sense perception. For example, that there is a computer screen in front of you when you are typing out a sentence such as the present one is evident. This proposition isn't certain: you might be hallucinating, after all; but it's—as we might say—*close* to certain. You wouldn't want to say, for example, while spying a coffee cup in front of you, in perfect health and having not ingested any mind-altering drugs, that the proposition that there's a cup in front of you is merely beyond reasonable doubt: you want to say, instead, that you are well within your epistemic "rights" in holding that it's *extremely* likely that there's a cup before you. This, again, is the category of the evident.

But moving down another Chisholmian notch in strength, we do in fact hit *beyond reasonable doubt*—which of course famously coincides roughly with what it takes in the United States to legally convict someone of murder. That is, to

[9] For example, our conclusion about believers in The Singularity would be obtained by turning instead to (Pollock 1974). This is as good a place as any to mention that both Chisholm's scheme, and Pollock's, are "computing-machine friendly." One of us has made use of Chisholm's strength-factor scheme to ground software for engineering argumentation; see (Bringsjord et al. 2008). And Pollock himself built an artificial agent on the basis of his epistemology; see for example (Pollock 1989, 1995).

convict someone of this kind of crime the evidence must make some such proposition as Jones is guilty beyond reasonable doubt. Finally, note that to convict on this standard, it's not sufficient to know that it's merely *probable* that Jones did it. Some proposition *P* being probable is the last notch before we reach *counterbalanced*, which of course means that a purely rational agent wouldn't bet in favor of *P*, and wouldn't bet against it. A perfectly rational agent who is agnostic about some proposition *P* would regard *P* to be counterbalanced.

What about the "negative" side of Chisholm's continuum? Since neither the empiricist not the rationalist, if abiding by their respective programs for belief fixation, would assent to propositions on negative side of *counterbalanced*, we have no need here to explore this epistemic terrain. Of course, all bets are off when it comes to the fideist. Kierkegaard even went so far as to recommend embracing the logically incoherent; that is, to recommend embracing certain propositions that are, viewed intellectually, certainly false (such as that Abraham was obligated to refrain from killing Isaac, and obligated to kill him). But we have no need to discuss the four notches of strength on the negative side in any detail.

Armed with Chisholm's spectrum, we can now offer encapsulation of the three main standards for belief to be applied to belief in **S**:

- **Rationalism**: The view that belief in a wutr proposition *P* must be supported by deductive proofs or arguments, where the inferences in this reasoning are each formally valid, and the premises are at least probable.
- **Empiricism**: The view that belief in a wutr proposition *P* must be supported by direct, neurobiologically normal sense perception of the constituents (i.e., of the being or event in question) of the propositions in question (making *P*, as noted above, evident), perhaps augmented from there by *some* formally valid deductive proof or argument.
- **Fideism**: The view that one ought to believe a wutr proposition *P* despite having little or no evidence for *P* (i.e., put in terms of arguments, every argument for *P* has at least one proposition at or below the level of counterbalanced).

Each of these doctrines are partitioned in our comprehensive breakdown into at least a *strong*, *moderate*, and *weak* sub-forms. This more fine-grained breakdown is beyond our needs in the present essay, but we do need hear a significant portion of the breakdown for rationalism (for reasons that will soon become clear). Accordingly, we note here that *strong rationalism* is the view (and as it happens, *our* view) that any human person believing some wutr *P* ought to have on hand at least one outright proof of *P*; that is, have on hand a formally valid chain of deductive inference originating from premises that are each certain.[10] The doctrine

[10] Some readers will inevitably ask: "Is there any such thing?!" We are of course well aware of the fact that even some axioms in some axiomatic set theories are controversial, and hence perhaps not certain. (Even the power set axiom in ZFC has its detractors.) Nonetheless, whatever one can deduce in deductively valid fashion from, say, $1 = 1$, would be certain, and one would be well-advised to believe such a consequence. For instance, $1 = 1 \vee Q$, for any proposition Q, would be an acceptable disjunction for even a strong rationalist to believe.

of *moderate rationalism* holds that if Jones abides by this doctrine and believes P, then Jones must have on hand at least one formally valid argument for P whose premises $P_1, P_2, \ldots, P_n$ are each at least evident, where each P_i is evident. And following suit we can say that *weak rationalism* requires only that the premises involved in deductive reasoning for the wutr P in question are at least probable. Readers will no doubt get the driving idea from the foregoing; the story would continue on, all the way through not only a more fine-grained ontology of rationalism,[11] but empiricism and fideism. In the case of the latter, the "bravest" fideists are those who believe self-contradictory propositions; Kierkegaard, as noted above, is known for commending the absurd, or certainly false, for assent. On the other hand, the most "timid" fideists would be those who believe a wutr P despite the fact that one or more premises are counter-balanced. In this case, under the "weakest-link principle," there is still wishful thinking.[12]

It's important to note that the above R-E-F framework is erected under the assumption that the human beings we are talking about are neurobiologically normal (and indeed alert readers will have noticed that we employed this condition in our definition of empiricist belief) and have had sufficient nurturing and training to be able to reason at the level of first-order logic. This assumption does idealize the situation to some degree, but we have known since the experiments of Piaget and colleagues that such human beings are certainly among us (e.g. see Inhelder and Piaget 1958), and indeed you no doubt are one of them.[13]

We conclude this section by pointing out that a nice testbed for understanding and contrasting the three different schemes for belief in wutr events and propositions can be found in the case of the credal Christian miracles. Mcgrew (2010) provides a thorough, readable discussion of the various forms of argument in favor of the veridicality of the credal miracles, some of which (in connection, e.g. with the Resurrection) are rationalist (e.g. Paley 2010), and some of which are empiricist (e.g. Habermas 1984). In addition, Swinburne (2010) has recently provided a formidable empiricist argument for the miracle of the Incarnation.

[11] For example, we could distinguish between the strength of inferential links in the argument for wutr P.

[12] Barbarically put, the principle states that an argument for Q is only as strong, overall, as the weakest inferential link in that argument. We leave aside the fascinating subject of fideism "forced" by decision–theoretic considerations. One who for example agrees with Pascal's Wager may decide to believe even if the best propositional evidence is counter-balanced, just because the potential disutility of not believing is infinitely large.

[13] That there are such humans in no way is inconsistent with results (e.g. those produced by the ingenious experimentation of Johnson-Laird 2000) showing that most humans fail to reason at the level of FOL. For additional evidence that some people are pretty darn good at deductive reasoning that coincides with FOL, see (Rips 1994).

Belief in The Singularity is Fideistic

We now articulate and defend our claim that belief in The Singularity is fideistic, and hence that such belief, while perhaps acceptable in the realm of religion, is not acceptable in the realm of science, where rationalism and empiricism together reign justifiably supreme. The basic line of reasoning in the argument is quickly and easily stated: We examine the main line of serious argument in support of **S**, and observe that by rationalist and empiricist standards this reasoning fails to fall under either umbrella. By disjunctive syllogism, the proponent of **S** is a fideist.

Without further ado, then, what is the argument? It's the one alluded to above, first given by Good (1965), and polished considerably by Chalmers (2010). The kernel of the argument, expressed in prose:

> Let an ultraintelligent machine be defined as a machine that can far surpass all the intellectual activities of any man however clever. Since the design of machines is one of these intellectual activities, an ultraintelligent machine could design even better machines; there would then unquestionably be an 'intelligence explosion,' and the intelligence of man would be left far behind. Thus the first ultraintelligent machine is the last invention that man need ever make. (Good 1965)

Chalmers reasonably takes Good to be here arguing for the second premise, that is, (P2), in the following overarching argument ($\mathcal{A}$). In this argument, 'HI' is human intelligence, 'AI' is artificial intelligence at the level of human persons, 'AI^+' is artifical intelligence above the level of human persons, and 'AI^{++}' refers to super-intelligence constitutive of $\mathcal{S}$. Note that we have labeled the conclusion in line with previously introduced notation.

$\mathcal{A}$

(P1) There will be AI (created by HI).
(P2) If there is AI, there will be AI^+ (created by AI).
(P3) If there is AI^+, there will be AI^{++} (created by AI^+).
$\therefore$ There will be AI^{++} ($= \mathcal{S}$ will occur).

Of course, $\mathcal{A}$ is deductive in form, and formally valid. Unfortunately, that's about where the good news ends for the proponent of The Singularity. To see this, we reason as follows. In order for belief that **S** to qualify as rationalist, the premises in question must be in Chisholm's continuum either probable, beyond reasonable doubt, certain, or evident. There can be no denying that (P1) isn't certain; in fact, all of us can be quite certain that (P1) isn't certain. Our certainty in the lack of certainty here can be established by showing, formally, that the denial of (P1) is consistent, since if not-(P1) is consistent, it follows that (P1) doesn't follow from any of the axioms of classical logic and mathematics (for example, from a standard axiomatic set theory, such as ZF). How then do we show that not-(P1) is consistent? We derive it from a set of premises which are themselves consistent. To do this, suppose that human persons are information-processing machines more powerful than standard Turing machines, for instance the infinite-time Turing machines specified and explored by Hamkins and Lewis (2000), that

AI (as referred to in $\mathcal{A}$) is based on standard Turing-level information processing, and that the process of creating the artificial intelligent machines is itself at the level of Turing-computable functions. Under these jointly consistent mathematical suppositions, it can be easily proved that AI can *never* reach the level of human persons (and motivated readers with a modicum of understanding of the mathematics of computer science are encouraged to carry out the proof). So, we know that (P1) isn't certain.

But as a matter of fact the reasoning we have just summarized suffices to show that (P1), and for that matter (P2) and (P3) as well, cannot be classified as *beyond reasonable doubt* or *evident*. Why? The answer is straightforward, and water-tight: It's not beyond reasonable doubt that those who hold that the human mind processes information in a manner above the Turing Limit are wrong. This point is made for example in the brief (Bringsjord and van Heuveln 2003), and made again in the sustained, book-length (Bringsjord and Zenzen 2003).

But there are also out-of-the-armchair reasons why (P1) isn't evident. Recall that we said evident propositions are typically those recommended by direct sense perception. But what is it that we perceive which provides reason to believe that human-level machine intelligence is coming, on the strength of human engineering? The answer is: "Nothing." For the fact of the matter is that a sharp toddler of today makes a mockery of any computing machine with designs on natural-language communication. And even if we leave natural-language communication out of the picture, and refer instead to human-level problem solving specifically in areas that would seem to be positively ideal for computing machines, we perceive not the steady advance of computing machines, but their paralysis when stacked against the capability of humans. For example, consider automatic programming, which is one of the original dreams of AI. Today, in 2012, we quite literally have no computing machines that can, having been supplied with a standard mathematical specification of an arbitrary number-theoretic function f (from the natural numbers to the natural numbers), supply as output a new computing machine that computes f—even when the input functions are as simple as those given to students in introductory programming classes!

We come then to the last possible escape from fideism available, at least in principle, to the believer in $\mathcal{S}$: weak rationalism, in the specific instantiation of this doctrine consisting in the claim that (P1)–(P3) are merely each probable. This move makes for an epistemic humility that we haven't seen among those proclaiming the arrival of superintelligent machines. Nonetheless, the point is that it seems to be a move available to the believer in $\mathcal{S}$. In addition, there is no denying that while in philosophy of religion the vast majority of cases made for the propositions central to the Christian brand of monotheism accept the burden of strong or at least moderate rationalism, there are instances of *weak* rationalism. Swimburne (1991), for example, argues only that the existence of God is more probable than not.

So, is the trio (P1)–(P3) probable? We don't think so. AI and the computational conception of mind, following Glymour (1992), can be said to have begun over two millennia ago with Aristotle's knowledge representation and reasoning frameworks; and yet, again, here we are, with hardware that moves information in silicon at a rate that makes the transmission speed of the brain seem as slow as a disoriented caterpillar by comparison, and we still don't have a machine that can problem-solve, even in highly formal domains like computer programming, at the level of a mediocre novice. It seems to us that at this point it's looking highly *un*likely that HI will produce AI, and moreover we have no reason to think that AI would be able to produce AI^+ at any rate. We concede that this isn't much of an argument. Is there any more principled philosophical reason for holding that one or more of the trio are less than probable, and hence that believing in The Singularity is to slip into fideism? Yes.

We give an argument based, first, on the observation that $\mathcal{A}$ is itself based on the concept of ever-increasing intelligence. More specifically, we note that it follows deductively from the trio in question that, where $L(M)$ yields the level of intelligence of a machine (or class of machines; we assume for the sake of argument that humans are bio/carbon-based computing machines), $L(\text{ HI}) < L(\text{ AI}^+)$. In fact, it follows deductively from the three propositions in question, and the defensive move that we have invoked on behalf of the proponent of **S** seeking to avoid falling into fideism, that

(P4) Proponents of the case $\mathcal{A}$ for **S** at this stage in the present dialectic know that it's probable that machines AI^+ will arrive such that $L(\text{ HI}) < L(\text{ AI}^+)$.

But if the proponents of the case in question know this, then surely they must know what the difference in intelligence between HI and AI^+ consists in. If they don't know what the difference consists in, then they aren't within their epistemic rights in asserting (P4). In fact, in that state of ignorance, asserting (P4) certainly has the look and feel of the core spirit of fideism, which is to forge ahead and believe, in the absence of the normal prerequisites. Do those who believe in The Singularity understand what the difference in question is? Apparently not. We have scoured the writings of pro-$\mathcal{S}$ thinkers for even an atom of an account of the difference, and have come up utterly empty. In fact, these writings, merged, yield self-refutation. For example, Chalmers (2010) understands that mere processing speed of hardware, ever increasing in conformity with Moore's Law, *contra* Kurzweil (2000), is insufficient to support the claim that super-human machine intelligence will probably arrive. As Chalmers notes, speed is one thing; that which is computed *by* that speed is quite another. What the proponent of $\mathcal{S}$ thus needs to supply in order to dodge the descent into fideism is the difference between HI and AI^+ *cashed out in a differential between the respective functions computable by each class of machine*. These details cannot be found in the literature—anywhere. Of course, the proponent of The Singularity could retort that ultraintelligent computing machines have super-human intelligence because, for example, they can: play better chess than any humans, or push further into complexity-intractable spaces (e.g. can solve in a reasonable amount of time more of the space of problems in the general propositional satisfiability problem) than humans have managed, or out-score any human on general tests of intelligence, and so on.[14]

Unfortunately, there are two fatal problems with this response. First, the response runs afoul of Chalmers' observation that speed in and of itself is in the end nothing worth writing home about. We've known for a long time, for example, that we have an algorithm for playing perfect chess. So if outerspace aliens landed tomorrow and proudly proclaimed that they can play invincible chess via this algorithm, because they have faster hardware than ours (perhaps implanted in their bodies), we really wouldn't be that impressed—and in analogy we wouldn't be inclined to say that if these aliens are just computing machines, ultraintelligent machines had arrived on our planet. The second fatal problem is that the current upward march of AI research is gradually producing precisely the sort of machines touted by the believer in S in the rejoinder under review at the moment; but these machines aren't in any way regarded to be a quantum leap beyond HI. We expect that soon enough computing machines will be able to process in real time all data relevant to the coördinated automated driving of every vehicle on our planet. Do these machines deserve to be called 'ultraintelligent'? No. They are just fast processors; their core functionality is rather trivial.

Conclusion

We conclude, then, that proponents of **S** are indeed fideists. This in no way implies that **S** is false. We have friends, and suspect you do as well, who assert the wutr propositions of this and that religion in the absence of a rationalist or empiricist basis for such assertions, and we wisely resist declaring that therefore they have put their faith in falsehoods. The most that can be said (unless of course disproofs of the propositions in question are on hand) is that faith of the fideistic sort is certainly in operation, and this is so whether or not the targets of that faith are real. So it is as well for the believer in The Singularity at the close of the investigation carried out herein.

---

[14] In our experience, the concept of intelligence as it's used in communication between those believing in S comes at least close to be conflated with the concept of *power*, or more precisely, *information-acquisition* power, conjoined with processing speed a la Moore's Law. Once this conflation occurs, the notion that machines of the future will be ultraintelligent quickly arrives on scene. Why? The point can be put in sci-fi terms: We imagine a *Terminator 3*-like event in which unmanned machines hooked into all digital information on the planet suddenly break through any and all privacy restrictions on use of this data, and proceed to exploit it. These machines are now able to do things that are unprecedentedly "intelligent." For example, the machines may now be able to prevent human crimes before they happen. (E.g. machines with access to everyone's email, and the processing power to check them for plans of foul play, could thwart criminals.) Needless to say, while this notion of information-theoretic super-intelligence is coherent, and may in fact even be likely to materialize, no fundamentally new functionality is in play, and hence, while in our interaction with believers in The Singularity we witness the conflation in question, the case for **S** isn't insulated from our counter-argumentation.

References

Bringsjord, S. (1992). *What robots can and can't be*. Dordrecht: Kluwer.

Bringsjord, S. (2010). Meeting Floridi's challenge to artificial intelligence from the knowledge-game test for self-consciousness. *Metaphilosophy 41*(3), 292–312. http://kryten.mm.rpi.edu/sb_on_floridi_offprint.pdf

Bringsjord, S., Taylor, J., Shilliday, A., Clark, M. Arkoudas, K. (2008). Slate: An argument-centered intelligent assistant to human reasoners. In: F. Grasso, N. Green, R. Kibble C. Reed (Eds.), *Proceedings of the 8th International Workshop on Computational Models of Natural Argument (CMNA 8)*, (pp. 1–10). Greece: Patras. http://kryten.mm.rpi.edu/Bringsjord_etal_Slate_cmna_crc_061708.pdf

van Bringsjord, S., & Heuveln, B. (2003). The mental eye defense of an infinitized version of Yablo's paradox. *Analysis, 63*(1), 61–70.

Bringsjord, S., & Zenzen, M. (2003). *Superminds: People harness hypercomputation, and more*. Dordrecht: Kluwer Academic Publishers.

Chalmers, D. (2010). The singularity: A philosophical analysis. *Journal of Consciousness Studies, 17*, 7–65.

Chesterton, G. (2009). *Orthodoxy*. Chicago: Moody Publishers.

Chisholm, R. (1977). *Theory of knowledge* . Englewood Cliffs: Prentice-Hall.

Dennett, D. (2007). *Breaking the spell: Religion as a natural phenomenon*. New York: Penguin.

Floridi, L. (2005). Consciousness, agents and the knowledge game. *Minds and Machines, 15*(3–4), 415–444. http://www.philosophyofinformation.net/publications/pdf/caatkg.pdf

Glymour, C. (1992). *Thinking things through*. Cambridge: MIT Press.

Good, I.J. (1965). Speculations concerning the first ultraintelligent machines. In F. Alt& M. Rubinoff (Eds.), *Advances in Computing* (Vol. 6, pp. 31–38). Academic: New York.

Habermas, G. (1984). *The resurrection of Jesus: An apologetic*. Lanham: University Press of America.

Hamkins, J. D., & Lewis, A. (2000). Infinite time turing machines. *Journal of Symbolic Logic, 65*(2), 567–604.

Inhelder, B., & Piaget, J. (1958). *The growth of logical thinking from childhood to adolescence*. New York: Basic Books.

Johnson-Laird, P. N., Legrenzi, P., Girotto, V., & Legrenzi, M. S. (2000). Illusions in reasoning about consistency. *Science, 288*, 531–532.

Kierkegaard, S. (1986). *Fear and trembling*. New York: Penguin.

Kurzweil, R. (2000). *The age of spiritual machines: When computers exceed human intelligence*. New York: Penguin USA.

Leibniz, G. (1998). *Theodicy*. Chicago: Open Court.

Lewis, C. S. (1960). *Mere christianity*. New York: Macmillan.

McGrew, T. (2010). Miracles, Stanford Encyclopedia of Philosophy. http://plato.stanford.edu/entries/miracles

Paley, W. (2010). Evidence of christianity, qontro classic books. Paley's apology was first published in 1794. The book is available through Project Gutenberg. The full official title of the book is A View of the Evidences of Christianity.

Pollock, J. (1974). *Knowledge and justification*. Princeton: Princeton University Press.

Pollock, J. (1989). *How to build a person: A prolegomenon*. Cambridge: MIT Press.

Pollock, J. (1995). *Cognitive carpentry: A blueprint for how to build a person*. Cambridge: MIT Press.

Rips, L. (1994). *The psychology of proof*. Cambridge: MIT Press.

Swinburne, R. (1981). *Faith and reason*. Oxford: Clarendon Press.

Swinburne, R. (1991). *The existence of god*. Oxford: Oxford University Press.

Swinburne, R. (2010). *Was Jesus god?*. Oxford: Oxford University Press.

Turing, A. (1950). Computing machinery and intelligence. *Mind LIX*, (*59*)(236), 433–460.

Vinge, V. (1993). *The coming technological singularity: How to survive in the post-human era.* Whole Earth Review.

Yudkowsky, E. (1996). *Starting into the singularity.* http://yudkowsky.net/obsolete/singularity.html

Chapter 19A
Vernor Vinge on Bringsjord et al.'s "Belief in the Singularity is Fideistic"

It's no surprise that pure rationalism is useless for discussing the possibility of the Singularity. Pure rationalism is not much use outside of mathematics. (And in computer science, _pace_ Edsger Dijkstra, it's not really useful outside of very simple situations.) In the sciences, the goal to strive for is rationalism combined with a focused empiricism consisting of cleverly planned observations and experiments that disprove as much as possible as quickly as possible.

Unfortunately, such a combination of rationalism and empiricism is rarely attainable in discussing future progress in science and engineering. (When it can be achieved, it amounts to Alan Kay's famous advice that "The best way to predict the future is to invent it.") For many planning environments, we must instead consider a variety of scenarios (e.g. (Vinge 1993, 2007). Risks and symptoms and benchmarks can then be watched for and used to support further plans and action. In this process, some of the players may be somewhat fideistic. That's fine. Without an element of fideism in our entrepreneurs, we'd have fewer failures, but we'd also lose or postpone many wonderful innovations.

As for Bringsjord et al.'s WUTR (weighty, unseen, and temporally removed) assessment of the Technological Singularity:

- Weighty:The possibility of the Singularity is certainly weighty. Progress along all the different paths to the Singularity is bringing into focus (and perhaps stark immediacy) a number of questions that have been endlessly debated over the last few thousand years (identity, consciousness, intelligence, mortality). Whether or not the Singularity happens, the technological interrogation of these issues has put us in a different playing field than all the philosophers of the past.

- Unseen:That there are no current examples of super-intelligence is not a surprise. On the other hand, the milestones already passed are not trivial, except as claimed to be so after they were attained. Bringsjord et al. propose an interesting milestone of their own, the problem of automatic program generation where the input is a simple function described in standard mathematical notation. Tell me more! This sounds like something that is doable with 2012-era computers/ software, at least competitive with human performance.)

Bringsjord raise a much broader complaint in saying that Singularity enthusiasts don't even specify the difference between human level intelligence and machine superhuman intelligence: "We have scoured the writings of pro-S thinkers for even an atom of an account of the difference, and have come up utterly empty."

In discussing this point, they raise the possibility that superintelligence might be claimed as simply the running of a computer very fast—and they dismiss that possibility as irrelevant. I agree that 2012 software running very very fast would be an absurd contender, but that is the wrong comparison. For myself (and I expect most people) the really hard thing to accept is that human equivalent intellects could run on a computer. But that is a goal we have a moderately good criterion for, namely Turing's Test (especially in the extended sense that Penrose describes in "The Emperor's New Mind", at the end of his generally skeptical discussion of the topic). Now imagine that such a Turing Test winner is run at much higher speed. In (Vinge 1993), I called such an achievement "weak superhumanity". In fact, I used the word "weak" because I believe there would be lot more to superhuman intelligence (Vinge 1993, 2010). Nevertheless, it provides a goal as specific as Turing's Test for the discussion of superhuman intelligence.

- Temporally removed:Until it actually happens, the Singularity will have this characteristic. But in the absence of technological surprises and classical disasters (e.g. nuclear war), I expect to see automation gradually achieving more and more of what has been human-only capabilities. At the same time, I expect that human/computer teams will be ever more powerful; they may in fact guide the Singularity into being. The Teens should be interesting years.

References

Vinge, V. (1993). The Coming Technological Singularity: How to Survive in the Post-Human Era, VISION-21 Symposium sponsored by NASA Lewis Research Center and the Ohio Aerospace Institute, Mar 30–31, 1993. NASA CP-10129. Available http://www-rohan.sdsu.edu/faculty/vinge/misc/singularity.html

Vinge, V. (2007). What If the Singularity Does NOT Happen. Talk presented at Seminars about Long-Term Thinking, 15 Feb 2007. Available http://www-rohan.sdsu.edu/faculty/vinge/longnow/index.htm

Vinge, V. (2010). Species of Mind. Talk presented at IAAI-10, 15 July 2010. http://www-rohan.sdsu.edu/faculty/vinge/misc/iaai10/

Chapter 19B
Michael Anissimov on Bringsjord et al.'s "Belief in The Singularity is Fideistic"

The substance of Brinsjord et al's critique is in a single paragraph of pages 10–11 of their essay, P1 referring to Chalmers' first assumption, "there will (eventually, barring defeaters) be Artificial Intelligence (of the human level)":

> There can be no denying that (P1) isn't certain; in fact, all of us can be quite certain that (P1) isn't certain. [. . .] ...suppose that human persons are information-processing machines more powerful than standard Turing machines, for instance the infinite-time Turing machines specified and explored by Hamkins and Lewis (2000), that AI (as referred to in A) is based on standard Turing-level information processing, and that the process of creating the artificial intelligent machines is itself at the level of Turing-computable functions. Under these jointly consistent mathematical suppositions, it can be easily proved that AI can never reach the level of human persons (and motivated readers with a modicum of understanding of the mathematics of computer science are encouraged to carry out the proof). So, we know that (P1) isn't certain.

It is difficult to ascertain on what basis Brinsjord et al are making the claim that human persons are information—processing machines "more powerful" than standard Turing machines. Occam's razor, along with decades of evidence from cognitive science, seem to imply that the human brain and mind can be viewed as a massively parallel Turing machine.

Supposing that artificial intelligences will "never reach the level of human persons" is a claim with few academic citations. Generally, such statements appear to be appeals to intuitions of human exceptiónalism—the notion that humans have something deeply special about them that could never be duplicated in a machine. Given this intuition, human exceptionalists are forced to retroactively search for supporting arguments. The notion that human brains somehow utilize extra-Turing information processing is one such argument.

The Church-Turing thesis is the idea that anything algorithmically computable is computable by a Turing machine. Given the nearly universally accepted supposition in the cognitive sciences that intelligence is made up of a collection of mental routines that are fuzzy algorithms, plus the Church-Turing thesis, we get the conclusion that intelligence is indeed computable by standard Turing machines. Acceptance of these two ideas is not universal in cognitive science and computer science, but the ideas are broadly accepted, with extensive discussions in the literature.

The history of science is filled with various examples of human exceptionalism that were proven wrong. For instance, the notion that human beings are animated by an immaterial soul has been replaced by the scientific notion of the brain as the director of behavior. Another example would be the pre-scientific notion of humans as separate from the animal kingdom, replaced by the idea of humans as a part of the animal kingdom. The notion that human beings are the only agents that can implement intelligence is being supplanted by the notion that intelligence is a bundle of algorithms that can be implemented by any suitable computer, whether carbon-based or silicon-based.

Chapter 20
A Singular Universe of Many Singularities: Cultural Evolution in a Cosmic Context

Eric J. Chaisson

Abstract Nature's myriad complex systems—whether physical, biological or cultural—are mere islands of organization within increasingly disordered seas of surrounding chaos. Energy is a principal driver of the rising complexity of all such systems within the expanding, ever-changing Universe; indeed energy is as central to life, society, and machines as it is to stars and galaxies. Energy flow concentration—in contrast to information content and negentropy production—is a useful quantitative metric to gauge relative degree of complexity among widely diverse systems in the one and only Universe known. In particular, energy rate densities for human brains, society collectively, and our technical devices have now become numerically comparable as the most complex systems on Earth. Accelerating change is supported by a wealth of data, yet the approaching technological singularity of 21st century cultural evolution is neither more nor less significant than many other earlier singularities as physical and biological evolution proceeded along an undirectional and unpredictable path of more inclusive cosmic evolution, from big bang to humankind. Evolution, broadly construed, has become a powerful unifying concept in all of science, providing a comprehensive worldview for the new millennium—yet there is no reason to claim that the next evolutionary leap forward beyond sentient beings and their amazing gadgets will be any more important than the past emergence of increasingly intricate complex systems. Nor is new science (beyond non-equilibrium thermodynamics) necessarily needed to describe cosmic evolution's interdisciplinary milestones at a deep and empirical level. Humans, our tools, and their impending messy interaction possibly mask a Platonic simplicity that undergirds the emergence and growth of complexity among the many varied systems in the material Universe, including galaxies, stars, planets, life, society, and machines.

E. J. Chaisson (✉)
Harvard-Smithsonian Center for Astrophysics,
Harvard University, Cambridge, MA, USA
e-mail: ejchaisson@cfa.harvard.edu
www.cfa.harvard.edu/~ejchaisson

A. H. Eden et al. (eds.), *Singularity Hypotheses*, The Frontiers Collection,
DOI: 10.1007/978-3-642-32560-1_20, © Springer-Verlag Berlin Heidelberg 2012

Introduction: My Philosophy of Approach

About a decade ago, a book of mine was co-reviewed along with another in the *Boston Globe* (Raymo 2002), both of them in the context of humanity's future prospects. *Cosmic Evolution* (Chaisson 2001) sought to explicate, from a strictly scientific viewpoint, the natural rise of complex systems throughout the nearly 14 billion year history of the Universe, including sentient humans and our useful yet disturbing technical devices. The other book, of which I was unaware at the time, *The Age of Spiritual Machines* (Kurzweil 2000) argued that the speed and volume of information processing are increasing so rapidly that computers will soon surpass humans as an event of singular importance—a cultural tipping point termed by some the Singularity—is fast approaching. Although our scholarship partly overlapped, Kurzweil's book seemed speculative and even passionate, so I never did critically assess the idea of a technological singularity until I was invited to contribute the present article to this Frontiers Collection.

For many years, my scientific agenda has aimed to go beyond mere words and speculation about humankind and its technological aids. I have striven to place human society into a cosmological framework and to quantitatively analyze just how complex we, our brains, and our machines really are. Frankly, as a confirmed empiricist, I am skeptical of forecasting our future because all such exercises entail much qualitative guesswork; nor do I regard future evolutionary events to be accurately predictable given that an element of chance always accompanies the necessity of natural selection. That said, it does seem inevitable, indeed quite ordinary, that new forms of complexity are destined to supplant humanity as the most complex system known, just as surely as people took precedence over plants and reptiles, and in turn even earlier life on Earth complexified beyond that of the galaxies, stars, and planets that made life possible. There is nothing abnormal about the oncoming clash of men and machines—other than perhaps damaging our egos. The Universe has spawned many such grand evolutionary, even transcendent, events in deep time, the scale used to measure biological, geological, and cosmological changes throughout history writ large. That carbon-based humans are about to merge with, or concede to, silicon-based machines during a so-called "technological singularity" (Kurzweil 2005) is entirely reasonable—although a more benign outcome is that we might simply learn to live with them, to coexist. Data presented in this paper suggest that singularities are part of the natural scheme of things—normal, broadly expected outcomes when concentrated energy flows gave rise to increasingly complex systems throughout the expanding Universe. (Note that the expression "singularity" in this paper matches that commonly used to mean a major evolutionary milestone, of which there were many in cosmic history and thus the word singularity, oddly, implies plurality, not the technical term that puzzles mathematicians when sizes and scales near zero and densities approach infinity, as in black holes.)

My philosophy of approach, as an experimental physicist, seeks to interpret natural history over many billions of years, and to do so by embracing the leitmotif of energy flow through increasingly complex systems. By contrast, Kurzweil, among many other strong artificial-intelligence advocates, prefer information content to explain and predict humanity's recent and impending changes over much shorter periods of time. This is not a criticism of those who characterize complexity and evolution by means of information theory, or even entropy production, although I personally find these concepts overly abstract (with dubious meanings), hard to define (to everyone's satisfaction), and even harder to measure (on any scale). Regarding the latter, neither maximum nor minimum entropy principles are evident in the data presented below. Regarding the former, I sense, but cannot prove, that information is another kind of energy; both information storage and retrieval need energy, and greater information processing and calculation require greater energy density. While information content and entropy production are powerful terms that offer much theoretical insight, neither provides clear, unambiguous empirical metrics. My practical stance is that information may be useful to describe some systems, but energy is needed to make and operate them.

Where we do all agree (apparently) is that cultured humans and their invented machines are now in the process of transcending biology, a topic bound to be emotional if only because it rubs our human nerves and potentially dethrones our perceived cosmic primacy (Dick and Lupisella 2009; Kelly 2010). The roots of this evolutionary milestone—perhaps it is a technological singularity—probably extend as far back as the onset of agriculture when our forebears began manipulating their local environs, yet has recently advanced rapidly as we now alter both our globe environmentally as well as our being genetically. Even so, these changes—and their outcomes—are probably nothing more than the natural way that cultural evolution developed beyond biological evolution, which in turn built upon physical evolution before that, each of these evolutionary phases being an integral part of a more inclusive cosmic evolution that also operates naturally, as it always has and likely always will, with the irreversible march of time.

Cosmic Evolution: A Scientific Worldview for the New Millennium

The past few decades have seen the emergence of a unified scenario of natural history, including ourselves as sentient beings, based on the time-honored concept of change. Heraclitus may well have been right some 25 centuries ago when he offered perhaps the best observation of Nature ever: *πaντα ρει*—"all flows... nothing stays the same." From stars and galaxies to life and humanity, a loose community of liberal researchers is now weaving an intricate pattern of understanding using the fabric of all the sciences—an interdisciplinary rendering of the origin and evolution of every known class of object in our richly endowed Universe. Often called cosmic evolution, this uncommonly broad cosmology that includes life as an integral part can be defined as *the study of the many varied developmental and generational changes in the assembly and composition of radiation, matter, and life throughout the history of the Universe*. These are the changes that have produced our Galaxy, our Sun, our Earth, and ourselves, and as such include both evolution and development (Salthe 1993). A localized "big-history" version of this scenario that places into larger perspective specifically humankind on Earth (Christian 2004; Brown 2007; Spier 2010; Grinin et al. 2011) is part of a more universal cosmic-evolutionary narrative that addresses the Universe at large (Chaisson 2001, 2006, 2009a, b; Dick 2009; Vakoch 2009). The result is a grand evolutionary synthesis bridging a wide variety of scientific specialties—physics, astronomy, geology, chemistry, biology, anthropology, among others and including the humanities—a genuine epic of vast proportions extending from the very beginning of time to the present—and presumably beyond in both space and time.

While entering this new age of synthesis, we are beginning to decipher how all known systems—atoms and galaxies, cells and brains, people and society, among myriad others—are interrelated and constantly changing. Our appreciation for evolution now extends well beyond the subject of biology; the concept of evolution, generally considered (as in most dictionaries) as *any process of ascent with change in the formation, growth, and development of systems*, has become a potent unifying factor in all of science. Yet questions remain: How realistic is our quest for unification, and will the integrated result resemble science or philosophy? How have the magnificent examples of order on and beyond Earth arisen from chaos? Can the observed constructiveness of cosmic evolution be reconciled with the inherent destructiveness of thermodynamics? Most notably, we want to understand the emergence of diverse structures spanning the Universe, and especially the complexity of such systems as defined by *intricacy, complication, variety, or involvement among the interconnected parts of a system.* Particularly intriguing is the rise of complexity over the course of time, and dramatically so in the Phanerozoic during the past $\sim$540 million years—a rise that has reached a crescendo on Earth with conscious beings, adroit machines, and their likely future intermingling. Could a technological singularity be the next great advance in the scenario of cosmic evolution?

Recent empirically based research, guided by huge new databases describing a multitude of complex systems, suggests robust answers to some of the above queries. Islands of ordered complexity that include galaxies, stars, planets, life, and society are more than balanced by great seas of increasing disorder elsewhere in the environments beyond those systems. All quantitatively agrees with the valued precepts of thermodynamics, especially non-equilibrium thermodynamics. None of Nature's organized structures, not even life itself, is a violation (nor even a circumvention) of the celebrated 2nd law of thermodynamics. Both order and entropy can increase together—the former locally and the latter globally. Thus, we arrive at a central question lurking in the minds of some of today's eclectic thinkers (e.g. Mandelbrot 1982; Wolfram 2002): Might there be a kind of Platonism at work in the Universe—an underlying principle, a unifying law, or perhaps a

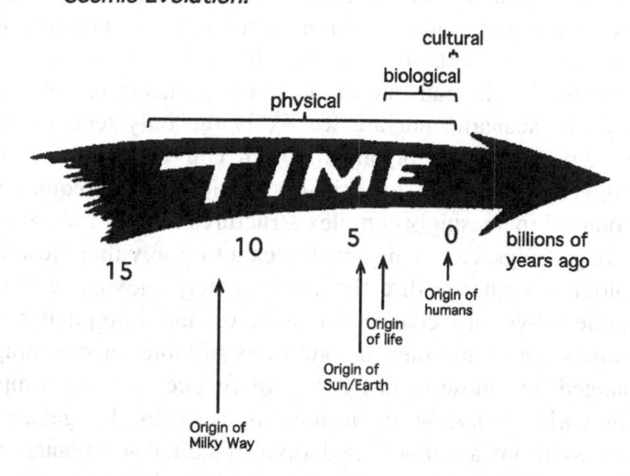

Fig. 20.1 An arrow of time, extending over nearly 14 billion years from the big bang at *left* to the present at *right*, symbolically represents the sweeping inclusiveness of cosmic evolution, an overarching subject that includes the three phases of physical, biological, and cultural evolution (*top* of figure). The arrow is not pointing at us; cosmic-evolutionary cosmology is not anthropcentric, yet it powerfully encapsulates the origin and evolution of our galaxy, star, and planet, as well as of life, humanity, and civilization (*bottom*)

surprisingly simple process that quite naturally creates, organizes, and maintains the form and function of complex systems everywhere?

Figure 20.1 depicts the archetypal illustration of cosmic evolution—the arrow of time. Regardless of its shape or orientation, such an arrow represents a symbolic guide to the sequence of events that have changed systems from simplicity to complexity, from inorganic to organic, from chaos to order. That sequence, as determined by a large body of post-Renaissance data, accords well with the idea that a thread of change links the evolution of primal energy into elementary particles, the evolution of those particles into atoms, in turn of those atoms into galaxies and stars, and of stars into heavy elements, the evolution of those elements into the molecular building blocks of life, of those molecules into life itself, and of intelligent life into the cultured and technological society that we now share. Despite the compartmentalization of today's academic science, evolution knows no disciplinary boundaries. As such, the most familiar kind of evolution—biological evolution, or neo-Darwinism—is just one, albeit important, subset of a broader evolutionary scheme encompassing much more than mere life on Earth. In short, what Darwinian change does for plants and animals, cosmic evolution aspires to do for all things. And if Darwinism created a revolution of understanding by helping to free us from the notion that humans differ from other life-forms on our planet, then cosmic evolution extends that intellectual revolution by treating matter on Earth and in our bodies no differently from that in the stars and galaxies beyond.

Anthropocentrism is neither intended nor implied by the arrow of time; it points toward nothing in particular, just the future generally. Anthropic principles notwithstanding, no logic supports the idea that the Universe was conceived to produce specifically us. We humans are unlikely the pinnacle or culmination of the cosmic-evolutionary scenario, nor are we likely the only technically competent beings to have emerged in the organically rich Universe. Time's arrow merely provides a convenient symbol, artistically depicting a ubiquitous flow that (somehow) produced increasingly complex structures from spiral galaxies to rocky planets to thinking beings. Nor is the arrow meant to imply that "lower," primitive life forms biologically change directly into "higher," advanced organisms, any more than galaxies physically change into stars, or stars into planets. Rather, with time—much time—the environmental conditions suitable for spawning simple life eventually changed into those favoring the emergence of more complex species; likewise, in the earlier Universe, environments were ripe for galactic formation, but now those conditions are more conducive to stellar and planetary formation. Changes in surrounding environments often precede change within ordered systems, and the resulting system changes have *generally* been toward greater amounts of diverse complexity, as numerically justified in the next section.

Energy Flows and Complexity Rises

Cosmic evolution as understood today is governed largely by the laws of physics, particularly those of thermodynamics. Note the adverb "largely," for this is not an exercise in traditional reductionism. Of all the known principles of Nature, thermodynamics perhaps best describes the concept of change—yet change dictated by a combination of randomness and determinism, of chance and necessity. Literally, thermodynamics, which tells us what can happen and not what does happen, means "movement of heat"; a more insightful translation (in keeping with the wider connotation in Greek antiquity of motion as change) would be "change of energy." Energy flows engendered largely by the expanding cosmos do seem to be as central in the origin of structured systems as anything yet found in Nature. Furthermore, the optimization of such energy flows might well act as a motor of evolution broadly conceived, thereby affecting all of physical, biological, and cultural evolution, the sum total of which constitutes cosmic evolution.

Energy does play a role in creating, ordering, and maintaining complex systems. Recognized decades ago at least qualitatively in words (Lotka 1922; von Bertalanffy 1932; Schroedinger 1944), the need for energy should now be embraced as an essential feature not only of biological systems such as plants and animals but also of physical systems such as stars and galaxies; energy's engagement is also widely recognized in cultural systems such as a city's inward flow of food and resources amidst its outward flow of products and wastes, indeed for all of civilization itself. All complex systems—whether alive or not—are open, organized, dissipative, non-equilibrated structures that acquire, store, and express energy.

In contrast to my enthusiasm for energy as an organizing principle, I acknowledge that entropy production (Kleidon and Lorenz 2005; Martyushev and Seleznev 2006) and information content (Hofkirchner 1999; Gleick 2011) are more often espoused in discussions of origin, evolution, and complexity. Yet, these alternative aspects of systems science are less encompassing and decidedly less empirical than many practitioners admit, their theoretical usefulness narrow, qualitative, and equivocal as general complexity metrics (Meyers 2009). Although yielding insightful properties of systems and their emergent and adaptive qualities unlikely to be understood otherwise, such efforts have reaped an unusual amount of controversy and only limited success to date (Mitchell 2009). Nor are information or negentropy useful in quantifying or measuring complexity, a slippery term for many researchers. In biology alone, much as their inability to reach consensus on a definition of life, biologists cannot agree on a complexity metric. Some (Maynard Smith 1995) use non-junk genome size, others (Bonner 1988) employ creature morphology and behavioral flexibility, still others chart the number of cell types in organisms (Kaufmann 1993) or appeal to cellular specialization (McMahon and Bonner 1983). All these attributes of life have qualitative worth, yet all are hard to quantify in practical terms. Cosmic evolutionists seek to push the analytical envelope beyond mere words, indeed beyond biology.

We thus return to the quantity having greatest appeal to physical intuition—energy—a term that is satisfactorily definable, understandable, and above all measurable. Not that energy has been overlooked in more recent discussions of systems' origin and assembly. Many researchers (e.g. Morrison 1964; Morowitz 1968; Dyson 1979; Odum 1988; Smil 1999; Lane and Martin 2010) have championed in different ways and limited contexts the cause of energy's organizational abilities. Even so, the quantity of choice cannot be energy alone, for a star is clearly more energetic than an amoeba, a galaxy much more than a single cell. Yet any biological system is surely more complicated than any inanimate entity. Absolute energies are not as indicative of complexity as relative values, which depend on a system's size, composition, and efficiency. To characterize complexity objectively—that is, to normalize all such structured systems in precisely the same way—a kind of energy density is judged most useful. Moreover, it is the *rate* at which free energy transits complex systems of given mass that seems especially constructive (as has long been realized for ecosystems: Lotka 1922; Ulanowicz 1972), thereby delineating energy *flow*. Hence, "energy rate density," symbolized by Φ_m, becomes an operational term whose meaning and measure are easily understood, indeed whose definition is clear: *the amount of energy passing through a system per unit time and per unit mass*. In this way, neither new science nor appeals to non-science are needed to explain the impressive hierarchy of the cosmic-evolutionary story, from quarks to quasars, from microbes to minds.

Experimental data and detailed computations of energy rate densities are reported elsewhere (Chaisson 2011a, b), most of them culled or calculated from values found in widely scattered journals over many years. In the briefest of compact summaries:

- For physical systems, stars and galaxies generally have energy rate densities (10^{-3}–10^2 erg/s/g) that are among the lowest of known organized structures. Galaxies show clear temporal trends in rising values of Φ_m while clustering hierarchically, such as for our Milky Way, which increased from $\sim 10^{-2}$ to 0.1 erg/s/g while changing from primitive dwarf status to mature spiral galaxy. Stars, too, adjust their states while evolving during one or more generations, their Φ_m values rising while complexifying with time as their interior thermal and chemical gradients steepen and differentiate; for the Sun, Φ_m increases from ~ 1 to 120 erg/s/g from young protostar to aged red giant.
- In turn, among biological systems, plants and animals regularly exhibit intermediate values of $\Phi_m = 10^3$–10^5 erg/s/g. For plant life on Earth, energy rate densities are well higher than those for normal stars and typical galaxies, as perhaps best demonstrated by the evolution of photosynthesizing gymnosperms, angiosperms, and C_4 plants, which over the course of a few hundred million years increased their Φ_m values nearly an order of magnitude to $\sim 10^4$ erg/s/g. Likewise, as animals evolved from fish and amphibians to reptiles, mammals, and birds, their Φ_m values rose from $\sim 10^{3.5}$ to 10^5 erg/s/g, here energy conceivably acting as a fuel for change, partly selecting systems able to utilize increased power densities, while forcing others to destruction and extinction—all likely in accord with neo-Darwinian principles.
- Furthermore, for cultural systems, advances in technology are comparable to those of society itself, each of them energy-rich and having $\Phi_m \geq 10^5$ erg/s/g—hence plausibly among the most complex systems known. Social evolution can be tracked, again in terms of normalized energy consumption, for a variety of human-related cultural advances among our ancestral forebears, from early agriculturists ($\sim 10^5$ erg/s/g) to modern technologists ($\sim 10^{6.5}$). Machines, too, and not just computers, but also ordinary engines that drove the 20th century economy, show the same trend from primitive devices of the industrial revolution ($\sim 10^5$ erg/s/g) to today's jet aircraft ($\sim 10^{7.5}$).

Of special note often neglected, although the absolute energy in astronomical systems is vastly larger than in our human selves, and although the mass densities of stars, planets, bodies, and brains are all comparable, the energy rate density for people and our society are upwards of a million times greater than for stars and galaxies. That's because the quantity Φ_m is an energy rate *density*. Although, for example, the Sun emits a vast luminosity, 4×10^{33} erg/s (equivalent to nearly a billion billion billion Watt light bulb), it also has an unworldly large mass, 2×10^{33} g; thus each second an amount of energy equaling only 2 ergs passes through each gram of this star. Many colleagues are likewise surprised to realize that, despite its huge size and scale, the Sun's mass density is small enough (well less than a rock) that this star would almost float if we could get it into a bathtub. By contrast, more energy flows through each gram of a plant's leaf during photosynthesis, and much more radiates through each gram of gray matter in our brains while thinking—which is why we have a hope of deciphering who we are and the Sun cannot!

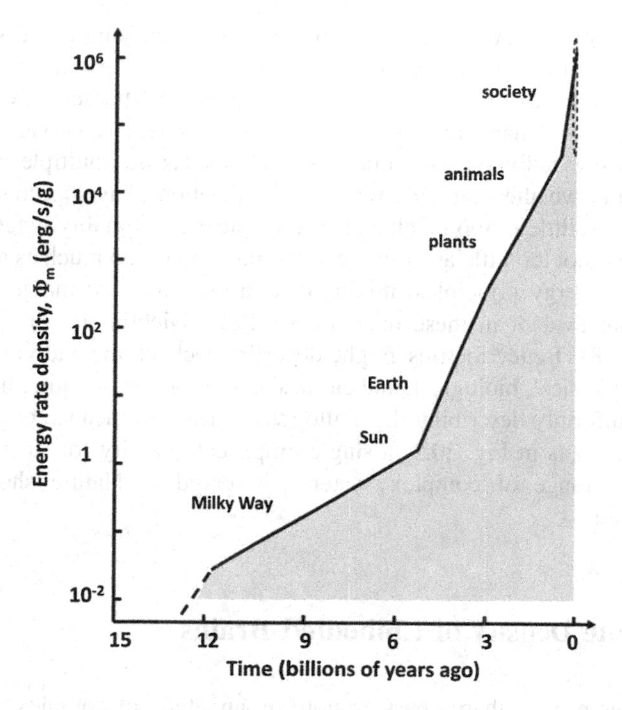

Fig. 20.2 Energy rate density, Φ_m, for a wide spectrum of systems observed throughout Nature displays a clear increase during ~14 billion years of cosmic history—in fact, an exponential rise whereby cultural evolution (steep slope at *upper right*) acts faster than biological evolution (moderate slope in middle part of curve), and even faster than physical evolution (smallest slope at *lower left*). The shaded area includes a huge ensemble of Φ_m values as individual types of localized systems continued changing and complexifying within the wider Universe that has become increasingly disordered. The Φ_m values and historical dates plotted here are estimates, each with outliers and uncertainties; yet it is not their absolute magnitudes that matter most as much as the perceived trend of Φ_m with the passage of time. The thin *dashed oval* at *upper right* outlines the magnitude of Φ_m and the duration of time plotted in Fig. 20.3

Figure 20.2, which is plotted on the same temporal scale as in Fig. 20.1, graphically compiles those data compactly presented in the three bullets above, depicting how physical, biological, and cultural evolution have transformed homogeneous, primordial matter of the early Universe into organized systems of increased intricacy and energy rate density—and it has done so with increasing speed, hence the exponentially rising curve. The graph shows the increase of Φ_m as measured or computed for representative systems having approximate evolutionary times at which they emerged in natural history. (For specific power units of W/kg, divide by 10^4.) Values given are typical for the general category to which each system belongs, yet variations and outliers are inevitable, much as expected for any simple, unifying précis of an imperfect Universe.

Energy is likely a common currency for all complex, ordered systems. Even for structures often claimed to be "self-assembled" or "self-organized," energy is inexorably involved. Energy flow is among the most unifying processes in all of science, helping to provide cogent explanations for the origin, evolution, and complexification of a whole array of systems spanning >20 orders of magnitude in scale and nearly as many in time—notably, how systems emerge, mature, and terminate during individual lifetimes as well as across multiple generations. Robust systems, whether stars, life forms, or civilizations, have optimum ranges of energy flow; too little or too much and systems abort. Optimality is favored in the use of energy—not too little as to starve a system, yet not too much as to destroy it; no maximum energy principles, minimum entropy states, or maximum entropy production are evident in these data (Lotka 1922; Nicolis and Prigogine 1977; Prigogine 1978). Better metrics might describe each of the individual systems governed by physical, biological, and cultural evolution, but no other metric seems capable of uniformly describing them altogether. The significance of plotting "on the same page" (as in Fig. 20.2) a single empirical quantity for such an extraordinarily wide range of complex systems observed in Nature should not be underestimated.

Energy Rate Density of Embodied Brains

Humans deserve more than a passing note in any study of complex systems, not because we are special but because we are them. Each individual adult normally consumes ~2,700 kcal/day in the form of food to fuel our metabolism. This energy, gained directly from that stored in other (plant and animal) organisms and only indirectly from the Sun, is sufficient to maintain our body structure and temperature as well as drive our physiological functions and tetrapodal movements. (Note that the thermodynamical definition of a calorie, 1 cal = 4.2×10^7 erg—the amount of heat needed to raise 1 g of H_2O by 1 °C—does not equal a dietician's large Calorie with a capital "C," which is 10^3 times more energetic than a physicist's calorie.) Therefore, with a body mass of 70 kg, a typical adult maintains $\Phi_m \approx 2 \times 10^4$ erg/s/g while in good health. Humans have mid-range mammalian metabolic values because our bodies house average complexity among endothermic mammals, all of which harbor comparable intricacy, including hearts, livers, kidneys, lungs, brains, muscles, and guts. Despite our manifest egos, our bodily beings do not have the highest energy rate density among animals (birds do, probably because they operate in 3 dimensions; Chaisson 2011b), nor are we more demonstrably complex than many other mammalian species.

The energy budget derived here for humans assumes today's average, sedentary citizen, who consumes ~65 % more than the basal metabolic rate of 1,680 kcal/day (or $\Phi_m \approx 1.2 \times 10^4$ erg/s/g) for an adult fasting while lying motionless. However, our metabolic rates increase substantially when performing occupational tasks or recreational activities—that's function, not structure. Even so, Φ_m once again scales with the degree of complexity of the function. For example, leisurely fishing, violin playing, tree cutting, and bicycle riding require about 3×10^4,

5×10^4, 8×10^4, and 2×10^5 erg/s/g, respectively (Ainsworth 2011). Clearly, jamming a musical instrument or balancing a moving bicycle are complex functions, and therefore more energetically demanding events, than waiting patiently for fish to bite. Thus, in the biological realm, the value-added quality of functionality does indeed count, in fact quantitatively so. Complex tasks actively performed by humans on a daily basis are typified by values of Φ_m that are often larger than those of even the metabolically imposing birds.

Nearly all zoological Φ_m values for bodies are tightly confined to within hardly more than an order of magnitude of one another—the great majority of specific metabolic rates for animals vary between 3×10^3 and 10^5 erg/s/g, despite their masses ranging over $\sim$11 orders of magnitude from fairy flies to blue whales (Makarieva et al. 2008)—all of them midway between smaller botanical values for photosynthesizing plants and higher neurological ones for central nervous systems. This, then, is how humankind, like all of the animal world, contributes to the rise of entropy in the Universe: We consume high-quality energy in the form of ordered foodstuffs and then radiate away as body heat (largely by circulating blood near the surface of the skin, by exhaling warm, humidified air, and by evaporating sweat) an equivalent amount of energy as low-quality, disorganized infrared photons. Like the stars and galaxies, we too among all other life forms are wasteful, dissipative structures (in our case glowing warmly in the infrared as a 130 W bulb), thereby connecting with earlier thermodynamic arguments that some researchers might (wrongly) think pertinent only to inanimate systems.

Regarding brains, which nuclear magnetic resonance (fMRI) imaging shows are always electrically active regardless of the behavioral posture (even while resting) of their parent animal bodies, they too derive nearly all their energy from the aerobic oxidation of glucose in blood; thus, for brains, basal and active rates are comparable. Similar trends in rising complexity noted above for bodies are also evident for brains, though with higher Φ_m brain values for each and every animal type—much as expected since cerebral structure and function are widely considered among the most complex attributes of life (Jerison 1973; Allman 1999). Here, some quantitative details are compiled from many sources, again treating brains as open, non-equilibrium, thermodynamic systems, and once more casting the analysis of energy flow through them in terms of energy rate density. (While several other potentially useful neural metrics exist—cortical neuron numbers, encephalization quotients, and brain/body ratios (Roth and Dicke 2005)—I have evaluated brains here in terms of their Φ_m values in order to be scrupulously consistent with the complexity metric used above for all inanimate and animate systems.) Caution is advised since brain metabolic values taken from the literature often suffer from a lack of standard laboratory methods and operational units; many reported brain masses need correction for wet (live) values (by multiplying measured in vitro dry masses by a factor of 5 since in vivo life forms, including brains, are $\sim$80 % H_2O). Note also that the ratio of brain mass to body mass (used by some neuroscientists as a sign of intelligence) differs from the ratio of brain power to brain mass (which equals Φ_m); nor is "brain power" the same as that used in colloquial conversation, rather here it literally equals the rate of energy flowing through the cranium.

No attempt is made here to survey brain Φ_m values comprehensively, a task seemingly impossible in any case given the primitive state of neurological data to date; rather, representative mean values suffice for a spectrum of extant animals. Comparing mammals and reptiles, $\Phi_m \approx 10^5$ erg/s/g for mice brains (in contrast to $\sim 4 \times 10^4$ for their whole bodies) exceeds $\sim 5 \times 10^4$ erg/s/g for lizard brains ($\sim 3 \times 10^3$ for their bodies) (Hulbert and Else 1981); this is generally the case for all such animal taxa as Φ_m values are somewhat greater for mammalian brains than those for reptilian brains by factors of 2–4, and those for mammal bodies by roughly an order of magnitude (Hofman 1983). The great majority of vertebrate fish and amphibians show much the same 5–10 times increase in brain over body Φ_m values (Freeman 1950; Itazawa and Oikawa 2005), with, as often the case in biology, some outliers (Nilsson 1996). Even many invertebrate insects show several factors increase in Φ_m values for their brains ($\sim 5 \times 10^4$) compared to their bodies ($\sim 10^4$), most notably the flying insects (Kern 1985). Among mammals alone, primates have not only high brain/body mass ratios but also relatively high Φ_m brain values ($\sim 2 \times 10^5$ erg/s/g). Although primates allocate for their brains a larger portion (8–12 %) of their total bodily (resting) energy budget than do non-primate vertebrates (2–8 %) (Armstrong Armstrong 1983; Hofman 1983; Leonard and Robertson 1992), average primate brains' Φ_m values tend to be comparable to those of brains of non-primates; Φ_m brain values remain approximately constant across 3 orders of magnitude in mammalian brain size (Karbowski 2007). As with bodies above, brains do not necessarily confer much human uniqueness; brains are special, but all animals have them, and our neural qualities seem hardly more than linearly scaled-up versions of those of other primates (Azevedo et al. 2009). Even so, brain function and energy allocation are revealing: Among living primates, adult humans ($\sim 1.5 \times 10^5$ erg/s/g for brains and $\sim 2 \times 10^4$ for bodies) seem to have the highest brain power per unit mass—that is, not merely ~ 10 times higher Φ_m than for our bodies, but also slightly higher than for the brains of our closest, comparably massive, ape relatives, including chimpanzees. This substantial energy–density demand to support the unceasing electrical activity of myriad neurons within our human brains, which represent only ~ 2 % of our total body mass yet account for 20–25 % of the total bodily energy intake (Clarke and Sokoloff 1999), testifies to the disproportionate amount of worth Nature has invested in evolved brains—and is striking evidence of the superiority of brain over brawn.

The tendency for complex brains to have high Φ_m values, much as for complex whole animal bodies, can be tentatively correlated with the evolution of those brains among major taxonomic groups (Allman 1999). Further, more evolved brains tend to be larger relative to their parent bodies, which is why brain-to-body-mass ratios also increase with evolution generally—mammals more than reptiles, primates notable among mammals, and humans foremost among the great apes (Hofman 1983; Roth and Dickey 2005). Part of the reason is that relatively big brains are energetically expensive. Neurons use energy as much as 10 times faster than average body tissue to maintain their (structural) neuroanatomy and to support their (functional) consciousness; the amount of brain devoted to network connections increases disproportionately with brain size and so does the clustering and layering of cells within the higher-processing neocortex of recently evolved vertebrates (Stevens 2001; Jarvis 2005). Much of this accords with the "expensive tissue hypothesis" (Aiello and Wheeler 1995; Isler and van Schaik 2006; Navarrete et al. 2011), which posits that high brain/body ratios are indeed more energetically costly, at least for mammals and many birds, that energy flow through brains is central to the maintenance of relatively large brains, especially for primates, and that relatively large brains evolve when either brain energy input increases or energy allocation shifts to the brain from other bodily organs or fat reserves. Although the human brain's metabolic rate is not much greater than for selected organs, such as the stressed heart or active kidneys, regional energy flux densities within the brain greatly exceed (often by an order of magnitude) most other organs at rest. The pressures of social groups and social networking might also drive growth in brain size, cognitive function, and neurophysiological complexity along insect, bird, and primate lineages (Dunbar 2003; Smith et al. 2010); evolving societies require even more energy to operate, at least for humankind advancing (cf. next section). Throughout Earth's biosphere, the high-energy cost of brains might reasonably limit brain size and constrain natural selection's effect on an animal's survival or reproductive success; indeed, the brain is the first organ to be damaged by any reduction in O_2. This, then, is the observed, *general* trend for active brains in vivo: not only are brains voracious energy users and demonstrably complex entities, but evolutionary adaptation also seems to have favored for the brain increasingly larger allocations of the body's total energy resources.

Among more recent prehistoric societies of special relevance to humanity, the genus *Homo*'s growing encephalization during the past $\sim$2 million years may be further evidence of natural selection acting on those individuals capable of exploiting energy- and protein-rich resources as their habitats expanded (Foley and Lee 1991). By deriving more calories from existing foods and reducing the energetic cost of digestion, cooking was likely central among cultural innovations that allowed humans to support big brains (Wrangham 2009). Energy-based selection would have naturally favored those hominids who could cook, freeing up more time and energy to devote to other things—such as fueling even bigger brains, forming social relationships, and creating divisions of labor, all of which arguably advanced culture. As with many gauges of human intelligence, it's not absolute brain size that apparently counts most; rather, brain size normalized by body mass is more significant, just as the proposed Φ_m complexity metric is normalized by mass, here for brains as well for all complex systems at each and every stage along the arrow of time, from big bang to humankind.

The net finding for brains, broadly stated though no less true for the vast majority of animals, is that their Φ_m values are systematically higher than those for the bodies that house them. Nearly all such brain values fall within a rather narrow range of Φ_m between lower biological systems (such as plants) and higher cultural ones (such as societies). Although absolute brain masses span $\sim$6 orders of magnitude, from insects to whales, their Φ_m brain values cluster within a few factors, more or less depending upon their absolute size and evolutionary provenance, of $\sim 10^5$ erg/s/g.

Energy Rate Density of Humankind Advancing

For cosmic evolution to qualify as a comprehensive scientific worldview, human society and its many cultural achievements should be included, anthropocentric criticisms notwithstanding. Nature, alone and without sentient, technological beings, could not have built the social systems and technological devices characterizing our civilization today. Humankind itself is surely a part of Nature and not apart from it; schemes that regard us as outside of Nature, or worse atop Nature, are misguided. To examine how well, and consistently so, cultural systems resemble physical and biological systems—and thus to explore cultural evolution in a cosmic context—this section explores the evolution of cultural complexity as quantified by the same heretofore concept of energy rate density. (Some colleagues prefer to relabel long-term cultural evolution as "post-biological evolution," especially as regards clever machines that may someday outwit flesh-and-blood humans (Dick 2003); they assert that technological civilization is guided by intelligence and knowledge, yet both these factors resemble the earlier-abandoned information theory. By contrast, I aim to skirt the vagueness of social studies while embracing once again empirical-based energy flow as a driver of cultural evolution—especially, in the interest of unification, if that driver manifests the same common process that governs physical and biological evolution as well.)

Consider modern civilization *en masse*, which can be deemed the totality of all humanity comprising a (thermodynamically) open, complex society going about its usual business. Today's ~ 7 billion inhabitants utilize ~ 18 TW to keep our global culture fueled and operating, admittedly unevenly distributed in developed and undeveloped regions across the world (U.N. 2008). The cultural ensemble equaling the whole of humankind then averages $\Phi_m \approx 5 \times 10^5$ erg/s/g. Here human society is taken to mean literally the mass of humanity, not its built infrastructure (of buildings, roadways, etc.), for what matters is the flow of energy through the aggregated human social network. Unsurprisingly, a group of brainy organisms working collectively is more complex than all of its individual human components (who each consume an order of magnitude less energy, lest our bodies fry), at least as regards the complexity criterion of energy rate density—a good example of the "whole being greater than the sum of its parts," a common characteristic of emergence fostered by the flow of energy through organized, and in this case social, systems.

Rising energy expenditure per capita has been a hallmark in the origin, development, and evolution of humankind, an idea dating back decades (White 1959; Adams 1975). Culture itself is often defined as a quest to control greater energy stores (Smil 1994). Cultural evolution occurs, at least in part, when far-from-equilibrium societies dynamically stabilize their organizational posture by responding to changes in flows of energy through them. A quantitative treatment of culture, peculiar though it may be from a thermodynamic viewpoint, need be addressed no differently than for any other part of cosmic evolution (Nazaretyan 2010). Values of Φ_m can be estimated by analyzing society's use of energy by our relatively recent hominid ancestors, and the answers illustrate how advancing peoples increasingly supplemented their energy budgets beyond the 2–3,000 kcal/day that each person actually eats as food (Cook 1976; Bennett 1976; Simmons 1996; Spier 2005; Chaisson 2008; 2011a): Hunter-gatherers ~300,000 years ago used $\sim 3 \times 10^4$ erg/s/g, agriculturists ~10,000 years ago increased energy expenditure to $\sim 10^5$, industrialists beginning nearly two centuries ago utilized $\sim 5 \times 10^5$, and today's technologists in the most developed countries use $\sim 2 \times 10^6$. Underlying, and quite possibly driving, all this cultural advancement was not only greater energy usage but also greater energy usage per capita (i.e., per unit mass) at each and every step of the way.

Much of this social advancement is aided and abetted by culturally acquired knowledge accumulated from one generation to the next, including client selection, rejection, and adaptation, a decidedly Lamarckian process. Cultural inventiveness enabled our immediate ancestors to evade some environmental limitations: Hunting and cooking allowed them to adopt a diet quite different from that of the australopithecines, while clothing and housing permitted them to colonize both drier and colder regions of planet Earth. Foremost among the cultural advances that helped make us technological beings were the invention and utilization of tools, which require energy to make and use, all the while decreasing entropy within those social systems employing them and increasing it elsewhere in wider environments beyond. The 2nd law demands that as any system complexifies—even "smart" human-centered systems—its surroundings necessarily degrade. Thermodynamic terminology may be unfamiliar to anthropologists or historians, but the fundamental energy-based processes governing the cultural evolution of technological society are much the same, albeit measurably more complex, as for the evolution of stars, galaxies, and life itself (Adams 2010). As for biological organisms before them, specialization permits social organizations to process more energy per unit mass and this is reflected in increased Φ_m values over the course of time.

Notable among social practices widespread on Earth today, not only in developed countries but also intensifying rapidly in undeveloped countries, is technology. Advancement of machines is a premier feature of cultural evolution—and also one that increases order in manufactured products mainly by means of energy expenditures that inevitably ravage the larger environment of raw materials used to make those goods. Of today's many cultural icons, surely one of the most prominent is the automobile, which for better or worse has become an archetypical symbol of technological innovation worldwide. Values of Φ_m can be calculated for today's average-sized automobiles, whose typical properties are ~1.6 tons of mass and $\sim 10^6$ kcal of gasoline consumption per day; the result, $\sim 10^6$ erg/s/g (assuming 6 h of daily operation), is likely to range higher or lower by several factors, given variations among vehicle types, fuel grades, and driving times, yet this average value accords well with that expected for a cultural invention of considerable magnitude. Put another way to further illustrate evolutionary trends and using numbers provided by the U.S. government (U.S. Highway Traffic Safety Administration 2005) for the past quarter-century, the horsepower-to-weight ratio (in English units of hp/100 lb) of American passenger cars has increased steadily from 3.7 in 1978 to 4.1 in 1988 to 5.1 in 1998 to 5.5 when last compiled in 2004; converted to the units of Φ_m used here, these values equal 6.1, 6.7, 8.4, and 9.1, all times 10^5 erg/s/g respectively. Not only in and of themselves but also when compared to less powerful and often heavier autos of >50 years ago (whose Φ_m values are less than half those above), the trend of these numbers confirms once again the general correlation of Φ_m with complexity, for who would deny that modern automobiles, with their electronic fuel injectors, computer-controlled turbochargers, and a multitude of dashboard gadgets are more culturally complex than Ford's model-T predecessor of a century ago? The bottom line is that more energy is required per unit mass to operate the newer vehicles—a rise in Φ_m that will almost certainly continue as machines soon fundamentally switch their inner workings by substituting lightweight electrons for burning fuel and fast computers for mechanical linkages.

The connection between complexity and the advance of cultural evolution can be more closely probed by tracing the changes in internal combustion engines that power automobiles among many other machines such as gas turbines that propel aircraft (Smil 1999). To be sure, the brief history of machines can be cast in evolutionary terms, replete with branching, phylogeny, and extinctions that are strikingly similar to billions of years of biological evolution—though here, cultural change is again less Darwinian than Lamarckian, hence quicker too. Energy remains a driver for these cultural evolutionary trends, reordering much like physical and biological systems from the simple to the complex, as engineering improvement and customer selection over generations of products made machines more elaborate and efficient. Modern automobiles are better equipped and mechanically safer than their simpler, decades-old precursors, not because of any self-tendency to improve, but because manufacturers constantly experimented with new features, keeping those that worked while discarding the rest, thereby acquiring and accumulating successful traits from one generation of cars to the next. For example, the pioneering 4-stroke, coal-fired Otto engine of 1878 had a Φ_m value ($\sim 4 \times 10^4$ erg/s/g) that surpassed earlier steam engines ($\sim 10^4$ erg/s/g), but it too was quickly bettered by the single-cylinder, gasoline-fired Daimler engine of 1899 ($\sim 2.2 \times 10^5$ erg/s/g), more than a billion of which have been installed to date in cars, trucks, aircraft, boats, lawnmowers, etc., thereby acting as a signature force in the world's economy for more than a century. Today's mass-produced automobiles, as noted in the previous paragraph, average several times the Φ_m value of the early Daimler engine, and some intricate racing cars can reach an order of magnitude higher still. Among aircraft, the Wright brothers' 1903 homemade engine ($\sim 10^6$ erg/s/g) was superseded by the Liberty engines of World War I ($\sim 7.5 \times 10^6$ erg/s/g) and then by the Whittle-von Ohain gas turbines of World War II ($\sim 10^7$ erg/s/g). Boeing's 707 airliner inaugurated intercontinental jet travel in 1959 when Φ_m reached $\sim 2.3 \times 10^7$ erg/s/g, and civilian aviation evolved into perhaps the premier means of global mass transport with today's 747-400 wide-body, long-range jet whose engines create up to 110 MW to power this 180 ton craft to just below supersonic velocity (Mach 0.9) with $\Phi_m \approx 2.7 \times 10^7$ erg/s/g.

The cultural rise of Φ_m can be traced particularly well over several generations of jet-powered fighter aircraft of the U.S. Air Force (though here engine thrust must be converted to power, and for unloaded military jets operating nominally without afterburners typically 1 N $\approx$ 500 W, for which Φ_m values then relate to thrust-to-weight ratios). First-generation subsonic aircraft of the late 1940, such as the F-86 Sabre, gave way to 2nd-generation jets including the F-105 Thunderchief and then to the 3rd-generation F-4 Phantom of the 1960s and 1970s, reaching the current state-of-the-art supersonic F-15 Eagle now widely deployed by many western nations; 5th-generation F-35 Lightning aircraft will soon become operational. (Fighter F-number designations do not follow sequentially since many aircraft that are designed never get built and many of those built get heavily redesigned.) These aircraft not only have higher values of Φ_m than earlier-era machines, but those energy rate densities also steadily rose for each of the 5 generations of military aircraft R&D during the past half century—2.6, 4.7, 5.7, 6.1, and 8.2, all times 10^7 erg/s/g respectively, and all approximations for their static engine ratings (U.S. Air Force 2010).

Stunning advances in computer technology can also be expressed in the same quantitative language—namely, the rate of energy flowing through computers made of densely compacted chips. In all cases, Φ_m values reveal, as for engines above, not only cultural complexity but also evolutionary trends. (To make the analysis manageable, I have examined only computers that I personally used in my career, except for the earliest such device.) The ENIAC of the 1940s, a room-sized, 8.5 ton, 50 kW behemoth, transformed a decade later into the even larger and more powerful (125 kW) UNIVAC with $\sim$5,200 vacuum tubes within its 14.5 ton mainframe. By the 1970s, the fully transistorized Cray-1 supercomputer managed within each of its several (<1 ton, $\sim$22 kW) cabinets less energy flow yet higher energy rate density as computers began shrinking. By 1990 desktop computers used less power but also amassed less bulk ($\sim$250 W and $\sim$13 kg), making Φ_m still high. And now, MacBook laptops need only $\sim$60 W to power a 2.2 kg chassis to virtually equal the computational capability and speed of early supercomputers. During this half-century span, Φ_m values of these cultural systems changed respectively: 6.4, 9.5, 32, 20, and 28, all times 10^4 ergs/s/g. Although the power consumed per transistor decreased with the evolution of each newer, faster, and more efficient computer generation, the energy rate *density* increased because of progressive miniaturization—not only for the transistors themselves, but also for the microchips on which they reside and the computers that house them all. This growth of Φ_m parallels Moore's law (Moore 1965)—whereby transistor numbers etched on silicon chips double roughly every 18 months—and may be the underlying reason for it.

Although these and other cultural Φ_m values often exceed biological ones, machines are not claimed here to be "smarter" than we humans. Values of Φ_m for today's computers approximate those for human brains largely because they number-crunch much faster than do our neurological networks; even laptops now have central-processing units with immense computational features and not surprisingly, in cultural terms, high Φ_m values. That doesn't make microelectronic devices more intelligent than humans, but it does arguably make them more complex, given the rapid rate at which they functionally process data—and not least consume energy per unit mass. Accordingly, our most advanced aircraft have even higher Φ_m values than our most sophisticated computers. Modern flying machines rely on computers but also possess many additional, technologically advanced widgets that together require even more energy density, making them extraordinarily complex. That computers per se are amazingly complex machines, but not amazing enough for them to fly on their own, does suggest that perhaps there is something significant—and inherently more complex—about both living species and technical devices that can operate in 3-D environments on Earth; whether insects, birds, or cutting-edge aircraft, airborne systems exhibit higher values of Φ_m within each of their respective categories, more so to execute their awesome functions than to support their geometrical structures.

Much of this cultural advancement has been refined over many human generations, transmitted to succeeding offspring not by genetic inheritance but by use and disuse of acquired knowledge and skills. Again a mostly Lamarckian process whereby evolution of a transformational type proceeds via the passage of adopted traits, cultural evolution, like physical evolution, involves neither DNA chemistry nor genetic selection that characterize biological evolution. Culture enables animals to transmit modes of living and survival to their descendants by non-genetic, meme-like routes; communication passes behaviorally, from brain to brain and generation to generation, and that is what causes cultural evolution to act so much faster than biological evolution (Dennett 1996; Blackmore 1999; Denning Denning 2009). Even so, a kind of selection acts culturally, arguably guided by energy use (Chaisson 2011a); the ability to start a fire or sow a plant, for example, would have been major selective advantages for those hominids who possessed them, as would sharpening tools or manipulating materials. The result is that selection yielded newer technologies and systematically cast older ones into extinction, often benefiting humanity over the ages. It is this multitude of cultural advancements in recent times that has escalated and complexified change—advancements which, in turn with the scientific method that derives from them, enable us to explore, test, and better probe the scenario of cosmic evolution.

Figure 20.3 collates all of the above-cited human- and machine-related values of Φ_m, noting that these data pertain only to the uppermost part of the graph in Fig. 20.2. That's because modern society and our technological inventions are, in the cosmic scheme of things, only very recent advances in the rising complexity of generally evolving systems in the Universe.

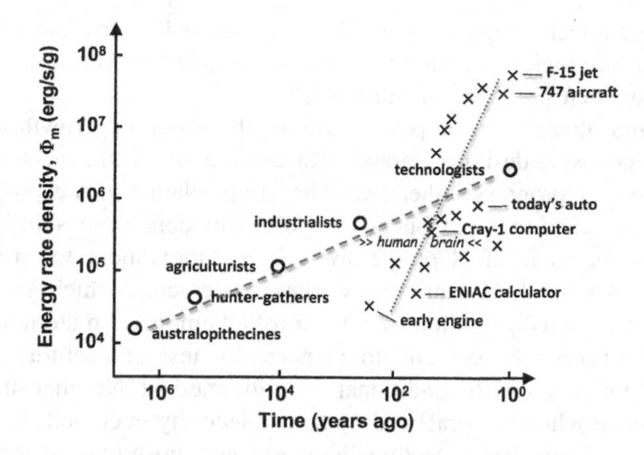

Fig. 20.3 Machines of the fast-paced 21st century not only evolve culturally, but are also doing so more quickly than humans evolve, either culturally or biologically—hence the reality, numerically delineated here, of a technological singularity. This graph shows some representative cultural systems that populate the uppermost part of the Φ_m curve plotted in Fig. 20.2. The time scale here covers only the past few million years, which is merely 0.02 % of the entire temporal scale of cosmic history illustrated in Figs. 20.1 and 20.2. This is a log–log plot, allowing meaningful display of data for society (plotted as Os linked by a *dashed line*) and for machines (Xs linked by a *dotted line*) over millions and hundreds of years, respectively, in the same figure. The value of Φ_m for the human brain is also indicated—but note well that Φ_m is a proposed measure of complexity, not necessarily of intelligence

Discussion: The Technological Singularity in Perspective

Today's civilization runs on energy for the simple reason that all complex, functioning systems need energy to survive and prosper. Whether aging stars, twirling galaxies, buzzing bees or redwood trees, it is energy that keeps open, non-equilibrium systems ordered and operating—to help them, at least locally and temporarily, avoid a disordered state (of high entropy) demanded by the 2nd law of thermodynamics. Whether living or non-living, dynamical systems need flows of energy to endure. If stars do not fuse matter into heat and light, they collapse; if plants fail to photosynthesize sunlight, they shrivel and decay; if humans stop eating, they die. Likewise, human society's fuel is energy: Resources come in and wastes go out while civilization conducts its daily business.

Throughout the long and storied, yet meandering, path of cosmic evolution, many complex systems have come and gone. Most have been selected out of Nature by Nature—destroyed and gone extinct—probably and partly because they were unable to utilize optimum amounts of energy per unit time and per unit mass; in all aspects of evolution, there are few winners and mostly losers. Is humankind among the preponderance of systems destined for extinction—owing perhaps to environmental degradation, societal collapse, or loss of control to machines? Will machines dominate us in the future, or might we merge with them to our mutual benefit? Would a technological singularity be good, bad, or irrelevant for us? Just what is the technological singularity and can we quantitatively assess its implications in ways that go beyond mere words?

To my mind, there is no purpose to any of the observed growth in universal complexity—no overt design or grand plan evident in cosmic evolution. Nor is there any obvious progress either; we who study Nature make progress while deciphering this grand scenario, but no compelling evidence exists that the cosmic-evolutionary process itself is progressive (as in "movement toward a goal or destination"). Admittedly I cannot prove these statements, which are themselves hardly more than squishy opinions. As a confirmed empirical materialist, my forte is to closely observe Nature and to numerically test conjectures about it—a mainstream application of the traditional scientific method. Not that subjectivity is absent in science while it's practiced; rather, objectivity eventually emerges only after much quantitative probing of qualitative ideas. Those ideas that pass the test of time survive—and those that don't are discarded; theoretical ideas are subject to selection and adaptation much like the complex material systems featured in this article. Hence my skepticism of parts of this volume that entail merely, mostly, and often exclusively beliefs, pronouncements, and speculations.

In the interest of full disclosure, I am also skeptical of much of what constitutes frontier physics these days. Progress toward a unified understanding of Nature need not postulate metaphysical schemes in abstract cosmology or untestable ideas in theoretical physics; nor does it necessarily require multiple universes, extra dimensions, or string theories for which there is no direct evidence (Greene 2011; Kragh 2011). A coherent, phenomenological explication of what is actually observed in our singular, four-dimensional universe populated mainly with galaxies, stars, planets, and life comprises a useful advance in comprehending, and to some extent unifying, the extended, diverse world around us. That is the intellectual stance from which I prefer to examine the idea of a technological singularity.

Figure 20.1 places cultural evolution on Earth during the past ~50 thousand years into the larger perspective of the more inclusive scenario of cosmic evolution that spans ~14 billion years. The arrow of time is an artistic graphic, not a numerical graph per se; it need not be examined closely. Figure 20.2 is that numerical graph and one that merits focused scrutiny, indeed one for which the key factor of this article—energy rate density—is plotted against precisely the same linear temporal scale as in Fig. 20.1. It compactly displays the rise of Φ_m for a wide array of systems throughout universal history to date. It rank-orders complex systems from the early Universe to civilization on Earth. And it shows, during each of the physical, biological, and cultural phases, how Φ_m rose increasingly rapidly—the growth of Φ_m accelerated. That, then, is what accelerates—Φ_m, the rate at which increasingly complex systems utilize energy—and it puts meat on the bones of all those soft and airy claims over the years that "something" is accelerating in our sophisticated world today. To be clear, on a linear plot as in Fig. 20.2, the whole graph taken together shows an exponentially rising trend; the slope of the curve is steeper for cultural evolution than for biological evolution, which in turn is steeper than for physical evolution. At least in terms of the Φ_m diagnostic discussed here, it seems unequivocal that the central mechanism of cosmic evolution, and the complexity products derived therefrom, have indeed accelerated with the march of time over billions of years.

Furthermore, though not shown here as much as elsewhere in detail (Aunger 2007; Chaisson 2011a, b), Φ_m rises exponentially for each type of complex system only for limited periods of time, after which their sharp rise often tapers off. Caution is warranted in order not to over-interpret these data, yet some but not all complex systems seem to slow their rate of growth while following a classic S-shape curve—much as microbes do in a petri dish while replicating unsustainably or as human population is expected to plateau later this century. That is, Φ_m values for a whole array of physical, biological, and cultural systems grow quickly during their individual evolutionary histories and then level off throughout the shaded area of Fig. 20.2 (whose drawn curve is then the compound sum of multiple S-curves); Φ_m for viable, complex systems show no noticeable decrease, rather often depict decreased rates of growth and S-shaped inflection perhaps once those systems have matured (Chaisson 2012). Some colleagues assume that means Φ_m decreases—it does not, at least not for surviving systems able to command optimal energy; others interpret that as complexity declining—but it also does not. The rate of change of Φ_m—which is itself a rate—might eventually decrease, but that means only that complexity's growth rate is lessening, not the magnitude of complexity per se.

Figure 20.3 allows a closer, numerical examination of the notion of a technological singularity—an occasion of some significance now probably underway during Earth's cultural evolution, which surely does transcend biological evolution. Note that the graph in Fig. 20.3 pertains only to the uppermost part of the curve in Fig. 20.2 and furthermore that this plot is not temporally linear; it is fully logarithmic. As such, the (dashed and dotted) straight lines exhibit exponential growth—as indicated individually for society advancing (plotted as Os, topped by modern technologists in developed countries today) and for machines rising (plotted as Xs, topped by 3-D, computer-controlled, military aircraft). *Prima facie*, the plotted graph does literally seem to display transcendence, as commonly defined "going beyond, surpassing, or cutting across," of machines over humankind. This is often claimed to be an event beyond which human affairs cannot continue—akin to mathematical singularities beset by values that transcend finite limitations—one for which humankind and the human mind as we currently know them are superseded and perhaps supplanted by strong, runaway, even transhuman artificial intelligence (Von Neumann and Ulam 1958; Kurzweil 2005). Alas, data in this paper are not accurate enough to test this unsettling fate.

The sum of the two curves for today's dominant cultural systems *en toto* results in faster-than-exponential growth—that is, the combined curve, dashed plus dotted in Fig. 20.3, sweeps upward on a log–log plot. Cultural change is indeed rapidly accelerating and the Φ_m data prove it. However, the data of Fig. 20.3 imply no evidence for a singularity of singular import or uniqueness. The technological singularity, which seems real and oncoming, may be central (and even threatening) to beings on Earth, yet is only one of many exceptional events throughout natural history, and unlikely more fundamental than many other profound evolutionary developments among complex systems over time immemorial. The cosmic-evolutionary narrative comprises innumerable transcendent phenomena that can be regarded as singularities all across the arrow of time in Fig. 20.1 and all the way up the rising curve of Φ_m in Fig. 20.2, including but by no means solely the birth of language (transcending symbolic signaling), the Cambrian explosion (land life transcending sea life), the onset of multicells (clusters transcending unicells), the emergence of life itself (life transcending matter), and even before that the origin and merger of stars and galaxies, among scores of prior and significant evolutionary events that led to humankind and its current existential crisis. Singulartarians need to think bigger and broader, thereby embracing the transformative concept of singularity in wider, cosmic settings extending all the way back along the arrow of deep time in reverse.

All things considered, this much seems evident from Fig. 20.3:

- Φ_m is increasing for humans and for machines, with the latter system rising faster
- Φ_m for humans and machines individually might each be slowing their rates of growth
- Φ_m for both humans and machines collectively accelerates hyper-exponentially
- a technological singularity, viewed as an evolutionary milestone, is indeed near.

Must we fear machines? Will they dominate or displace us, or merely aid us? The Φ_m data to date are not reliable enough to extrapolate an answer to these fateful questions, and in any case evolution is not a predictive science. Random chance always works in tandem with deterministic necessity, the two comprising natural selection that acts as a ruthless editor or pruning device to delete those systems unable to command energy in optimal ways; that is why "non-random elimination" is perhaps a better term for natural selection broadly applied to all complex systems (Mayr 1997). Thus, and sadly for those who agonize about future outcomes, Φ_m analyses cannot presently determine if humans will merge with machines or be overwhelmed by them in the coming years—although the data of Fig. 20.3 do imply that some machines are already more complex (higher Φ_m) than the humans and their brains who created them. Given that so many aspects of Nature are neither black nor white, rather shades of grey throughout, it is not inconceivable that humankind could survive while becoming more machine-like, all the while machines become more human-like—these two extremely complex systems neither merging nor dominating, as much as coexisting. After all, earlier evolutionary milestones that could easily have been considered transcendent singularities at the time—such as galaxies spawning complex stars, primitive life emerging on hostile Earth, or plants and animals adapting for the benefit of each—did not result in dominance, but rather coexistence.

Men and machines need not compete, battle, or become mutually exclusive; they might well join into a symbiotically beneficial relationship as have other past complex systems, beyond which even-higher Φ_m systems they—and we—may already be ascending with change, that is, evolving a whole new complex state that again becomes greater than the sum of its parts. The technological singularity—one of many other singularities among a plethora of evolutionary milestones in natural history and not likely the pinnacle or culmination of future cosmic evolution—fosters controversy because it potentially affects our human selves, and even elicits calls for ethical constraints and regulatory restrictions on technological innovation and advancement. Should we strive to preserve our essential humanity and halt the growth of machines? To my mind, given the natural rise in an expanding Universe of the curves in Figs. 20.2 and 20.3, we should not and could not.

The culturally increasing Φ_m values reported here—whether slow and ancestral such as for controlling fire and tilling lands by our provincial forebears, or fast and modern as with operating engines or programming computers in today's global economy—relate to evolutionary events in which energy flow and cultural selection played significant roles. All of this complexification, which has decidedly bettered the quality of human life as measured by health, education, and welfare, inevitably came—and continues to come—at the expense of greatly increased demand for more and enriched energy, which now drives us toward a fate on Earth that remains unknown.

Summary

Cosmic evolution is more than a subjective, qualitative narration of one unrelated event after another from big bang to humankind. This extensive scientific scenario provides an objective, quantitative framework that supports much of what comprises material Nature. It addresses the coupled topics of system change and rising complexity—the temporal advance of the former having apparently led to the spatial growth of the latter, yet the latter feeding back to make the former increasingly productive. It implies that basic differences both within and among the many varied complex systems in the Universe are of degree, not of kind. And it contends that evolution, broadly construed, is a universal concept, indeed a unifying principle throughout modern science.

More than perhaps any other single factor, energy plays a central role throughout the physical, biological, and cultural sciences. Energy seems to be an underlying, universal driver like no other in the evolution of all things, serving as a common currency in the potential unification of much of what is actually observed in Nature. Energy rate density, in particular, is an unambiguous, weighted measure of energy flow, enabling assessment of all complex systems in like manner—one that gauges how over the course of natural history writ large some systems optimally commanded energy and survived, while others apparently could not and did not.

Human society and its invented machines are among the most energy-rich systems known, hence plausibly the most complex yet encountered in the

Universe. Cultural innovations, bolstered by increased energy allocation as numerically tracked by rising Φ_m values, enable 21st century *H. sapiens* not only to circumvent the degrading environment on Earth but also to challenge it, indeed manipulate it. Technological civilization and its essential energy usage arguably act as catalysts, speeding the course of cultural change, which like all of cosmic evolution itself is unceasing, uncaring, and unpredictable.

Whatever our future portends—whether a whole new phase of cosmic evolution or merely the next, gradual step in cultural evolution, be it complex survival or simple termination—it will be a normal, natural outcome of cosmic evolution itself. For humanity, too, is part of Nature—and however humbling, we are likely just another chapter in a meta-story yet unfinished. Grand evolutionary events such as the oncoming technological singularity of human–machine interplay have occurred in the past many billions of years, and they will likely continue occurring indefinitely, forevermore yielding creativity and diversity in a Universe that expands, accelerates, and evolves. Think big, accept change, use energy wisely, adapt and prosper.

Acknowledgments I thank colleagues at the Smithsonian Astrophysical Observatory and students at Harvard University for discussions of this topic, and la Fondation Wright de Geneve for research support.

References

Adams, R. (1975). *Energy and structure*. Austin: University of Texas Press.

Adams, R. N. (2010). Energy, complexity and strategies, as illustrated by Maya Indians of Guatemala. *World Futures. Journal of General Evolution, 66*, 470.

Aiello, L.C., Wheeler, P. (1995). The expensive-tissue hypothesis. *Current Anthropology, 36*, 199.

Ainsworth, B.E., Compendium of physical activities tracking guide. http://prevention.sph.sc.edu/tools/docs/documents_compendium.pdf17Aug2011.

Allman, J. M. (1999). *Evolving brains*. New York: Scientific American Books.

Armstrong, E. (1983). Relative brain size and metabolism in mammals. *Science, 220*, 1302.

Aunger, R. (2007). *Technological Forecasting and Social Change, 74*, 1164.

Azevedo, F., Carvalho, L., Grinberg, L., Farfel, J., Feretti, R., Filho, W., et al. (2009). Equal numbers of neuronal and nonneuronal cells make the human brain an isometrically scaled-up primate brain. *Journal of Comparative Neurology, 513*(5), 532.

Bennett, J. W. (1976). *Ecological transition*. New York: Pergamon.

Blackmore, S. (1999). *The meme machine*. Oxford: Oxford University Press.

Bonner, J. T. (1988). *The evolution of complexity*. Princeton: Princeton University Press.

Brown, C. S. (2007). *Big history*. New York: The New Press.

Chaisson, E. J. (2001). *Cosmic evolution: The rise of complexity in nature*. Cambridge: Harvard University Press.

Chaisson, E. (2006). *Epic of evolution: Seven ages of the cosmos*. New York: Columbia University Press.

Chaisson, E. J. (2008). Long-term global heating from energy usage. *EOS Transactions of the American Geophysical Union, 89*, 253.

Chaisson, E. J. (2009a). Cosmic evolution: State of the science. In *Cosmos and Culture*, p. 3, S. Dick and M. Lupisella (Eds.). Washington: NASA Press.
Chaisson, E. J. (2009b). Exobiology and complexity. In: *Encyclopedia of Complexity and Systems Science*, p. 3267, R. Meyers (Ed.). Berlin: Springer.
Chaisson, E. J. (2011a). Energy rate density as a complexity metric and evolutionary driver. *Complexity*, *16*, 27. doi: 10.1002/cplx.20323.
Chaisson, E.J. (2011b). Energy rate density II: probing further a new complexity metric. *Complexity*, *17*, 44. doi: 10.1002/cplx.20373.
Chaisson, E. J. (2012). Using complexity science to search for unity in the natural sciences. In: *The self-organizing universe: Cosmology, biology, and the rise of complexity*, C. Lineweaver, P. Davies, M. Ruse (Eds.). Cambridge: Cambridge University Press.
Christian, D. (2004). *Maps of time*. Berkeley: University of California Press.
Dennett, D. (1996). *Darwin's dangerous idea*. New York: Simon and Shuster.
Denning, K. (2009). Social evolution. In: *Cosmos and Culture*, p. 63, S. Dick & M. Lupisella (Eds.). Washington: NASA Press.
Dick, S. J. (2003). Cultural evolution, the postbiological universe and SETI. *International Journal of Astrobiology, 2*, 65.
Dick, S. J. (2009). Cosmic evolution: History, culture, and human destiny. In: *Cosmos and Culture*, p. 25, S. Dick & M. Lupisella (Eds.). Washington: NASA Press.
Dick, S. J., Lupisella, M. L. (Eds.) (2009). *Cosmos and culture*: *Cultural evolution in a cosmic context*. NASA SP-2009 4802, Washington.
Clarke, D. D., Sokoloff, L. (1999). Circulation and energy metabolism of the brain. In: *Basic Neurochemistry*, 6th edition, G.J. Siegel (Ed.), p. 637. New York: Lippincott-Raven.
Cook, E. (1976). *Man, energy, society*. San Francisco: W. H. Feeman.
Dunbar, R. I. M. (2003). The social brain: mind, language, and society in evolutionary perspective. *Annual Review of Anthropology, 32*, 163.
Dyson, F. (1979). Time without end: physics and biology in an open universe. *Reviews of Modern Physics, 51*, 447.
Foley, R. A., & Lee, P. C. (1991). Ecology and energetics of encephalization in hominid evolution. *Philosophical Transactions of the Royal Society London Series B, 334*, 223.
Freeman, J.A. (1950). Oxygen consumption, brain metabolism and respiration movements of goldfish during temperature acclimatization. *Biological Bulletin*, *99*, 416.
Gleick, J. (2011). *The information*. New York: Pantheon.
Greene, B. (2011). *The hidden reality*. New York: Knopf.
Grinin, L. E., Korotayev, A. V., & Rodrigue, B. H. (Eds.). (2011). *Evolution: A big history perspective*. Volgograd: Uchitel Publishing House.
Hofkirchner, W. (Ed.). (1999). *The quest for a unified theory of information*. Amsterdam: Gordon and Breach.
Hofman, M. A. (1983). Energy metabolism, brain size and longevity in mammals. *Quarterly Review of Biology, 58*, 495.
Hulbert, A. J., & Else, P. L. (1981). Comparison of the 'mammal machine' and the 'reptile machine'. *American Journal of Physiology Regulatory, Integrative and Comparative Physiology, 241*, 350.
Isler, K., van Schaik, C.P. (2006). Metabolic costs of brain size evolution. *Biology Letters*, *2*, 557.
Itazawa, Y., & Oikawa, S. (2005). Metabolic rates in excised tissues of carp. *Cellular and Molecular Life, 39*, 160.
Jarvis, E.D. et al. (28 coauthors) (2005). Avian brains and a new understanding of vertebrate brain evolution. *Nature Reviews Neuroscience*, *6*, 151.
Jerison, H. J. (1973). *Evolution of the brain and intelligence*. New York: Academic.
Karbowski, J. (2007). Global and regional brain metabolic scaling and its functional consequences. *BMC Biology, 5*, 18.
Kaufmann, S. (1993). *Origins of order*. Oxford: Oxford University Press.
Kelly, K. (2010). *What technology wants*. New York: Viking.

Kern, M. J. (1985). Metabolic rate of the insect brain in relation to body size and phylogeny. *Comparative Biochemistry and Physiology, 81A*, 501.
Kleidon, A., & Lorenz, R. D. (Eds.). (2005). *Non-equilibrium thermodynamics and production of entropy*. Berlin: Springer.
Kragh, H. (2011). *Higher speculations*. Oxford: Oxford University Press.
Kurzweil, R. (2000). *The age of spiritual machines*. New York: Penguin.
Kurzweil, R. (2005). *The singularity is near: When humans transcend biology*. New York: Penguin.
Lane, N., & Martin, W. (2010). Energetics of genome complexity. *Nature, 467*, 929.
Leonard, W. R., & Robertson, M. L. (1992). Nutritional requirements and human evolution: A bioenergetics model. *American Journal of Human Biology, 4*, 179.
Lotka, A. J. (1922). Contributions to the energetics of evolution. *Proceedings of the National academy of Sciences of the United States of America, 8*, 147.
Makarieva, A. M., Gorshkov, V. G., Li, B.-L., Chown, S. L., Reich, P. B., & Gavrilov, V. M. (2008). Mean mass-specific metabolic rates are strikingly similar across life's major domains. *Proceedings of the National academy of Sciences of the United States of America, 105*, 16994.
Mandelbrot, B. (1982). Mean mass-specific metabolic rates are strikingly similar across life's major domains. *Fractal geometry of nature*. San Francisco: W.H. Freeman.
Martyushev, L. M., & Seleznev, V. D. (2006). Maximum entropy production principle in physics, chemistry and biology. *Physics Reports, 426*, 1.
Maynard Smith, J. (1995). Non-junk DNA. *Nature, 374*, 227.
Mayr, E. (1997). *This is biology*. Cambridge: Harvard University Press.
McMahon, T., & Bonner, J. T. (1983). *On size and life*. San Francisco: Freeman.
Meyers, R. (Ed.) (2009). Encyclopedia of complexity and systems science, 11 vol. Berlin: Springer.
Mitchell, M. (2009). *Complexity: A guided tour*. Oxford: Oxford University Press.
Moore, G. E. (1965). Cramming more components onto integrated circuits. *Electronics, 38*, 8.
Morowitz, H. J. (1968). *Energy flow in biology*. New York: Academic.
Morrison, P. (1964). A thermodynamic characterization of self-reproduction. *Reviews of Modern Physics, 36*, 517.
Navarrete, A., Isler, K., & van Schaik, C. P. (2011). Energetics and the evolution of human brain size. *Nature, 480*, 91.
Nazaretyan, A. P. (2010). *Evolution of non-violence: studies in big history, self-organization, and historical psychology*. Saarbrucken: Lambert Publishing.
Nicolis, G., & Prigogine, I. (1977). *Self-organization in non-equilibrium systems*. New York: Wiley.
Nilsson, G. E. (1996). Brain and body oxygen requirements of *Gnathonemus petersii*. *Journal of Experimental Biology, 199*, 603.
Odum, H. (1988). Self-organization, transformity, and information. *Science, 242*, 1132.
Prigogine, I. (1978). Time, structure, and fluctuations. *Science, 201*, 777.
Raymo, C. (2002). Will evolution leave humanity behind? *Boston Globe*, 17 Sep 2002.
Roth, G., Dicke, U. (2005). Evolution of the brain and intelligence. *Trends in Cognitive Science, 9*, 250.
Salthe, S. N. (1993) *Development and Evolution*, Cambridge: MIT Press.
Schroedinger, E. (1944). *What is life?*. Cambridge: Cambridge University Press.
Simmons, I. G. (1996). *Changing the face of the earth*. London: Blackwell.
Smith, A.R., Seid, M.A., Jimenez, L.C. & Wcislo, W.T. (2010). Wernicke's area homologue in chimpanzees. *Proceedings of the Royal Society B*, 277, 2157.
Spier, F. (2005). How big history works: energy flows and the rise and demise of complexity. *Social Evolution and History, 4*, 87.
Spier, F. (2010). *Big history and the future of humanity*. New York: Wiley.
Smil, V. (1994). *Energy in world history*. New York: Westview.
Smil, V. (1999). *Energies*. Cambridge: MIT Press.
von Bertalanffy, L. (1932). *Theoretische biologie*. Berlin: Borntraeger.

Stevens, C. F. (2001). An evolutionary scaling law for the primate visual system and its basis in cortical function. *Nature, 411*, 193.

Ulanowicz, R. E. (1972). Mass and energy flow in closed ecosystems. *Journal of Theoretical Biology, 34*, 239.

U. N. (2008). Department of Economic Social Affairs, Population. Div., *World Population Prospects*, New York.

U.S. Air Force, Factsheets available for various aircraft, such as for F-15: http://www.af.mil/information/factsheets/factsheet.asp?id=101. Accessed 17 Sept 2010.

U.S. National Highway Traffic Safety Administration, U.S. Dept. of Transportation, DOT HS 809 512, 2005.

Vakoch, D. A. (2009). Encoding our origins. In *Cosmos and Culture*, p. 415, S. Dick and M. Lupisella (Eds.), Washington: NASA Press.

Von Neumann, J., cf. Ulam, S. (1958). *Bulletin of the AmericanMathematical Society, 64.3*, 1.

White, L. (1959). *Evolution of culture*. New York: McGraw-Hill.

Wolfram, S. (2002). *A new kind of science*. Champaign: Wolfram Media.

Wrangham, R. (2009). *Catching fire: How cooking made us human*. New York: Basic Books.

Chapter 20A Theodore Modis on Chaisson's "A Singular Universe of Many Singularities: Cultural Evolution in a Cosmic Context"

The concept of Φ_m is the best attempt at rigorously quantifying complexity that I have seen, albeit with shortcomings, e.g. no one will accept that bicycle riding is ten times more complex than violin playing or that a jet engine is 1000 times more complex than a mammalian organism! My attempt to quantify complexity (discussed in the second part of my essay) is only in relative terms and is based on data that may be subject to subjective judgment. Of course there must have also been some subjective estimates in Chaisson's data, for example, in the calculation of Energy Rate Densities of hunter-gatherers, agriculturists, industrialists, etc., which may mask a leveling-off of the straight-line trend of the O data points in Fig. 20.3, similar to the visible leveling-off of the X data points. These leveling-offs are evidence that we are dealing with S-curves and combined with the acknowledged leveling-off of the two early curves in Fig. 20.2, reinforces the general conclusion that exponential trends of Phi are in fact early parts of S-curves.

Chaisson is being conservative. He modestly says that "I sense, but cannot prove, that information is another kind of energy" while he could have easily argued that information content is proportional to entropy which is equal to Q/T (heat over temperature), which IS energy. He also says that the drawn curve of the shaded area of Fig. 20.2 is the compound sum of multiple S-curves, but stops short of using S-curves to extrapolate it into the future. In fact he refrains from committing himself to any future eventuality one way or another. (One would have welcomed at least an educated guess from such an expert!)

Having spent most of my career with S-curves I can see in Chaisson's Fig. 20.3 that the two "S-curves" depicted by the dashed and dotted lines determine the shape of the late part of the third "S-curve" labeled society on Fig. 20.2. Furthermore, these two curves in Fig. 20.3 have life cycles that become shorter with time (acceleration effect). Life cycles getting shorter is evidence for saturation. As I mention in my essay there is a fractal aspect to S-curves. A large-scale S-curve can be decomposed to smaller constituent S-curves the life cycles of which become shorter as we approach the ceiling of the envelope curve (see also publication http://www.growth-dynamics.com/articles/Fractal.pdf). I can then conjecture that the line labeled society in Fig. 20.2 is an S-curve that presently finds itself beyond its midpoint, i.e. experiences a progressive slowdown of its rate of growth. An imminent slowdown in the rate of growth of Phi (and complexity) corroborates a similar conclusion in my essay.

Printed in the United States
By Bookmasters